Cognitive Science and Technology

Series editor

David M.W. Powers, Adelaide, Australia

More information about this series at http://www.springer.com/series/11554

Before the class started, Sidney wanted to go over the text one more time

James K. Peterson

Calculus for Cognitive Scientists

Derivatives, Integrals and Models

Springer

James K. Peterson
Department of Mathematical Sciences
Clemson University
Clemson, SC
USA

ISSN 2195-3988 ISSN 2195-3996 (electronic)
Cognitive Science and Technology
ISBN 978-981-13-5719-0 ISBN 978-981-287-874-8 (eBook)
DOI 10.1007/978-981-287-874-8

Printed on acid-free paper

This Springer imprint is published by SpringerNature
The registered company is Springer Science+Business Media Singapore Pte Ltd.

I dedicate this work to all of my students who have been learning this material from courses taught over the last 10 years on the interplay between mathematics, biology, computation, and cognitive science. I am also indebted to the practicing scientists who have helped an outsider think more clearly and to my family who have listened to my ideas in the living room and over dinner for many years. I hope that this text helps inspire everyone to consider mathematics and computer science as indispensable tools in their own work in the cognitive and biological sciences.

Acknowledgments

We would like to thank all the students who have used the various iterations of our mathematics, biology/cognitive science, and computation notes as they have evolved to the fully typed version here. We have been pleased by the enthusiasm you have brought to this interesting combination of ideas from many disciplines.

For our new text on how to start learning calculus with a cognitive and biological science slant, we would like to thank all of the students from the past size years who have listened to similar lectures on these matters. Their interest and plentiful feedback and comments are helping this new endeavor be successful. I am looking forward to hearing from all of my new students on how to improve my presentations!

We wanted a new painting for this new book and decided on a vision of what the future will certainly hold: the squids of today are phenomenally intelligent and will undoubtedly evolve to the point where studying calculus, mathematics, and computation is an absolute must. The details of how to do that in a water-based environment are easily solved. In our painting, the squid Sydney is getting ready to go to class and has decided to read the lesson of the day one more time or perhaps Sydney is simply working on these notes as part of a program of self-study as we hope many of you will be.

Finally, we gratefully acknowledge the support of Hap Wheeler in the Department of Biological Sciences in the years 2006–2014 for believing that courses of this sort would be useful to biology students.

Contents

Part III Using Multiple Variables

List of Figures

List of Table

List of Code Examples

Abstract

This book attempts to give a program of self-study on how mathematics, computer science, and science can be usefully and pleasurably intertwined. You must all learn to allow ideas from mathematics and computation to be part of the way you approach understanding cognitive and biological science. A wonderful quote from **Mathematical Models of Social Evolution: A Guide For the Perplexed** by Richard McElreath and Robert Boyd published by the University of Chicago Press in 2007 gives you a good perspective on this.

> Imagine a field in which nearly all important theory is written in Latin, but most researchers can barely read latin and certainly cannot speak it. Everyone cites the Latin papers, and a few even struggle through them for implications. However, most rely upon a tiny cabal of fluent Latin-speakers to develop and vet important theory.
>
> Things aren't quite so bad as this in evolutionary biology [think our combination of mathematics, biology and computation here instead!]. Like all good caricature, it exaggerates but also contains a grain of truth. Most of the important theory developed in the last 50 years is at least partly in the form of mathematics, the "Latin" of the field, yet few animal behaviorists and behavioral ecologists understand formal evolutionary models. The problem is more acute among those who take an evolutionary approach to human behavior, as students in anthropology and psychology are customarily even less numerate than their biology colleagues.
>
> We would like to see the average student and researcher in animal behavior, behavioral ecology, and evolutionary approaches to human behavior [think biological sciences in general!] become more fluent in the "Latin" of our fields. Increased fluency will help empiricists better appreciate the power and limits of the theory, and more sophisticated consumers will encourage theorists to address issues of empirical importance. Both of us teach courses to this end, and this book arose from our experiences teaching anxious, eager, and hard-working students the basic tools and results of theoretical evolutionary ecology.

These authors offer much good advice to you too. You are starting on a new journey where we will try hard to get you to enjoy this triad of mathematics, computer work, and science. So to get the most out of this class, take the advice of Richard McElreath and Robert Boyd to heart.

While existing theory texts are generally excellent, they often assume too much mathematical background. Typical students in animal behavior, behavioral ecology, anthropology or psychology [think biological and cognitive scientists in general!] had one semester of calculus long ago. It was hard and didn't seem to have much to do with their interests, and, as a result, they have forgotten most of it. Their algebra skills have atrophied from disuse, and even factoring a polynomial is only an ancient memory, as if from a past life. They have never had proper training in probability or dynamical systems. In our experience, these students need more hand holding to get them started.

This book does more hand-holding than most, but ultimately learning mathematical theory is just as hard as learning a foreign language. To this end, we have some suggestions. In order to get the full use of this book, the reader should

1. Read the book in order. Each chapter builds on the last....
2. Work through the examples within each chapter. Math is like karate: it must be learned by doing. If you really want to understand this stuff, you should work through every derivation and understand every step. [So] read the chapters at two levels, for general understanding and for skill development.
3. Work all the problems....
4. Be patient, both with the material and yourself. Learning to understand and build formal evolutionary models is much like learning a language. This book is a first course, and it can help the reader understand the grammar and essential vocabulary. It is even a passable pocket dictionary. However, only practice and use of the language will build fluency. Students sometimes jump to the conclusion that they are stupid because they do not immediately understand a theoretical paper of interest. This is not justified. No one quickly learns to comprehend and produce mathematical arguments, any more than one can quickly become fluent in Latin.

For our part, our contract to you is to work very hard at getting you through this interesting mixture of complicated stuff! So let us begin this journey together!

History

Some of the materials for this course are based on:
Calculus Two for Biology: More Models and Their Analysis
Notes On MTHSC 108 for Biologists developed during the Spring 2006 and Fall 2006, Spring 2007 and Fall 2007 courses
The first edition of this text was used in
Spring 2008 and Fall 2008 courses
The course was then relabeled as MTHSC 111 and the text was used in the Spring 2009 and Fall 2009 courses
The second edition of this text was used in
Spring 2010 and Fall 2010 courses.
The third edition was used in the
Spring 2011 and Fall 2011,
Spring 2012 and Summer Session I courses
The fourth edition was used in
Fall 2012 and Spring 2013 courses.
Also, we have used material from notes on Partial Differential Equation Models which has been taught to small numbers of students since 2008.
This text specialized for biology was used in the first semester calculus course for biologists
Fall 2013 and Spring 2014
Fall 2014 and Spring 2015

Part I
Introduction

Chapter 1
Introductory Remarks

In this course, we want to learn how to use mathematics and what is called computation to help us understand how to study cognitive sciences. The problem, of course, is the mathematics and computer science part seem to overwhelm all of us and keep us from seeing how helpful it is to allow those extra points of view. Now Physics is a big field which has been traditionally very mathematical and physics students have similar complaints about losing sight of their science by being narrowly focused on the quantitative underpinnings. We are going to try hard to avoid that trap, but make no mistake. We expect you are all willing to go on this journey eagerly and work hard to achieve the goals we set out.

Our goals are deceptively simple.

- Learn enough **Calculus** to allow us to think clearly about biological and cognitive science models. This means we have to learn about how to use functions to model the data we find in our experiments. This is the area of study we call Biological Modeling. Then we have to make assumptions about how the *things* in our data that vary depend on each other. We typically call these *things* **variables**. For this course, we will limit ourselves to two variables most of the time, but towards the end we will look at relationships between more than two things. But remember, we are the ones who look at the data we measure and make this call. Our intellect makes the decisions on how our **variables** depend on one other.
  Usually how variables change over time is important to us and that leads us to think about that idea more abstractly. This leads to the idea of the **derivative** of a function. This idea is a nice abstraction of stuff we do all the time in our physical world. We notice how things change! In fact, we tend to *not* notice things that stay the same. Think about the shirt or blouse you are wearing. When you first put it on, you might notice how it feels on your skin, but our sensory subsystems for skin pressure and so forth quickly stop paying attention. Instead, it is change in pressure we notice. So we need ways to model such changes and that is the idea of a derivative.
  If we were looking at data on how an animal was eating over a period of weeks, we might want to add up how the animal has eaten over that time period. It is

J.K. Peterson, *Calculus for Cognitive Scientists*, Cognitive Science and Technology, DOI 10.1007/978-981-287-874-8_1

pretty easy to see that if we added up the food intake each day that would in general give an answer a bit different from what we would get it we measured food intake every twelve hours and then added. And what about measurements taken each hour and added? Or measurements taken each minute and added? This progression of additions over different time scales leads naturally to a new idea: that of the **integration** of a function.
Of course, these three concepts: **function**, **differentiation** and **integration** have lots of nuance we have to work our way through. But these are all tools we can use to help us model the data we measure.
We are going to intertwine modeling ideas with our learning of these basic Calculus ideas so that you always see why we are learning all these new things.
We will also introduce computational ideas early using MatLab so that you can all get comfortable right away with tools that can help us in cases where we just can't solve our equations by hand. We typically use a mixture of computer work and mathematics to gain insight. This is a very profitable way of gaining insight and we hope to train you so well in this that you carry it forward to other work you will do in your future.

- Once we have basic calculus ideas down, we can start modeling biology in earnest. We typically do this in the beginning with what are called **differential equations** which are just equations where some of the variables involve derivatives. We can model protein synthesis in various ways with these ideas quite nicely. We can also use our tools to model large scale ideas such as domestication rates in wild wheat and barley. And many others, of course. We will learn enough in this course to get you started and you will find you can build models of your own soon enough using what you learn in this one semester course.
- We also want to be able to *build* models of biologically relevant stuff even if we can't solve them with our calculus tools. This leads us to use computational techniques. In this course, we will be using **Matlab** throughout and we will show how to use it to help you solve problems that are way too hard to solve by hand techniques. Using Matlab we can run what are called **simulations** to solve our differential equation models and help us understand how the variables we choose to focus on interact over appropriate time scales even when we can't solve the models using pencil and paper.
- Finally, we allow our focus to broaden to look at more that two variables of interest. This will lead us to look at what is called **partial derivatives** which are really quite simple. You just take our usual derivative thinking of one variable at a time as being allowed to change. We will find the mechanics of this is very simple and we will do just enough theoretical stuff to help you understand what you are really doing. Having more than two variables will allow us to look at our biological data in more sophisticated ways. We can only scratch the surface of this here, but you can take more courses and learn more. And we encourage you to do so! Biology, Mathematics and Computation are merging more and more and studying all three areas is a wise thing to do for your future.

1.1 Our Design Philosophy

Our philosophy is then that all parts of the course must be integrated, so we don't want to use mathematics, science or computer approaches for their own intrinsic value. Experts in these separate fields must work hard to avoid this. This is the time to be generalists and always look for connective approaches. Also, models are carefully chosen to illustrate the basic idea that we know far too much detail about virtually any biologically based system we can think of. Hence, we must learn to throw away information in the search of the appropriate abstraction. The resulting ideas can then be phrased in terms of mathematics and simulated or solved with computer based tools. However, the results are not useful, and must be discarded and the model changed, if the predictions and illuminating insights we gain from the model are incorrect. We must always remember that throwing away information allows for the possibility of mistakes. This is a hard lesson to learn, but important. Note that models from population biology, genetics, cognitive dysfunction, regulatory gene circuits and many others are good examples to work with. All require massive amounts of abstraction and data pruning to get anywhere, but the illumination payoffs are potentially quite large.

Now many of us are trying to build an interface between biology, mathematics and computation. There is a recent article which discusses the issues involved and we encourage you to read it ("Interfacing Mathematics and Biology: A Discussion on Training, Research, Collaboration, and Funding" by Laura Miller and Silas Alben published in *Integrative and Comparative Biology* in 2012 (Miller and Alben 2012)). The paper discusses what happened at a workshop on this interface which was held at the 2012 annual meeting of the Socicty for Integrative and Comparative Biology. We have our own thoughts on how to teach this material too and we offer our perspectives in "Some Thoughts on BioMathematics Education" published in 2013 (Peterson 2013).

> The broad goal of much of this recent activity has been to promote new collaborations between biologists, physical scientists, and mathematicians, to inspire the next generation of biology students to use quantitative approaches and to use applications in the life sciences to fuel the development of new mathematical and numerical techniques. The common challenges that have emerged from these discussions center on the best practices for training students and the need to develop mechanisms that foster multidisciplinary collaborations.

Ideally, students such as yourselves who are interested in studying computational cognitive science would learn the material of this text and follow it up with the additional volumes (Peterson 2015a, b, c). We have taught versions of this text for the last two years and versions of Peterson (2015a) for about eight years to quite a few students. The last two texts have been taught to small numbers of students over the past five years as well but we always hope for more interest! The point is to take the tools we are teaching and use them in your own work. Unfortunately, you won't find enough of your traditional biology and cognitive scientists using these tools. However, that does not mean their use isn't the future. It just means it is always hard to integrate interdisciplinary material into traditional research. Your science teachers and colleagues are probably as afraid of mathematics and computation as you are so it is only natural that it is hard to get the ball rolling. But we think it is very important to try to get this off the ground as future work in biologically based science is clearly tied to increased use of mathematics and computation.

A standard criticism of being more interdisciplinary in approach is that if you teach students that way, they lose depth. So the challenges here are deep and require focus and work from professionals in all areas. As the paper says

> [t]o address these challenges, a variety of educational initiatives, funding programs, and institutes focused on mathematics and biology have been established over the past decade....An important theme in the discussion was the training and mentoring of researchers at the interface of mathematics, engineering and biology. The panel discussed the ideal timing of a switch from a more traditional disciplinary education to a more interdisciplinary education – at the undergraduate level, the graduate level, or beyond? There was a general concern expressed about interdisciplinary students gaining breadth at the expense of breadth. Less depth in a traditional field can be a significant disadvantage when applying to, and completing the requirements for a graduate degree in a particular field. However, faculty pursuing interdisciplinary research are more likely to admit students who have shown an interest in previous interdisciplinary work. Ultimately, success in coursework and research is strongly enhanced by following one's intellectual interests. Pursuing research that overlaps with the priority research areas of faculty and funding opportunities is one of the practical realities of training students.

As you can see, we are still working out how to do this. For example, one of my former students who is just now graduating was interested in interdisciplinary work and ended up doing a dual major in biochemistry and biology. He took all of our biological mathematics courses and did additional work with us on modeling brain networks and programming. He had to work very hard at finding the right graduate school. Most graduate schools are still not so keen on interdisciplinary work. He found that he had to be upfront about his need to wear multiple hats and not be *typecast* into one area or another. He found three likely candidates and is going to start in the Fall at the University of Washington in their NIH funded training program for medical researchers. He will be able to direct how his research focuses with a team's backing which is what he wanted. To finish with this student's story, here is an example of how the mathematics, computation and biology worked together for him. At one point in his senior research, he needed primers to bind to the DNA he was working on. It was hard to find and order the right ones as the organism he was working on was just in the process of having its genome annotated. He contacted the group doing that research and they shared the program there were using with him. He was able to modify the code of that program which was written in MatLab to search for the primers he needed. So these tools we are going to be talking about are useful!

The training we are trying to do here is also being done at other places. Indeed, as the paper says

> At the undergraduate level, colleges and universities are offering increasing numbers of courses and degree programs that cross the interface of mathematics and biology. The panelists and workshop attendees generally agreed that these courses were excellent mechanisms to pique student interest in interdisciplinary work and that undergraduate education

> often allows students the flexibility for students to develop a solid background in more than one field. One of the most common courses being developed is BioCalculus (i.e., Calculus for Life Scientists). This course typically focuses on covering standard calculus material using applications to biology. More and more colleges and universities are also offering upper level courses such as Quantitative Biology (from the biology side) or Mathematical Biology (from the mathematics side).

So this book is part of a more general phenomenon. However, there is much more work to be done. One of the biggest problems is that the mathematics teachers usually do not have biological training and the biology teachers lack the mathematical training so it is hard for most teachers to be comfortable in using ideas and tools outside of their discipline. Here, as the paper says

> [t]he panel and workshop group generally agreed that more curriculum development is needed at the interface of mathematics and biology. Parallels can be drawn to some degree between physics and mathematics. These fields have a long history of intersecting research and education. Physic majors take a significant number of mathematics courses, and substantial mathematics is included in many physics courses. Applied mathematics courses often are motivated by examples from physics. A limitation in mathematical training is due to the fact that while students may take a course in mathematical or quantitative biology, these ideas are not reinforced in other courses. If a freshman takes a course in BioCalculus, for example, but does not see many applications of mathematics in the courses that follow, this training likely will not be retained.
>
> A consensus was reached during the discussion that publicly available lesson plans that easily could be incorporated into standard classes by non experts have the potential to greatly enhance training at the interface. Repositories of biological examples in mathematics should be available for mathematicians with little of no biology training. There is a similar need for straightforward applications of mathematics and computation to standard material in the biological sciences.

Then, they note

> One approach to this problem is to design courses that generally address how to develop mathematical models in biology. The construction of original mathematical models for new biological problems can be both tractable and challenging to many students at all levels. Of course the study of biological models gives us the background to study cognitive models as well, so this is a win - win plan!

and this is one of our guiding principles!

1.2 Insights from Other Fields

A very mathematical part of Physics is called **Quantum Field Theory** or just **QFT** for short and it is very hard to train new physicists in this area as it requires hard and difficult mathematics to work through the problems. However, it is easy for the teachers and masters of this area to get lost in the beauty and wonder of the mathematical manipulations and forget that all the models come from physics. Just to give you a feeling for this, we thought you should see some reviews of the new textbook **Quantum Field Theory in a Nutshell: Second Edition** by A. Zee published by

Princeton University Press in 2013. These reviews can be found on the Amazon web page you see when you look at Zee's book. Zee's book is well received because it places the Physics first and the mathematical detail in a subservient position. As the first review says

Review by M. Haque

One problem with learning QFT is that it is so easy to get lost in the mathematical details that the core physics concepts often get obscured. In my opinion, Tony Zee overcomes this particular problem quite successfully. He keeps algebra to a bare minimum, and tries to find the shortest route to the physics ideas. He chooses examples that illustrate concepts in the fastest possible way. Wholeheartedly recommended.

This is why we are trying hard to illustrate how to use the mathematics and computation in biology with carefully chosen examples and models. They give insight! Now Zee does have one advantage we don't have. By the time students get to QFT, they have had years of mathematical training so Zee can assume a lot of basic expertise. Of course, we don't have that but the basic principles are the same. Focus on the science and let the mathematics and computation serve that. The next review states how Zee's emphasis on ideas is really helpful.

Review by Alexander Scott

From my experiences in quantum field theory: [there are different types of books on QFT:] The kind that you can read, the kind that work out examples, and the kind that your professors want you to understand. The last are [texts] that dummies like me in a QFT class will never be able to use ("dummies in a QFT class" may sound like an oxymoron, but we're not all geniuses...). The kind of texts that works out examples ... have been invaluable to me, but I still have not always been able to understand the IDEAS contained in the mathematics. "QFT in a Nutshell" heralds the introduction of a book on quantum field theory that you can sit down and read. My professor's lectures made much more sense as I followed along in this book, because concepts were actually EXPLAINED, not just worked out. I still recommend having all three types of texts, but I am glad that now I have three types and not just the last two.

So, explanation is key! In our course, we will always work hard at getting you to understand thoroughly the ideas. We think memorizing formulae and thinking of the mathematics as some sort of *black box* you just plug things into is a very bad way to train scientists and critical thinkers. Note how the reviewers of the QFT book, who are all physicists in training, repeatedly state it is the underlying ideas that count. That point is just as important in biology as it is in physics. Let us leave you with one final review

Review by Mobius

I was tempted to give this book four stars, simply to stand out among the sea of five star reviews, but I cannot, for this book truly is deserving of five stars. This is indeed a wonderful book, though it is not the mythic "one field theory text you will ever need" or the book that can make Sarah Palin understand instantons.

This book covers quite a bit of ground, but that does not mean it is shallow. I've read some crap textbooks whose authors try to cram every topic under the sun into the table of contents, but do nothing to convey any real understanding... This book is at the other end of the spectrum.

> In physics identifying the truly interesting questions usually proves to be more difficult than performing the calculations, and what this book does really well is show what the interesting questions are and why they are interesting. If the calculational details Zee presents are too sparse, and I think they are in a few places, you can always find more information on the interwebs.

Hold onto that thought: identifying the truly interesting questions is very difficult. Another way of looking at this is that all interesting questions in science require complicated mathematics that can't be easily solved without concomitant computational work. Our aim in this course is to get you started on learning how to do this.

Now let's switch gears and go from Physics to Climate Modeling. Here are some quotes from a new textbook on this subject: **Principles of Planetary Climate** by Raymond Pierrehumbert published by Cambridge University Press in 2010. Climate models are extremely interdisciplinary. As he says

> When it comes to understanding the whys and wherefores of climate, there is an infinite amount one needs to know, but life affords only a finite time in which to learn it; the time available before ones fellowship runs out and a PhD thesis must be produced affords still less. Inevitably, the student who wishes to get launched on significant interdisciplinary problems must begin with a somewhat hazy sketch of the relevant physics, and fill in the gaps as time goes on. This book [our book too!] is an attempt to provide the student with a sturdy scaffolding which a deeper understanding may be built on.
>
> The climate system is made up of building blocks which in themselves are based on elementary physical principles, but which have surprising and profound collective behavior when allowed to interact on the planetary scale. In this sense, the "climate game" is rather like to game of Go, where interesting structure emerges from the interaction of simple rules on a big playing field, rather than complexity in the rules themselves. ...
>
> A guiding principle is that new ideas come from profound analysis of simple models – thinking deeply about simple things. The goal is to teach students how to build simple models of diverse planetary phenomena, and to provide the tools necessary to analyze their behavior.

We completely agree. The focus should be on a deep analysis of relatively simple models that we think very carefully about. Always remember that our goal is explanatory insight!

1.3 How Should You Study?

Since this course is probably a lot different from other mathematics courses you have taken, we thought it would be nice for you to hear what the authors of another text that tries to combine mathematics and biology has to say. The text is **Mathematical Models of Social Evolution: A Guide For the Perplexed** by Richard McElreath and Robert Boyd published by the University of Chicago Press in 2007. They are teaching evolutionary models to biologists, anthropologist and psychologists who at best have had a calculus course a long time in the past. They state the problem very nicely

Imagine a field in which nearly all important theory is written in Latin, but most researchers can barely read latin and certainly cannot speak it. Everyone cites the Latin papers, and a few even struggle through them for implications. However, most rely upon a tiny cabal of fluent Latin-speakers to develop and vet important theory.

Things aren't quite so bad as this in evolutionary biology [think our combination of mathematics, biology and computation here instead!]. Like all good caricature, it exaggerates but also contains a grain of truth. Most of the important theory developed in the last 50 years is at least partly in the form of mathematics, the "Latin" of the field, yet few animal behaviorists and behavioral ecologists understand formal evolutionary models. The problem is more acute among those who take an evolutionary approach to human behavior, as students in anthropology and psychology are customarily even less numerate than their biology colleagues.

We would like to see the average student and researcher in animal behavior, behavioral ecology, and evolutionary approaches to human behavior [think biological scientists] become more fluent in the "Latin" of our fields. Increased fluency will help empiricists better appreciate the power and limits of the theory, and more sophisticated consumers will encourage theorists to address issues of empirical importance. Both of us teach courses to this end, and this book arose from our experiences teaching anxious, eager, and hard-working students the basic tools and results of theoretical evolutionary ecology.

They note the most common problem is that textbooks are too mathematical at the expense of the science. So the key thing is many examples and worked out steps. They have suggestions for you!

While existing theory texts are generally excellent, they often assume too much mathematical background. Typical students in animal behavior, behavioral ecology, anthropology or psychology had one semester of calculus long ago. It was hard and didn't seem to have much to do with their interests, and, as a result, they have forgotten most of it. Their algebra skills have atrophied from disuse, and even factoring a polynomial is only an ancient memory, as if from a past life. They have never had proper training in probability or dynamical systems. In our experience, these students need more hand holding to get them started.

This book does more hand-holding than most, but ultimately learning mathematical theory is just as hard as learning a foreign language. To this end, we have some suggestions. In order to get the full use of this book, the reader should

1. Read the book in order. Each chapter builds on the last. ...
2. Work through the examples within each chapter. Math is like karate: it must be learned by doing. If you really want to understand this stuff, you should work through every derivation and understand every step. [So] read the chapters at two levels, for general understanding and for skill development.
3. Work all the problems. ...
4. Be patient, both with the material and yourself. Learning to understand and build formal evolutionary models is much like learning a language. This book is a first course, and it can help the reader understand the grammar and essential vocabulary. It is even a passable pocket dictionary. However, only practice and use of the language will build fluency. Students sometimes jump to the conclusion that they are stupid because they do not immediately understand a theoretical paper of interest. This is not justified. No one quickly learns to comprehend and produce mathematical arguments, any more than one can quickly become fluent in Latin.

They let you know that you need a lot of practice at working through all the steps in a derivation so that you really understand.

> This book has lots of algebra....The reason is that beginning students need it. If you really want to learn this material, you have to follow the derivations one mathematical step at a time. Students that haven't done much math is a while often have a shaky grasp on how to do the algebra; they make lots of mistakes, and even more important, don't have much information about how to get from point A to point B is a mathematical argument. All this means that they can find themselves stuck, unable to derive the next result, and unsure of whether they don't understand something fundamental or the obstacle is just a mathematical trick or algebraic error. Many years of teaching this material convinces us that the best remedy is to show lots of intermediate steps in derivations.

We will follow this plan too but we expect that you work hard to understand all the steps so that your growth path is on target to helping you become a modern biologist.

Finally, they talk about how we use computation in our modeling work. Their comments are very true. We know many mathematicians who study biology, chemistry, ecology or physics by having the equations that some other group has decided provide a model handed to them without understanding how the model is derived. And they don't want to understand that! They simply put a thin veneer of science into their mathematics. Their work is usually impossible for scientists to use as it is divorced from the science. The same is true about computational approaches. All of these things need to be fully integrated with the science. As they say

> There is a growing number of modelers who know very little about analytic methods. Instead, these researchers focus on computer simulations of complex systems. When computers were slow and memory was tight, simulation was not a realistic option. Without analytic methods, it would have taken years to simulate even moderately complex systems. With the rocketing ascent of computer speed and plummeting price of hardware, it has become increasingly easy to simulate very complex systems. This make it tempting to give up on analytic methods, since most people find them difficult to learn and understand.
>
> There are several reasons why simulations are poor substitutes for analytic models.
>
> **Equations Talk** Equations – given the proper training – really do speak to you. They provide intuitions about the reasons an evolutionary [think more general biological models!] system behaves the way it does, and these reasons can be read from the expressions that define the dynamics and resting states of the system. Analytic models therefore tell us things that we must infer, often with great difficulty, from simulation results. Analytic models can provide proofs, while simulations only provide a collection of examples.
>
> **Sensitivity Analysis** It is difficult to explore the sensitivity of simulations to changes in parameter values. Parameters are quantities that specify assumptions of the model for a given run of the simulation – things like population size, mutation rate, and the value of a resource. In analytic models, the effects of these changes can be read directly from equations or by using various analytic techniques. In simulations, there are no analogous results. Instead the analyst has to run large numbers of simulations, varying the parameters in all combinations. For a small number of parameters, this may not be so bad. But let's assume a model has four parameters of interest, each of which has only 10 interesting values. Then we require 10^4 simulations. If there are any stochastic effects in the model, we will need maybe 100 or 1000 Or 10,000 times as many....[M]anaging and interpreting the large amounts of data generated from the rest of the combinations can be a giant project, and this data-management problem

will remain no matter how fast computers become in the future. Technology cannot save us here. When simple analytic methods can produce the same results, simulation should be avoided, both for economy and sanity.

Computer Programs are hard to communicate and verify There is as yet no standard way to communicate the structure of a simulation, especially a complicated "agent-based" simulation, in which each organism or other entity is kept track of independently. Often, key aspects of the model are never mentioned at all. Subtle and important details of how organisms reproduce or interact have benefited from generations of notational standardization, and even unmentioned assumptions can be read from expressions in the text. Thus it is much easier for other researchers to verify and reproduce modeling results in the analytic case. Bugs are all too common, and simulations are rarely replicated, so this is not a minor virtue.

Overspecification The apparent ease of simulation often tempts the modeler to put in every variable which might matter, leading to complicated and uninterpretable models of an already complicated world. Surprising results can emerge from simulations, effects we cannot explain. In these cases, it is hard to tell what exactly the models have taught us. We had a world we didn't understand and now we have added a model we don't understand.

If the temptation to overspecify is resisted, however, simulation and analytic methods complement each other. Each is probably most useful when practiced along side each other. There are plenty of important problems for which it is simply impossible to derive analytic results. In these cases, simulation is the only solution. And many important analytic expressions can be specified entirely as mathematical expressions but cannot be solved, except numerically. For these reasons, we would prefer formal and simulation models be learned side by side.

We hope by the end of this course, to convince you that equations do talk and to help you understand what they are saying! Enjoy the journey!

1.4 Code

All of the code we use in this book is available for download as individual files from **Starting Calculus for Cognition** (http://www.ces.clemson.edu/~petersj/startingcalcforcognition.html). Just save each file to your working directory on your computer and you are good to go. If you want to avoid saving each file one at a time, you can instead download the file **CompCogOne.tar.gz** from the same site.

In addition, although you are just in book one, if you are interested, all of the code we use for all four books is available for download from the **BioInformation Web Site** (http://www.ces.clemson.edu/~petersj/CognitiveModels.html). The code samples can then be downloaded as the zipped tar ball **CognitiveCode.tar.gz** and unpacked where you wish. If you have access to MatLab, just add this folder with its sub folders to your MatLab path. If you don't have such access, download and install **Octave** on your laptop. Now Octave is more of a command line tool, so the process of adding paths is a bit more tedious. When we start up an Octave session, we use the following trick. We write up our paths in a file we call `MyPath.m`. For us, this code looks like this

Listing 1.1: How to add paths to Octave

```
function MyPath()
%
s1 = '/home/petersj/MatLabFiles/BioInfo/:';
s2 = '/home/petersj/MatLabFiles/BioInfo/GSO:';
s3 = '/home/petersj/MatLabFiles/BioInfo/HH:';
s4 = '/home/petersj/MatLabFiles/BioInfo/Integration:';
s5 = '/home/petersj/MatLabFiles/BioInfo/Interpolation:';
s6 = '/home/petersj/MatLabFiles/BioInfo/LinearAlgebra:';
s7 = '/home/petersj/MatLabFiles/BioInfo/Nernst:';
s8 = '/home/petersj/MatLabFiles/BioInfo/ODE:';
s9 = '/home/petersj/MatLabFiles/BioInfo/RootsOpt:';
s10 = '/home/petersj/MatLabFiles/BioInfo/Letters:';
s11 = '/home/petersj/MatLabFiles/BioInfo/Graphs:';
s12 = '/home/petersj/MatLabFiles/BioInfo/PDE:';
s13 = '/home/petersj/MatLabFiles/BioInfo/FDPDE:';
s14 = '/home/petersj/MatLabFiles/BioInfo/3DCode';
s = [s1,s2,s3,s4,s5,s6,s7,s8,s9,s12];
addpath(s);
end
```

The paths we want to add are setup as strings, here called **s1** etc., and to use this, we start up Octave like so. We copy **MyPath.m** into our working directory and then do this

Listing 1.2: Setting the path in Octave

```
octave>> MyPath();
```

We agree it is not as nice as working in MatLab, but it is free! You still have to think a bit about how to do the paths. For example, in Peterson (2015c), we develop two different ways to handle graphs in MatLab. The first is in the directory **GraphsGlobal** and the second is in the directory **Graphs**. They are not to be used together. So if we wanted to use the setup of **Graphs** and nothing else, we would edit the **MyPath.m** file to set **s = [s11];** only. If we wanted to use the **GraphsGlobal** code, we would edit **MyPath.m** so that **s11 = '/home/petersj/MatLabFiles/BioInfo/GraphsGlobal:';** and then set **s = [s11];**. Note the directories in the **MyPath.m** are ours: the main directory is **'/home/petersj/MatLabFiles/BioInfo/** and of course, you will have to edit this file to put your directory information in there instead of ours.

All the code will work fine with **Octave**. So pull up a chair, grab a cup of coffee or tea and let's get started.

1.5 Some Glimpses of Modeling

To whet your appetite, let's look at some typical modeling problems that are much more than just solving a mathematical problem.

1.5.1 West Nile Virus Models

Flaviviruses are a family of viruses and, with the exception of hepatitis virus, are transmitted by mosquitoes and ticks with an impact that is important for varied sociological and economic reasons. These diseases include dengue, yellow fever, tick-borne encephalitis and West Nile fever. They are widely distributed throughout the world with the exception of the polar regions, although a specific flavivirus may be geographically restricted to a continent or a particular part of it. These single-stranded RNA viruses are expanding in geographical distribution. West Nile virus, for example, has emerged in recent years in temperate regions of Europe and North America, presenting a threat to public and animal health. The most serious manifestation of West Nile virus (WNV) infection is fatal encephalitis (inflammation of the brain) in humans and horses, as well in certain domestic and wild birds. The virus is maintained in nature through a transmission cycle involving mosquitoes and birds. Most individuals will usually experience an inapparent or a mild febrile illness; others, a dengue-like illness while a minority, often including the elderly may develop an encephalitis which can be fatal. The diagnosis is usually made by serology although the virus can be detected in the blood by molecular techniques, or in tissue culture. No vaccine for the virus is available and there is no specific therapy.

This virus exhibits important behaviors within the host during the pathogenesis of the disease. Infection features a substantial upregulation of cell surface molecules on a variety of cell types which are in the G_0 or resting state of cell division. Curiously, cells that are progressing through the cell cycle, (e.g. in the G_1 state) exhibit this upregulation, but at much lower levels. Although many cell surface molecules are expressed at a much higher concentrations in infected quiescent cells, For the moment, let's focus primarily on the increase in the major histocompatibility antigen-I (MHC-I) complex and intercellular adhesion molecule-1 (ICAM-1) and the upregulatory effects on both these molecules by induced interferons (IFN).

To build a model of this infection requires many compromises and many levels of abstraction. We model the infection within a single *abstract* host. We choose initial cell population sizes and immune system components so that our *abstract* version of a host exhibits reasonable responses closely aligned to what we find in the literature.

A lot of data on West Nile virus infections have been collected by cellular biologists that involve multiple hosts. In a survival experiment, a population of mice of size N are all infected with the same amount of virus, and the number of animals that die are determined. The amount of virus used is measured in plaque forming units (pfu). This determines the concentration of infectious virus by virtue of the number of areas of cell death in a cell monolayer in vitro. Hence, it is a statistical measurement and somewhat imprecise. The experiment is then repeated, using N mice per group for a range of virus concentrations. If we did this for 12 pfu levels, we could then graph the number of mice that survive versus the virus concentration. This a survival experiment is expensive to undertake, requiring $12N$ mice. If we used $N = 100$ for statistical accuracy, this would be 1200 mice at perhaps \$20.00 per mouse. At \$24, 000 for the mice alone, this is costly. With most viruses, a traditional survival

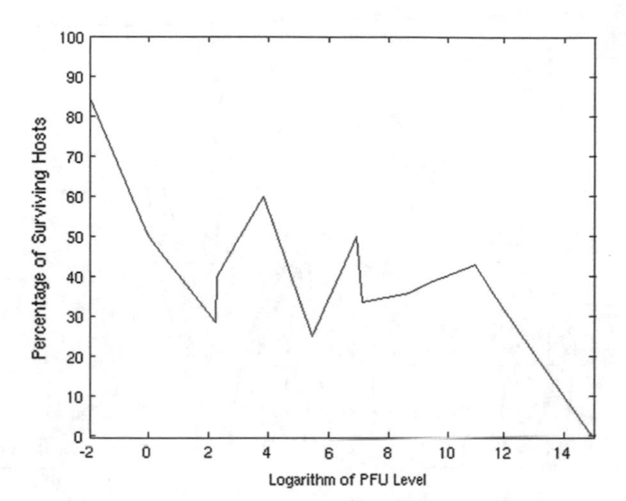

Fig. 1.1 Actual survival curve

experiment in which the virus is titrated in this way gives a classical dose-response curve which progressively and smoothly decays down to a survival of 0 at high virus concentrations. WNV has a peculiar survival curve which is shown in Fig. 1.1.

Note that survival sometimes increases with increasing virus concentration, in contrast to the data seen in most other viral infections. Our big question is how can understand this survival curve? In our model, we have a variable that represents the number of cells that are not infected and that number changes with time as the infection progresses for a given choice of the parameter in our model that represents pfu level. How do we decide a host dies in our simulation? This is certainly not easy to do and how we do it effects the kind of simulated survival curve we generate in our simulation. However, if we can build a model which faithfully generates a simulated survival curve qualitatively similar to what we see in the real data, we have come a little closer to understanding how the Wet Nile Virus infects hosts. And shed a little light, perhaps, on how this virus kills a certain number of infected people. This infection exhibits what is called immunopathology which is a kind of self-attack by the immune system, so it is important to study of that level too. We may learn insights into other auto immune problems. Learning how to answer this question is a mix of mathematics, computation and the science of infectious diseases. The code is written in MatLab and we can perform the simulations on any reasonably fast laptop.

1.5.2 Simple Brain Models

A cortex module is built from cortical cans assembled into columns which are then organized into one or two dimensional sheets. So the process of building a cortex module is pretty intense! However, we can use the resulting module as an isocortex building block as we know all cortex prior to environmental imprinting is the same.

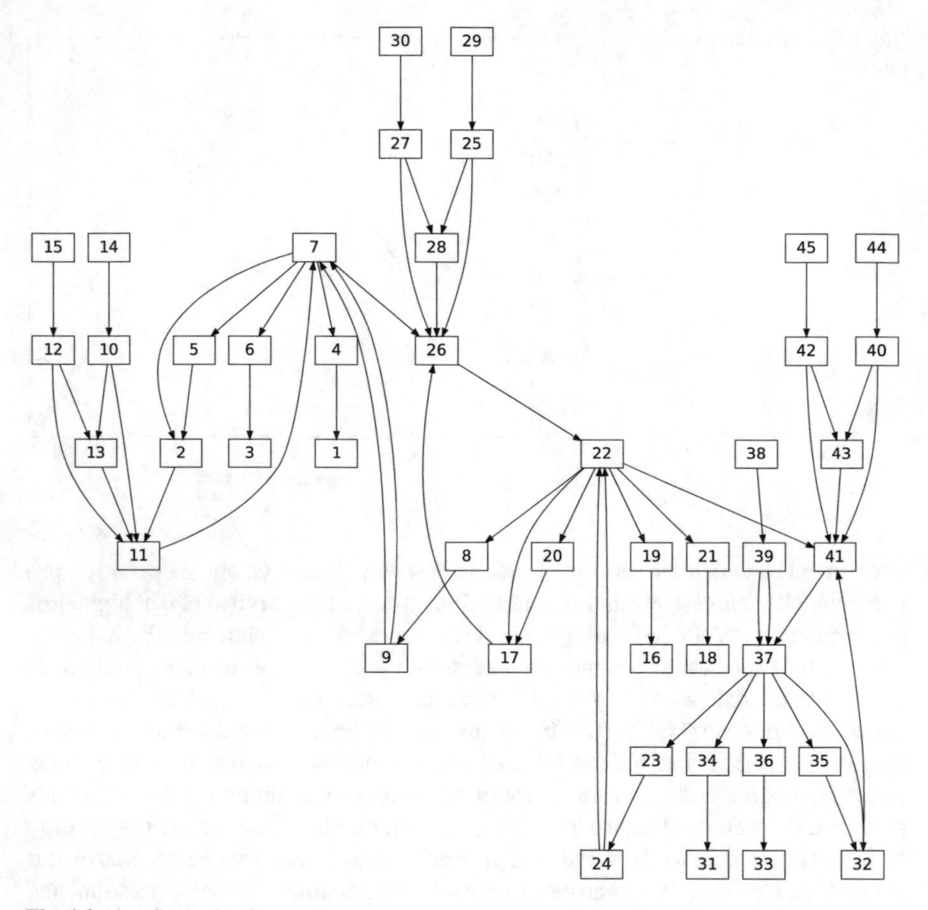

Fig. 1.2 A column circuit

Hence, we can build a cortex module for visual cortex, auditory cortex etc. and simply label their respective global vector addresses appropriately. To build the model of cortex, we then glue together as many cortex columns as we want.

We can build a model of a typical column easily in MatLab which is shown in Fig. 1.2. Then we glue together as many of these columns as we like to make a cortical sheet. Each node is a neuron and edges between nodes are synaptic connections which may be inhibitory or excitatory. This type of model is called a *graph* model and we want to use it to ask various questions. For example, if we use inputs to certain nodes and want other nodes to have specified values which are considered outputs, how do we *train* our cortex model to *map* these inputs to the desired outputs? Learning how to do this is now a mix of mathematics, computation and neurobiology. The code is again written in MatLab and we can perform the simulations on any reasonably fast laptop.

1.5.3 A Cancer Model

We discuss this carefully in Chap. 19, but here is a teaser. In colon cancer, there is a special gene called a tumor suppressor gene (TSG) which comes in two forms called alleles. If the first allele of a TSG is inactivated by a point mutation while the second allele is inactivated by a loss of one parent's contribution to part of the cell's genome, the second allele is inactivated because it is lost in the copying process. This is an example of chromosomal instability or CIN Let's look now at colon cancer itself. Look at Fig. 1.3. In this figure, you see typical colon crypts. Stem cells at the bottom of the crypt differentiate and move up the walls of the crypt to the colon lining where they undergo apoptosis.

At the bottom of the crypt, a small number of *stem* cells slowly divide to produce *differentiated* cells. These differentiated cells divide a few times while migrating to the top of the crypt where they undergo apoptosis. This architecture means only a small subset of cells are at risk of acquiring mutations that become fixed in the permanent cell lineage. Many mutations that arise in the differentiated cells will be removed by apoptosis. Colon rectal cancer is thought to arise as follows. A mutation inactivates the **Adenomatous Polyposis Coli** or **APC** TSG pathway. The crypt in which the **APC** mutant cell arises has abnormal growth and produces a *polyp*. Large *polyps* seem to require additional oncogene activation. Then 10–20 % of these large polyps progress to cancer. A general model of cancer based on TSG inactivation is as follows. The tumor starts with the inactivation of a **TSG** called **A**, in a small compartment of cells. A good example is the inactivation of the **APC** gene in a colonic crypt, but it could be another gene. Initially, all cells have two active alleles of the **TSG**.

We will denote this by $\boldsymbol{A}^{+/+}$ where the superscript "$+/+$" indicates both alleles are active. One of the alleles becomes inactivated at mutation rate u_1 to generate a cell type denoted by $\boldsymbol{A}^{+/-}$. The superscript $+/-$ tells us one allele is inactivated. The second allele becomes inactivated at rate $\hat{u}_2$ to become the cell type $\boldsymbol{A}^{-/-}$. In addition, $\boldsymbol{A}^{+/+}$ cells can also receive mutations that trigger **CIN**. This happens at the rate u_c resulting in the cell type $\boldsymbol{A}^{+/+CIN}$. This kind of a cell can inactivate the first allele of the **TSG** with normal mutation rate u_1 to produce a cell with one

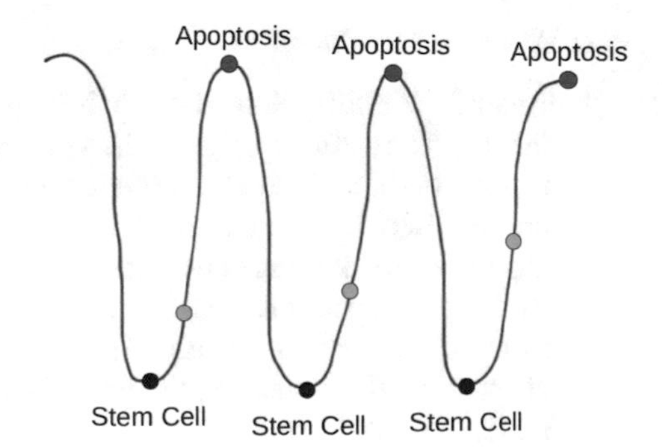

Fig. 1.3 Typical colon crypts

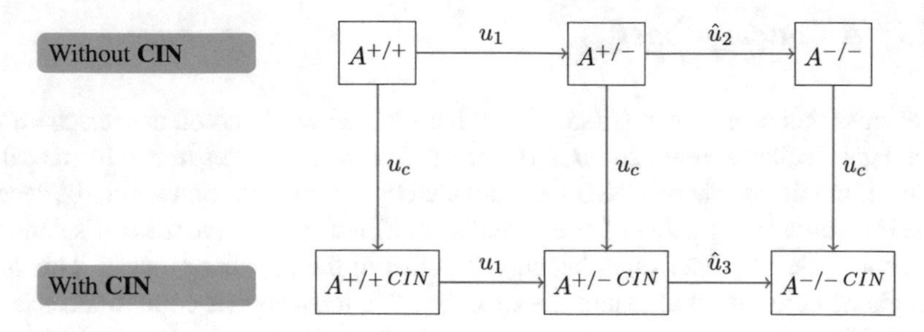

Fig. 1.4 The pathways for the **TSG** mutations

inactivated allele (i.e. a $+/-$) which started from a CIN state. We denote these cells as $A^{+/- CIN}$. We can also get a cell of type $A^{+/- CIN}$ when a cell of type $A^{+/-}$ receives a mutation which triggers **CIN**. We will assume this happens at the same rate u_c as before. The $A^{+/- CIN}$ cell then rapidly undergoes **LOH** at rate $\hat{u}_3$ to produce cells of type $A^{-/- CIN}$. Finally, $A^{-/-}$ cells can experience **CIN** at rate u_c to generate $A^{-/- CIN}$ cells. We show this information in Fig. 1.4.

Our question is which pathway to colon cancer is dominant? The top pathway through pointwise mutation or the bottom pathway through CIN? This is not answered by solving a simple model using calculus. It requires a lot of thought to get the insight we need. Hence, again, a blend of mathematics, computation and science is needed to get insight!

1.6 A Roadmap to the Text

In this course, we introduce enough relevant mathematics and the beginnings of useful computational tools so that you can begin to understand some interesting biological models. We present a selection of nonlinear biological models and slowly build you to the point where you can begin to have a feel for the model building process. Here is the plan so you can either be excited by all the cool stuff or horrified that you are going to have to read a lot as well as listen to us. Oh well, all interesting things require effort. Remember that.

Part II: Using One Variable

Chapter 2: Viability Selection In this chapter, we discuss how we might model the long term effects of genes in a population. We carefully chose this model to start with because it has interesting science, the questions we are asking are very high level and difficult to model well and we need nothing other than algebra to work things out. Many of you are probably a bit afraid of algebra partially because it seems so abstract and useless to real world stuff. But be assured, it is a skill set that you can use very effectively to gain insight. In this chapter, we develop some things and use some mathematics we can return to multiple times for profit.

Chapter 3: Limits and Basic Smoothness Here, we are going to try something new. Usually in a book on calculus, we carefully go through the big ideas in separate chapters—a bit tedious and boring perhaps. So let's try something new. In this chapter, we are going to talk about the smoothness of the functions we see in modeling in a sneaky way. We will explore the ideas of limits first with some examples and then hit you with continuity. From there, we show you the special kind of limit that leads to a derivative and end with an nice example of the very strange limiting process that leads to what is called a Riemann integral. So after this chapter, you have some concrete things to hold in your mind about where we are going. We then get to the nitty gritty of it all and look at the details in the later chapters.

Chapter 4: Continuity and Derivatives Now we hit the idea of derivative in earnest. Lots of detail now but we will use a lot of examples and try hard to convince you this is a useful tool. We learn how to take simple derivatives very quickly and go over all the cool rules: the product rule, the quotient rule and the chain rule. All are pretty self evident things when all is said and done and you need to realize that and not let the mathematical detail get your goat as they say!

Chapter 5: Sin, Cos and All That Now we add sine and cosine functions to the mix. The limits that define the derivative of the sine function are tricky and we certainly can't do them directly. So we show you how to sneak up on them more easily with some geometric arguments. And yes, we do expect you remember some of your basic trigonometry from high school. Not a lot, but some. We finish with a very nice example of the kind of recursive processing we see in neural circuits. Yep, the chain rule applies but we try to find other ways to get the rate of change information we need as the chain rule is too hard! Remember, there are always multiple ways to get where you want to go.

Chapter 6: Antiderivatives Next, we work out how to do a lot of antiderivatives which we essentially can guess since we know how to find a lot of derivatives now from the previous chapters.

Chapter 7: Substitutions Once we know about antiderivatives, we can use the idea of a variable substitution to see that many antiderivatives we want to find are really disguised forms of a simpler pattern. Learning how to reach into the heart of a thing and dredge out its hidden core is a hugely important idea in doing science. Buried in data are underlying patterns and possibly laws which the relationships between variables follow. Believe it or not, the simple act of learning how to do substitution prepares your mind to dig deep like this. Plus it is a cool party trick and you can amaze friends for hours with this kind of skill.

Chapter 8: Riemann Integration In this chapter, we discuss the Riemann integral we introduced in the smoothness chapter much more carefully. It is hard to believe, but we find the connection between this particular limit process of sums and our previously discussed antiderivatives! With those results in hand, we are poised to solve a lot of problems. Of course, some of this is a bit intense, but we need these tools so hang in there.

Chapter 9: The Logarithm and Its Inverse With the Riemann integral firmly established in our minds and the connection between it and antiderivatives found, we can use those ideas to define the natural logarithm function and its inverse called the exponential function. Most of this follows quite naturally from the previous chapter and we finish with a lot of practice with our new derivative friends: the derivative of the logarithm and exponential function is there many disguises.

Chapter 10: Exponential and Logarithm Function Properties Of course, the logarithm and exponential function have useful properties which are completely non intuitive. We will use these properties a lot so you need to be very comfortable with them. We show you how these properties come about in detail.

Chapter 11: Simple Rate Equations Now we are ready to solve our first models which are called simple rate equations. We also learn about half lives. We find that the exponential function is the fundamental building block of our biological modeling.

Chapter 12: Simple Protein Models We are now ready to some biology. We build a simple model of protein transcription. Yes, it misses some detail but it gives us great insight and we can see how the tools we have been building really begin to pay off now.

Chapter 13: Logistics Models The next model concerns growth when there are limited resources and is easy for us to solve with our new tools. At this point, you should be beginning to get a lot more comfortable with all of this machinery.

Chapter 14: Function Approximation Many times the functions we want to work with are just so complicated we want a way to approximate them so we can get good approximate answers, seat of the pants or back of the envelope calculations, to help us build insight into something really hard. In this chapter, we learn how to build simple approximations to functions using straight lines or quadratic functions. This requires some new abstraction to pull off, but the gain we get in our ability to model is well worth the work. We also use our new ideas to approximate things like the difference of two closely related decay exponentials.

Chapter 15: Extreme Values We need to know more about when functions have maxima and minima. Here we give you some tools for finding where a function can have such extreme values. Our tools are derivative based.

Chapter 16: Numerical Methods Order One ODEs At this point, we show you how to solve our models using numerical tools. The first one is the Euler method and it is pretty easy to understand. It is based on approximating our models using tangent line method. Eventually we also discuss better ways called Runge-Kutta methods.

Chapter 17: Advanced Protein Models We finish the first part with a nice discussion of how we can understand the rate of error in protein transcription. We only use simple derivative ideas really and we show how our first model doesn't match experimental results but a better second model—necessarily more complicated!—does. Remember interesting biology is going to require more sophisticated mathematics and computational approaches. And that is the fun of it!

Part III: Using Multiple Variables

Chapter 18: Matrices and Vectors We are now ready to begin thinking about functions built with more that one variable. We start with some preliminaries on vectors and matrices because some of the language we need now uses those thing.

Chapter 19: A Cancer Model We then discuss a simple model of cancer which is based on six variables and show you how to solve half of it. We also show you the solutions we find are impossible to really understand—grok if you want (read Heinlein's *Stranger in a Strange Land*)—without our previously worked out approximations for differences of exponential functions.

Chapter 20: First Order Multivariable Calculus Now we add in the idea of derivatives and continuity for a function of two variables. We start with nice visuals courtesy of MatLab and then work out how to define partial derivatives and the tools we need to work with them. There is indeed a new version of the chain rule here!

Chapter 21: Second Order Multivariable Calculus We need to understand how to approximate a function of two variables with a tangent plane and how much error we make doing so. This is where we discuss that. Parts of this discussion are pretty heavy, but this is necessary. We get to extremal theory for functions of two variables and finish with a nice discussion of regression lines which uses these tools for the derivation.

Chapter 22: Hamilton's Rule In Evolutionary Biology We finish the book with a chapter on a big problem in biology: how does altruism spread throughout a population so that it is retained? This has been identified as being a problem of huge importance and it is very hard to model as it is a question it is difficult to make quantitative. We show a variety of models that slowly build insight into this question. We finish with a description of Hamilton's Rule. Hamilton postulated a parameter r, the coefficient of relatedness, in his work but it is pretty hard to understand the meaning of r. This chapter works it way up to describing r as the slope of particular regression line. We figure this out a couple of ways and one use the chain rule! So all things come full circle here. Most of this chapter uses a lot of thinking and very little mathematics beyond algebra!!

Part IV: Summing It All Up

Chapter 23: Final Thoughts We have some final thoughts on the triad of mathematics, computation and science we insist on writing down!

Part V: Advise to the Beginner

Chapter 24: Background Reading Here, we suggest some outside reading you might enjoy after you take this class!

1.7 Final Thoughts

We want you to continue to grow in these areas and we encourage you to learn more. Every time you try to figure something out in science, you will find there is a lot of stuff you don't know and you have to go learn new tricks. That's ok and you shouldn't be afraid of it. You will probably find you need more mathematics, statistics and so forth in your work, so don't forget to read more! We have written some companion texts that continue to discuss this blend of mathematics, computation and science which you can read and learn from: **Calculus for Cognitive Scientists: Higher Order Models and Their Analysis**, Peterson (2015a) and **Calculus for Cognitive Scientists: Partial Differential Equation Models**, Peterson (2015b). In addition, there is a fourth volume on bioinformation processing, **BioInformation Processing: A Primer On Computational Cognitive Science**, Peterson (2015c) which this material will prepare you for. Enjoy your journeys!

References

L. Miller, S. Alben, Interfacing mathematics and biology: a discussion on training, research, collaborations, and funding. Integr. Comp. Biol. **52**(5), 616–621 (2012)

J. Peterson, Some thoughts on biomathematics education. Biol. Int. **54**, 130–179 (2013)

J. Peterson, *Calculus for Cognitive Scientists: Higher Order Models and Their Analysis*, Springer Series on Cognitive Science and Technology (Springer Science+Business Media Singapore Pte Ltd., Singapore, 2015a in press)

J. Peterson, *Calculus for Cognitive Scientists: Partial Differential Equation Models*, Springer Series on Cognitive Science and Technology (Springer Science+Business Media Singapore Pte Ltd., Singapore, 2015b in press)

J. Peterson, *BioInformation Processing: A Primer On Computational Cognitive Science*, Springer Series on Cognitive Science and Technology (Springer Science+Business Media Singapore Pte Ltd., Singapore, 2015c in press)

Part II
Using One Variable

Chapter 2
Viability Selection

Let's start you on your journey towards an understanding of the ideas of basic calculus by just looking at how far we can go with applying algebra alone. We are going to look carefully at a model of natural selection called **viability selection**. We are going to build some nice models using lots of letters to represent things of interest and all we need to understand this is your basic algebra skills from high school. Now in our experience, most of you probably dislike mathematical things because they seem so divorced from something real. So our tact here is to build something useful and along the way, we have to do various manipulations which will help you gently remember all your basic skills. Now these discussions come from the first chapter of a great book on evolutionary biology (McElreath and Boyd 2007) which we encourage you to pick up and study at some point. We are not going to cover these ideas in the great detail there; instead we will focus on a few portions. We also use a bit different language as our target is people who are just seeing calculus perhaps although we are also hoping to get you to read this as a return to old things. At any rate, remember, the purpose of modeling is insight into complicated stuff and mathematics is a great tool to help us achieve that. Now let's get to it.

2.1 A Basic Evolutionary Model

We are interested in understanding the long term effects of genes in a population. Obviously, it is very hard to even frame questions about this. One of the benefits of our use of mathematics is that it allows us to build a very simplified model which nevertheless helps us understand general principles. These are biological versions of the famous Einstein *gedanken* experiments: i.e. thought experiments which help develop intuition and clarity.

We will explicitly use *letters* to represent quantities in the biology we want to keep track of. We will then want to make some assumptions about how these quantities depend and interact with one another. Of course, in doing so, we always make error

J.K. Peterson, *Calculus for Cognitive Scientists*, Cognitive Science
and Technology, DOI 10.1007/978-981-287-874-8_2

which is called **model error** as there is no way we can specify everything in a real biological scenario. We must make assumptions.

We start by assuming we have a population of N individuals at a given time. It also seems reasonable to think of our time unit as **generations**. So we would say the population at generation t is given by $N(t)$. Hence, the number of individuals in our population is not necessarily fixed but can change from one generation to the next. To understand this change, we need to know how our population reproduces so as to create the next generation. We want a very simple model here, so we will also assume each individual in our population is **haploid** which means new individuals are produced without sex or any sort of genetic recombination. Wait a minute, you say! What could this mean? Usually, an adult has a certain number of chromosomes; call this number $2P$. The gametes are the cells with half of the genetic material and therefore have only P chromosomes. Then in sexual reproduction there is a complicated process by which the sperm and the egg interact to create a new cell called a *zygote* which has $2P$ chromosomes. The gametes are considered **haploid** as they each have half of the chromosomes of the adult. Sexual reproduction allows a mixing of the chromosomes from two adults to form a new individual having $2P$ chromosomes. The cell formed by the union of the sperm and egg is called a **zygote** and it is diploid as it has $2P$ chromosomes. So we are making a very big simplifying assumption. Essentially we are saying each adult has Q chromosomes and the reproduction process does not mix genetic information from another adult and hence the zygote formed by what is evidently some form of asexual reproduction also has Q chromosomes. Note, calling the cell formed by this reproductive process a zygote is a bit odd as usually that word is reserved for the cell formed by a sexual reproduction. So we have a really simplistic population dynamic here! Hence, we use the term *zygote* here loosely!

With this said, we also assume in each generation individuals go through their life cycle exactly the same: all individuals are born at the same time and all individuals reproduce at the same time. We will call this a **discrete** dynamic. Note a zygote does not have to live long enough to survive to an adult.

Since we want to develop a very simple model we assume there are only two genotypes, type **A** and type **B**. We also assume **A** is more likely to survive to an adult. We need to start defining quantities of interest: i.e. variables now.

- N_A is the number of individuals of phenotype A and N_B is the number of individuals of phenotype B in a given generation.
- N is the number of individuals in the population and this number changes each generation as

$$N = N_A + N_B.$$

This is our first equation and note it is pretty simple. It simply counts things. Now if we really wanted to be careful, we might let t represent the generation we are in: i.e we start at generation 0 so $t = 0$ initially. Then the first generation is $t = 1$ and so forth. Eventually, we will want to know what happens in the population as

t increases, but that is for later. So if we wanted to be really explicit, we would add the variable t to our equation above to get

$$N(t) = N_A(t) + N_B(t).$$

It is a bookkeeping device really and not terribly essential. However, without adding (t) in there variables, we have to remember that each one does depend on the generation we are in. Also, note generations here are integers not numbers like $1/2$ or 2.4. Nothing but integers for now!

- It is also convenient to keep track of the fraction of individuals in the population that are genotype **A** or **B**. This fraction is also called the **frequency** of type **A** and **B** respectively. We use new variables for this.

$$P_A = \frac{N_A}{N_A + N_B}$$
$$P_B = \frac{N_B}{N_A + N_B}$$

We can also explicitly add the generation variable t to get

$$P_A(t) = \frac{N_A(t)}{N_A(t) + N_B(t)}$$
$$P_B(t) = \frac{N_B(t)}{N_A(t) + N_B(t)}$$

- Finally, as we said, individuals of each phenotype do not necessarily survive to adulthood. We will assume each phenotype survives to adulthood with a certain probability, V_A and V_B, respectively.

2.1.1 Examples

Let's do some calculations.

Example 2.1.1 If $N_A(0) = 25$ and $N_B(0) = 15$, find $P_A(0)$ and $P_B(0)$.

Solution *This is an easy computation.*

$$P_A(0) = \frac{N_A(0)}{N_A(0) + N_B(0)} = \frac{25}{25 + 15} = \frac{25}{40}$$

$$\begin{aligned}P_B(t) &= \frac{N_B(0)}{N_A(0) + N_B(0)}\\ &= \frac{15}{25 + 15} = \frac{15}{40}.\end{aligned}$$

2.1.2 *Homework*

Exercise 2.1.1 *If* $N_A(0) = 35$ *and* $N_B(0) = 60$*, find* $P_A(0)$ *and* $P_B(0)$.

Exercise 2.1.2 *If* $N_A(0) = 1500$ *and* $N_B(0) = 900$*, find* $P_A(0)$ *and* $P_B(0)$.

Exercise 2.1.3 *If* $N_A(0) = 82$ *and* $N_B(0) = 47$*, find* $P_A(0)$ *and* $P_B(0)$.

2.2 The Next Generation

We now have enough of a setup to look more carefully at how our population moves from one generation to the next. Let's do this very carefully. We will keep track of how the frequency P_A changes. Let $P_A(t)$ be the frequency for **A** at generation t. What is the frequency at the next generation $t + 1$? Here is how we figure it out.

- The number of zygotes from individuals of genotype **A** at generation t is assumed to be $z\,N_A(t)$ where z is the number of zygotes each individual of type **A** produces. Note that z plays the role of the **fertility** of individuals of type **A**. We will assume that individuals of type **B** also create z zygotes. Hence, their fertility is also z. We could also have assumed these fertilities are different and labeled them as z_A and z_B, respectively, but we aren't doing that here. So the number of zygotes of **B** type individuals is $z\,N_B(t)$.
- The frequency of **A** zygotes at generation t is then

$$P_{AZ}(t) = \frac{z\,N_A(t)}{z\,N_A(t) + z\,N_B(t)}$$

where we add an additional subscript to indicate we are looking at zygote frequencies. Note the z's cancel to show us that the frequency of **A** zygotes in the population does not depend on the value of z at all. We have

$$P_{AZ}(t) = \frac{N_A(t)}{N_A(t) + N_B(t)} = P_A(t).$$

- But not all zygotes survive to adulthood. If we multiply numbers of zygotes by their probability of survival, V_A or V_B, the number of **A** zygotes that survive to adulthood is $V_A\,N_A(t)$ and the number of **B** zygotes that survive to adulthood is

$V_B \, N_B(t)$. We see the frequency of **A** zygotes that survive to adulthood to give the generation $t+1$ must be

$$P_{AZS}(t+1) = \frac{V_A \, N_A(t)}{V_A \, N_A(t) + V_B \, N_B(t)}.$$

where we have added yet another subscript S to indicate survival. Note we add the generation label $t+1$ to P_{AZS} because this number is the frequency of adults that start generation $t+1$. Also, note this fraction is exactly how we define our usual frequency of **A** at generation $t+1$. Hence, we can say

$$P_A(t+1) = \frac{V_A \, N_A(t)}{V_A \, N_A(t) + V_B \, N_B(t)}.$$

Now for the final step. From the way we define stuff, notice that

$$P_A(t) = \frac{N_A(t)}{N_A(t) + N_B(t)}$$

Now replace the denominator by $N(t)$ and multiply both sides by this denominator to get

$$N(t) \, P_A(t) = N_A(t).$$

Then consider the frequency for **B**. Note

$$\begin{aligned} 1 - P_A(t) &= 1 - \frac{N_A(t)}{N_A(t) + N_B(t)} \\ &= 1 - \frac{N_A(t)}{N(t)}. \end{aligned}$$

Getting a common denominator, we find

$$1 - P_A(t) = \frac{N(t) - N_A(t)}{N(t)}.$$

But we can simplify $N(t) - N_A(t)$ to simply $N_B(t)$. Using this in the last equation, we have found that

$$1 - P_A(t) = \frac{N_B(t)}{N(t)}$$

which leads to the identity we wanted: $N_B(t) = (1 - P_A(t)) \, N(t)$. This analysis works just fine at generation $t+1$ too, but at that generation, we have

$$\begin{aligned} N_A(t+1) &= N(t)\, V_A\, P_A(t) \\ N_B(t+1) &= N(t)\, V_B\, P_B(t) \\ &= N(t)\, V_B\, (1 - P_A(t)) \end{aligned}$$

So we can say

$$\begin{aligned} P_A(t+1) &= \frac{N_A(t+1)}{N_A(t+1) + N_B(t+1)} \\ &= \frac{P_A(t)\, N(t)\, V_A}{P_A(t)\, N(t)\, V_A + (1 - P_A(t)) N(t)\, V_B}. \end{aligned}$$

Since $N(t)$ is common in both the numerator and denominator, we can cancel them to get

$$P_A(t+1) = \frac{V_A\, P_A(t)}{P_A(t)\, V_A + (1 - P_A(t))\, V_B}.$$

This tells us how the frequency of the **A** genotype changes each generation.

2.2.1 Examples

Example 2.2.1 Let $V_A = 0.7$ and $V_B = 0.45$ and assume $P_A(0) = 0.03$. Find $P_A(1)$.

Solution *We know*

$$\begin{aligned} P_A(1) &= \frac{V_A\, P_A(0)}{P_A(0)\, V_A + (1 - P_A(0))\, V_B} \\ &= \frac{0.7\,(0.03)}{0.7\,(0.03) + 0.45\,(0.97)} = \frac{0.0210}{0.0210 + 0.4365} = 0.0459. \end{aligned}$$

2.2.2 Homework

Exercise 2.2.1 *Let* $V_A = 0.8$ *and* $V_B = 0.25$ *and assume* $P_A(0) = 0.02$. *Find* $P_A(1)$.

Exercise 2.2.2 *Let* $V_A = 0.6$ *and* $V_B = 0.1$ *and assume* $P_A(0) = 0.01$. *Find* $P_A(1)$.

Exercise 2.2.3 *Let* $V_A = 0.85$ *and* $V_B = 0.3$ *and assume* $P_A(0) = 0.05$. *Find* $P_A(1)$.

2.3 A Difference Equation

We can also derive a formula for the change in frequency at each generation by doing a subtraction. We consider

$$P_A(t+1) - P_A(t) = \frac{V_A\, P_A(t)}{P_A(t)\, V_A + (1 - P_A(t))\, V_B}. - P_A(t).$$

Now the algebra gets a bit messy, but bear with us. Get a common denominator next.

$$P_A(t+1) - P_A(t) = \frac{V_A\, P_A(t)}{P_A(t)\, V_A + (1 - P_A(t))\, V_B}. - P_A(t)\, \frac{P_A(t)\, V_A + (1 - P_A(t))\, V_B}{P_A(t)\, V_A + (1 - P_A(t))\, V_B}.$$

Multiply everything out and put into one big fraction.

$$\begin{aligned} P_A(t+1) - P_A(t) &= \frac{V_A\, P_A(t) - P_A(t)\left(P_A(t)\, V_A + (1 - P_A(t))\, V_B\right)}{P_A(t)\, V_A + (1 - P_A(t))\, V_B} \\ &= \frac{V_A\, P_A(t) - (P_A(t))^2\, V_A - \left(P_A(t)\, (1 - P_A(t))\, V_B\right)}{P_A(t)\, V_A + (1 - P_A(t))\, V_B}. \end{aligned}$$

Whew! Messy indeed. Now factor a bit to get

$$P_A(t+1) - P_A(t) = \frac{\left(P_A(t) - {P_A}^2(t)\right) V_A - \left(P_A(t)\, (1 - P_A(t))\, V_B\right)}{P_A(t)\, V_A + (1 - P_A(t))\, V_B}.$$

Almost done. As a final step note that $P_A(t) - {P_A}^2(t)$ is the same as $P_A(t)(1 - P_A(t))$. So in the numerator of this complicated fraction, we can factor that term out to give

$$P_A(t+1) - P_A(t) = \frac{\left(P_A(t)\, (1 - P_A(t))\right)\left(V_A - V_B\right)}{P_A(t)\, V_A + (1 - P_A(t))\, V_B}.$$

This is what is called a **recursion** equation. It is called that because it gives us a formula which tells us how to find the next generation results give the previous generation's frequency. For example, if the frequency of **A** in the population started at $P(0) = p_0$, the change in frequency of **A** in the next generation is given by

$$P_A(1) - P_A(0) = \frac{\left(P_A(0)\, (1 - P_A(0))\right)\left(V_A - V_B\right)}{P_A(0)\, V_A + (1 - P_A(0))\, V_B}.$$

Substituting in the value p_0 we find we can solve for $\boldsymbol{P_A}(1)$ as follows:

$$\boldsymbol{P_A}(1) = p_0 + \frac{p_0\,(1-p_0)\left(\boldsymbol{V_A} - \boldsymbol{V_B}\right)}{p_0\,\boldsymbol{V_A} + (1-p_0)\,\boldsymbol{V_B}}$$

and continuing in this vein, the frequency at generation 2 would be

$$\boldsymbol{P_A}(2) = p_1 + \frac{p_1\,(1-p_1)\left(\boldsymbol{V_A} - \boldsymbol{V_B}\right)}{p_1\,\boldsymbol{V_A} + (1-p_1)\,\boldsymbol{V_B}}$$

where we denote $\boldsymbol{P_A}(1)$ more simply by p_1. We can then continue and calculate as many of these as we want. Note, this will give us a table of numbers: for each generation t, we calculate a frequency of **A** in the population $\boldsymbol{P_A}(t)$. We can then plot these ordered pairs on a standard set of axis: the horizontal **generation** axis and the vertical **frequency** of **A** axis. Of course, these means a lot of computation! So as time goes on, we will use a tool called **MatLab** to do this more easily. We also notice that an interesting question is what happens to this frequency as the number of generations increase? Doe the frequency plateau at some level or does it rise forever? A little thought should show you that the frequency can't go above 1 as frequency is a fraction between 0 and 1 at each generation. So the question is does the frequency plateau at 1 or something less? With this question, we are really asking about what is called the **limiting** behavior of our frequency model. Our studies in Calculus will give us a variety of tools and ideas to let us handle this question, but already you should see that it is an important thing to think about.

Another important thing to notice is that the idea of **generation** is a fluid concept as **generation** means very different things in different species. So when we use t to represent a generation, the time interval to get to generation $t+1$ can be years, months, days, hours and even less. Our severely odd thought experiment creature of this model can be replaced by other creatures with sexual reproduction and all sorts of other more accurate assumptions. But the basic questions will still be the same. We use our model to find how the frequencies change in each generation and we use calculation to generate plots so we can see the behavior visually. And we ask the big question: what is the limiting behavior?

2.3.1 Examples

We should do some examples of how to use these ideas. First, some computations.

Example 2.3.1 Find the frequency of type **A** individuals at generation 1 given that initially the frequency of **A** individuals is $p_0 = 0.01$ and the probabilities of **A** and **B** are $\boldsymbol{V_A} = 0.8$ and $\boldsymbol{V_B} = 0.3$.

Solution *We use the equation above. We have*

$$\begin{aligned}
\boldsymbol{P_A}(1) &= p_0 + \frac{p_0\,(1-p_0)\left(\boldsymbol{V_A} - \boldsymbol{V_B}\right)}{p_0\,\boldsymbol{V_A} + (1-p_0)\,\boldsymbol{V_B}} \\
&= 0.01 + \frac{0.01\,(0.99)\,(0.8-0.3)}{0.01\,(0.8) + 0.99\,(0.3)} \\
&= 0.01 + \frac{0.00495}{0.305} = 0.01 + 0.01623 = 0.0262.
\end{aligned}$$

So the change in frequency of **A** *is* $\boldsymbol{P_A}(1) - \boldsymbol{P_A}(0) = 0.0262 - 0.01 = 0.0162$ *with the new frequency given by* $\boldsymbol{P_A}(1) = 0.0262$. *We can clearly calculate as many of these as we wish.*

2.3.2 Homework

Exercise 2.3.1 *Find the frequency of type* **A** *individuals at generation* 1 *given that initially the frequency of* **A** *individuals is* $p_0 = 0.02$ *and the probabilities of* **A** *and* **b** *are* $\boldsymbol{V_A} = 0.85$ *and* $\boldsymbol{V_B} = 0.25$.

Exercise 2.3.2 *Find the frequency of type* **A** *individuals at generation* 1 *given that initially the frequency of* **A** *individuals is* $p_0 = 0.03$ *and the probabilities of* **A** *and* **b** *are* $\boldsymbol{V_A} = 0.75$ *and* $\boldsymbol{V_B} = 0.45$.

Exercise 2.3.3 *Find the frequency of type* **A** *individuals at generation* 1 *given that initially the frequency of* **A** *individuals is* $p_0 = 0.04$ *and the probabilities of* **A** *and* **b** *are* $\boldsymbol{V_A} = 0.65$ *and* $\boldsymbol{V_B} = 0.2$.

Exercise 2.3.4 *Find the frequency of type* **A** *individuals at generation* 1 *and generation* 2 *given that initially the frequency of* **A** *individuals is* $p_0 = 0.02$ *and the probabilities of* **A** *and* **b** *are* $\boldsymbol{V_A} = 0.75$ *and* $\boldsymbol{V_B} = 0.3$.

Exercise 2.3.5 *Find the frequency of type* **A** *individuals at generation* 1 *and generation* 2 *given that initially the frequency of* **A** *individuals is* $p_0 = 0.01$ *and the probabilities of* **A** *and* **b** *are* $\boldsymbol{V_A} = 0.6$ *and* $\boldsymbol{V_B} = 0.2$.

2.4 The Functional Form of the Frequency

Let's derive a functional form for $\boldsymbol{P_A}(t)$. We will start with the number of **A** in the population initially, $\boldsymbol{N_A}(0)$. From our discussions, we know that at the next generation, the number of **A** is

$$N_A(1) = \left(\text{Fertility of } \mathbf{A}\right) \times \left(\text{The number of } \mathbf{A} \text{ at generation } 0\right)$$
$$\times \left(\text{The probability an } \mathbf{A} \text{ type lives to adulthood}\right)$$
$$= z\, N_A(0)\, V_A.$$

Then, it is easy to see that $N_A(2)$ is given by

$$N_A(2) = \left(\text{Fertility of } \mathbf{A}\right) \times \left(\text{The number of } \mathbf{A} \text{ at generation } 1\right)$$
$$\times \left(\text{The probability an } \mathbf{A} \text{ type lives to adulthood}\right)$$
$$= z\, N_A(1)\, V_A = z\,(z\, N_A(0)\, V_A)\, V_A = N_A(0)\,(z\, V_A)^2.$$

We can do this over and over again. We find

$$N_A(3) = N_A(0)\,(z\, V_A)^3$$
$$N_A(4) = N_A(0)\,(z\, V_A)^4$$
$$\vdots$$

We can easily extrapolate from this to see that, in general,

$$N_A(t) = N_A(0)\,(z\, V_A)^t$$

and a similar analysis shows us that

$$N_B(t) = N_B(0)\,(z\, V_B)^t$$

We are now ready to figure out a functional form for $P_A(t)$. From the definition of the frequency for **A**, we have

$$P_A(t) = \frac{N_A(t)}{N_A(t) + N_B(t)}.$$

Now divide top and bottom of this fraction by $N_A(t)$ to get

$$P_A(t) = \frac{\frac{N_A(t)}{N_A(t)}}{\frac{N_A(t)+N_B(t)}{N_A(t)}}$$
$$= \frac{1}{1 + \frac{N_B(t)}{N_A(t)}}.$$

Now plug in our formulae for $N_A(t)$ and $N_B(t)$. Note the fraction

$$\begin{aligned}\frac{N_B(t)}{N_A(t)} &= \frac{N_B(0)\,(z\,V_B)^t}{N_A(0)\,(z\,V_A)^t} \\ &= \frac{N_B(0)}{N_A(0)}\left(\frac{V_B}{V_A}\right)^t.\end{aligned}$$

Using this in our frequency formula, we have

$$P_A(t) = \frac{1}{1 + \frac{N_B(0)}{N_A(0)}\left(\frac{V_B}{V_A}\right)^t}.$$

2.4.1 Examples

Let's redo our previous example using this new formula.

Example 2.4.1 Find the frequency of type **A** individuals at generation 1 given that initially the frequency of **A** individuals is $p_0 = 0.01$ and the probabilities of **A** and **B** are $V_A = 0.8$ and $V_B = 0.3$.

Solution *Since* $p_0 = 0.01$, *we can use that to find the initial values of* $N_A(0)$ *and* $N_B(0)$. *We have*

$$P_A(0) = 0.01 = \frac{N_A(0)}{N_A(0) + N_B(0)}.$$

Now rewrite this as by dividing top and bottom of the fraction by $N_A(0)$ *to get*

$$P_A(0) = 0.01 = \frac{1}{1 + \frac{N_B(0)}{N_A(0)}}.$$

We can then solve for

$$\begin{aligned}\frac{1}{1 + \frac{N_B(0)}{N_A(0)}} &= 0.01 \\ 1 + \frac{N_B(0)}{N_A(0)} &= \frac{1}{0.01} = 100 \\ \frac{N_B(0)}{N_A(0)} &= 99.\end{aligned}$$

Using the probabilities, we thus have

$$
\begin{aligned}
\boldsymbol{P_A}(1) &= \frac{1}{1 + \frac{N_B(0)}{N_A(0)} \frac{V_B}{V_A}} \\
&= \frac{1}{1 + 99.0\,(\frac{0.3}{0.8})^1} \\
&= \frac{1}{1 + 99.0\,\frac{3}{8}} \\
&= \frac{1}{38.125} = 0.0262
\end{aligned}
$$

So the change in frequency of **A** *is* $\boldsymbol{P_A}(1) - \boldsymbol{P_A}(0) = 0.0262 - 0.01 = 0.0162$ *with the new frequency given by* $\boldsymbol{P_A}(1) = 0.0264$*. This is just like we calculated before. However, this formula makes it easy for us to find out what happens after many generations! Note,* $\boldsymbol{P_A}(5)$ *would be*

$$\boldsymbol{P_A}(5) = \frac{1}{1 + 99.0\,(\frac{0.3}{0.8})^5}$$

and $\boldsymbol{P_A}(15)$ *would be*

$$\boldsymbol{P_A}(15) = \frac{1}{1 + 99.0\,(\frac{0.3}{0.8})^{15}}$$

2.4.2 Homework

Now use our new frequency formula to find some frequencies of **A** in the population. Note how, although we are not doing it yet, we can now do all the calculations to generate a plot of $\boldsymbol{P_A}(t)$ versus generation t!

Exercise 2.4.1 *Find the frequency of type* **A** *individuals at generation* 5 *given that initially the frequency of* **A** *individuals is* $p_0 = 0.02$ *and the probabilities of* **A** *and* **b** *are* $\boldsymbol{V_A} = 0.85$ *and* $\boldsymbol{V_B} = 0.25$.

Exercise 2.4.2 *Find the frequency of type* **A** *individuals at generation* 6 *given that initially the frequency of* **A** *individuals is* $p_0 = 0.03$ *and the probabilities of* **A** *and* **b** *are* $\boldsymbol{V_A} = 0.75$ *and* $\boldsymbol{V_B} = 0.45$.

Exercise 2.4.3 *Find the frequency of type* **A** *individuals at generation* 10 *given that initially the frequency of* **A** *individuals is* $p_0 = 0.04$ *and the probabilities of* **A** *and* **b** *are* $\boldsymbol{V_A} = 0.65$ *and* $\boldsymbol{V_B} = 0.2$.

Exercise 2.4.4 *Find the frequency of type* **A** *individuals at generation* 5*, generation* 10 *and generation* 20 *given that initially the frequency of* **A** *individuals is* $p_0 = 0.02$ *and the probabilities of* **A** *and* **b** *are* $V_A = 0.75$ *and* $V_B = 0.3$*. What do you think is happening as the number of generations increases?*

Exercise 2.4.5 *Find the frequency of type* **A** *individuals at generation* 10*, and generation* 20 *and generation* 30 *given that initially the frequency of* **A** *individuals is* $p_0 = 0.01$ *and the probabilities of* **A** *and* **b** *are* $V_A = 0.6$ *and* $V_B = 0.2$*. What do you think is happening as the number of generations increases?*

2.4.3 Biology and the Model

Let's step back a minute and ask what we are doing. We have been interested in exploring how a population of two phenotypes spread throughout a population. Since the size of the population can change at each generation, it is much more useful to look at the frequencies of each phenotype in the population. We have developed a model that gives us insight into how this happens. Recall our recursion equation for the change in frequency for type **A**:

$$P_A(t+1) - P_A(t) = \frac{\left(P_A(t)\,(1 - P_A(t))\right)\left(V_A - V_B\right)}{P_A(t)\,V_A + (1 - P_A(t))\,V_B}.$$

It is traditional in evolutionary biology to look at the denominator term $P_A(t)\,V_A + (1 - P_A(t))\,V_B$ and try to understand what it means biologically. A common way to interpreting it is to tag the term $P_A(t)\,V_A$ as the **fitness** of type **A** in the population and the other term, $(1 - P_A(t))\,V_B$ as the fitness of type **B**. Hence, the denominator term is like a weighted or average fitness of the entire population. We often use a **bar** above a variable to indicate we are looking at that quantities *average value*. The fitness is usually denoted by the variable w and hence, this denominator term is $\bar{w}$.

To get additional insight, let's let the symbol ΔP_A represent the change $P_A(t+1) - P_A(t)$; we read the symbol Δ as **change in**. And let's drop all the (t) labels so we can rewrite our recursion as

$$\Delta P_A = \frac{P_A\,(1 - P_A)\,(V_A - V_B)}{\bar{w}}.$$

Finally, we could do this sort of analysis for any phenotype not just type **A**. So let's replace the specific term P_A by just p. Also the term $V_A - V_B$ is the difference in *fitness* between **A** and **B**. Let's denote that by Δf where f represents **fitness**. Then we can rewrite our recursion again as

$$\Delta p = p\,(1 - p)\,\frac{\Delta f}{\bar{w}}.$$

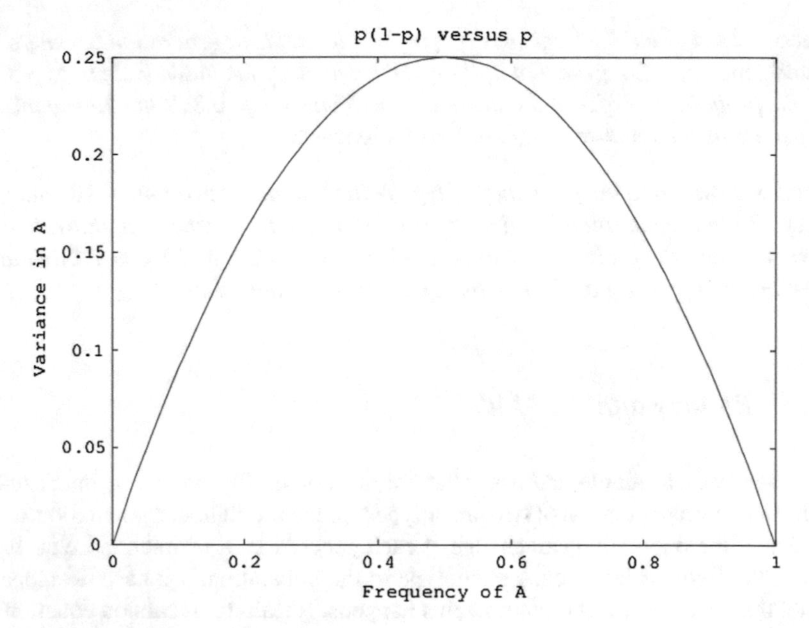

Fig. 2.1 The product $\boldsymbol{p}\,(1 - \boldsymbol{p})$

If we assume the difference in fitness is constant, then the amount of change in the frequency of **A** is determined by the leading term $\boldsymbol{p}\,(1 - \boldsymbol{p})$. To see what this means, consider the graph of $\boldsymbol{p}$ on the interval [0, 1]. We don't need to use any other values of $\boldsymbol{p}$ as it is a frequency as so ranges from 0 to 1. This graph is shown in Fig. 2.1.

The largest value the product $\boldsymbol{p}\,(1 - \boldsymbol{p})$ can be seen by looking at the plot. This maximum value occurs at $\boldsymbol{p} = 0.5$. Hence, the change in **fitness** is largest when the product $\boldsymbol{p}\,(1 - \boldsymbol{p})$ is largest. If $\boldsymbol{p}$ is small or close to 1, the product $\boldsymbol{p}\,(1 - \boldsymbol{p})$ is also small. We usually interpret our equation

$$\boldsymbol{\Delta p} = \boldsymbol{p}\,(1 - \boldsymbol{p})\,\frac{\boldsymbol{\Delta f}}{\bar{\boldsymbol{w}}}.$$

as the magnitude of **natural selection** acting on our population. Hence, we can see that **natural selection** is made largest (we say **maximized**) when the product $\boldsymbol{p}\,(1 - \boldsymbol{p})$ is maximized. It turns out the product $\boldsymbol{p}\,(1 - \boldsymbol{p})$ is known as the **variance** in genotypes in the population which is why we used the term *variance* in the title of Fig. 2.1! So our thought experiment has shed light on an interesting principle: **fitness is maximized when the variance of the genotypes in the population is maximal**!!

Remember, our goal with the mathematics and the computations we do is to gain insight into how to understand biological processes.

2.5 A Gentle Introduction to MatLab

In this book, we will use a computational tool called **MatLab** (see http://www.mathworks.com/) so that you can learn a little about a typical interactive program that many scientists use for their work. Once you have it installed, if you are using your laptop or computer, just click on the Matlab icon to get started. Next, create a folder or directory on your laptop for your work in this book. Something like **neanderthal**. All of our stuff for this book will then go into that folder. When you start up MatLab, you see the **command** screen shown in Fig. 2.2. You screen image might be a bit different but it will be similar.

Next, we have to set your path. With Matlab running do this:

- On the left, there is a **File** option you can click on. Click on that and go to the **Save Path** option.
- When the program Matlab is searching for its instructions on how to do things, it first looks for them in all of the folders that were setup for Matlab as part of its installation. We don't want to put any of our stuff in those folders.
- We want our stuff in our personal folder **neanderthal**. So scroll down to find this folder, choose it and then click on the **Add Path** button to add this folder to the *search path*. Then choose **Save** and you are done.

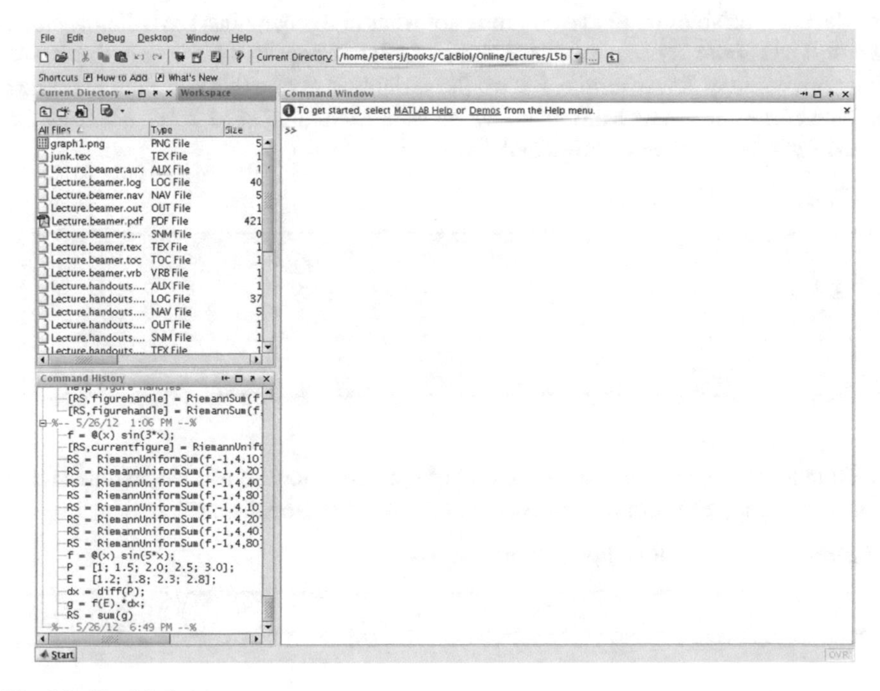

Fig. 2.2 The MatLab startup screen

- Now at the top of the running Matlab, choose the folder **neanderthal** so that Matlab is running in that folder. You can check this by typing **pwd** in the big middle window. It should spit back this folder name.

2.5.1 Matlab Vectors

First, set up our function. MatLab allows us to define a function inside the MatLab environment as follows

Listing 2.1: Defining a Matlab Function

```
f = @(x) (x.^2);
```

This defines the function $f(x) = x^2$. If we had wanted to define $g(x) = 2 * x^2 + 3$, we would have used

Listing 2.2: Defining a second Matlab Function

```
g= @(x) (2*x.^2+3);
```

In MatLab, variables that have columns are what in mathematics are called *vectors*. We will talk about *vectors*, *matrices* and so forth in Chap. 18 a bit later in this book. For now, consider this example. Set up the variable *X* as one that has columns. The **;** between the numbers let MatLab know we want each number to start a new row. So *X* is a variable which has 3 rows and 1 column.

Listing 2.3: A column vector

```
X = [1; 2; 3]

X =

      1
      2
      3
```

Note there is no semicolon at the end of the line below so Matlab displays what *X* is after we type the command. Adding the **;** turns off the display.

Listing 2.4: A column vector with no display

```
X = [1; 2; 3];
```

If we had used **,** between the numbers instead of the **;** we would have made a variable which consists of 1 row and 3 columns:

Listing 2.5: A row vector

```
Y = [1, 2, 3]
Y =

       1     2     3
```

Adding the **;** turns off the display.

Listing 2.6: A row vector with no display

```
Y = [1, 2, 6, -8];
```

Now let Z be another column vector the same *size* as X.

Listing 2.7: New Vector Z

```
Z = [4;-2;6]

Z =

     4
    -2
     6
```

The MatLab notation $X.*Z$ means to multiple the components of X and Z as follows

$$\begin{bmatrix} 1 \\ 2 \\ 3 \end{bmatrix} \cdot * \begin{bmatrix} 4 \\ -2 \\ 6 \end{bmatrix} = \begin{bmatrix} (1)(4) \\ (2)(-2) \\ (3)(6) \end{bmatrix} = \begin{bmatrix} 4 \\ -4 \\ 18 \end{bmatrix}$$

So in MatLab, we have

Listing 2.8: Component wise Multiplication of Vectors

```
X.*Z

ans =

     4
    -4
    18
```

If we wanted to square everything in X, we would write $X.^2$ to square each *component* creating a new *vector* with each entry squared.

Listing 2.9: A Vector Squared

```
X.^2
ans =

     1
     4
     9
```

The way we set up the function

Listing 2.10: The function f(x)

```
f = @(x) (x.^2);
```

makes use of this. We don't know if the variable x is a *vector* or not. So we write $x.^2$ so that if x is a vector, we handle the squaring of each component properly. Otherwise, MatLab *ignores* the extra **.** in front of the multiplication symbol * when the variable x is just a number. So for our function, to find $f(2)$, we just type

Listing 2.11: f(2)

```
f(2)

ans =

     4
```

as the argument sent into `f` is just a number. But if we had sent in a vector, like `X`, it would still be handled properly because of the way we wrote the code for `f`. So for our function, to find f for all the values in X, we just type

Listing 2.12: f(X)

```
f(X)
ans =
     1
     4
     9
```

2.5.1.1 Examples

Example 2.5.1 Let's set up some vectors. Write the Matlab code to set up the row vector $A = [1, 2, -5, 8]$ and the column vector

$$B = \begin{bmatrix} -4 \\ 2 \\ -5 \\ 10 \end{bmatrix}$$

Solution *We write*

Listing 2.13: Set up some vectors

```
A = [1,2,-5,8]
B = [-4;2;-5;10]
```

2.5.1.2 Homework

Exercise 2.5.1 *Write the Matlab code to set up the row vector $A = [-2, 4, 3, -18]$ and the column vector*

$$B = \begin{bmatrix} 6 \\ 10 \\ 7 \\ 1 \end{bmatrix}$$

Exercise 2.5.2 *Write the Matlab code to set up the row vector $A = [4, 0, -50, 80]$ and the column vector*

$$B = \begin{bmatrix} -40 \\ 22 \\ 35 \\ 8 \end{bmatrix}$$

Exercise 2.5.3 *Write the Matlab code to set up the row vector $A = [1, 2, 0, -28]$ and the column vector*

$$B = \begin{bmatrix} -6 \\ -32 \\ 43 \\ 1 \end{bmatrix}$$

2.5.2 *Graphing a Function*

To graph f we need to set up a variable which tells us how many data points to use in the plot. This variable is different from our partition variable. The **linspace** command below sets up a variable y to be a vector with 21 points in it. The first point is 1 and the last point is 3 and the interval $[1, 3]$ is divided into 20 equal size pieces. So this command **`linspace(1,3,21)`** creates y values spaced 0.1 apart:

$$\{y_1 = 1, y_2 = 1.1, y_3 = 1.2, \ldots, y_{20} = 2.9, y_{21} = 3.0\}.$$

We use the pairs $(y_i, f(y_i))$ to make a plot by connecting the dots determined by the pairs using lines. To do the plot in Matlab is easy

Listing 2.14: Setting up a function plot

```
y = linspace(1,3,21);
plot(y,f(y));
```

We can add stuff to this bare bones plot.

Listing 2.15: Adding labels to the plot

```
xlabel('x axis');
ylabel('y axis');
legend('x^2','location','best');
title('Plot of f(x) = x^2 on [1,3]');
```

where

- **xlabel** sets the name printed under the horizontal axis.
- **ylabel** sets the name printed next to the vertical axis.
- **legend** sets a blurb printed inside the graph explaining the plot. Great when you plot multiple things on the same graph.
- **title** sets the title of the graph.

The graph pops up in a separate window as you can see. Using the **file** menu, select **save as** and scroll through the choices to save the graph as a **.png** file—a Portable Network Graphics file. You'll need to give the file a name. We chose **graph1.png** and it is shown in Fig. 2.3.

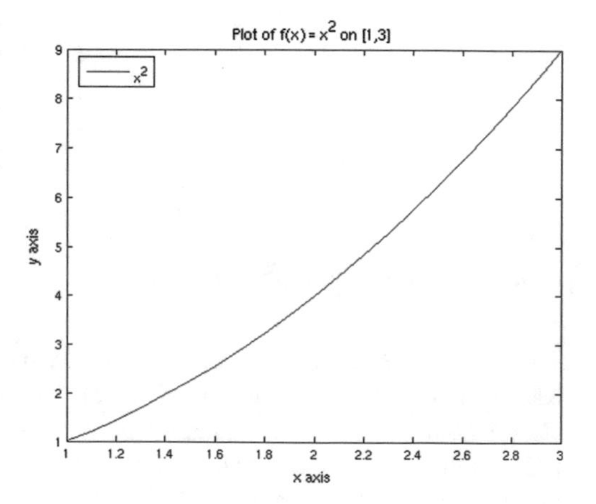

Fig. 2.3 Graph of f

2.5.2.1 Examples

You should type in these commands in Matlab on your laptop or some other computer and make sure you know how to do all the steps. To turn in the homework, make a screen print of your Matlab session. Later we will take the session and put it into a word document but that is to come.

Example 2.5.2 Let's plot $f(t) = t^3 + 2t + 3$ on the interval $[-1, 1]$.

Solution *Type in these commands to see the plot—we don't show it here but you should see it!*

Listing 2.16: Graphing $f(t) = t^3 + 2t + 3$

```
f = @(t) (t.^3 + 2*t + 3);
T = linspace(-1,1,41);
plot(T,f(T));
xlabel('t axis');
ylabel('y axis');
legend('t^3+2t=3','location','best');
title('Plot of f(t) = t^3+2t+3 on [-1,1]');
```

2.5.2.2 Homework

Graph the following functions on the given interval nicely with labels and so forth. You'll probably have to play with the **linspace** command to get a nice plot.

Exercise 2.5.4 *Graph* $f(x) = 2x + x^4$ *on the interval* $[-2, 3]$.

Exercise 2.5.5 *Graph* $f(t) = 2t - 5t^2$ *on the interval* $[-1, 2]$.

Exercise 2.5.6 *Graph* $h(y)) = 12y - 6y^3$ *on the interval* $[-4, 4]$.

2.5.3 A Simple Virus Infection Model

Let's look now at a sample use of MatLab to plot some data. We will use some data arising from computational models of West Nile Virus infection. West Nile Virus is in the Flavivirus family which is a family of viruses transmitted by mosquitoes and ticks with an impact that is important for varied sociological and economic reasons. These diseases include dengue, yellow fever, tick-borne encephalitis and West Nile fever. They are widely distributed throughout the world with the exception of the polar regions, although a specific flavivirus may be geographically restricted to a continent or a particular part of it. With global warming, these single-stranded RNA viruses are entering the radars of more regions of the world than ever. West Nile virus, for example, has emerged in recent years in temperate regions of Europe and

North America, presenting a threat to public and animal health. The most serious manifestation of the West Nile virus infection is fatal encephalitis (inflammation of the brain) in humans and horses, as well as mortality in certain domestic and wild birds. The virus is maintained in nature through a transmission cycle involving mosquitoes and birds. Children will usually experience an apparent or a mild febrile illness. Adults may experience a dengue-like illness while the elderly may develop an encephalitis which is sometimes fatal. The diagnosis is usually made by serology although the virus can be isolated from the blood in tissue culture. No vaccine for the virus is available and there is no specific therapy. The West Nile Virus infections feature a substantial up-regulation of cell surface molecules of a variety of cell types which are in the G_0 resting state of cell division. Curiously, cells that are dividing (i.e. in the G_1 state) do not have this up-regulation. Although many cell surface molecules are expressed at a much higher rate in the quiescent cells, the model developed in Peterson et al. (2015) so far focuses primarily on the increase in the MHC-I complex. A simulation model was built as closely tied to the epidemiological and biological literature as possible (although there is a mathematical framework as well). The model was then used to run simulations to help us understand the data on West Nile Virus infections that have been collected by cellular biologists that involve multiple hosts. In a survival experiment, a population of mice of size N are all infected with the same amount of antigen. In the laboratory, the level of antigen used is measured in *pfu* or *plaque forming units*. This level is actually measured visually by looking at a slide infected with antigen and seeing how many "plaques" form when the virus is exposed to a phage. Hence, it is a somewhat imprecise measurement even in the cell biologist's laboratory setting. After these N mice are infected, we wait to see how many die. We then repeat the experiment using N mice for a reasonable number of different pfu levels. Say we did this for 12 pfu levels on 12 different populations of size N. We could then graph the number of mice that survive versus the pfu level. This is called a survival experiment and as you can see, it is fairly expensive to mount, requiring $12N$ mice. If we used $N = 100$, this would be 1200 mice at perhaps $20.00 per mouse. The experiment thus costs over $24,000 for the mice alone. Needless to say, this is costly. Now a traditional survival experiment gives a survival plot which smoothly decays down to a survival of 0 at high pfu level. West Nile Virus has a peculiar survival curve which is shown in Fig. 2.4. Note that survival actually increases sometimes with increasing pfu. This is quite different from the data that we see in other viral infections. The computer simulations are used to generate this kind of survival data for 10 hosts infected at 18 different pfu levels. This data is shown in Fig. 2.6b and has been entered into a file called **survival.dat** (Fig. 2.5). We want to generate a plot of the number of mice that survive versus the logarithm of the pfu level. We can use MatLab to do this easily.

We want to load the information contained in the file **survival.dat** into a MatLab variable. You want to make sure the file is a simple text file—in a Windows machine, use *Notepad* and in a Mac, use *textedit*. Just make sure you save the file in a text format. Also, as we said above, make sure you set the path in MatLab so that MatLab can find your data file. Now, note that we have entered the data in a three column format.

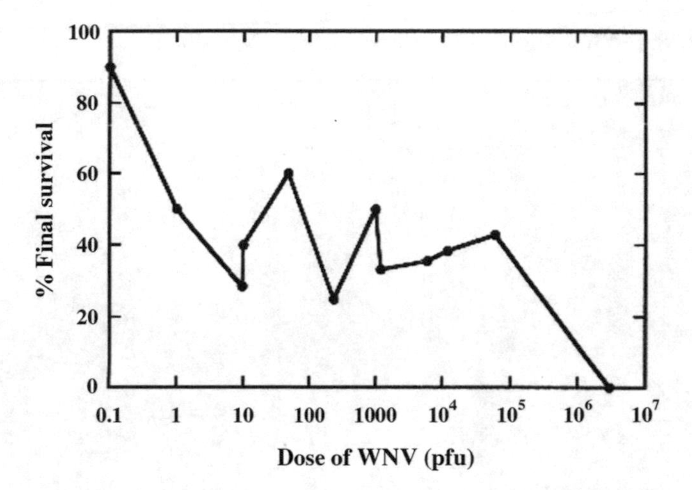

Fig. 2.4 Actual survival curve

Surviving Mice	Infection Level	Surviving Fraction	Surviving Mice	Infection Level	Surviving Fraction
10	100	0.999990	4	25000	0.439213
10	250	0.999982	2	50000	0.358584
10	500	0.808539	2	75000	0.367035
8	750	0.589315	3	90000	0.379991
10	1000	0.739119	4	120000	0.359158
8	2500	0.497502	0	360000	0.296874
5	5000	0.454120	1	600000	0.301097
3	7500	0.425226	2	900000	0.356034
3	10000	0.370141	0	1200000	0.297576

Fig. 2.5 Simulation WNV survival data

The numbers in each row are entered separated by spaces. You can't use comma's in numbers here. For example, if you entered 25000 as 25,000, the, would be interpreted in MatLab as the second entry in the row is 25, the third entry in 000 and the fourth entry is 0.439213. Of course, there should be only 3 entries in each row, so this would cause MatLab to get very confused. To load the data in this file, we use the following command.

Listing 2.17: Loading Data From a File

```
> Data = load('survival.dat');
```

The information in the file is now placed into a MatLab variable called **Data** which consists of 18 rows with 3 columns each. We take all the numbers in column one and store them in a new variable called **Survival** with this line.

Listing 2.18: Putting Survival Data into a Matlab Variable

```
Survival = Data(:,1)

Survival =

    10
    10
    10
     8
    10
     8
     5
     3
     3
     4
     2
     2
     3
     4
     0
     1
     2
     0
```

You see the data is echoed to your screen. This is because we did not put a final ; after the line. If we had, there would have been no additional output. Note the syntax here. The command **`Data(:,1)`** tells MatLab to use all the data in column 1 to load into the variable **`Survival`**. We could also have used the command **`Data(1:18,1)`**, but this is a little more work as we need to know exactly how many rows of data there are. The first way, using a : is a lot easier! Next, we load the second column of data into a variable called **`Pfu`** where we have the ellipsis to indicate data we are not showing because it takes up so many lines! Of course, in the MatLab environment, you would see all the data printed out.

Listing 2.19: Loading data into MatLab variables

```
Pfu = Data(:,2)

Pfu =

         100
         250
         500
         ...
      900000
     1200000
```

Finally, we load the last or third column into a variable called **`HealthyPercent`**.

Listing 2.20: Loading Healthy Percent

```
HealthyPercent = Data(:,3)

HealthyPercent =

    1.0000
    1.0000
    0.8085
     ...
    0.3560
    0.2976
```

Now we will look at plots involving the logarithm of the data. You have seen the logarithm function in high school and we are going to explain what logarithms are very carefully in a bit, but for now, let's assume you know about them. So, if we wanted to graph **Survival** versus the logarithm of **Pfu**, we first compute the natural logarithm of each number in **Pfu** with the line

Listing 2.21: Logarithm of Pfu

```
logPfu = log(Pfu);
```

This takes the column of information in the variable *Pfu* and applies the natural logarithm to each entry. Now we haven't discussed natural logarithms yet, but you probably remember them from high school classes. Our discussion will be in Chap. 10. But you should be able to remember enough to get the gist of what we are doing here. Now we put the **;** on the end of this line and so the new variable *logPfu* is not printed out. If we had left off **;**, we would have seen

Listing 2.22: Listing logPFU Values

```
logPfu = log(Pfu)

logPfu =

    4.6052
    5.5215
    6.2146
    6.6201
    6.9078
    7.8240
    8.5172
    8.9227
    9.2103
   10.1266
   10.8198
   11.2252
   11.4076
   11.6952
   12.7939
   13.3047
   13.7102
   13.9978
```

The plot is then generated with the lines

Listing 2.23: Simple logPfu versus Surviving Hosts Plot

```
plot(logPfu,Survival);
xlabel('Logarithm Of PFU Level');
ylabel('Surviving Hosts');
axis([4.0 14.5 -0.5 10.5]);
title('Survival Experiment: 10 Hosts, final time = 960');
```

The plot command here will use default colors and will be generated with no axis labels, title and so forth. The **xlabel**, **ylabel** commands above set the axis labels to the string we want to use. The **title** command allows us to pick the title for our graph. Finally, the **axis** command allows us to override the default minimum x, maximum x, minimum y and maximum y values used in the plot. MatLab automatically chooses these for you, but the **axis** command lets you choose more pleasing settings if you want. The first line, **plot**, generates the figure right away and you will see it pop up. As each of the other lines is typed and you hit the carriage return key, the strings are added to the existing figure. When all is done, you can go to the figure and save it as a graphics file with an appropriate extension. For us, since we want to add these files to word or open office documents, we choose **.png** or **.jpg** files. You will have to choose where you save the file also before you save it.

A similar set of lines generates the plot of **HealthyPercent** versus **logPfu**.

Listing 2.24: Healthy Percent versus logPfu Plot

```
plot(logPfu,HealthyPercent);
xlabel('Logarithm Of PFU Level');
ylabel('Percentage of Healthy Cells Left');
axis([4.0 14.5 -0.5 10.5]);
title('Survival Experiment: 10 Hosts, final time = 960');
```

The plots we have generated are shown in Fig. 2.6a, b.
Note, that the simulated survival curve is qualitatively similar to the real data shown in Fig. 2.4. To end this section, note that the entire MatLab session to build both plots is quite compact. Here it is without commentary.

Listing 2.25: Entire Matlab Survival Data Session

```
Data = load('survival.dat');
Survival = Data(:,1);
Pfu = Data(:,2);
HealthyPercent = Data(:,3);
logPfu = log(Pfu);
plot(logPfu,Survival);
xlabel('Logarithm Of PFU Level');
ylabel('Surviving Hosts');
axis([4.0 14.5 -0.5 10.5]);
title('Survival Experiment: 10 Hosts, final time = 960');
plot(logPfu,HealthyPercent);
xlabel('Logarithm Of PFU Level');
ylabel('Percentage of Healthy Cells Left');
axis([4.0 14.5 -0.1 1.05]);
title('Survival Of Healthy Cells : 10 Hosts, final time = 960');
```

Not bad, eh?

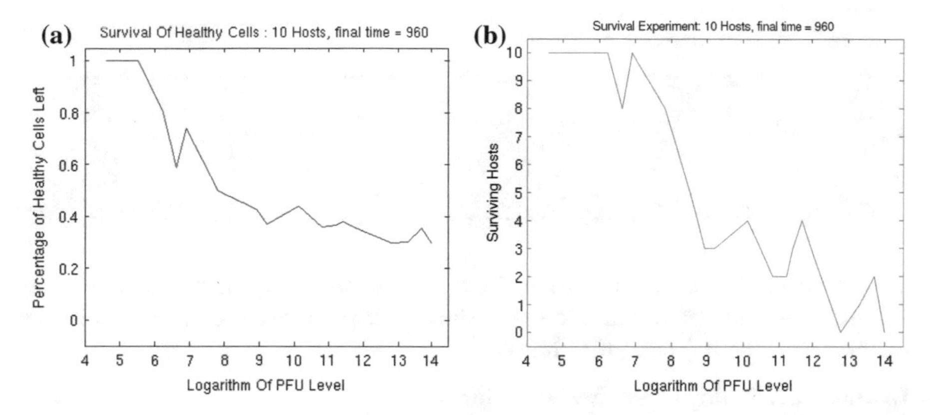

Fig. 2.6 Survival experiment results for 10 hosts and 18 fpu Levels. **a** Healthy cells left in survival experiment: we infect 10 hosts with 18 different pfu levels. The simulation is run for 960 time units. **b** Survival curve: we clearly see up and down variability in survival as pfu levels increase

2.6 Long Term Consequences

Now, let's go back to our original question about long term behavior of the frequency of **A**. Let's generate the plot in Matlab. Recall the frequency formula for **A** is given by

$$P_A(t) = \frac{1}{1 + \frac{N_B(0)}{N_A(0)} \left(\frac{V_B}{V_A} \right)^t}.$$

So we need to setup variables in Matlab for various parameters. We will use

- **`VA`** to represent V_A.
- **`VB`** to represent V_B.
- **`p0`** to represent $p0$.
- **`r`** to represent the ratio $N_B(0)/N_A(0)$. It is easy to see that when we solve for this ratio, we find $r = 1 - 1/p0 = (1 - p0)/p0$. In Matlab, this is **`r = (1-p0/p0)`** also!
- **`v`** to represent the ratio V_B/V_A.
- **`N`** to represent the last generation we are interested in looking at.

We set up these initializations like this

Listing 2.26: Initialize our constants

```
VA = .8;
VB = .3;
p0 = 0.01;
r = (1-p0)/p0;
v = VB/VA;
N = 25;
```

We then define the **P** function. Since Matlab starts everything at the counter 1, we will use $t - 1$ in our formula to make sure the Matlab **P(1)** corresponds to the mathematical $\boldsymbol{P}(0)$. The Matlab line is

Listing 2.27: Setting up the frequency function

```
P = @(t) 1./(1+r*v.^(t-1));
```

Note we use the Matlab **./** and **.^** to allow our function to deal with data that is a vector (we talked about this!). Then, we set up our **linspace** to give us generations 0 to 25 and generate the plot. We add labels and a title too.

Listing 2.28: Plotting the frequency of A

```
T = linspace(0,N,N+1);
plot(T,P(T),'o');
xlabel('Generation');
ylabel('Frequency of A');
title('The Frequency of Genotype A vs Generation');
```

This shows an open circle at each generation. We could let Matlab connect the open circles to show a plot that is smooth, but of course, this plots data points for generations like 2.75 which don't really make sense. Still, generalizing from data points each generation to data points at in between times is something we often do. We show the generated plot in Fig. 2.7.

2.7 The Domestication of Wheat

Have you ever thought about how the varieties of wheat we use today came about? We are going to tell you a plausible story which we found in the fantastic book by Steve Mithen on what we know about humanity in the post ice age time (Mithen 2004). Here is the tale paraphrased from Mithen's words—you should really read his book though!

In wild wheat the **ear** that contains the **spikelets** is very brittle and when ripe, the ear falls apart spontaneously. This scatters the seed on the ground and it is difficult to harvest it from there. However, about 1 or 2 out of a million wheat plants are

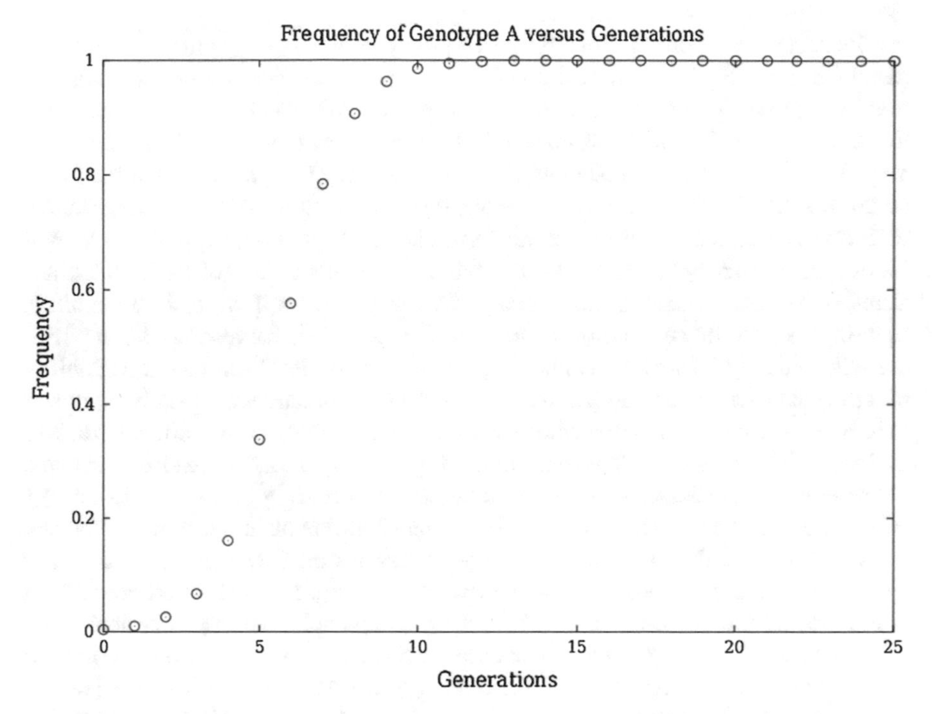

Fig. 2.7 The frequency of **A**

genetic mutants whose ears arc not very brittle. When ripe, these ears do not shatter spontaneously and so they are easier to collect. As Mithen says on pp. 37–38

> Imagine the situation in which a small party of Natufians [the peoplc Mithen is talking about here] arrived to begin cutting a stand of wild cereals. If the wheat was already ripe, then much of the grain from the brittle plants would already have been scattered. But the rare non-brittle plants would still be intact. So when the stalks were cut, the grain from those plants would still be intact. So when the stalks were cut, the grain from those plants would have been relatively more abundant within the harvest than it had been in the woodlands or on the steppe. … The next step is to imagine what would happen if the Natufian people began to reseed the wild stands of cereal by scattering grain saved from a previous harvest....

When this began to happen, the frequency of the non-brittle variant would have been enhanced. Over many generations, this frequency would have continued to climb until eventually the non-brittle variety was dominant. This was a fundamental change in how humans lived for many reasons. One is that the wheat now required human intervention to grow as the non-brittle ears. We are going to use our nice viability model to understand how this might happen. If you look at Hillman and Davies (1990), you can find out a lot more about how to make educated guesses about how long it might take for the domesticated (i.e. the non-brittle variety) wheat to become dominant. Hillman and Davies suggest somewhere between 20 and 50 cycles of sowing and harvesting for this to happen.

Our model will work like this. We will let the **A** variant correspond to the domesticated wheat variety and start its probability as $V_A = 1.0\text{e}{-6}$ which corresponds to 1 non-brittle wheat plant per one million brittle plants. This matches the frequencies we think are realistic. We won't assume the Natufians planted their own wheat plots every year. They might have harvested wild stands and every few years planted their own stands because they had settled down to one region for awhile. Whenever they planted their own stands, the probability of the domestic variety would go up a bit. So we will let **domfrequency** be our MatLab variable which tells us how often the Natufians planted their own wheat; perhaps every 5 years or so. So if we had 20 Nautifian plantings, we would be looking at a total of $5 \times 20 = 100$ harvestings, 90 of which are wild and only 10 are human planting events. We will let **Vdom** be the probability that the wild wheat generates the domesticated wheat mutant: we typically set this as **Vdom = 1.0e-6** and then set the probability of the domesticated variety **VB** to be a multiple of **Vdom**. We will keep our model simple: every **domfrequency** harvests, we reset the domesticated probability to be **VB = Vdom * (1+ epsilon)^j** where **j** is our counter which tells us how many Nautifian plantings there have been. So for example, if the Natufian plant every 5 years, for the first 5 years, the domesticated variety has probability **VB = Vdom * (1+ epsilon)** where **epsilon** is the amount the domesticated probability goes up each planting. A typical value might be **epsilon = 0.8** in which case the domesticated probability is **Vdom** for the first 5 years, is **Vdom*1.8** for the next 5, **Vdom*1.8^2** for the next five and so on. Even though **Vdom** is very small—say 1.0e – 6, after 20 Nautifian plantings, the probability is **Vdom * 1.8^(20)= 0.127**. We reset the probability for the brittle variety to be **1 - VB** at each planting so after 20 plantings, the probability of brittle has fallen to 0.873. And of course, we use a given domesticated probability for say 5 years, so it increases for 5 years following our usual P_A equation during those years.

When we do this simulation, eventually, **Vdom*(1+epsilon)^j** will exceed the wild probability. For example, after a number of Nautifian harvests, we might find **Vdom*(1+epsilon)^j = 0.51**. At that point, **VA = 0.51** and **VB = 0.49**. The ratio **VB/VA > 1** and in the P_A equation, the value of P_A rapidly switches from 0 to 1! This switch point is what we are looking for as it tells us when the domesticated variety becomes dominant. So we will implement this with a **while loop** like this:

Listing 2.29: Finding the switching point

```
% set our simulation counter to 1: this is our first harvesting
i = 0;
% as long as quitloop is 0, we will keep going.
% in the calculations inside the loop as soon
% VAnew exceeds VBnew, we know domesticated
% wheat is dominant.  So we set quitloop = 1
% so we will stop our calculations there.
quitloop = 0;
while  quitloop < 1
    % find the probability ratio
    v = VBnew/VAnew;
    probratio(i) = v;
    % now calculate the new Pdom
    % until a Nautifian planting is done which
    % will increase the domesticated probability
    % this loop is for domfrequency harvests
    for j = 1:domfrequency
        k = i*domfrequency + j;
        PWild(k) = P(k,v);
        Pdom(k) = 1- PWild(k);
       if(Pdom(k) > PWild(k) )
          % print out some variables
          v
          VAnew
          VBnew
          k
          i
          % setting quitloop = 1 here will stop our
          % calculations once we go to the top
          % and see the test while quitloop < 1
          quitloop = 1;
          break;
        end
    end
    % As long as we are ok to continue
    % we reset the probabilities because of  the
    % Nautifian planting
    VBnew = VBnew*(1+epsilon);
    VAnew = 1-VAnew;
    % we increase the counter for the number of
    % Nautifian plantings
    i = i+1;
end
```

Now we want to wrap all this up into a function and do some plots to, so we take the basic loop we just looked at and at some stuff. The basic code as a function returns a number of things: **`Pdom`**, the vector of the domesticated variety frequencies, **`PWild`**, the vector of brittle variety frequencies, **`VDom`**, the vector of domesticated probabilities, **`VWild`**, the vector of brittle variety probabilities and two more. For convenience, we return the ratio of the domesticated and brittle variety probabilities, **`probratio`** and the vector of harvest times **`T`** so we can construct additional plots easily.

Listing 2.30: Our domesticated wheat model

```
function [probratio ,Pdom,PWild ,VWild ,Vdom,T] = domesticatedwheat(Vdom,domfrequency
    ,epsilon)
%
% Vdom = probability of domesticated wheat
% domfrequency = how often harvest strategy leads to
%                              an increase in domestic wheat probability
% epsilon = the change in the probability of domesticated wheat which is
%                    modeled as (1 + epsilon) VA
%
% set the initial domesticated probability
VA = 1-Vdom;
% set the initial brittle probability
VB = Vdom;
% set the initial frequeny of the domesticated variety
p0 = VA;
% find r = NB(0)/NA(0) = ( 1- p0)/p0
r = (1-p0)/p0;
r
% define the population function for domesticated wheat
P = @(t,v) 1./(1+r*v.^(t-1));
%
% every  domfrequency generations, the probability of
% domesticated wheat is increased by the multiplier 1 + epsilon
%
% setup vectors to hold wild and domesticated probalities
% as well as the ratios VB/VA at each step
PWild = [];
Pdom = [];
probratio = [];
% initial VAnew and VBnew
VAnew = VA;
VBnew = VB;
% we start our model with counter i = 1
i = 0;
%
quitloop = 0;
while  quitloop < 1
      v = VBnew/VAnew;
    probratio(i) = v;
     for j = 1:domfrequency
          k = (i-1)*domfrequency + j;
          PWild(k) = P(k,v);
          Pdom(k) = 1- PWild(k);
          VWild(k) = VAnew;
          Vdom(k) = VBnew;
        if(Pdom(k) > PWild(k) )
            k
            i
            quitloop = 1;
            break;
          end
      end
      VBnew = VBnew*(1+epsilon);
      VAnew = 1-VBnew;
      i = i+1;
end
% clear any old plot and plot the results
clf;
N = k;
T = linspace(0,N,N);
plot(T,PWild,'o');
xlabel('Generations');
ylabel('Frequency of Wild Wheat');
title('The Frequency of Wild Wheat vs Generation');
end
```

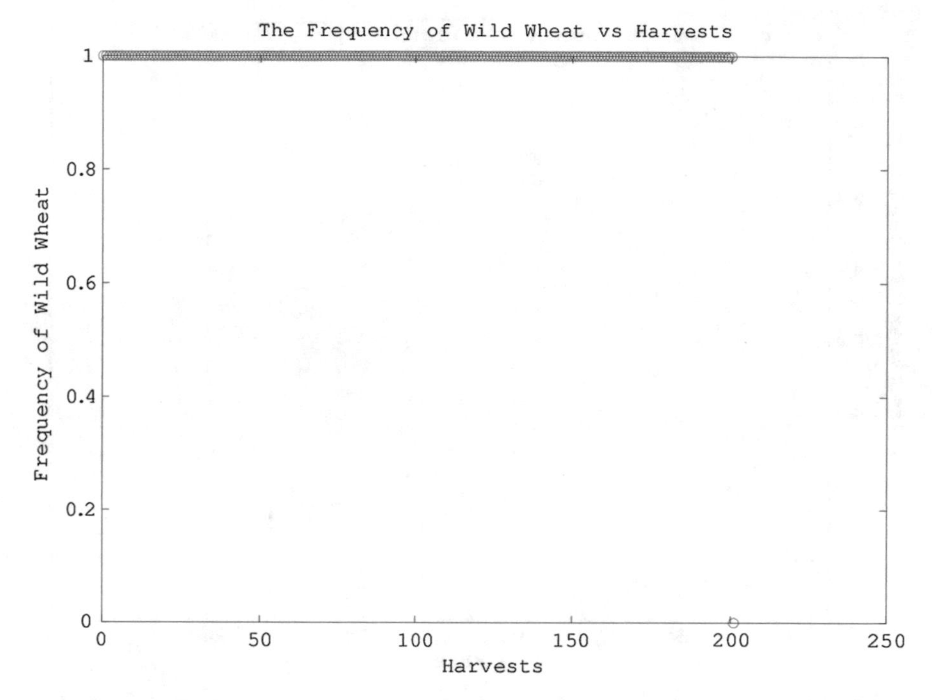

Fig. 2.8 Domesticated wheat dominance: Nautifian harvests every 5 years, domesticated wheat probability multiplier is 0.4 and the initial domesticated wheat probability is 1.0e − 6

Now let's try a simulation. Let's set **`epsilon = 0.4`** and **`domfrequency = 5`** and use an initial domestic wheat probability **`Vdom = 1.0e-6`**. Then in MatLab run this:

Listing 2.31: Sample Domesticated Wheat Simulation

```
[probratio ,Pdom, PWild , VWild , Vdom, T] = domesticatedwheat (1.0 e −6 ,5 ,.4) ;
```

This returns the plot we see in Fig. 2.8. If you look at this graph, you'll see all the points are at the top—where the frequency of the brittle variety is 1. It isn't until you get to harvest 200 that you see the jump down to 0.
So we can clearly see that the switch occurs after $k = 200$ harvests; i.e about 200 years which matches well what Himman and Davies say in their paper. However, It is even easier to see the switch if we plot the domesticated wheat probability versus harvests. This is easily done as shown below.

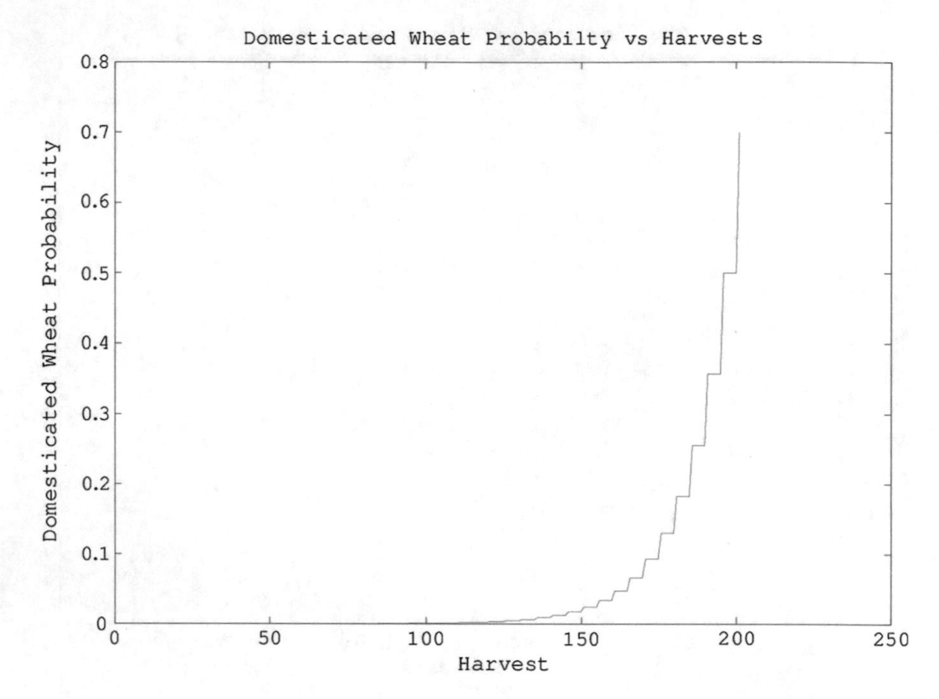

Fig. 2.9 Domesticated wheat probabilities: Natufians plant every 5 years, domesticated wheat probability multiplier is 0.4 and the initial domesticated wheat probability is 1.0e − 6

Listing 2.32: The Domesticated Wheat Probability versus Harvests

```
plot(T,Vdom);
xlabel('Harvest');
ylabel('Domesticated Wheat Probability');
title('Domesticated Wheat Probability vs Harvests');
```

We can see the resulting plot in Fig. 2.9. Again, it is easy to see the domesticated wheat variety becomes dominant at harvest 200. Note how the probability of our domesticated variety jumps up every 5 years as that is the interval between the Nautifian plantings. We could do a similar graph for the probability for the brittle variety too. Cool beans!

2.7.1 Project

Your project is do use our viability models to develop a domesticated wheat dominance model just as we have done in this section. First, download the MatLab code, **domesticatedwheat.m** from the course website as usual. Make sure you start up

MatLab in your working directory and, if you do, after you download the file, you will see it in the left panel when MatLab is up and running. For our project, use the following basic template for our simulations: in this template, we are showing the same values we used in the example we worked out above.

Listing 2.33: Sample Domesticated Wheat Simulation

```
% this is the basic simulation
Vdom = 1.0e-6;
domfrequency = 5;
epsilon = 0.4;
[probratio ,Pdom,PWild ,VWild ,Vdom,T] = domesticatedwheat(Vdom,domfrequency ,epsilon)
  ;
% this is the plot for the domesticated wheat variety probabilities
plot(T,Vdom);
xlabel('Harvest');
ylabel('Domesticated Wheat Probability');
title('Domesticated Wheat Probability vs Harvests');
```

Then, all you have to do is plug in the values you are supposed to use for the project. This is what we want from you. The project report is in Word using singlespaced format. This is a **50 Point** project.

Introduction: We want three pages on a careful discussion of wild versus domesticated wheat using primary sources. Make sure you reference them using any standard scheme: i.e. reference them in the text with a number and list them at the end of the report in a section called **References**. This is **12 Points**.

Annotated discussion of the code domesticatedwheat.m: Here you explain very carefully what the code **domesticatedwheat.m** is doing. Think of explaining this code to a literate person who does not know MatLab—a friend or family member. We are not interested in you simply rehashing the explanations in the text. Try to be fresh and original. MatLab code should be bold font and explanations should be in italic font prefaced by the usual % sign. This is **10 points**.

Simulation One: Use **`Vdom = 9.0e-7`**, **`domfrequency = 6`** and **`epsilon = 0.6`**. Generate two plots: one is the frequency of wild wheat and the other is the domesticated wheat probabilities. Write a short introduction to the simulation problem and then insert the MatLab code to run the simulation. Just annotate this code lightly as the other section did the main work. Then insert the two plots and comment on what they show. This is **10 Points**.

Simulation Two: Use **`Vdom = 1.5e-6`**, **`domfrequency = 4`** and **`epsilon = 0.3`**. Generate two plots: one is the frequency of wild wheat and the other is the domesticated wheat probabilities. Organize the simulation results just like the first one. This is **10 Points**.

Conclusion: Here explain what the simulations are showing. Go back to how we derive the viability model and discuss why it is reasonable to use this model for this problem. For example, the viability model assumes the fertility of both phenotypes is the same. Is that reasonable? Think carefully about all the assumptions! This is **8 Points**.

References: This will show the references you used to prepare your report; make sure you reference this text as you use the code and so forth from it.

References

G. Hillman, M. Davies, Measured domestication rates in wild wheats and barley under primitive cultivation, and their archaeological implications. J. World Prehistory **4**, 157–222 (1990)

R. McElreath, R. Boyd, *Mathematical Models of Social Evolution: A Guide for the Perplexed* (University of Chicago Press, Chicago, 2007)

S. Mithen, *After The Ice: A Global Human History, 20,000–50,000 BC* (Harvard University Press, Cambridge, 2004)

J. Peterson, A.M. Kesson, N.J.C. King, A simulation for flavivirus infection decoy responses. Adv. Microbiol. **5**(2), 123–142 (2015)

Chapter 3
Limits and Basic Smoothness

We are now ready to start discussing an important topic which is the **smoothness** of a function. We have seen nice examples of functions in Chap. 2 on viability selection. Recall, the frequency of type **A** individuals at generation t was given by

$$\boldsymbol{P_A}(t) = \frac{1}{1 + \frac{N_B(0)}{N_A(0)} \left(\frac{V_B}{V_A}\right)^t}.$$

and the change from generation t to $t+1$ was governed by the formula

$$\boldsymbol{P_A}(t+1) - \boldsymbol{P_A}(t) = \frac{\Big(\boldsymbol{P_A}(t)\,(1 - \boldsymbol{P_A}(t))\Big)\Big(\boldsymbol{V_A} - \boldsymbol{V_B}\Big)}{\boldsymbol{P_A}(t)\,\boldsymbol{V_A} + (1 - \boldsymbol{P_A}(t))\,\boldsymbol{V_B}}.$$

The length of a generation can vary widely. So far we are just letting t denote the generation number and we have not paid attention to the average time a generation lasts. Another way of looking at it is that we are implicitly thinking of our time unit as being a **generation**. The actual duration of the generation was not needed in our thinking. For convenience, let's switch variables now. You need to get comfortable with the idea that we can choose to name our variables of interest as we see fit. It is usually better to name them so that they mean something to you for the biology you are trying to study. So in viability selection, variables and parameters like $\boldsymbol{P_A}$ and $\boldsymbol{N_B}$ were chosen to remind us of what they stood for. However, that name choice is indeed arbitrary. Switch $\boldsymbol{P_A}$ to x and the ratio $\boldsymbol{N_B}(0)/\boldsymbol{N_A}(\mathbf{0})$ to simply a and the ratio $\boldsymbol{V_B}/\boldsymbol{V_A}$ to just b and we have

$$x(t) = \frac{1}{1 + a\,b^t}.$$

which looks a lot simpler even though it says the same thing. Our biological understanding of the viability selection problem tells us that b is not really arbitrary. We

J.K. Peterson, *Calculus for Cognitive Scientists*, Cognitive Science and Technology, DOI 10.1007/978-981-287-874-8_3

know $V_B/V_A < 1$ because V_B is smaller than V_A. We also know these values are between 0 and 1 as they are frequencies. The same can be said for a; the initial values of N_A and N_B are positive and so a is some positive number. So should say a bit more

$$x(t) = \frac{1}{1 + a\, b^t}, \quad \text{where } a > 0 \text{ and } 0 < b < 1.$$

Currently, $x(t)$ is defined for a time unit that is generations, so it is easy to see that this formula defines a function which generates a value for any generation t whether t is measured in seconds, days, hours, weeks or years! It is really quite general. So our initial viability selection model gives us something more general: a function which models the frequency of a choice of action for any t regardless of which unit is used to measure t. In fact, we could take a step towards abstraction and say we could use this formula for any value of t on the real line. That is, we can stop thinking of t measured in terms of finite numbers of a time unit and extend our understanding of this formula by letting it be valid for any t. Make no mistake, this is an abstraction as our arguments only worked for generations! We will often make this step into abstraction as it allows us to bring powerful tools to bear on understanding our models that we can't use if we are restricted to finite time measurements. So we have a general modeling principle:

- Build a model using units that are relevant for the biology. Use generations, seasonal cycles and so forth. Work hard to make the model realistic for your choice of units. This is a hard but important step.
- From your discrete time unit model, make the abstract jump to allow time to be any value at all. That is, you let your model formula be valid for any time value at all instead of only being valid for your discrete values.
- With the model extended to all time values, apply tools that allow you to manipulate these models to gain insight. These tools from Calculus include things like **limiting behavior**, **smoothness issues called continuity** and **rate of change smoothness issues called differentiability**. The tools from computation include learning how to use MatLab to do calculations when we either can't use mathematical techniques or it is just too hard to use the mathematics.

Now **continuity** for our model roughly means the graph of our model with respect to time doesn't have jumps. Well, of course, our model for generational time has jumps! The $x(t)$ value simply jumps to the new value $x(t + 1)$ when we apply the formula. The continuity issue arises when we pass to letting t be any number at all. Another way of looking at it is that the generation time becomes smaller and smaller. For example, the generation time for a virus or a bacteria is very small compared to the generation time of a human! So letting our model handle shrinking generation time seems reasonable and as the generation time shrinks, it makes sense that the jumps we see get smaller. Continuity is roughly the idea that as the generation time shrinks to zero (yeah, odd concept!) the model we get has no jumps in it at all!

To get a feel for the **differentiability** thing, again we resort to passing to an abstract model from a discrete time one. The frequency update law can be rewritten in terms of x also as

$$x(t+1) - x(t) = \frac{x(t)\,(1 - x(t))\,(c - d)}{c\,x(t) + d\,(1 - x(t))}$$

where we let $c = \boldsymbol{V_A}$ and $d = \boldsymbol{V_B}$. We know $c > d$ from our viability discussions too, so we should say

$$x(t+1) - x(t) = \frac{x(t)\,(1 - x(t))\,(c - d)}{c\,x(t) + d\,(1 - x(t))}, \quad \text{with } c > d > 0.$$

Now our time units are generations here, so t is really thought of as an integer. So $(t+1) - t$ is always 1! Now rewrite what we have above as

$$\frac{x(t+1) - x(t)}{(t+1) - t} = \frac{x(t)\,(1 - x(t))\,(c - d)}{c\,x(t) + d\,(1 - x(t))}.$$

Now let's start thinking about the size of that generation. Now think of $(t+1) - t$ as **One** time unit. Call this unit h just for fun. So our general t is really some multiple of the generation time unit h. Call it Nh for some N. Then $t + 1 = (N+1)h$. But it is easier to think of $t + 1$ as this: $t + 1 = Nh + h = t + h$! So we can rewrite again as

$$\frac{x(t+h) - x(t)}{h} = \frac{x(t)\,(1 - x(t))\,(c - d)}{c\,x(t) + d\,(1 - x(t))}.$$

Now we are being a lot more clear about generation time. An obvious question is what happens as the generation time, now measures as h, gets smaller and smaller. If as we let h get really tiny, we find the fraction $\frac{x(t+h)-x(t)}{h}$ approaches some stable number, we can be pretty confident that this stable number somehow represents an abstraction of the rate of change of x over a generation. This shrinking of h and our look at the behavior of the fraction $\frac{x(t+h)-x(t)}{h}$ as h shrinks is the basic idea of what having a derivative at a point means. Of course, our original model is very discrete and possesses jumps and this fraction makes no sense for time units smaller than the particular generation time we were using for our model. Nevertheless, since we want our models be essentially independent of the actual generation time unit, we are led to thinking about looking at this ratio as the generation time unit shrinks and shrinks.

With these introductory remarks out of the way, it is time to start studying these ideas of continuity and differentiability in earnest. Smoothness for our functions means we add usually at least being continuous and having a derivative to our requirements for the function we are using to model the biology. So for awhile, to study these ideas more carefully, we will move away from biology and just do some mathematical stuff. But remember, our reasons for this are that these abstractions will give us powerful tools to bring to bear on gaining insight from the models we want to build.

3.1 Limits

We are all familiar with the idea of seeing what happens as we approach something. The idea of a **limit** is similar to that. Let's start with some examples you have all seen even if you didn't know it!

3.1.1 The Humble Square Root

Let's consider real numbers. We all know that there are fractions which are ratios of two integers and then the others which can't be written that way. For example $\sqrt{2}$ can't be written as fraction although we can approximate it nicely with one. Pull out your calculator and punch in the commands to see $\sqrt{2}$. You'll see something like 1.414214 or perhaps even more digits depending on your calculator. This is not $\sqrt{2}$ but it is an approximation to it using fractions. The number we have above is really a fraction built out of powers of 10. We have

$$\sqrt{2} \approx 1 + \frac{4}{10} + \frac{1}{100} + \frac{4}{1{,}000} + \frac{2}{10{,}000} + \frac{1}{100{,}000} + \frac{4}{1{,}000{,}000}$$

which we can also express in terms of powers of 10:

$$\sqrt{2} \approx 1 + \frac{4}{10^1} + \frac{1}{10^2} + \frac{4}{10^3} + \frac{2}{10^4} + \frac{1}{10^5} + \frac{4}{10^6}$$

This sum of fractions is exactly the same as the fraction 1,414,214/1,000,000 so as we said, we are approximating $\sqrt{2}$ by *this* fraction. But we can easily add more terms to this approximation and if your calculator can't supply them, we have other tools such as computer programs to do that for us. The point is that $\sqrt{2}$ is not *exactly* any fraction even though we can approximate it as closely as we like by fractions by just adding more terms of the form $a/10^p$ where a is the next digit which will range for 0 to 9 and p is the next power. The next two digits are actually 36 and $\sqrt{2} \approx 1.4142136$. Note the last digit we had before which was a 4 has now become a 36 as our calculator had rounded the 36 up to a 4 on our display. So our approximation of $\sqrt{2}$ has replaced the $4/10^6$ with the terms $3/10^6$ and $6/10^7$. And we can go on and on with these approximations.

Let's introduce another notation called the *summation* notation. We can rewrite our approximation to $\sqrt{2}$ like this

$$\sqrt{2} \approx 1 + \sum_{n=1}^{6} \frac{a_n}{10^n}$$

where $a_1 = 4$, $a_2 = 1$ and so forth are the digits we have for our first approximation to six digits. Our second approximation would then be

$$\sqrt{2} \approx 1 + \sum_{n=1}^{7} \frac{a_n}{10^n}$$

where the new $a_6 = 3$ and $a_7 = 6$. These new forms are just a different way of expressing our standard decimal notation which is $\sqrt{2} \approx 1.414214$ to six digits or $\sqrt{2} \approx 1.4142136$. Since we can carry on this approximation idea to as many digits as we like, we need a new notation to say that. We could say $\sqrt{2} = 1.414214\ldots$ where the ... indicates we are not going to stop. This idea of going on forever is the idea of a limiting process. Using our a_n notation, we could also say

$$\sqrt{2} = 1 + \sum_{n=1}^{6} \frac{a_n}{10^n} + \ldots$$

but the standard way in mathematics to say this is to use the symbol ∞ to indicate we are doing something forever. So instead of ... in the formula above we would write

$$\sqrt{2} = 1 + \sum_{n=1}^{\infty} \frac{a_n}{10^n}$$

to tells us that this process of finding more digits never stops. Of course, we would need a way to keep finding these digits and there are such calculational tools available. But that is not our purpose here. You have all seen real number, fractions or otherwise, so whether you realized it or not, that means you have seen **limits**!

3.1.1.1 Example

Example 3.1.1 Express $\sqrt{3}$ to 6 digits as a fraction and as sum of fractions.

Solution

$$\begin{aligned}
\sqrt{3} &\approx 1.732051 \\
&\approx \frac{1{,}732{,}051}{1{,}000{,}000} \\
&\approx 1 + \frac{7}{10^1} + \frac{3}{10^2} + \frac{2}{10^3} + \frac{0}{10^4} + \frac{5}{10^5} + \frac{1}{10^6}
\end{aligned}$$

3.1.1.2 Homework

Exercise 3.1.1 *Express* $\sqrt{5}$ *to 6 digits as a fraction and as sum of fractions.*

Exercise 3.1.2 *Express* $\sqrt{7}$ *to 6 digits as a fraction and as sum of fractions.*

Exercise 3.1.3 *Express* $\sqrt{11}$ *to 6 digits as a fraction and as sum of fractions.*

3.1.2 A Cool Polynomial Trick

Here is another example. Look at the following polynomials and their factored form. Now we are assuming you will get out paper and pencil and actually check that these work! This is not a spectator sport here!

$$1 - x^2 = (1 - x)\,(1 + x)$$
$$1 - x^3 = (1 - x)\,(1 + x + x^2)$$
$$1 - x^4 = (1 - x)\,(1 + x + x^2 + x^3)$$

A little thought and some common sense tells you that this factoring is a general principle which tells us that for any positive integer n, we should have

$$1 - x^{n+1} = (1 - x)\,(1 + x + \ldots + x^n)$$

The notation ... just means there are powers of x in between that we have left out. So already we are being a bit abstract and symbolic. Hold onto your hats as there is more! Now if x was strictly between -1 and 1, it is easy to see using a calculator that x^p for increasing powers of p gets smaller and smaller. As a matter of fact, as p increases, we can easily see the x^p gets as close to 0 as we want. To get closer at any time, we simply have to increase p. This is another idea of a limit. We would say

$$x^p \to 0 \quad \text{as } p \text{ gets larger if } -1 < x < 1.$$

But we usually say this differently. The phrase p gets large is replaced by $\lim_{p\to\infty}$ and we write

$$x^p \to 0 \quad \text{as } p \to \infty \text{ if } -1 < x < 1.$$

But we usually just say $\lim_{p\to\infty} x^p = 0$ if $-1 < x < 1$. Now powers of numbers don't have to be this nicely behaved. Take $x = 2$. then the powers 2^p grow larger for ever—check this out on your calculator—and we would say $\lim_{p\to\infty} 2^p = \infty$ as a shorthand. The behavior can be worse. Take $x = -1$. Then the powers of -1 are $(-1)^p$ whose values flip between $+1$ and -1 depending on whether the power p is even or odd. So we would have to say $\lim_{p\to\infty} (-1)^p$ just doesn't exist.

Anyway, back to our polynomials. If x is strictly between -1 and $+1$, we can divide by $1 - x$ and get the nice formula

$$1 + x + x^2 + \cdots + x^n = \frac{1 - x^{n+1}}{1 - x}$$

which we can also write in our summation notation as

$$\sum_{i=0}^{n} x^i = \frac{1 - x^{n+1}}{1 - x}$$

The choice of i here is not important. It is the summation variable and it can be replaced by any other letter and the summation would be the same. It is really a reminder of what we should do. Also, note the summation starts at $i = 0$ because the first term, 1 is really the same as x^0.

What happens as n gets larger and larger? By what we have said above, in this case we know $x^{n+1} \to 0$ since x is strictly between -1 and 1. So the fraction $\frac{1-x^{n+1}}{1-x} \to \frac{1}{1-x}$. We would say

$$\lim_{n\to\infty}\left(\sum_{i=0}^{n} x^i\right) = \frac{1}{1 - x}.$$

But we usually just write this as

$$\sum_{i=0}^{\infty} x^i = \frac{1}{1 - x}.$$

So there you have it: another example of this thing called a **limit**.

3.1.2.1 Examples

Example 3.1.2 Evaluate $\sum_{i=0}^{\infty}(1/3)^i$.

Solution

$$\sum_{i=0}^{\infty}(1/3)^i = \frac{1}{1 - 1/3} = \frac{3}{2}.$$

Example 3.1.3 Evaluate $\sum_{i=1}^{\infty}(1/3)^i$.

Solution *This summation starts at* $i = 1$ *instead of* $i = 0$ *so we are missing the first term. So we will sum the whole thing and then subtract the first term.*

$$\sum_{i=1}^{\infty}(1/3)^i = \sum_{i=0}^{\infty}(1/3)^i - 1$$
$$= \frac{1}{1-1/3} - 1 = \frac{3}{2} - 1 = \frac{1}{2}.$$

Example 3.1.4 Evaluate $\sum_{i=2}^{\infty}(1/5)^i$.

Solution *This summation starts at i $= 2$ instead of i $= 0$ so we are missing the first two terms. So we will sum the whole thing and then subtract the first two terms.*

$$\sum_{i=2}^{\infty}(1/5)^i = \sum_{i=0}^{\infty}(1/5)^i - \left(1 + \frac{1}{5}\right)$$
$$= \frac{1}{1-1/5} - \frac{6}{5} = \frac{5}{4} - \frac{6}{5} = \frac{25-24}{20} = \frac{1}{20}.$$

Example 3.1.5 Determine, if possible, the following limits. If the limit is $+\infty$ or $-\infty$ say that and if the limit is oscillating like $(-1)^n$ does say that. Determine

- $\lim_{n\to\infty} 3^n$
- $\lim_{n\to\infty}(-3)^n$
- $\lim_{n\to\infty}(1/4)^n$
- $\lim_{n\to\infty}(-1/4)^n$

Solution • $\lim_{n\to\infty} 3^n$ *is* ∞.
- $\lim_{n\to\infty}(-3)^n$ *is oscillating so does not exist.*
- $\lim_{n\to\infty}(1/4)^n$ *is* 0.
- $\lim_{n\to\infty}(-1/4)^n$ *is* 0.

3.1.2.2 Homework

Exercise 3.1.4 *Evaluate* $\sum_{i=0}^{\infty}(1/7)^i$.

Exercise 3.1.5 *Evaluate* $\sum_{i=1}^{\infty}(1/4)^i$.

Exercise 3.1.6 *Evaluate* $\sum_{i=2}^{\infty}(1/2)^i$.

Exercise 3.1.7 *Evaluate* $\sum_{i=1}^{\infty}(1/6)^i$.

Exercise 3.1.8 *Determine, if possible, the following limits. If the limit is* $+\infty$ *or* $-\infty$ *say that and if the limit is oscillating like* $(-1)^n$ *does say that. Determine*

- $\lim_{n\to\infty} 4^n$
- $\lim_{n\to\infty}(-1.5)^n$
- $\lim_{n\to\infty}(1/7)^n$
- $\lim_{n\to\infty}(-2/3)^n$

3.1.3 Change and More Change!

Let's look at the rate problem. Earlier, we found the change in frequency of gene of type **A** could be expressed as the formula

$$\frac{x(t+1)-x(t)}{(t+1)-t} = \frac{x(t)\,(1-x(t))\,(c-d)}{c\,x(t)+d\,(1-x(t))}.$$

and we argued that we could re express this in terms of time units of h as the generation time as

$$\frac{x(t+h)-x(t)}{h} = \frac{x(t)\,(1-x(t))\,(c-d)}{c\,x(t)+d\,(1-x(t))}.$$

Given our discussions so far, a reasonable thing to do is to ask what happens to this ratio as h gets smaller and smaller. Now this particular ratio is way to hard to handle directly, so let's make up a very simple example—strictly mathematical!—and see how it would go. Let's simplify our life and choose the function $f(x) = x^2$. Let's choose an h and consider the ratio

$$\begin{aligned}\frac{f(x+h)-f(x)}{(x+h)-(x)} &= \frac{f(x+h)-f(x)}{h}\\ &= \frac{(x+h)^2-x^2}{h}\end{aligned}$$

Now we need to do some algebra. We have

$$\begin{aligned}(x+h)^2-x^2 &= x^2+2\,x\,h+h^2-x^2\\ &= 2\,x\,h+h^2.\end{aligned}$$

Plugging this into our fraction, we find

$$\begin{aligned}\frac{f(x+h)-f(x)}{(x+h)-(x)} &= \frac{f(x+h)-f(x)}{h}\\ &= \frac{(x+h)^2-x^2}{h}\\ &= \frac{2\,x\,h+h^2}{h}\\ &= 2\,x+h.\end{aligned}$$

As h gets smaller and smaller, we see the *limiting value* is simply $2x$. We would say

$$\lim_{h\to 0}\left(\frac{f(x+h)-f(x)}{h}\right) = 2\,x.$$

This kind of limit works well for many functions f, including our function for the frequency of the gene of type **A** in the population if we assume our frequency function can be extended from being defined for just discrete generation times to all time (which we routinely do). It is such an important limiting process, it is given a special name: the **derivative** of f with respect to x. We often write this difference of function values of f divided by differences in x values as $\frac{\Delta f}{\Delta x}$ which of course hides a lot and say

$$\lim_{\Delta x \to 0} \left(\frac{\Delta f}{\Delta x} \right) = 2\,x.$$

The traditional symbol we use for this special limiting process is taken from the notation Δ. As h gets smaller and smaller, we use the symbol $\frac{df}{dx}$ to indicate we are taking this limit and write

$$\frac{df}{dx} = \lim_{\Delta x \to 0} \left(\frac{\Delta f}{\Delta x} \right) = 2\,x.$$

Congratulations! You have found your first derivative! More to come naturally, but this is all there is too it. We will add some graphical interpretations and other stuff but you have the idea now!

3.1.3.1 Examples

Example 3.1.6 Write down the limit formula for the derivative of $f(x) = x^3$ at $x = 2$ using both the h and Δx forms.

Solution

$$\begin{aligned} \frac{df}{dx}(2) &= \lim_{\Delta x \to 0} \left(\frac{(2 + \Delta x)^3 - (2)^3}{\Delta x} \right) \\ &= \lim_{h \to 0} \left(\frac{(2 + h)^3 - (2)^3}{h} \right) \end{aligned}$$

Example 3.1.7 Write down the limit formula for the derivative of $f(x) = x^4$ at $x = 1$ using both the h and Δx forms.

Solution

$$\begin{aligned} \frac{df}{dx}(1) &= \lim_{\Delta x \to 0} \left(\frac{(1 + \Delta x)^4 - (1)^4}{\Delta x} \right) \\ &= \lim_{h \to 0} \left(\frac{(1 + h)^4 - (1)^4}{h} \right) \end{aligned}$$

Example 3.1.8 Write down the limit formula for the derivative of $f(x) = 2x^3$ at $x = 4$ using both the h and Δx forms.

Solution

$$\frac{df}{dx}(4) = \lim_{\Delta x \to 0} \left(\frac{2\,(4 + \Delta x)^3 - 2\,(4)^3}{\Delta x} \right)$$
$$= \lim_{h \to 0} \left(\frac{2\,(4 + h)^4 - 2\,(4)^3}{h} \right)$$

3.1.3.2 Homework

Exercise 3.1.9 *Write down the limit formula for the derivative of* $f(x) = 5x^2$ *at* $x = 1$ *using both the h and* Δx *forms.*

Exercise 3.1.10 *Write down the limit formula for the derivative of* $f(x) = x^6$ *at* $x = 2$ *using both the h and* Δx *forms.*

Exercise 3.1.11 *Write down the limit formula for the derivative of* $f(x) = 7x^2$ *at* $x = 3$ *using both the h and* Δx *forms.*

Exercise 3.1.12 *Write down the limit formula for the derivative of* $f(x) = x^{10}$ *at* $x = 5$ *using both the h and* Δx *forms.*

3.1.4 How Many Do We Have?

What if we wanted to add up the frequency total for our gene of type $\mathbf{A}$ for some number of generations? We would want to add up the following sum which we will call S. The sum is

$$S = \boldsymbol{P_A}(1) + \boldsymbol{P_A}(2) + \boldsymbol{P_A}(3)$$

Think of the units of $\boldsymbol{P_A}(1)$ which is really **frequency per generation time**. So the term $\boldsymbol{P_A}(1)$ can be thought of as $\boldsymbol{P_A}(1)$ frequency per time unit $\times$ 1 time unit. But we could be more general and think of the time unit as perhaps smaller or larger and call the time unit h like before. Then our terms should be of the form $\boldsymbol{P_A}(1)$ frequency per time unit $\times$ h time units. The time 1 should be written as 1 but time 2 should now be $1 + h$ and time 3 should be $1 + 2h$. So our sum should be written as

$$S = \boldsymbol{P_A}(1)\,h + \boldsymbol{P_A}(1 + h)\,h + \boldsymbol{P_A}(1 + 2h)\,h$$

Now suppose we wanted to add up the frequencies for a time of 2 years when each generation h is a lot smaller than a year. So we need to add up $2/h$ pieces to get from 1 to 3 years. Let N be the closest integer to the fraction $2/h$. Then, our sum becomes

$$S = \boldsymbol{P_A}(1)\, h + \boldsymbol{P_A}(1 + h)\, h + \boldsymbol{P_A}(1 + 2h)\, h + \boldsymbol{P_A}(1 + N\, h)\, h$$

We could then write this in a summation notation as

$$S = \sum_{i=0}^{N} \boldsymbol{P_A}(1 + i\, h)\, h$$

Since this sum is for the time range 1 to 3, let's change the notation of the sum to

$$S_1^3 = \sum_{i=0}^{N} \boldsymbol{P_A}(1 + i\, h)\, h$$

and since h divides 2 into N pieces, let's add the integer N too. This gives

$$(S_1^3)^N = \sum_{i=0}^{N} \boldsymbol{P_A}(1 + i\, h)\, h$$

We have already discussed that it is easy to imagine smaller and smaller generation times h. So as h gets smaller, N gets larger and we add up more pieces. What is the **limiting value** here? That is, does $\lim_{N\to\infty}(S_1^3)^N$ go to some stable value? Clearly our $\boldsymbol{P_A}(t)$ function is way to hard to work with, so let's drop back to a real simple example which illustrates all of this too.

Let $f(x) = x^2$ again. Then the sum becomes

$$(S_1^3)^N = \sum_{i=0}^{N} f(1 + i\, h)\, h$$

and this sum measures some sort of adding up of the values of f times the interval they act on over a chunk of the x axis. The actual interpretation would depend on what we are trying to model. Here, for the moment, it is just a mathematical thing. Let's expand the term $f(1 + i\, h) = (1 + i\, h)^2$. We have

$$(1 + i\, h)^2 = 1 + 2\, i\, h + i^2\, h^2,$$

Now it is hard to handle this at the moment because our h is not written in terms of N. So let's flip it around. Instead of starting with h, let's set $h = 2/N$; that is, choose the h to be a size that depends on N. This gives our same result: $N = 2/h$! Why $2/h$? Well, we are trying to find out what happens between year 1 and year 3 and so there are $(3-1)/h$ generations we go through between those years. Now rewrite the expansion of f using this to get

$$\left(1+i\,\frac{2}{N}\right)^2 = 1+2\,i\,\frac{2}{N}+i^2\left(\frac{2}{N}\right)^2,$$

Now plug it all into the summation to get

$$\left(S_1^3\right)^N = \sum_{i=0}^{N}\left(1+2\,i\,\frac{2}{N}+\left(\frac{2}{N}\right)^2\right)\frac{2}{N}$$

Now don't get disheartened by all this mess. We will now manipulate this to make it easier to understand. First, it seems reasonable we can break this sum into three parts, so let's do so now.

$$\left(S_1^3\right)^N = \sum_{i=0}^{N} 1\,\frac{2}{N} + \sum_{i=0}^{N} 2\,i\frac{2}{N}\,\frac{2}{N} + \sum_{i=0}^{N} i^2\left(\frac{2}{N}\right)^2\frac{2}{N}$$

Now look at each piece separately.

- $\sum_{i=0}^{N} 1\,\frac{2}{N}$ just means to add up the constant $\frac{2}{N}$ $N+1$ times (it is $N+1$ because we start counting at 0 instead of 1). So the answer is $2\frac{N+1}{N} = 2\,(1+1/N)$.
- $\sum_{i=0}^{N} 2\,i\frac{2}{N}\,\frac{2}{N}$ can be made simpler be moving all the constant terms out of the sum to give $2\left(\frac{2}{N}\right)^2 \sum_{i=0}^{N} i$. Now mathematicians have known for a long time a formula for $\sum_{i=0}^{N} i$; $\sum_{i=0}^{N} i = \frac{N\,(N+1)}{2}$ and it is not terribly important how we would prove this. So let's just take it as given. So

$$\begin{aligned}
2\left(\frac{2}{N}\right)^2\sum_{i=0}^{N} i &= 2\left(\frac{2}{N}\right)^2\frac{N\,(N+1)}{2}\\
&= \frac{2\times 4}{2}\,\frac{N\,(N+1)}{N\,N}\\
&= 4\,\frac{N+1}{N} = 4\,(1+1/N).
\end{aligned}$$

- The last piece is handled like the second piece. We pull the constant stuff out and write

$$\sum_{i=0}^{N} i^2\left(\frac{2}{N}\right)^2\frac{2}{N} = \left(\frac{2}{N}\right)^3\sum_{i=0}^{N} i^2.$$

It turns out mathematicians know how to handle the sum $\sum_{i=0}^{N} i^2$ too. It is $\sum_{i=0}^{N} i^2 = \frac{N\,(N+1)\,(2N+1)}{6}$—a totally cool formula. So we have

$$\begin{aligned}\sum_{i=0}^{N} i^2 \left(\frac{2}{N}\right)^2 \frac{2}{N} = \left(\frac{2}{N}\right)^3 \sum_{i=0}^{N} i^2 \\ &= \frac{8}{N^3}\,\frac{N\,(N+1)\,(2N+1)}{6} = \frac{8}{6}\,\frac{N\,(N+1)\,(2N+1)}{N\,N\,N} \\ &= \frac{4}{3}\,\frac{N+1}{N}\,\frac{2N+1}{N} = \frac{4}{3}\,(1+1/N)\,(2+1/N).\end{aligned}$$

Wow, what a lot of work! We will work hard on learning new tools that will let us avoid this kind of stuff later, but for now let's soldier through. With all this done, we have

$$\begin{aligned}(S_1^3)^N &= \sum_{i=0}^{N} \left(1 + 2\,i\,\frac{2}{N} + \left(\frac{2}{N}\right)^2\right) \frac{2}{N} \\ &= 2\,(1+1/N) + 4\,(1+1/N) + \frac{4}{3}\,(1+1/N)\,(2+1/N).\end{aligned}$$

and so as h goes to zero or equivalently as N goes to ∞ we have

$$\lim_{N\to\infty} (S_1^3)^N = 2 + 4 + 2\,\frac{4}{3} = \frac{26}{3}.$$

Now since we are still doing summations although of course using more and more terms as h gets smaller, it would be nice to have a new notation for this limiting value. The standard one is to use a stretched out **S** and write it as $\int$—see how it looks like an **S** sort of? So we use

$$\int_1^3 = \lim_{N\to\infty} (S_1^3)^N$$

But we also want our notation to help us remember that we are using the function x^2 and that we are using smaller and smaller chunks h. If you think about it a bit, you can see that the chunk h is the same as a change in x; $h = x + h - x$ which we usually call Δx. We typically use the notation dx to remind us that we are letting $h \to 0$ or $\Delta x \to 0$ here. So we add the x^2 and the dx to our notation and get

$$\int_1^3 x^2\,dx = \lim_{N\to\infty} (S_1^3)^N$$

and we end up with having calculated this

$$\begin{aligned}\int_1^3 x^2\,dx &= \lim_{N\to\infty} \sum_{i=0}^{N} \left(1 + 2\,i\,\frac{2}{N} + \left(\frac{2}{N}\right)^2\right) \frac{2}{N} \\ &= \lim_{N\to\infty} \sum_{i=0}^{N} \left(1 \ + i\ \Delta x\right)^2 \ \Delta x\end{aligned}$$

Note how weird this limit is! There is an N on the top of the summation symbol ($\sum_{n=0}^{N}$) as well as an N in the thing we are summing (the $\frac{2}{N}$'s). Also note one term is getting bigger while the other is getting smaller. These kind of limits always are like that and it always requires a lot of thinking to figure out if the limit exists and is some nice value. You can probably imagine that this process might fail for some functions f even though it certainly worked for our $f(x) = x^2$. In general, for an arbitrary function f for terms between $x = 1$ and $x = 3$, we would want to look at the limit

$$\begin{aligned}\int_1^3 f(x)\,dx &= \lim_{N\to\infty} \sum_{i=0}^{N} f(1 \ + i\ \Delta x)\ \Delta x \\ &= \lim_{N\to\infty} \sum_{i=0}^{N} f\left((1 \ + i\left(\frac{2}{N}\right)\right) \frac{2}{N}\end{aligned}$$

This is another special limit then. This one is called the **Riemann Integral** of f on the interval $[1, 3]$. We certainly don't want to have to find these limits this way! So we are going to learn tools that help us to this very quickly. That, of course, means a bit of theory and mind-stretching, but it is worth it!

3.1.4.1 Examples

Example 3.1.9 Write down the summation for the $\int_2^5 x^2 dx$ in limit form.

Solution *This is easy. The interval is* $[2, 5]$ *now so we should use* $\Delta x = h = (5-2)/N = 3/N$.

$$\int_2^5 x^2\,dx = \lim_{N\to\infty} \sum_{i=0}^{N} \left(2 \ + i\ \frac{3}{N}\right)^2 \frac{3}{N}.$$

Example 3.1.10 Write down the summation for the $\int_1^6 x^3 dx$ in limit form.

Solution *This is easy. The interval is* $[1, 6]$ *now so we should use* $\Delta x = h = (6 - 1)/N = 5/N$.

$$\int_1^6 x^3\, dx = \lim_{N \to \infty} \sum_{i=0}^{N} \left(1 + i\,\frac{5}{N}\right)^3 \frac{5}{N}.$$

3.1.4.2 Homework

Exercise 3.1.13 *Write down the summation for the* $\int_3^9 x^4 dx$ *in limit form.*

Exercise 3.1.14 *Write down the summation for the* $\int_2^4 x^2 dx$ *in limit form.*

Exercise 3.1.15 *Write down the summation for the* $\int_3^{10} x^5 dx$ *in limit form.*

3.1.5 This Function Is Smooth!

A function f defined on an interval $[a, b]$ where a and b are finite numbers can be quite strange. For example, here is a legitimate function f defined on $[-1, 1]$:

$$f(t) = \begin{cases} 1 & \text{if } t \text{ is a rational number} \\ -1 & \text{if } t \text{ is an irrational number} \end{cases} \tag{3.1}$$

Now don't get upset here. Rational numbers are just ratios of integers and everything else is irrational. A little thought should help you see that you can get as close as you want to any irrational number like $\sqrt{2}$ using as many terms in its decimal expansion as needed and flipping it around, we can get as close as we want to any fraction with an irrational! In other words, we can't draw a circle about a rational number and say there are no irrationals in there or vice-versa. So this function is horribly odd: it is not possible to graph it at all. It flips between -1 and $+1$ all the time. Since we can't find a circle of only rationals, we can't find a circle where the function is $+1$ always and by the same reasoning, we can't find a circle full of only irrationals where the function is -1 always. We can't really get a function that behaves much worse! The idea of **smoothness** or more technically, **continuity** is that we can fully predict what a function will do near a point on the basis of its behavior around that point. We naturally want to study and use functions that are much better behaved than this. We want our functions to have derivatives and this notion of continuity too! Now these ideas of *continuity* and *differentiability* (i.e. having a derivative at a point) are *pointwise* concepts. So to do anything useful for biological modeling, we want the function we use in our models to to satisfy the requirements for continuity and differentiability at all points in whatever interval we are interested in. For example, the function f defined by

$$f(t) = 2t^3 + 32\,t + 16$$

is nicely defined on any interval we want. Sketch its graph on a piece of paper or using MatLab. We usually draw some points and connect the dots to make a nice *smooth* curve between the points we have. We can choose as many points to plot as we want and they will always fit nicely in between other points we have already plotted. So for example, we would say

$$\begin{aligned}
\lim_{t \to 2} f(t) &= \lim_{t \to 2} \left(2\,t^3 + 32\,t + 16 \right) \\
&= 2\,(2)^3 + 32\,(2) + 16 \\
&= 16 + 64 + 16 = 96
\end{aligned}$$

because we fully expect that the value we get for this function is found by simply plugging in the number 2 into its formula. In fact, we could do this at any point and we could always say

$$\lim_{t \to c} f(t) = f(c)$$

for any number c. Make no mistake there is a lot of sophisticated stuff going on here—limits and more limits!

Chapter 4
Continuity and Derivatives

We have discussed a variety of the special limits we will no doubt need in our biological modeling and we know approximately what limits, the notion of continuity, the idea of a derivative and the idea of a Riemann integral mean. We have even been exposed to the needed notation and we have tried to explain why we use the notation we do. But with our preliminary tour done, it is time to do this stuff in more detail. I know you are all beginning to shudder and think to yourselves, "Ok, here it comes. Time to be blown away..." but we think you will be pleasantly surprised at how we do this. So just be patient and do a lot of scratch work of your own on paper while you read. Remember, learning is not a spectator sport!!

4.1 Continuity

Let's start with this smoothness notion called **continuous**. Continuity is a very important concept to us and so far we have approached it through example. However, we can be more precise. From what we have said, this is how we would define continuity formally.

Definition 4.1.1 (*Continuity Of A Function At A Point: Limit Version*)
f is said to be continuous at a point p in its domain if several conditions hold:

1. f is actually defined at p
2. The limit as t approaches p of f exists
3. The value of the limit above matches the value $f(p)$.

This is usually stated more succinctly as $f(p)$ exists and $\lim_{t \to p} f(t) = f(p)$, You have already seen that the polynomial $f(t) = 2t^3 + 32\, t + 16$ is continuous at all t values using these definitions. Now we are sort of tacitly using the idea of limits a lot here. If we really wanted to be careful we would have to define that idea more carefully too. Here is an example. Take this step function:

J.K. Peterson, *Calculus for Cognitive Scientists*, Cognitive Science and Technology, DOI 10.1007/978-981-287-874-8_4

$$f(x) = \begin{cases} 1, & \text{if } 0 \leq x \leq 1 \\ 2 & \text{if } 1 < x \leq 2. \end{cases}$$

It is pretty easy to see that as we approach 1 from the left, the function values are always 1 and so our expectation is that the limiting value is 1 from the left. We would write that as $\lim_{x \to 1^-} f(x) = 1$ where the superscript $^-$ is to remind us that the limit is from the left only. Similarly, if we approach 1 for the right, the function values are always 2 and so we would expect the limiting value to be 2 from the right. We would write this as $\lim_{x \to 1^+} f(x) = 2$ where the superscript $^+$ reminds us that the limit is from the right only. It seems reasonable to agree that the actual limit can't exist unless the right and left hand limit both exist and match. Since $1 \neq 2$, we don't have a match and since we have a jump from 1 to 2 in function value at 1, this function is clearly not continuous at 1. Note what we have here

- The $\lim x \to 1\ f(x)$ does not exist because the right and left hand limits don't match.
- The function value $f(1)$ is fine and $f(1) = 1$.
- So using our more formal definition of continuity we know f is not continuous at 1 because the $\lim x \to 1\ f(x)$ does not exist and that failure in the conditions is enough make continuity at 1 fail.

Now do it again; this time define f as follows:

$$f(x) = \begin{cases} 1, & \text{if } 0 \leq x < 1 \\ 10, & \text{if } x = 1 \\ 1 & \text{if } 1 < x \leq 2. \end{cases}$$

Note what happens now.

- The $\lim x \to 1\ f(x)$ does exist and equals 1 because the right and left hand limits match and equal 1.
- The function value $f(1)$ is fine and $f(1) = 10$.
- So using our more formal definition of continuity we know f is not continuous at 1 because the $\lim x \to 1\ f(x)$ does exist but the value of that limit does not match the function value and that is enough to make the function fail to be continuous at 1. Of course, this failure is pretty stupid as it is caused by a bad definition of the function value at 1. Hence, the lack of continuity is due to a flaw in the way we defined our function and it can easily be removed by simply redefining the function to be 1 at 1. So this kind of lack of continuity is called a **removable discontinuity** to reflect that.

But let's go back to the first f which was

$$f(x) = \begin{cases} 1, & \text{if } 0 \leq x \leq 1 \\ 2 & \text{if } 1 < x \leq 2. \end{cases}$$

Let's pick a number called a **tolerance** which is represented by the Greek letter ϵ. Let's make $\epsilon = 1/2$. Can we find a circle about $x = 1$ so that $f(x) = 1$ in that whole circle? Such a circle would be of the form $1 - \delta < x < 1 + \delta$ for any radius δ. The problem is on the left side, $f(x) = 1$ for sure, but on the right side $f(x) = 2$ instead. So we will never be able to find such a circle. If we wanted instead to find a circle about $x = 1$ so that $f(x) = 2$ in the circle, we would fail again as on the left we would get 1 and on the right 2. The only two reasonable choices for the value of $\lim_{x \to 1} f(x)$ are 1 and 2 and for the choice $\epsilon = 1/2$ we find

$$1 - \delta < x < 1 + \delta \quad \text{does not imply} \quad -\epsilon < f(x) - 1 < \epsilon$$

because on the left it works and we have $-1/2 < 0 < 1/2$ as $f(x) = 1$ there and the subtraction is 0 but on the right, we get $f(x) - 1 = 2 - 1 = 1$ which is bigger than our epsilon of $1/2$. Now we know this is a hard slog and quite technical, but it is the simplest example which shows the following. A limit exists at $x = p$ with value L if given any tolerance ϵ we can find some radius δ so that

$$p - \delta < x < p + \delta \implies -\epsilon < f(x) - L < \epsilon$$

Now again, we don't have to think about limits this carefully most of the time, but it is a great exercise to work it through in your mind. This leads to another more formal definition for limits.

Definition 4.1.2 (*Limit Of A Function At A Point: $\epsilon - \delta$ Version*)
f is said to be have a limit at a point p in its domain if given any tolerance ϵ there is a restriction δ on the values of x so that

$$p - \delta < x < p + \delta \implies -\epsilon < f(x) - L < \epsilon$$

We can also write this using absolute values as

$$|x - p| < \delta \implies |f(x) - L| < \epsilon$$

Using these more formal ideas of limits, we can also define continuity differently in terms of tolerance. Let's write that one down too. This time we'll use t instead of x so you can get used to the idea that the choice of letter is up to us!

Definition 4.1.3 (*Continuity Of A Function At A Point: $\epsilon - \delta$ Version*)
f is said to be continuous at a point p in its domain if given any tolerance ϵ there is a restriction δ on the values of t so that $\mid f(t) - f(p) \mid < \epsilon$ if t is in the domain of f and t satisfies $\mid t - p \mid < \delta$.

Now, this second definition is written in the very formal language of mathematics and is almost fussily precise. We will need that level of fussiness a few times in our discussions but not too often. Now let's do some more examples of limits and continuity but this time let's work with something more interesting that a step function. Consider these three versions of a function f defined on $[0, 2]$.

$$f(x) = \begin{cases} x^2, & \text{if } 0 \le x < 1 \\ 10, & \text{if } x = 1 \\ 1 + (x-1)^2 & \text{if } 1 < x \le 2. \end{cases}$$

The first version is not continuous at $x = 1$ because although the $\lim_{x \to 1} f(x)$ exists and equals 1 ($\lim_{x \to 1^-} f(x) = 1$ and $\lim_{x \to 1^+} f(x) = 1$), the value of $f(1)$ is 10 which does not match the limit. Hence, we know f here has a removable discontinuity at $x = 1$. Note continuity failed because the limit existed but the value of the function did not match it. The second version of f is given below.

$$f(x) = \begin{cases} x^2, & \text{if } 0 \le x \le 1 \\ (x-1)^2 & \text{if } 1 < x \le 2. \end{cases}$$

In this case, the $\lim_{x \to 1^-} = 1$ and $f(1) = 1$, so in a sense, we could say f is continuous from the left. However, $\lim_{x \to 1^+} = 0$ which does not match $f(1)$ and so f is not continuous from the right! Also, since the right and left hand limits do not match at $x = 1$, we know $\lim_{x \to 1}$ does not exist. Here, the function fails to be continuous because the limit does not exist. The final example is below:

$$f(x) = \begin{cases} x^2, & \text{if } 0 \le x \le 1 \\ x + (x-1)^2 & \text{if } 1 < x \le 2. \end{cases}$$

Here, the limit and the function value at 1 both match and so f is continuous at $x = 1$.

Whew! This has been hard stuff and right now it seems pretty far removed for biological modeling. But these notions of smoothness which have buried in them the ideas of limiting actions are important to help us build the tools we need to study our models later.

4.1.1 Example

Example 4.1.1 For this function

$$f(x) = \begin{cases} 3, & \text{if } 0 \le x \le 1 \\ 5 & \text{if } 1 < x \le 2. \end{cases}$$

- find the right and left hand limits as $x \to 1$.
- Show that for $\epsilon = 1$, there is no radius δ so that the definition of the limit is satisfied.
- explain why f is not continuous at 1.

Solution • *The left hand limit is* 3 *and the right hand limit is* 5 *so we know the limit at* $x = 1$ *does not exist.*

- *The only choices for the limit are* 3 *or* 5*. For the choice* 3*, to show that for* $\epsilon = 1$*, there is no radius* δ *so that the definition of the limit is satisfied, note that on the left,* $1 - \delta < x$ *implies* $f(x) - 3 = 0$ *which is less than* $\epsilon = 1$ *but on the right,* $x < 1 + \delta$ *implies* $f(x) - 3 = 5 - 3 = 2$ *which is larger than* $\epsilon = 1$*. Since we can't find a* δ *that works, the limit can't be* 3*. The other argument is similar. For the choice* 5*, to show that for* $\epsilon = 1$*, there is no radius* δ *so that the definition of the limit is satisfied, note that on the left,* $1 - \delta < x$ *implies* $f(x) - 5 = 3 - 5 = -2$ *which is below* $-\epsilon = -1$ *but on the right,* $x < 1 + \delta$ *implies* $f(x) - 5 = 5 - 5 = 0$ *which is smaller than* $\epsilon = 1$*. Since we can't find a* δ *that works, the limit can't be* 5*.*
- f *is not continuous at* 1 *because the limit at* 1 *does not exist.*

Example 4.1.2 For this function

$$f(x) = \begin{cases} 2, & \text{if } 0 \leq x \leq 2 \\ 6 & \text{if } 2 < x \leq 4. \end{cases}$$

- find the right and left hand limits as $x \to 2$.
- Show that for $\epsilon = 1$, there is no radius δ so that the definition of the limit is satisfied.
- explain why f is not continuous at 2.

Solution • *The left hand limit is* 2 *and the right hand limit is* 6 *so we know the limit at* $x = 2$ *does not exist.*

- *The only choices for the limit are* 2 *or* 6*. For the choice* 2*, to show that for* $\epsilon = 1$*, there is no radius* δ *so that the definition of the limit is satisfied, note that on the left,* $2 - \delta < x$ *implies* $f(x) - 2 = 0$ *which is less than* $\epsilon = 1$ *but on the right,* $x < 2 + \delta$ *implies* $f(x) - 2 = 6 - 2 = 4$ *which is larger than* $\epsilon = 1$*. Since we can't find a* δ *that works, the limit can't be* 2*. The other argument is similar. For the choice* 6*, to show that for* $\epsilon = 1$*, there is no radius* δ *so that the definition of the limit is satisfied, note that on the left,* $2 - \delta < x$ *implies* $f(x) - 6 = 2 - 6 = -4$ *which is below* $-\epsilon = -1$ *but on the right,* $x < 2 + \delta$ *implies* $f(x) - 6 = 6 - 6 = 0$ *which is smaller than* $\epsilon = 1$*. Since we can't find a* δ *that works, the limit can't be* 6*.*
- f *is not continuous at* 1 *because the limit at* 2 *does not exist.*

Example 4.1.3 For the function

$$f(x) = \begin{cases} x^2 + 4x, & \text{if } 0 \leq x \leq 1 \\ -3x & \text{if } 1 < x \leq 2. \end{cases}$$

- Find the right and left hand limits at $x = 1$.
- Decide if f is continuous at 1 and explain your decision.

Solution • *The left hand limit is* 5 *and the right hand limit is* −3.
• *Since the limit does not exist at* 1*, the function is not continuous at* $x = 1$.

Example 4.1.4 For the function

$$f(x) = \begin{cases} x^2 + 3x + 2, & \text{if } 0 \le x < 2 \\ x^3 + 4 & \text{if } 2 \le x \le 5. \end{cases}$$

• Find the right and left hand limits at $x = 2$.
• Decide if f is continuous at 2 and explain your decision.

Solution • *the left hand limit is* 12 *and the right hand limit is* 12 *so the limit at* 2 *does exist and has the value* 12.
• *Since* $f(2) = 12$*, the function value at* 2 *matches the limit value at* 2 *and so the function is continuous at* 2.

4.1.2 Homework

Exercise 4.1.1 *For this function*

$$f(x) = \begin{cases} 1, & \text{if } 0 \le x \le 1 \\ 5 & \text{if } 1 < x \le 2. \end{cases}$$

• *find the right and left hand limits as* $x \to 1$.
• *Show that for* $\epsilon = 1$*, there is no radius* δ *so that the definition of the limit is satisfied.*
• *explain why* f *is not continuous at* 1.

Exercise 4.1.2 *For this function*

$$f(x) = \begin{cases} 4, & \text{if } 0 \le x \le 2 \\ 7 & \text{if } 2 < x \le 4. \end{cases}$$

• *find the right and left hand limits as* $x \to 2$.
• *Show that for* $\epsilon = 1$*, there is no radius* δ *so that the definition of the limit is satisfied.*
• *explain why* f *is not continuous at* 2.

Exercise 4.1.3 *For this function*

$$f(x) = \begin{cases} -2, & \text{if } 0 \le x \le 3 \\ 1 & \text{if } 3 < x \le 5. \end{cases}$$

- *find the right and left hand limits as* $x \rightarrow 3$.
- *Show that for* $\epsilon = 1$, *there is no radius* δ *so that the definition of the limit is satisfied.*
- *explain why* f *is not continuous at* 3.

Exercise 4.1.4 *For the function*

$$f(x) = \begin{cases} 2x + 4, & \text{if } 0 \leq x \leq 2 \\ x^2 + 4 & \text{if } 2 < x \leq 4. \end{cases}$$

- *Find the right and left hand limits at* $x = 2$.
- *Decide if* f *is continuous at* 2 *and explain your decision.*

Exercise 4.1.5 *For the function*

$$f(x) = \begin{cases} 2x^2, & \text{if } 0 \leq x < 3 \\ x^3 & \text{if } 3 \leq x \leq 6. \end{cases}$$

- *Find the right and left hand limits at* $x = 3$.
- *Decide if* f *is continuous at* 3 *and explain your decision.*

4.2 Differentiability

We have already discussed what a derivative means in our opening salvo on limits. I'm sure you remember the pain and the agony. Still, we do need to get proper tools to build our models, so let's revisit this again but this time with a bit more detail. Differentiability is an idea we need to discuss more carefully. From what we have said before, it seems a formal definition of a derivative at a point would be this.

Definition 4.2.1 (*Differentiability of A Function At A Point*)
f is said to be differentiable at a point p in its domain if the limit as t approaches $p, t \neq p$, of the quotients $\frac{f(t) - f(p)}{t - p}$ exists. When this limit exists, the value of this limit is denoted by a number of possible symbols: $f'(p)$ or $\frac{df}{dt}(p)$. This can also be phrased in terms of the right and left hand limits $f'(p^+) = \lim_{t \rightarrow p^+} \frac{f(t) - f(p)}{t - p}$ and $f'(p^-) = \lim_{t \rightarrow p^-} \frac{f(t) - f(p)}{t - p}$. If both exist and match at p, then $f'(p)$ exists and the value of the derivative is the common value.

All of our usual machinery about limits can be brought to bear here. For example, all of the limit stuff could be rephrased in the $\epsilon - \delta$ framework but we seldom need to go that deep. However, the most useful way of all to view the derivative is to use an error term. Let

$$E(x - p) = f(x) - f(p) - f'(p)\,(x - p).$$

If the derivative exists, we can use the $\epsilon - \delta$ formalism to get some important information about the error. Choose $\epsilon = 1$. Then by definition, there is a radius δ so that

$$|x - p| < \delta \Longrightarrow \left| \frac{f(x) - f(p)}{x - p} - f'(p) \right| < 1.$$

We can rewrite this by getting a common denominator as

$$|x - p| < \delta \Longrightarrow \left| \frac{f(x) - f(p) - f'(p)(x - p)}{x - p} \right| < 1.$$

But the numerator here is the error, so we have

$$|x - p| < \delta \Longrightarrow \left| \frac{E(x - p)}{x - p} \right| < 1.$$

This tells us

$$|x - p| < \delta \Longrightarrow |E(x - p)| < |x - p|.$$

So as $x \to p$, $|x - p| \to 0$ and the above tells us $E(x - p) \to 0$ as well. Good to know as we will put this fact to use right away.

Here is our new definition of the derivative existing at a point that uses error.

Definition 4.2.2 (*Error Form for Differentiability of A Function At A Point*)
Let the value of the derivative of f at p be denoted by $f'(p)$ and let the error term $E(x - p)$ be defined by

$$E(x - p) = f(x) - f(p) - f'(p)\,(x - p).$$

Then if $f'(p)$ exists, the arguments above show us

- $E(0) = 0$ (i.e. $x = p$ and everything disappears).
- $E(x - p) \to 0$ as $x \to p$ and
- $E(x - p)/(x - p) \to 0$ as $x \to p$ also. Note when we say this, we are assuming $x \neq p$ of course.

So the error acts a lot like $(x - p)^2$! Going the other way, if there is a number L so that

$$E(x - p) = f(x) - f(p) - L\,(x - p).$$

satisfies the same two conditions, then f has a derivative at p whose value is L. We usually say this faster like this: f has a derivative $f'(p)$ if and only if the error $E(x - p)$ and $E(x - p)/(x - p)$ goes to 0 as $x \to p$.

Here is an example which should help. We will take our old friend $f(x) = x^2$. Let's look at the derivative of f at the point x. We have

$$\begin{aligned} E(\Delta x) &= f(x + \Delta x) - f(x) - f'(x)\,\Delta x \\ &= (x + \Delta x)^2 - x^2 - 2x\,\Delta x \\ &= 2x\,\Delta x + (\Delta x)^2 - 2x\,\Delta x \\ &= (\Delta x)^2. \end{aligned}$$

See how $E(\Delta x) = (\Delta x)^2 \to 0$ and $E(\Delta x)/\Delta x = \Delta x \to 0$ as $\Delta x \to 0$? This is the essential nature of the derivative. Replacing the original function value $f(x + \Delta x)$ by the value given by the straight line $f(x) + f'(x)\,\Delta x$ makes an error that is roughly proportional to $(\Delta x)^2$. Good to know.

Also, note using the error definition, if we can find a number L so that $E(x - p) = f(x) - f(p) - L\,(x - p)$ goes to zero in the two ways above, we know $f'(p) = L$. This is a fabulous tool.

A fundamental consequence of the existence of a derivative of a function at a point t is that it must also be continuous there. This is easy to see using the error form of the derivative. If f has a derivative at p, we know that

$$f(x) = f(p) + f'(p)\,(x - p) + E(x - p)$$

and as $x \to p$, we get $\lim_{x \to p} f(x) = f(p)$ because the other terms vanish. So having a derivative implies continuity. Think of continuity as a first level of smoothness and having a derivative as the second level of smoothness. So things at the second level should get the first level for free! And they do. Remember this as it will be important in a lot of things later. We state this as Theorem 4.2.1.

Theorem 4.2.1 (Differentiability Implies Continuity)
Let f be a function which is differentiable at a point t in its domain. Then f is also continuous at t.

Proof We just did this argument! ■

4.2.1 Example

Example 4.2.1 We can show the derivative of $f(x) = x^3$ is $3x^2$. Using this, write down the definition of the derivative at $x = 1$ and also the error form at $x = 1$. State the two conditions on the error too.

Solution *The definition of the derivative is*

$$\frac{dy}{dx}(1) = \lim_{h \to 0} \frac{(1 + h)^3 - (1)^3}{h}.$$

The error form is

$$x^3 = (1)^3 + 3(1)^2(x-1) + E(x-1)$$

where $E(x-1)$ and $E(x-1)/(x-1)$ both go to zero as $x \to 1$.

Example 4.2.2 We can show the derivative of $f(x) = \sin(x)$ is $\cos(x)$. Using this, write down the definition of the derivative at $x = 2$ and also the error form at $x = 2$. State the two conditions on the error too.

Solution *The definition of the derivative is*

$$\frac{dy}{dx}(2) = \lim_{h \to 0} \frac{\sin(2+h) - \sin(2)}{h}.$$

The error form is

$$\sin(x) = \sin(2) + \cos(2)(x-2) + E(x-2)$$

where $E(x-2)$ and $E(x-2)/(x-2)$ both go to zero as $x \to 2$.

Example 4.2.3 We know $f(x) = x^5$ has a derivative at each x and equals $5x^4$. Explain why $f(x) = x^5$ must be a continuous function.

Solution *x^5 is continuous since it has a derivative.*

Example 4.2.4 Suppose a function has a jump at the point $x = 5$. Can this function have a derivative there?

Solution *No. If the function did have a derivative there, it would have to be continuous there which it is not since it has a jump at that point.*

4.2.2 Homework

Exercise 4.2.1 *Suppose a function has a jump at the point $x = -2$. Can this function have a derivative there?*

Exercise 4.2.2 *We know $f(x) = \cos(x)$ has a derivative at each x and equals $-\sin(x)$. Explain why $f(x) = \cos(x)$ must be a continuous function.*

Exercise 4.2.3 *We can show the derivative of $f(x) = x^7$ is $7x^6$. Using this, write down the definition of the derivative at $x = 1$ and also the error form at $x = 1$. State the two conditions on the error too.*

Exercise 4.2.4 *We can show the derivative of $f(x) = x^2 + 5$ is $2x$. Using this, write down the definition of the derivative at $x = 4$ and also the error form at $x = 4$. State the two conditions on the error too.*

4.3 Simple Derivatives

We need some fast ways to calculate these derivatives. Let's start with constant functions. These never change and since derivatives are supposed to give rates of change, we would expect this to be zero. Here is the argument. Let $f(x) = 5$ for all x. Then to find the derivative at any x, we calculate this limit

$$\begin{aligned}\frac{dy}{dx}(x) &= \lim_{h \to 0} \frac{f(x+h) - f(x)}{h} \\ &= \lim_{h \to 0} \frac{5 - 5}{h} \\ &= \lim_{h \to 0} 0 = 0.\end{aligned}$$

A little thought shows that the value 5 doesn't matter. So we have a general result which we dignify by calling it a theorem just because we can!

Theorem 4.3.1 (The Derivative of a constant)
If c is any number, then the function $f(t) = c$ gives a constant function. The derivative of c with respect to t is then zero.

Proof We just hammered this out! ■

Let's do one more, the derivative of $f(x) = x$. This one is easy too. We calculate

$$\begin{aligned}\frac{dy}{dx}(x) &- \lim_{h \to 0} \frac{f(x+h) - f(x)}{h} \\ &= \lim_{h \to 0} \frac{x + h - x}{h} \\ &= \lim_{h \to 0} \frac{h}{h} \\ &= \lim_{h \to 0} 1 = 1.\end{aligned}$$

So now we know that $\frac{d}{dx}(x) = 1$ and it wasn't that hard. To find the derivatives of more powers of x, we are going to find an easy way. The easy way is to prove a general rule and then apply it to the two examples we know. This general rule is called the **Product Rule**. Here it is.

Theorem 4.3.2 (The Product Rule)
If f and g are both differentiable at a point x, then the product $f\,g$ is also differentiable at x and has the value

$$\Big(f(x)\,g(x)\Big)' = f'(x)\,g(x) + f(x)\,g'(x)$$

Proof We need to find this limit at the point x:

$$\frac{d}{dx}(fg)(x) = \lim_{h \to 0} \frac{f(x+h)\,g(x+h) \;-\; f(x)\,g(x)}{h}.$$

Looks forbidding doesn't it? Let's add and subtract just the right term:

$$\frac{d}{dx}(fg)(x) = \lim_{h \to 0} \frac{f(x+h)\,g(x+h) \;-\; f(x)\,g(x+h) \;+\; f(x)\,g(x+h) \;-\; f(x)\,g(x)}{h}.$$

Now group the pieces like so:

$$\frac{d}{dx}(fg)(x) = \lim_{h \to 0} \frac{\Big(f(x+h)\,g(x+h) \;-\; f(x)\,g(x+h)\Big) \;+\; \Big(f(x)\,g(x+h) \;-\; f(x)\,g(x)\Big)}{h}.$$

Factor out common terms:

$$\frac{d}{dx}(fg)(x) = \lim_{h \to 0} \frac{\Big(f(x+h) \;-\; f(x)\Big)\,g(x+h) \;+\; \Big(g(x+h) \;-\; g(x)\Big)\,f(x)}{h}.$$

Now rewrite as two separate limits:

$$\frac{d}{dx}(fg)(x) = \lim_{h \to 0} \left(\frac{f(x+h) \;-\; f(x)}{h}\right) g(x+h) \;+\; \lim_{h \to 0} \left(\frac{g(x+h) \;-\; g(x)}{h}\right) f(x)$$

Almost there. In the first limit, the first part goes to $f'(x)$ and the second part goes to $g(x)$ because since g has a derivative at x, g is continuous at x. In the second limit, the $f(x)$ doesn't change and the other piece goes to $g'(x)$. So there you have it:

$$\frac{d}{dx}(f(x)\,g(x)) = f'(x)\,g(x) \;+\; f(x)\,g'(x).$$

■

We are now ready to *blow your mind* as Jack Black would say.

- Let $f(x) = x^2$. Let's apply the product rule. The first function is x and the second one is x also.

$$\begin{aligned}\frac{d}{dx}(x^2) &= \frac{d}{dx}(x)\,x \;+\; (x)\,\frac{d}{dx}(x)\\ &= (1)\,(x) \;+\; (x)\,(1) \;=\; 2\,x.\end{aligned}$$

  This is just what we had before!

- Let $f(x) = x^3$. Let's apply the product rule here. The first function is x^2 and the second one is x.

$$\frac{d}{dx}(x^3) = \frac{d}{dx}(x^2)\, x \;+\; (x^2)\, \frac{d}{dx}(x)$$
$$= (2\,x)\,(x) \;+\; (x^2)\,(1) \;=\; 3\,x^2.$$

- Let $f(x) = x^4$. Let's apply the product rule again. The first function is x^3 and the second one is x.

$$\frac{d}{dx}(x^4) = \frac{d}{dx}(x^3)\, x \;+\; (x^3)\, \frac{d}{dx}(x)$$
$$= (3\,x^2)\,(x) \;+\; (x^3)\,(1) \;=\; 4\,x^3.$$

We could go on, but you probably see the pattern. If P is a positive integer, then $\frac{d}{dx}\, x^P \;=\; P\, x^{P-1}$. That's all there is too it. Just to annoy you, we'll state it as a Theorem:

Theorem 4.3.3 (The Simple Power Rule: Positive Integers)
If f is the function x^P for any positive integer P, then the derivative of f with respect to x satisfies

$$\left(x^P\right)' = P\, x^{P-1}$$

Proof We just reasoned this out! ■

Next, we don't want to overload you with a lot of tedious proofs of various properties, so let's just get to the chase. From the way the derivative is defined as a limit, it is pretty clear the following properties hold: first **the derivative of a sum of functions is the sum of the derivatives**:

$$\frac{d}{dx}\,(f(x) \;+\; g(x)) = \frac{d}{dx}(f(x)) \;+\; \frac{d}{dx}(g(x))$$

Second, **the derivative of a constant times a function is just constant times the derivative**.

$$\frac{d}{dx}\,(c\; f(x)) = c\, \frac{d}{dx}(f(x)).$$

Armed with these tools, we can take a lot of derivatives!

4.3.1 Examples

First, derivatives of various polynomials.

Example 4.3.1 Find $\left(t^2 + 4\right)'$.

Solution

$$\left(t^2 + 4\right)' = 2t.$$

Example 4.3.2 Find $\left(3\,t^4 + 8\right)'$.

Solution

$$\left(3\,t^4 + 8\right)' = 12\,t^3.$$

Example 4.3.3 Find $\left(3\,t^8 + 7\,t^5 - 2\,t^2 + 18\right)'$.

Solution

$$\left(3\,t^8 + 7\,t^5 - 2\,t^2 + 18\right)' = 24\,t^7 + 35\,t^4 - 4\,t.$$

Next, some product rules.

Example 4.3.4 Find $\left((t^2 + 4)\,(5t^3 + t^2 - 4)\right)'$.

Solution

$$\left((t^2 + 4)\,(5t^3 + t^2 - 4)\right)' = (2t)\,(5t^3 + t^2 - 4) + (t^2 + 4)\,(15t^2 + 2t)$$

and, of course, this expression above can be further simplified.

Example 4.3.5 Find $\left((2t^3 + t + 5)\,(-2t^6 + t^3 + 10)\right)'$.

Solution

$$\begin{aligned}\left((2t^3 + t + 5)\,(-2t^6 + t^3 + 10)\right)' &= (6t^2 + 1)\,(-2t^6 + t^3 + 10)\\ &\quad + (2t^3 + t + 5)\,(-12t^5 + 3t^2)\end{aligned}$$

4.3.2 Homework

Exercise 4.3.1 *Find* $\left(6\,t^7 + 4\,t^2\right)'$.

Exercise 4.3.2 *Find* $\left(3\,t^2 + 7\,t + 11\right)'$.

Exercise 4.3.3 *Find* $\left(-2\,t^4\,+\,7\,t^2\,-\,2\,t+1\right)'$.

Exercise 4.3.4 *Find* $\left(2\,+\,3\,t\,+\,7\,t^{10}\right)'$.

Exercise 4.3.5 *Find* $\left(3\,t\,+\,7\,t^8\right)'$.

Exercise 4.3.6 *Find* $\left((6t\,+\,4)\,(4t^2\,+\,t)\right)'$.

Exercise 4.3.7 *Find* $\left((5t^4\,+\,t^3\,+\,15)\,(4\,t\,+\,t^2\,+\,3)\right)'$.

Exercise 4.3.8 *Find* $\left((-8t^3\,+\,t^4)\,(4\,t\,+\,t^8)\right)'$.

Exercise 4.3.9 *Find* $\left((-10t^7\,+\,14t^5\,+\,8t^2\,+\,8)\,(5t^2\,+\,t^9)\right)'$.

Exercise 4.3.10 *Find* $\left((8t^{11}\,+\,t^{15})\,(-5t^2\,+\,8t\,-\,2)\right)'$.

4.4 The Quotient Rule

Derivatives of functions that have denominators are next. Let's assume we want to find the derivative of a function $1/g(x)$. We need to calculate

$$\left(1/g(x)\right)' = \lim_{h\to 0}\left(\frac{1}{g(x+h)} - \frac{1}{g(x)}\right)/h$$

where, of course, this doesn't make sense at any point where $g(x) = 0$! So as long as g is not zero at the point of interest, the first thing we do is to get a common denominator:

$$\left(1/g(x)\right)' = \lim_{h\to 0}\left(\frac{g(x) - g(x+h)}{g(x+h)\,g(x)}\right)/h$$

Now reorganize these fractions like so

$$\left(1/g(x)\right)' = \lim_{h\to 0}\left(\frac{g(x) - g(x+h)}{h}\right)\frac{1}{g(x+h)\,g(x)}$$

Now pull out a minus sign to get

$$\left(1/g(x)\right)' = \lim_{h\to 0}\; -\left(\frac{g(x+h) - g(x)}{h}\right)\frac{1}{g(x+h)\,g(x)}$$

We are done really! The first part goes to $g'(x)$ and since g is continuous at x since it is differentiable there, the second part goes to $g(x)\,g(x)$. So combining, we have $(1/g(x))' = -g'(x)/g(x)^2$. Now with that done, we can tackle the big question all of you are just dying to know: what is $(f(x)/g(x))'$? We will set this up as a theorem: this is a mathematics book after all!

Theorem 4.4.1 (The Quotient Rule)
If f and g are both differentiable at a point x in their domains and $g(x)$ is not zero, then the quotient f/g is also differentiable at x and has the value

$$\left(\frac{f(x)}{g(x)}\right)' = \frac{f'(x)\, g(x) \;-\; f(x)\, g'(x)}{(g(x))^2}$$

Proof This is not so hard. We just use the product rule that we know.

$$\begin{aligned}\left(\frac{f(x)}{g(x)}\right)' &= \left(f(x)\,\frac{1}{g(x)}\right)' \\ &= (f(x))'\,\frac{1}{g(x)} \;+\; f(x)\left(\frac{1}{g(x)}\right)' \\ &= (f(x))'\,\frac{1}{g(x)} \;-\; f(x)\left(\frac{(g(x))'}{(g(x))^2}\right).\end{aligned}$$

We usually write this a bit more simply.

$$\left(\frac{f(x)}{g(x)}\right)' = f'(x)\,\frac{1}{g(x)} \;-\; f(x)\,\frac{g'(x)}{g^2(x)}.$$

It is also common to get a common denominator to get this form:

$$\left(\frac{f(x)}{g(x)}\right)' = \frac{f'(x)\, g(x) \;-\; f(x)\, g'(x)}{g^2(x)}.$$

■

Most of us use this mnemonic too.

$$\left(\frac{\textbf{top}}{\textbf{bottom}}\right)' = \frac{\textbf{top}'\ \textbf{bottom} - \textbf{top}\ \textbf{bottom}'}{\textbf{bottom}^2}.$$

Now let's apply the magic. Rubbing our fingers together with glee we note

- Let $f(x) = x^{-1} = 1/x$. We know how to take this derivative. $(1/g)' = -g'/g^2$ and here $g(x) = x$. So we have

$$\frac{d}{dx}(1/x) = -1\ (1/x^2).$$

- Let $f(x) = x^{-2}$. Same idea. We know how to take this derivative. $(1/g)' = -g'/g^2$ and here $g(x) = x^2$. So we have

$$\frac{d}{dx}(1/x^2) = -2x\ (1/x^4)\ =\ -2/x^3.$$

- Let $f(x) = x^{-3}$. Same idea. We know how to take this derivative. $(1/g)' = -g'/g^2$ and here $g(x) = x^3$. So we have

$$\frac{d}{dx}(1/x^3) = -3x^2\ (1/x^6)\ =\ -3/x^4.$$

We could go on, but you probably see the pattern. If P is a positive integer, then $\frac{d}{dx}\ x^{-P} = -P\ x^{-P-1}$. That's all there is too it. We'll state it as a Theorem:

Theorem 4.4.2 (The Simple Power Rule: Negative Integers)
If f is the function x^{-P} for any positive integer P, then the derivative of f with respect to x satisfies

$$\left(x^{-P}\right)' = -P\ x^{-P-1}$$

Proof Duhh, look above dude. ■

4.4.1 Examples

First, derivatives of various polynomials.

Example 4.4.1 Find $\left(t^{-2}\ +\ 4\right)'$.

Solution

$$\left(t^{-2}\ +\ 4\right)' = -2/t^3.$$

Example 4.4.2 Find $\left(3\,t^4\ +\ 8t^{-5}\right)'$.

Solution

$$\left(3\,t^4\ +\ 8t^{-5}\right)' = 12\,t^3 - 40/t^6$$

Example 4.4.3 Find $\left(3\,t^{-18}\ +\ 3\,t^{-2} - 2\,t^2\right)'$.

Solution

$$\left(3\,t^{-18}\ +\ 3\,t^{-2} - 2\,t^2\right)' = -54/t^{19} - 6/t^3 - 4t$$

Next, some quotient rules.

Example 4.4.4 Find $\Big(t/(t+1)\Big)'$

Solution

$$\Big(t/(t+1)\Big)' = \frac{1\,(t+1) \;-\; (t)\,1}{(t+1)^2} = \frac{t+1-t}{(t+1)^2} = 1/(t+1)^2$$

Of course, simplifying these expressions is not always so easy!

Example 4.4.5 Find $\Big(t^2/(t^3+2t+5)\Big)'$

Solution

$$\Big(t^2/(t^3+2t+5)\Big)' = \frac{(2t)\,(t^3+2t+5) \;-\; (t^2)\,(3t^2+2)}{(t^3+2t+5)^2}$$

Ughh. The thought of simplifying this expression makes us weep. So let's leave this one alone.

Example 4.4.6 Find $\Big((t^2\,+\,4)/(5t^3\,+\,t^2\,-\,4)\Big)'$

Solution

$$\left(\frac{t^2\,+\,4}{5t^3\,+\,t^2\,-\,4}\right)' = \frac{2t\,\left(5t^3\,+\,t^2\,-\,4\right) \,-\, \left(t^2\,+\,4\right)\,\left(15t^2\,+\,2t\right)}{\left(5t^3\,+\,t^2\,-\,4\right)^2}$$

4.4.2 Homework

Exercise 4.4.1 $\Big(8t/(t^2+3t^3+50)\Big)'$

Exercise 4.4.2 $\Big((2t^4+5t^3)/(t+15)\Big)'$

Exercise 4.4.3 *Find* $\left(5\,t^{-7}\,+\,9\,t^{-4} - 12\,t^4\right)'$.

Exercise 4.4.4 *Find* $\left(2\,t^{-1}\,+\,3/t^4\right)'$.

Exercise 4.4.5 *Find* $\left(2/t^2\,+\,3/t^5+88t^3\right)'$.

Exercise 4.4.6 *Find* $\left(t^3\,+\,3/t^7\right)'$.

Exercise 4.4.7 *Find* $(4\,t^{-5} + 3t^6)'$.

Exercise 4.4.8 $\left((2t + 3t^2)/(t^5 + 15t^3) + 2t + 5\right)'$

Exercise 4.4.9 $\left((1/t)/(1/t + 6)\right)'$

Exercise 4.4.10 $\left((t + 1/t)/(t^2 + 2)\right)'$

4.5 Chain Rule

This one is harder to figure out, but it is a computational rule we use quite a bit. It is not too hard to see this. This one is about composition of functions. You probably didn't like that idea much in high school, but it is actually a simple concept. You shove one function into another and calculate the result. We know how to find x^2 and u^2 for any x and u. So what about $(x^2 + 3)^2$? This just means take $u = x^2 + 3$ and square it. That is if $f(u) = u^2$ and $g(x) = x^2 + 3$, the *composition* of f and g is simply $f(g(x))$. Compositions are really nice to know about and later we'll show you an example that comes out of how neurons compute their output based on the inputs that come into them. In that example, their output is a complicated composition!! So useful, yes it is.

Let's talk about continuity first and then we'll go on to the idea of taking the derivative of a composition. If f and g are both continuous, then if you think about it, another way to phrase the continuity of f is that

$$\lim_{y \to x} f(y) = f(x) \Longrightarrow f(x) = f(\lim_{y \to x} y).$$

So if f and g are both continuous, we can say

$$\begin{aligned}\lim_{y \to x} f(g(y)) &= f(\lim_{y \to x} g(y)\\ &= f(g(\lim_{y \to x} y))\\ &= f(g(x)).\end{aligned}$$

So the **composition of continuous functions is continuous**. Heave a big sigh of relief as smoothness has not been lost by pushing one smooth function into another smooth function! The next question is whether or not the composition of functions having derivatives gives a new function which has a derivative. And how could we calculate this derivative if the answer is yes? It turns out this is true but to see if requires a bit more work with limits.

So we want to know what $\Big(f(g(x))\Big)'$ is. First, if g was always constant, the answer is easy. It is $\Big(f(\textbf{constant})\Big)' = 0$ which is a special case of the formula we are going to develop. So let's assume g is not constant. Then, we want to calculate

$$\Big(f(g(x))\Big)' = \lim_{h \to 0} \frac{f(g(x+h)) \; - \; f(g(x))}{h}.$$

Now do a simple trick as follows. Rewrite by dividing and multiplying by $g(x+h) - g(x)$ which is ok to do as we assume g is not constant and so we don't divide by 0. We get

$$\Big(f(g(x))\Big)' = \lim_{h \to 0} \frac{f(g(x+h)) \; - \; f(g(x))}{g(x+h) - g(x)} \, \frac{g(x+h) - g(x)}{h}$$

Now as $h \to 0$, the second piece goes to $g'(x)$. Letting $u = g(x)$ and $\Delta u = g(x+h) - g(x)$, we see the first term looks like $\Big(f(u + \Delta u) - f(u)\Big)/\Delta u$. Since g is continuous because g has a derivative, as $h \to 0$, $\Delta u \to 0$ too. So the first term looks like $\lim \Delta u \to 0 \Big(f(u + \Delta u) - f(u)\Big)/\Delta u = f'(u)$. And so we have it

$$\Big(f(g(x))\Big)' = f'(g(x))\, g'(x).$$

Theorem 4.5.1 (Chain Rule For Differentiation)
If the composition of f and g is defined at a number x and if both $f'(g(x))$ and $g'(x)$ exist, then the derivative of the composition of f and g also exists and is given by

$$\Big(f(g(x))\Big)' = f'(g(x))\, g'(x)$$

Proof We reasoned this out above. We usually think about this as follows

$$\Big(f(\textbf{inside})\Big)' = f'(\textbf{inside})\, \textbf{inside}'(x)$$

■

It is easiest to see how this works with some examples:

4.5.1 Examples

Example 4.5.1 Find the derivative of $(t^3 + 4)^3$.

Solution *It is easy to do this if we think about it this way.*

$$\left((\textbf{thing})^{\textbf{power}}\right)' = \textbf{power} \times (\textbf{thing})^{\textbf{power}-1} \times (\textbf{thing})'$$

Thus,

$$\left((t^3 + 4)^3\right)' = 3\,(t^3 + 4)^2\,(3t^2)$$

Example 4.5.2 Find the derivative of $1/(t^2 + 4)^3$.

Solution *This is also*

$$\left((\textbf{thing})^{\textbf{power}}\right)' = \textbf{power} \times (\textbf{thing})^{\textbf{power}-1} \times (\textbf{thing})'$$

where **power** *is* -3 *and thing is* $t^2 + 4$. *So we get*

$$\left(1/(t^2 + 4)^3\right)' = -3\,(t^2 + 4)^{-4}\,(2t)$$

Example 4.5.3 Find the derivative of $(6t^4 + 9t^2 + 8)^6$.

Solution

$$\left((6t^4 + 9t^2 + 8)^6\right)' = 6\,(6t^4 + 9t^2 + 8)^5\,(24t^3 + 18t)$$

4.5.2 Homework

Exercise 4.5.1 *Find the derivative of* $(-2t^3 + 9t + 81)^7$.

Exercise 4.5.2 *Find the derivative of* $(4t^2 + 10t^8)^3$.

Exercise 4.5.3 *Find the derivative of* $(1 + 4t + 9t^2)^6$.

Exercise 4.5.4 *Find the derivative of* $\left((t^2)/(1 + t^2)\right)^3$.

Exercise 4.5.5 *Find the derivative of* $(2t + 1/(5t))^5$.

Chapter 5
Sin, Cos and All That

5.1 Sin, Cos and All That!

We sometimes need to work with the trigonometric functions.

5.1.1 The Sin and Cos Functions

We need to find the derivatives of the sin and cos functions. We will do this indirectly. Now we will assume you know about sin and cos and their usual properties. After all, you have seen this before! What we want to do it something new with them—find their derivatives. So if you are rusty about these two functions, crack open you're old high school book and refresh your mind.

Ok good, you're back. Look at Fig. 5.1 carefully.

From it, we can figure out three important relationships. First, from high school times, you should know a number of cool things about circles. The one we need is the area of what is called a sector. Draw a circle of radius r in the plane. Measure an angle x counterclockwise from the horizontal axis. Then look at the pie shaped wedge formed in the circle that is bounded above by the radial line, below by the horizontal axis and to the side by a piece of the circle. It is easy to see this in Fig. 5.1. Now we are going to be interested in the angle x being quite small, but the picture is drawn with x about 30°. If we tried to draw it with a very small angle x, it would be very hard to see the relationships we want to find. Looking at the picture, note there is a first sector or radius $\cos(x)$ and a larger sector of radius 1. It turns out the area of a sector is $1/2r^2\theta$ where θ is the angle that forms the sector. From the picture, the area of the first sector is clearly less than the area of the second one. So we have

$$(1/2)\ \cos^2(x)\, x < (1/2)\, x$$

J.K. Peterson, *Calculus for Cognitive Scientists*, Cognitive Science and Technology, DOI 10.1007/978-981-287-874-8_5

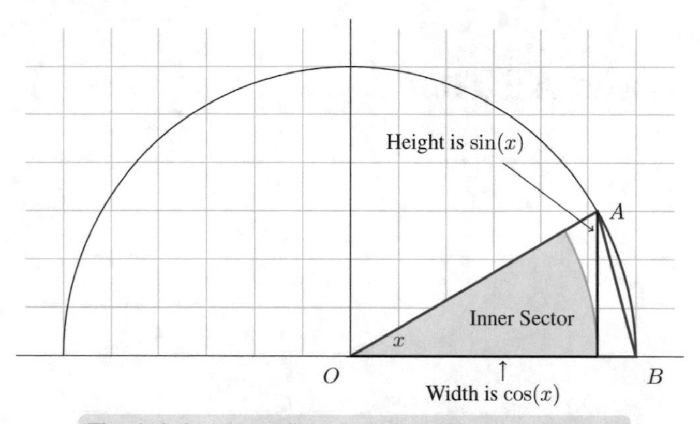

The circle here has radius 1 and the angle x determines three areas: the area of the inner sector, $\frac{1}{2}\cos^2(x)\,x$, the area of triangle Δ**OAB**, $\frac{1}{2}\sin(x)$ and the area of the outer sector, $\frac{1}{2}\,x$. We see the areas are related by

$$\frac{1}{2}\cos^2(x)\,x < \frac{1}{2}\sin(x) < \frac{1}{2}\,x.$$

Fig. 5.1 The critical area relationships that help us understand how to estimate $\sin(x)/x$

Now if you look at the picture again, you'll see a triangle caught between these two sectors. This is the triangle you get with two sides having the radial length of 1. The third side is the straight line right below the arc of the circle cut out by the angle x. The area of this triangle is $(1/2)\ \sin(x)$ because the height of the triangle is $\sin(x)$. This area is smack dab in the middle of the two sector areas. So we have

$$(1/2)\ \cos^2(x)\,x < (1/2)\ \sin(x) < (1/2)\,x.$$

These relationships work for all x and canceling all the $(1/2)$'s, we get

$$\cos^2(x)\,x < \sin(x) < x.$$

Now as long as x is not zero, we can divide to get

$$\cos^2(x) < \sin(x)/x < 1.$$

Almost done. From our high school knowledge about cos, we know it is a very smooth function and has no jumps. So it is continuous everywhere and so $\lim_{x\to 0}\cos(x) = 1$ since $\cos(0) = 1$. If that is true, then the limit of the square is 1^2 or still 1. So $\lim_{x\to 0}\sin(x)/x$ is trapped between the limit of the $\cos^2$ term and the limit of the constant term 1. So we have to conclude

$$\lim_{x\to 0}\sin(x)/x = 1.$$

Hah you say. Big deal you say. But what you don't know is the you have just found the derivative of sin at 0. Note

$$\left(\sin(x)\right)'(0) = \lim_{h \to 0} \frac{\sin(h) - \sin(0)}{h} = \lim_{h \to 0} \frac{\sin(h)}{h}$$

because we know $\sin(0) = 0$. Now if $\lim_{x \to 0} \sin(x)/x = 1$, it doesn't matter if we switch letters! We also know $\lim_{h \to 0} \sin(h)/h = 1$. Using this, we see

$$\left(\sin(x)\right)'(0) = \lim_{h \to 0} \frac{\sin(h)}{h} = 1 = \cos(0)$$

as $\cos(0) = 1$! (Review, Review....). This result is the key. Consider the more general result

$$\left(\sin(x)\right)' = \lim_{h \to 0} \frac{\sin(x+h) - \sin(x)}{h}.$$

Now dredge up another bad high school memory: the dreaded sin identities. We know

$$\sin(u+v) = \sin(u)\ \cos(v) + \cos(u)\ \sin(v)$$

and so

$$\sin(x+h) = \sin(x)\ \cos(h) + \cos(x)\ \sin(h).$$

Using this we have

$$\begin{aligned}
\left(\sin(x)\right)' &= \lim_{h \to 0} \frac{\sin(x+h) - \sin(x)}{h} \\
&= \lim_{h \to 0} \frac{\sin(x)\ \cos(h) + \cos(x)\ \sin(h) - \sin(x)}{h} \\
&= \lim_{h \to 0} \frac{\sin(x)\ (-1 + \cos(h)) + \cos(x)\ \sin(h)}{h}.
\end{aligned}$$

Now regroup a bit to get

$$\left(\sin(x)\right)' = \lim_{h \to 0} \sin(x)\ \frac{(-1 + \cos(h))}{h} + \cos(x)\ \frac{\sin(h)}{h}.$$

We are about done. Let's rewrite $(1 - \cos(h))/h$ by multiplying top and bottom by $1 + \cos(h)$. This gives

$$\begin{aligned}\frac{\sin(x)\,(-1 + \cos(h))}{h} &= \frac{\sin(x)\,(-1 + \cos(h))}{h}\,\frac{(1 + \cos(h))}{(1 + \cos(h))}\\ &= -\sin(x)\frac{(1 - \cos^2(h))}{h}\,\frac{1}{(1 + \cos(h))}.\end{aligned}$$

Now $1 - \cos^2(h) = \sin^2(h)$, so we have

$$\begin{aligned}\frac{\sin(x)\,(-1 + \cos(h))}{h} &= -\sin(x)\frac{\sin^2(h)}{h}\,\frac{1}{(1 + \cos(h))}\\ &= -\sin(x)\frac{\sin(h)}{h}\,\frac{\sin(h)}{1 + \cos(h)}.\end{aligned}$$

Now $\sin(h)/h$ goes to 1 and $\sin(h)/(1 + \cos(h))$ goes to $0/1 = 0$ as h goes to zero. So the first term goes to $-\sin(x)\ \times 1\ \times\ 0 = 0$. We also know the second limit is $\cos(x)\ (1)$. So we conclude

$$\left(\sin(x)\right)' = \cos(x).$$

And all of this because of a little diagram drawn in Quadrant I for a circle of radius 1 plus some high school trigonometry.

What about cos's derivative? The easy way to remember that sin and cos are shifted versions of each other. We know $\cos(x) = \sin(x + \pi/2)$. So by the chain rule

$$\left(\cos(x)\right)' = \left(\sin(x + \pi/2)\right)' = \cos(x + \pi/2)\ (1).$$

Now remember another high school trigonometry thing.

$$\cos(u + v) = \cos(u)\ \cos(v) - \sin(u)\ \sin(v)$$

and so $\cos(x + \pi/2) = \cos(x)\ \cos(\pi/2) - \sin(x)\ \sin(\pi/2)$. We also know $\sin(\pi/2) = 1$ and $\cos(\pi/2) = 0$. So we find

$$\left(\cos(x)\right)' = -\sin(x).$$

Let's summarize:

$$\begin{aligned}\boldsymbol{(\sin(x))' = \cos(x)}\\ \boldsymbol{(\cos(x))' = -\sin(x)}\end{aligned}$$

And, of course, we can use the chain rule too!! As you can see, in this book, the fun never stops (that's a reference by the way to an old Styx song...)

5.1.1.1 Examples

Example 5.1.1 Simple chain rule! Find $\left(\sin(4t)\right)'$.

Solution *This is easy:*

$$\left(\sin(4t)\right)' = \cos(4t) \times 4 = 4 \cos(4t).$$

Example 5.1.2 Chain rule! Differentiate $\sin^3(t)$

Solution *The derivative is*

$$\begin{aligned}(\sin^3(t))' &= 3 \sin^2(t) \left((\sin)'(t)\right)\\ &= 3 \sin^2(t) \cos(t)\end{aligned}$$

Example 5.1.3 Find $\left(\sin^3(x^2+4)\right)'$.

Solution

$$\left(\sin^3(x^2+4)\right)' = 3 \sin^2(x^2+4) \cos(x^2+4) \times (2x).$$

Example 5.1.4 Chain and product rule Find $\left(\sin(2x) \cos(3x)\right)'$.

Solution

$$\left(\sin(2x) \cos(3x)\right)' = 2 \cos(2x) \cos(3x) - 3 \sin(2x) \sin(3x).$$

Example 5.1.5 Quotient rule! Find $\left(\tan(x)\right)'$.

Solution *We usually don't simplify our answers, but we will this time as we are getting a new formula!*

$$\left(\tan(x)\right)' = \left(\frac{\sin(x)}{\cos(x)}\right)'$$

$$= \frac{\cos(x)\ \cos(x) - \sin(x)\ (-\sin(x))}{\cos^2(x)}$$
$$= \frac{\cos^2(x) + \sin^2(x)}{\cos^2(x)}.$$

Now if you remember, $1/cos^2(x)$ *is called* $\sec^2(x)$*. So we have a new formula:*

$$\mathbf{(\tan(x))' = \sec^2(x)}$$

5.1.1.2 Homework

Exercise 5.1.1 *Find* $\left(\sin^4(2x + 5)\right)'$.

Exercise 5.1.2 *Find* $\left(\cos(4t)\right)'$.

Exercise 5.1.3 *Find* $\left(\sin(8t)\right)'$.

Exercise 5.1.4 *Find* $\left(\sin(x^3)\ \cos(5x)\right)'$.

Exercise 5.1.5 *Find* $\left(\cos(4t)\ \sin(9t)\right)'$.

Exercise 5.1.6 *Find* $\left(\tan(6t)\right)'$.

Exercise 5.1.7 *Find* $\left(\frac{\sin(4t)}{\cos(3t)}\right)'$.

Exercise 5.1.8 *Find* $\left(\frac{1}{2+\tan^2(4t)}\right)'$.

5.2 A New Power Rule

Now we can do more chain rules!!! Now that we have more functions to work with, let's use the chain rule in the special case of a power of a function. We state this as a theorem giving our long standing tradition of trying hard to get you to see things more abstractly.

Theorem 5.2.1 (Power Rule For Functions)
If f is differentiable at the real number x, then for any integer p, f^p is also differentiable at the number x with

$$\left(f^p(x)\right)' = p\ \left(f^{p-1}(x)\right)\ f'(x)$$

5.2.1 Examples

Example 5.2.1 Differentiate $(1+\sin^3(2t))^4$

Solution *The derivative uses lots of chain rules.*

$$\begin{aligned}\left((1+\sin^3(2t))^4\right)' &= 4\ (1+\sin^3(2t))^3\ 3\ \sin^2(2t)\ \cos(2t)\ 2\\ &= 3\ \sin^2(t)\ \cos(t)\end{aligned}$$

Example 5.2.2 Find $\left((1+2x+9x^2)^{10}\right)'$.

Solution

$$\left((1+2x+9x^2)^{10}\right)' = 10\ (1+2x+9x^2)^9\ (2+18\ x)$$

5.2.2 Homework

Exercise 5.2.1 *Find* $\left((2+2\sin(3t))^3\right)'$.

Exercise 5.2.2 *Find* $\left((2x+5x^2+9x^3)^6\right)'$.

Exercise 5.2.3 *Find* $\left((1+sin(3t))^4\right)'$.

Exercise 5.2.4 *Find* $\left((\cos(2t)+\sin(5t))^3\right)'$.

Exercise 5.2.5 *Find* $\left((x^2+4)^{11}\right)'$.

Exercise 5.2.6 *Find* $\left((1 + \cos(6t))^9 \right)'$.

Exercise 5.2.7 *Find* $\left(\left(\frac{\sin(2t)}{\sin(3t)+\cos(4t)} \right)^3 \right)'$.

5.3 Derivatives of Complicated Things

We will often need to find the derivatives of more interesting things than simple polynomials. Finding rate of change is our mantra now! Let's look at a small example of how an excitable neuron transforms signals that come into it into output signals called **action potentials**. For now, think of an excitable neuron as a processing node which accepts inputs x and transforms them using a processing function we will call $\sigma(x)$ into an output y. As we know, neural circuits are built out of thousands of these processing nodes and we can draw them as a graph which shows how the nodes interact. Look at the simple example in Fig. 5.2.

In the picture you'll note there are four edges connecting the neurons. We'll label them like this: $E_{1\to2}$, $E_{1\to3}$, $E_{2\to4}$ and $E_{3\to4}$. When an excitable neuron generates an action potential, the action potential is like a signal. The rest voltage of the cell is about -70 millivolts and if the action potential is generated, the voltage of the cell rises rapidly to about 80 millivolts or so and then falls back to rest. The shape of the action potential is like a scripted response to the conditions the neuron sees as the input signal. In many ways, the neuron response is like a digital **on** and **off** signal, so many people have modeled the response as a curve that rises smoothly from 0 (the **off**) to 1 (the **on**). Such a curve looks like the one shown in Fig. 5.3.

The standard neural processing function has a high derivative value at 0 and small values on either side. You can see this behavior in Fig. 5.4.

We can model this kind of function using many approaches: for example, $\sigma(x) = \frac{x^2}{1+x^2}$ works and we can also build σ using exponential functions which we will get to later. Now the action potential from a neuron is fed into the input side of other neurons and the strength of that interaction is modeled by the edge numbers $E_{i\to j}$ for our various i and j's. To find the input into a neuron, we take the edges going in

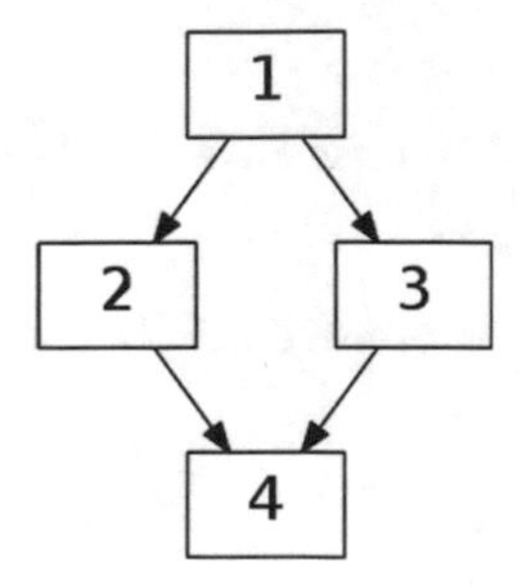

Fig. 5.2 A simple neural circuit: 1-2-1

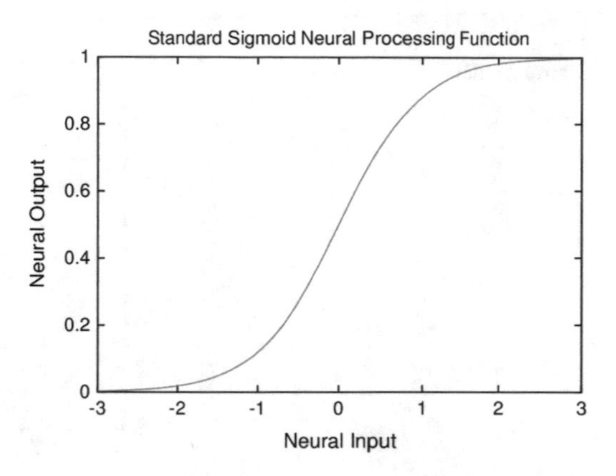

Fig. 5.3 A simple neural processing function

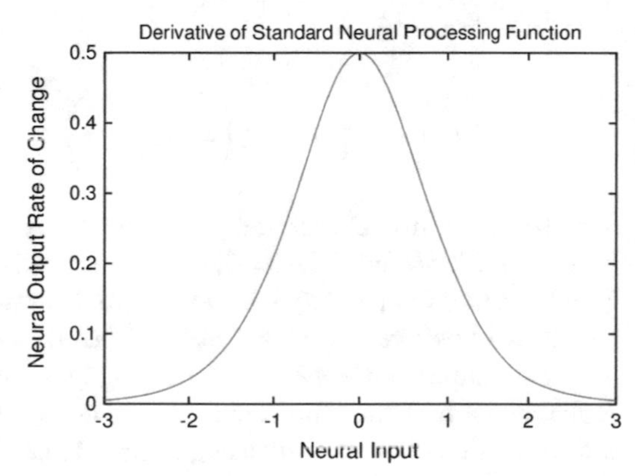

Fig. 5.4 A simple neural processing function's derivative

and multiply them by the output of the node the edge is coming from. If we let Y_1, Y_2, Y_3 and Y_4 be the outputs of our four neurons, then if x is the input fed into neuron one, this is what happens in our small neural model

$$
\begin{aligned}
Y_1 &= \sigma(x) \\
Y_2 &= \sigma(E_{1\to 2}\, Y_1) \\
Y_3 &= \sigma(E_{1\to 3}\, Y_1) \\
Y_4 &= \sigma(E_{2\to 4}\, Y_2 + E_{3\to 4}\, Y_3).
\end{aligned}
$$

Note that Y_4 depends on the initial input x in a complicated way. Here is the *recursive* chain of calculations. First, plug in for Y_2 and Y_3 to get Y_4 in terms of Y_1.

$$
Y_4 = \sigma\left(E_{2\to 4}\, \sigma(E_{1\to 2}\, Y_1) + E_{3\to 4}\, \sigma(E_{1\to 3}\, Y_1) \right).
$$

Now plug in for Y_1 to see finally how Y_4 depends on x.

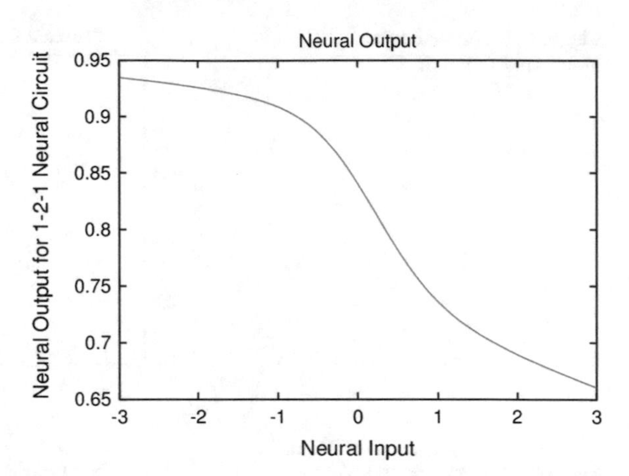

Fig. 5.5 Y_4 output for our neural model

$$Y_4(x) = \sigma\left(E_{2\to 4}\, \sigma\left(E_{1\to 2}\, \sigma(x) \right) + E_{3\to 4}\, \sigma\left(E_{1\to 3}\, \sigma(x) \right) \right).$$

For the randomly chosen edge values $E_{1\to 2} = -0.11622$, $E_{1\to 3} = -1.42157$, $E_{2\to 4} = 1.17856$ and $E_{3\to 4} = 0.68387$, we can calculate the Y_4 output for this model for all x values from -3 to 3 and plot them. Now negative values correspond to **inhibition** and positive values are **excitation**. As seen in Fig. 5.5, our simple model generates outputs between 0.95 for strong inhibition and 0.65 for strong excitation. Probably not realistic! But remember the edge weights were just chosen randomly and we didn't try to pick them using realistic biologically based values. We can indeed do better. But you should see a bit of how interesting biology can be illuminated by mathematics that comes from this book!

Note that this is essentially a $\sigma(\sigma(\sigma))$ series of function compositions! So the idea of a composition of functions is not just some horrible complication mathematics courses throw at you. It is really used in biological systems. Note, while we know very well how to calculate the derivative of this monster, $Y_4'(x)$ using the chain rule, it requires serious effort and the answer we get is quite messy. Fortunately, over the years, we have found ways to get the information we need from models like this without finding the derivatives by hand! Also, just think, real neural subsystems have hundreds or thousands or more neurons interacting with a vast number of edge connections. Lots of sigmoid compositions going on!

Let's do a bit more here. We know we can approximate the derivative using a slope term. Here that is

$$Y_4(x) = Y_4(p) + Y_4'(p)(x - p) + E(x - p).$$

So we can approximate $Y_4(x)$ at some point x by a straight line $Y_4(p) - Y_4'(p)(x - p)$ at a nearby point p. Since we know the error $E(x - p)$ is small close to p, **locally** (i.e. near p) we can really almost replace $Y_4(x)$ by the values of the straight line.

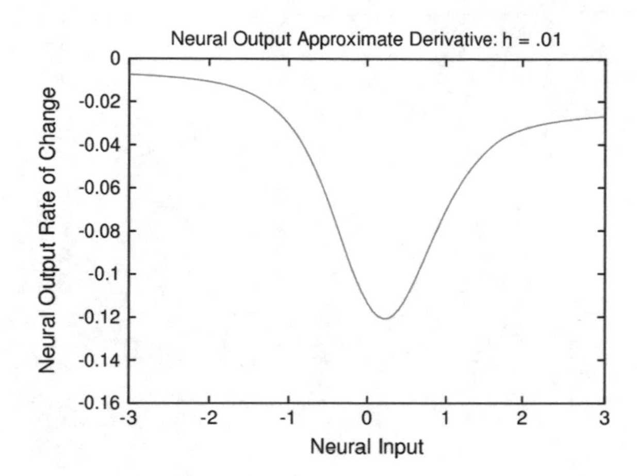

Fig. 5.6 Y_4's approximate derivative using $h = 0.01$

We call this straight line the **Tangent Line based at** p for our function. We will do more with this idea later, but it is nice to start using it. We usually say $Y_4(x) \sim Y_4(p) - Y_4'(p)(x - p)$ where the symbol $\sim$ means *approximately* and the burden is on us to understand that there is indeed error. But our feeling is that this error is *reasonably* small near each point p. We can go one step further. If we can't find the derivative easily, we can approximate the derivative's value at a point x by the fraction $(Y_4(x + h) - Y_4(x))/h$ as long as h is small. This is because we can solve the tangent line equation to get

$$\frac{Y_4(x+h) - Y_4(x)}{h} = Y_4'(p) + \frac{E(h)}{h}$$

and since $E(h)/(h)$ is small near x too, we can say

$$\frac{Y_4(x+h) - Y_4(x)}{h} \sim Y_4'(p)$$

near x. We can use this idea to calculate the approximate value of the derivative of Y_4 and plot it. As you can see from Fig. 5.6, the derivative is not that large, but it is always negative. Remember, derivatives are rates of change and looking at the graph of Y_4 we see it is always going down, so the derivative should always be negative.

Of course, we would have very different plots for different choices of the edge values. And, we could easily make up more complicated graphs of computational nodes.

Chapter 6
Antiderivatives

The idea of an Antiderivative or Primitive is very simple. We just guess! We say F is the **antiderivative** of f if $F' = f$. Since we know a group of simple derivatives, we can guess a group of simple antiderivatives! From this definition, we can see immediately that antiderivatives are not unique. The derivative of any constant is zero, so adding a constant to an antiderivative just gives a new antiderivative!

6.1 Simple Integer Power Antiderivatives

As we said, we can guess many antiderivatives. The symbol $\int$ we previously introduced as the symbol for a Riemann integral (be patient, we will be getting to that soon!) is also used to denote the **antiderivative**. This common symbol for the antiderivative of f has thus evolved to be $\int f$ because of the close connection between the antiderivative of f and the Riemann integral of f which is given in the **Cauchy Fundamental Theorem of Calculus**, Theorem 8.6.1 which we will get to in a bit. The usual Riemann integral, $\int_a^b f(t)\, dt$ of f on $[a, b]$ computes a *definite* value—hence, the symbol $\int_a^b f(t)\, dt$ is therefore usually referred to as the *definite integral of f on* $[a, b]$ to contrast it with the *family* of functions represented by the antiderivative $\int f$. Since the antiderivatives are arbitrary up to a constant, most of us refer to the antiderivative as the *indefinite integral of f*. Also, we hardly ever say "let's find the antiderivative of f"—instead, we just say, "let's *integrate* f". We will begin using this shorthand now!

Let's begin with antiderivative we can guess for the function t.

- We know the derivative of t^2 is $2t$ so it follows the derivative of $1/2\, t^2$ must be t. In fact, adding a constant doesn't change the result.
- In general, we have for any constant C that

$$\frac{d}{dt}\left(1/2t^2 + C\right) = t$$

J.K. Peterson, *Calculus for Cognitive Scientists*, Cognitive Science and Technology, DOI 10.1007/978-981-287-874-8_6

- We say that $1/2t^2 + C$ is the family of **antiderivatives** of t or more simply, just $1/2t^2 + C$ is the **antiderivative** of t.
- Note we could also say these are the **primitives** of t too.
- The usual symbol for the antiderivative is the Riemann integral symbol without the a and b. So we would say $\int t\,dt$ represents the antiderivative of t and

$$\int t\,dt = 1/2t^2 + C.$$

Next, let's look at the function t^2.

- We know the derivative of t^3 is $3t^2$ so it follows the derivative of $1/3\,t^3$ must be t^2. In fact, adding a constant doesn't change the result.
- In general, we have for any constant C that

$$\frac{d}{dt}\left(1/3t^3 + C\right) = t^2$$

- We say that $1/3t^3 + C$ is the family of **antiderivatives** of t^2 or more simply, just $1/3t^3 + C$ is the **antiderivative** of t^2.
- Note we could also say these are the **primitives** of t^2 too.
- Also, we would say for any constant C that

$$\int t^2\,dt = 1/3t^3 + C.$$

We can do a similar analysis for other powers. You should be able to convince yourself that for these positive powers, we have

- $$\int \mathbf{1}\,dt = t + C.$$
- $$\int \mathbf{t}\,dt = 1/2\,t^2 + C.$$
- $$\int \mathbf{t^2}\,dt = 1/3\,t^3 + C.$$
- $$\int \mathbf{t^3}\,dt = 1/4\,t^4 + C.$$

Further, we can guess for negative powers also. We can do a similar analysis for negative powers. You should be able to convince yourself that

- $$\int t^{-2}\,dt = -t^{-1} + C.$$
- $$\int t^{-3}\,dt = -1/2t^{-2} + C.$$
- $$\int t^{-4}\,dt = -1/3\,t^{-3} + C.$$
- $$\int t^{-5}\,dt = -1/4\,t^{-4} + C.$$

We can then glue together these antiderivatives to handle polynomials!

6.1.1 Examples

Example 6.1.1 Find $\int (2t+3)\,dt$.

Solution

$$\begin{aligned}\int (2t+3)\,dt &= \int (2t)\,dt + \int 3\,dt\\ &= 2\int t\,dt + 3\int 1\,dt\\ &= t^2 + 3t + C\end{aligned}$$

where the C indicates that we can add any constant we want and still get an antiderivative. C is often called the **integration constant**.

Example 6.1.2 Find $\int (5t^2 + 8t - 2)\,dt$.

Solution

$$\int (5t^2 + 8t - 2)\,dt = 5\left(t^3/3\right) + 8\left(t^2/2\right) - 2\left(t\right) + C.$$

Example 6.1.3 Find $\int (5t^{-3} + 8t^{-2} - 2t)\, dt$.

Solution

$$\int (5t^{-3} + 8t^{-2} - 2t)\, dt = 5\left(t^{-2}/(-2)\right) + 8\left(t^{-1}/(-1)\right) - 2\left(t^2/2\right) + C$$

Example 6.1.4 Find $\int (5t^5 + 4t^2 + 20)\, dt$.

Solution

$$\int (5t^5 + 4t^2 + 20)\, dt = 5\left(t^6/6\right) + 4\left(t^3/3\right) + 20t + C.$$

6.1.2 Homework

Exercise 6.1.1 *Find* $\int (15t^4 + 4t^3 + 9t + 7)\, dt$.

Exercise 6.1.2 *Find* $\int (6t^3 - 4t^2 - 12)\, dt$.

Exercise 6.1.3 *Find* $\int (50t^7 + 80t - 2/(t^2))\, dt$.

Exercise 6.1.4 *Find* $\int (12 + 8t^7 - 2t^{12})\, dt$.

Exercise 6.1.5 *Find* $\int (4 + 7t)\, dt$.

Exercise 6.1.6 *Find* $\int (6 + 3t^4)\, dt$.

Exercise 6.1.7 *Find* $\int (1 + 2t + 3t^2)\, dt$.

Exercise 6.1.8 *Find* $\int (-45 + 12t - 4t^5)\, dt$.

6.2 Simple Fractional Power Antiderivatives

Now we haven't yet discussed derivative of fractional powers of x. It is not that hard but it is easy to get blown away by a listing of too many mathy things, boom, one after the other. Here is a simple example to show you how to do it.

Consider $f(x) = x^{2/3}$. For convenience, let $y = f(x)$. Then we have $y = x^{2/3}$. Cube both sides to get $y^3 = x^2$. Now use the chain rule on the left hand side and a regular derivative on the right hand side to get $3\,y^2\,y' = 2\,x$. Now we just manipulate

$$3\,y^2\,y' = 2\,x \implies y' = \frac{2\,x}{3\,y^2}.$$

But, we can plug in for $y^2 = x^{4/3}$ to get

$$y' = \frac{2\,x}{3\,x^{4/3}} = \frac{2}{3}\,x^{1-4/3} = \frac{2}{3}\,x^{-1/3}.$$

This mix of chain rule and regular differentiation is an easy trick. You can see we can do this for any fraction p/q. We get another theorem!

Theorem 6.2.1 (The Simple Power Rule: Fractions!)
If f is the function $x^{p/q}$ for any integer p and q except $q = 0$, of course, then the derivative of f with respect to x satisfies

$$\left(x^{p/q}\right)' = \frac{p}{q}\,x^{p/q-1}$$

Proof We did the example for the power $2/3$ but the reasoning is the same for any fraction! ∎

So we can also find antiderivatives of fractional powers. You should be able to convince yourself that

- $$\int t^{\frac{1}{2}}\,dt = 2/3\,t^{3/2} + C.$$
- $$\int t^{\frac{4}{5}}\,dt = 5/9\,t^{9/5} + C.$$
- $$\int t^{\frac{5}{8}}\,dt = 8/13\,t^{13/8} + C.$$
- $$\int t^{-\frac{2}{3}}\,dt = 3\,t^{1/3} + C.$$

It is thus easy to guess the antiderivative of a power of t as we have already mentioned. We can state these results as Theorem 6.2.2.

Theorem 6.2.2 (Antiderivatives Of Simple Fractional Powers)
If u is any fractional power other than -1, then the antiderivative of $f(t) = t^u$ is $F(t) = t^{u+1}/(u+1) + C$. This is also expressed as $\int t^u\,dt = t^{u+1}/(u+1) + C$.

Since this result holds for any fractional power and fractions and irrational numbers can't be isolated from one another, we can show more. The rule above holds for any number other than -1*: even a number like* $\sqrt{2}$*. But we won't belabor this point now.*

Proof This is just a statement of all the results we have gone over. ■

6.2.1 Examples

Example 6.2.1 Find $\int t^{4/5}\, dt$.

Solution

$$\int t^{4/5}\, dt = t^{9/5}/(9/5) + C.$$

Example 6.2.2 Find $\int (t^{1/2} + 9t^{1/3})\, dt$.

Solution

$$\int (t^{1/2} + 9t^{1/3})\, dt = t^{3/2}/(3/2) + 9\left(t^{4/3}/(4/3)\right) + C.$$

Example 6.2.3 Find $\int (6t^{5/7})\, dt$.

Solution

$$\int (6t^{5/7})\, dt = 6\left(t^{12/7}/(12/7)\right) + C.$$

Example 6.2.4 Find $\int (8t^{-3/4} + 12t^{-1/5})\, dt$.

Solution

$$\int (8t^{-3/4} + 12t^{-1/5})\, dt = 8\left(t^{1/4}/(1/4)\right) + 12\left(t^{4/5}/(4/5)\right) + C.$$

6.2.2 Homework

Exercise 6.2.1 *Find* $\int (6\, t^{2/7})\, dt$.

Exercise 6.2.2 *Find* $\int (4\, t^{3/2} + 5t^{1/3})\, dt$.

Exercise 6.2.3 *Find* $\int (20\, t^{12/5})\, dt$.

Exercise 6.2.4 *Find* $\int (2\,t^{2/7} + 14t^{5/8})\,dt$.

Exercise 6.2.5 *Find* $\int (-22\,t^{11/3} + 6\,t^{1/4})\,dt$.

Exercise 6.2.6 *Find* $\int (3\,x^{9/8})\,dx$.

Exercise 6.2.7 *Find* $\int (6\,u^{4/3} + 5\,u^{7/2})\,du$.

Exercise 6.2.8 *Find* $\int (-19y^{1/6})\,dy$.

6.3 Simple Trigonometric Function Antiderivatives

The simple trigonometric functions $\sin(t)$ and $\cos(t)$ also have straightforward antiderivatives as shown in Theorem 6.3.1. Let's just state the results for the big two we are interested in!

Theorem 6.3.1 (Antiderivatives of Simple Trigonometric Functions)

1. *The antiderivative of* $\sin(t)$ *is* $-\cos(t) + C$
2. *The antiderivative of* $\cos(t)$ *is* $\sin(t) + C$

6.3.1 Examples

Example 6.3.1 Find $\int (2\cos(t) + 5\sin(t))\,dt$.

Solution

$$\int (2\cos(t) + 5\sin(t))\,dt = 2\sin(t) - 5\cos(t) + C.$$

Example 6.3.2 Find $\int (12\cos(t) + 15\sin(t))\,dt$.

Solution

$$\int (12\cos(t) + 15\sin(t))\,dt = 12\sin(t) - 15\cos(t) + C.$$

Example 6.3.3 Find $\int (-4\cos(t) - 5\sin(t))\,dt$.

Solution

$$\int (-4\cos(t) - 5\sin(t))\,dt = -4\sin(t) + 5\cos(t) + C.$$

Example 6.3.4 Find $\int (3\cos(x) + 8\sin(x))\, dx$.

Solution

$$\int (3\cos(x) + 8\sin(x))\, dx = 3\sin(x) - 8\cos(x) + C.$$

6.3.2 Homework

Exercise 6.3.1 *Find* $\int (2\cos(y) - 3\sin(y))\, dy$.

Exercise 6.3.2 *Find* $\int (\sin(t) - 5\cos(t))\, dt$.

Exercise 6.3.3 *Find* $\int (12\cos(t) - 16\sin(t))\, dt$.

Exercise 6.3.4 *Find* $\int (-\cos(x) + \sin(x))\, dx$.

Exercise 6.3.5 *Find* $\int (11\cos(x) + 9\sin(x))\, dx$.

Exercise 6.3.6 *Find* $\int (8\sin(t))\, dt$.

Exercise 6.3.7 *Find* $\int (4\cos(t))\, dt$.

Chapter 7
Substitutions

We can use the antiderivative ideas coupled with the chain rule to figure out how to *integrate* many functions that seem complicated but instead are just *disguised* versions of simple power function integrations. Let's get started with this. We will do this by example!

7.1 Simple Substitution Polynomials

Let's go through some in great detail.

Example 7.1.1 Compute $\int (t^2 + 1)\, 2t\, dt$

Solution *When you look at this integral, you should train yourself to see the simpler integral* $\int u\, du$ *where* $u(t) = t^2 + 1$*. Here are the steps:*

1. *We make the change of variable* $u(t) = t^2 + 1$*. Now differentiate both sides to see* $u'(t) = 2t$*. Thus, we have*

$$\int (t^2 + 1)\, 2t\, dt = \int u(t)\, u'(t)\, dt$$

2. *Now recall the chain rule for powers of functions, Theorem 5.2.1, we know*

$$\left((u(t))^2\right)'(t) = 2\, u(t)\, u'(t)$$

Thus,

$$u(t)\, u'(t) = \frac{1}{2}\left((u(t))^2\right)'(t)$$

J.K. Peterson, *Calculus for Cognitive Scientists*, Cognitive Science and Technology, DOI 10.1007/978-981-287-874-8_7

This then tells us that

$$\int (t^2+1)\,2t\,dt = \int u(t)\,u'(t)\,dt$$
$$= \int \frac{1}{2}\left((u(t))^2\right)'(t)dt$$

Now, the notation $\int \left((u(t))^2\right)'(t)dt$ *is just our way of asking for the* antiderivative *of the function behind the integral sign. Here, that function is* $(u^2)'$. *This antiderivative is, of course, just* u^2*! Plugging that into the original problem, we find*

$$\int (t^2+1)\,2t\,dt = \int u(t)\,u'(t)\,dt$$
$$= \int \frac{1}{2}\left((u(t))^2\right)'(t)dt$$
$$= \frac{1}{2}\,u^2(t) + C$$
$$= \frac{1}{2}\,(t^2+1)^2 + C$$

Whew!! That was awfully complicated looking. Let's do it again in a bit more streamlined fashion. Note all of the steps we go through below are the same as the longer version above, but since we write less detail down, it is much more compact. You need to get very good at understanding and doing all these steps!! Here is the second version:

Solution

1. *We make the change of variable* $u(t) = t^2+1$. *But we write this more simply as* $u = t^2+1$ *so that the dependence of* u *on* t *is implied rather than explicitly stated. This simplifies our notation already! Now differentiate both sides to see* $u'(t) = 2t$. *We will write this as* $du = 2t\,dt$, *again hiding the* t *variable, using the fact that* $\frac{du}{dt} = 2t$ *can be written in its differential form (you should have seen this idea in your first Calculus course). Thus, we have*

$$\int (t^2+1)\,2t\,dt = \int u\,du$$

2. *The antiderivative of* u *is* $u^2/2 + C$ *and so we have*

$$\int (t^2+1)\,2t\,dt = \int u\,du$$
$$= \frac{1}{2}\,u^2 + C$$

$$= \frac{1}{2}\,(t^2+1)^2 + C$$

Now let's try one a bit harder:

Example 7.1.2 Compute $\int (t^2+1)^3\, 4t\, dt$

Solution *When you look at this integral, again you should train yourself to see the simpler integral* $2 \int u^3\, du$ *where* $u(t) = t^2+1$*. Here are the steps: first, the detailed version*

1. *We make the change of variable* $u(t) = t^2 + 1$*. Now differentiate both sides to see* $u'(t) = 2t$*. Thus, we have*

$$\int (t^2+1)^3\, 4t\, dt = 2 \int u^3(t)\, u'(t)\, dt$$

2. *Now recall the chain rule for powers of functions, Theorem 5.2.1, we know*

$$\left((u(t))^4\right)'(t) = 4\, u^3(t)\, u'(t)$$

Thus,

$$2\, u^3(t)\, u'(t) = 2\, \frac{1}{4}\left((u(t))^4\right)'(t)$$

This then tells us that

$$\begin{aligned}\int (t^2+1)^3\, 4t dt &= 2 \int u^3(t)\, u'(t)\, dt \\ &= \int \frac{1}{2}\left((u(t))^4\right)'(t) dt\end{aligned}$$

Now, the notation $\int \left((u(t))^4\right)'(t)dt$ *is just our way of asking for the* antiderivative *of the function behind the integral sign. Here, that function is* $(u^4)'$*. This antiderivative is, of course, just* u^4*! Plugging that into the original problem, we find*

$$\begin{aligned}\int (t^2+1)^3\, 4t\, dt &= 2 \int u^3(t)\, u'(t)\, dt \\ &= \frac{1}{2}\, u^4(t) + C \\ &= \frac{1}{2}\,(t^2+1)^4 + C\end{aligned}$$

Again, this was awfully complicated looking the streamlined version is as follows:

1. *We make the change of variable* $u(t) = t^2 + 1$. *Now differentiate both sides to see* $u'(t) = 2t$ *and write this as* $du = 2t\ dt$. *Thus, we have*

$$\int (t^2+1)^3\ 4t\ dt = 2 \int u^3\ du$$

2. *The antiderivative of* u^3 *is* $u^4/4 + C$ *and so we have*

$$\begin{aligned}\int (t^2+1)^3\ 4t\ dt &= 2 \int u^3\ du \\ &= \frac{1}{2}\ u^4 + C \\ &= \frac{1}{2}\ (t^2+1)^4 + C\end{aligned}$$

Now let's do one the short way only.

Example 7.1.3 Compute $\int \sqrt{t^2+1}\ 3t\ dt$.

Solution *When you look at this integral, again you should train yourself to see the simpler integral* $3/2 \int u^{1/2}\ du$ *where* $u = t^2 + 1$. *Here are the steps: we know* $du = 2t\ dt$. *Thus*

$$\begin{aligned}\int \sqrt{t^2+1}\ 3t\ dt &= \frac{3}{2} \int u^{\frac{1}{2}}\ du \\ &= \frac{3}{2}\ \frac{1}{\frac{3}{2}}\ u^{\frac{3}{2}} + C \\ &= \frac{3}{2}\ \frac{2}{3}\ (t^2+1)^{\frac{3}{2}} + C\end{aligned}$$

Example 7.1.4 Compute $\int \sin(t^2+1)\ 5t\ dt$.

Solution *When you look at this integral, again you should train yourself to see the simpler integral* $5/2 \int \sin(u)\ du$ *where* $u(t) = t^2 + 1$. *Here are the steps: we know* $du = 2t\ dt$. *Thus*

$$\begin{aligned}\int \sin(t^2+1)\ 5t\ dt &= \frac{5}{2} \int \sin(u)\ du \\ &= \frac{5}{2}\ (-\cos(u)) + C \\ &= -\frac{5}{2}\ \cos(t^2+1) + C\end{aligned}$$

7.2 Substitution for Polynomials Quick and Dirty

Now let's do some problems in a more streamlined way.

Example 7.2.1 Find $\int (t^2 + 4)\, t\, dt$

Solution • *Let* $u(t) = t^2 + 4$.
• *Then* $\frac{du}{dt} = u'(t) = 2t$ *or* $1/2\, u'(t) = t$.
• *Write this in differential form* $1/2du = t\, dt$.
• *Then* **substitute** *into the antiderivative but this time write* u *instead of* $u(t)$*:*

$$\int (t^2 + 4)\, t\, dt = \int u\, 1/2\, du = 1/2 \int udu$$

• *But the* **simplest** *antiderivative of* u *is* $1/2u^2$. *We know our answer can have any constant added to it, so we add* C *at the end.*

$$\int (t^2 + 4)\, t\, dt = 1/2 \int u\, du = 1/4u^2 + C$$
$$= 1/4(t^2 + 4)^2 + C$$

Example 7.2.2 Find $\int (t^2 + 4)^2\, 5t\, dt$

Solution • *Let* $u(t) = t^2 + 4$.
• *Then* $\frac{du}{dt} = u'(t) = 2t$ *or* $1/2\, u'(t) = t$.
• *Write this in differential form* $1/2du = t\, dt$.
• *Then* **substitute** *these things into the antiderivative we want to find but this time just write* u *instead of* $u(t)$*:*

$$\int (t^2 + 4)^2\, 5t\, dt = \int u^2\, 5/2\, du$$
$$= 5/2 \int u^2 du$$

• *But the* **simplest** *antiderivative of* u^2 *is* $1/3u^3$ *and we add* C *at the end.*

$$\int (t^2 + 4)\, t\, dt = 5/2 \int u^2 du = 5/6u^3 + C$$
$$= 5/6(t^2 + 4)^3 + C$$

Example 7.2.3 Find $\int (t^3 + 6)^4\, 8t^2\, dt$

Solution • *Let* $u(t) = t^3 + 6$. *So* $u = t^3 + 6$.
• *The differential is* $du = 3t^2 dt$ *so write*

$$1/3du = t^2\, dt$$

- *Hence*

$$\begin{aligned}\int (t^3+8)^4\, 8t^2\, dt &= \int u^4\, 8/3\, du \\ &= 8/3 \int u^4 du \\ &= \frac{8}{3}\frac{1}{5}\, u^5 + C \\ &= 81/5\, (t^2+4)^5 + C.\end{aligned}$$

7.2.1 Homework

Time to do some yourself using the streamlined notation! Try these and write down all the steps.

- What is u?
- What is du?
- What is the new integral?
- What is the answer in u?
- What is the answer in t?

First, let's do polynomials of first order.

Exercise 7.2.1 *Find* $\int (2t+3)^3\, dt$.

Exercise 7.2.2 *Find* $\int (4t-2)^5\, dt$.

Exercise 7.2.3 *Find* $\int (-8t+1)^6\, dt$.

Exercise 7.2.4 *Find* $\int (19t+8)^{-2}\, dt$.

Exercise 7.2.5 *Find* $\int (5t+4)^{-4}\, dt$.

Second, let's do quadratics.

Exercise 7.2.6 *Find* $\int (2t^2+10)^3\, t\, dt$.

Exercise 7.2.7 *Find* $\int (5t^2+10t+3)^3\, (5t+5)\, dt$.

Exercise 7.2.8 *Find* $\int (6t^2+4t+8)^5\, (3t+1)\, dt$.

Exercise 7.2.9 *Find* $\int (-5t^2+15t+1)^4\, (-2t+3)\, dt$.

Exercise 7.2.10 *Find* $\int (t^2+t+19)^{-3}\, (2t+1)\, dt$.

Third, let's do cubics.

Exercise 7.2.11 *Find* $\int (4t^3 + 12)^4 \, (t^2) \, dt$.

Exercise 7.2.12 *Find* $\int (5t^3 - 6)^9 \, (10t^2) \, dt$.

Exercise 7.2.13 *Find* $\int (2t^3 + 4t - 8)^{10} \, (6t^2 + 4) \, dt$.

Exercise 7.2.14 *Find* $\int (9t^3 + 18t^2 + 25)^4 \, (3t^2 + 4t) \, dt$.

Exercise 7.2.15 *Find* $\int (10t^3 + 25t - 2)^6 \, (6t^2 + 5) \, dt$.

Lastly, let's just mix them all up!

Exercise 7.2.16 *Find* $\int (t^3 + 3t^2 + 10)^5 \, (3t^2 + 6t) \, dt$.

Exercise 7.2.17 *Find* $\int (t^3 + 3t^2 + 10)^5 \, (t^2 + 2t) \, dt$.

Exercise 7.2.18 *Find* $\int (t^4 + t^2 + 10t)^{-2} \, (4t^3 + 2t + 10) \, dt$.

Exercise 7.2.19 *Find* $\int (t^4 + t^2 + 10t)^{-2} \, (2t^3 + t + 5) \, dt$.

7.3 Sin's and Cos's

Now we can handle substitutions with trigonometric functions in then. We can do a lot with this and we would if we were preparing you for physics or engineering. But for our needs now, a few simple sin and cos substitutions is good enough.

7.3.1 Examples

Example 7.3.1 Find $\int \cos(5t + 18) \, dt$.

Solution *Here* $u = 5t + 18$ *and* $du = 5dt$. *So*

$$\int \cos(5t + 18) \, dt = \int \cos(u) \, (du/5) = (1/5) \, \sin(u) + C$$
$$= (1/5) \sin(5t + 18) + C.$$

Example 7.3.2 Find $\int \sin(8t) \, dt$.

Solution *Here* $u = 8t$ *and* $du = 8dt$. *So*

$$\int \sin(8t) \, dt = -(1/8) \, \cos(8t) + C.$$

Example 7.3.3 Find $\int \cos(t^2) \, t \, dt$.

Solution *Here $u = t^2$ and $du = 2tdt$. So*

$$\begin{aligned}\int \cos(t^2)\, t\, dt &= \int \cos(u)\, (du/2) = (1/2)\sin(u) + C \\ &= (1/2)\sin(t^2) + C.\end{aligned}$$

Example 7.3.4 Find $\int \sin(5t^4)\, t^3\, dt$.

Solution *Here $u = 5t^4$ and $du = 20t^3dt$. So*

$$\int \sin(5t^4)\, t^3\, dt = \int \sin(u)\, (du/20) = -(1/20)\, \cos(5t^4) + C.$$

7.3.2 Homework

Exercise 7.3.1 *Find $\int \cos(4t)\, dt$.*

Exercise 7.3.2 *Find $\int \sin(9t)\, dt$.*

Exercise 7.3.3 *Find $\int \cos(4t + 5)\, 8dt$.*

Exercise 7.3.4 *Find $\int \sin(-2t + 7)\, dt$.*

Exercise 7.3.5 *Find $\int \sin(1 + t^2)\, 4t\, dt$.*

Exercise 7.3.6 *Find $\int \sin(12t + 8)\, 10\, dt$.*

Exercise 7.3.7 *Find $\int \cos(5t^5)\, t^4\, dt$.*

Exercise 7.3.8 *Find $\int \sin(2t^{-1})\, (1/t^2)dt$.*

Exercise 7.3.9 *Find $\int \sin(t^4 + t^2 + 10t)\, (4t^3 + 2t + 10)\, dt$.*

Exercise 7.3.10 *Find $\int \cos(t^4 + t^2 + 10t)\, (2t^3 + t + 5)\, dt$.*

Chapter 8
Riemann Integration

We talked earlier about the special limiting process that leads to a Riemann integral in Chap. 3 and we have gone over derivatives and antiderivatives of a wide class of functions. But we need desperately a connection between these ideas. In this new chapter, we are going to find the **easy** way to calculate a Riemann integral. No limits at all! Here is the rule

$$\int_a^b f(t)\, dt = F(b) - F(a)$$

where F is **any** antiderivative of f! All we have to do is guess one and we have had a lot of practice doing just that in the last chapter. Of course, this is **not** magic and it follows from careful reasoning. And that is what we are going to show you now. So get ready for a nice long read. Remember the punch line: this new tool helps us and understanding how a tool is made helps us master it. This is the same principle we bring to the wood shop to make a complicated sleigh bed. We master our tools and learn to make them do our bidding. So here we go with another step on your road to becoming craftsman.

8.1 Riemann Sums

Let's go back and be formal for a bit. To study the integration of a function f, there are two intellectually separate ideas: the primitive or antiderivative and the Riemann integral. They seem quite different, don't they? We have

1. the idea of a *Primitive* or *Antiderivative*: of a function f. This is any function F which is differentiable and satisfies $F'(t) = f(t)$ at all points in the domain of f. Normally, the domain of f is a finite interval of the form $[a, b]$, although it could also be an infinite interval like all of $\Re$ or $[1, \infty)$ and so on. Note that an *antiderivative* does not require any understanding of the process of Riemann integration at all—only what differentiation is!

J.K. Peterson, *Calculus for Cognitive Scientists*, Cognitive Science and Technology, DOI 10.1007/978-981-287-874-8_8

2. The idea of the Riemann integral of a function.

Riemann integration is far more complicated to setup than the process of guessing a primitive or antiderivative. To define a Riemann integral properly, first, we start with a bounded function f on a finite interval $[a, b]$. This kind of function f need not be continuous! Then select a finite number of points from the interval $[a, b]$, $\{t_0,\ t_1, \ldots, t_{n-1},\ t_n\}$. We don't know how many points there are, so a different selection from the interval would possibly gives us more or less points. But for convenience, we will just call the last point t_n and the first point t_0. These points are not arbitrary—t_0 is always a, t_n is always b and they are ordered like this:

$$t_0 = a < t_1 < t_2 < \cdots < t_{n-1} < t_n = b$$

The collection of points from the interval $[a, b]$ is called a Partition of $[a, b]$ and is denoted by some letter—here we will use the letter **P**. So if we say P is a partition of $[a, b]$, we know it will have $n + 1$ points in it, they will be labeled from t_0 to t_n and they will be ordered left to right with strict inequalities. But, we will not know what value the positive integer n actually is. The simplest Partition P is the two point partition $\{a,\ b\}$. Note these things also:

1. Each partition of $n + 1$ points determines n subintervals of $[a, b]$
2. The lengths of these subintervals always adds up to the length of $[a, b]$ itself, $b - a$.
3. These subintervals can be represented as

$$\{[t_0, t_1],\ [t_1, t_2], \ldots, [t_{n-1}, t_n]\}$$

   or more abstractly as $[t_i, t_{i+1}]$ where the index i ranges from 0 to $n - 1$.
4. The length of each subinterval is $t_{i+1} - t_i$ for the indices i in the range 0 to $n - 1$.
5. The largest subinterval length is called the **norm** of the partition and we denote it by the symbol $\| P \|$.

Now from each subinterval $[t_i, t_{i+1}]$ determined by the Partition P, select any point you want and call it s_i. This will give us the points s_0 from $[t_0, t_1]$, s_1 from $[t_1, t_2]$ and so on up to the last point, s_{n-1} from $[t_{n-1}, t_n]$. At each of these points, we can evaluate the function f to get the value $f(s_j)$. Call these points an **Evaluation Set** for the partition P. Let's denote such an evaluation set by the letter E. If the function f was nice enough to be positive always and continuous, then the product $f(s_i) \times (t_{i+1} - t_i)$ can be interpreted as the area of a rectangle. Then, if we add up all these rectangle areas we get a sum which is useful enough to be given a special name: the Riemann sum for the function f associated with the Partition P and our choice of evaluation set $E = \{s_0, \ldots, s_{n-1}\}$. This sum is represented by the symbol $S(f, P, E)$ where the things inside the parenthesis are there to remind us that this sum depends on our choice of the function f, the partition P and the evaluation set E. So formally, we have the definition

Definition 8.1.1 (*Riemann Sum*)
The Riemann sum for the bounded function f, the partition P and the evaluation set $E = \{s_0, \ldots, s_{n-1}\}$ from $P\{t_0,\ t_1, \ldots, t_{n-1}, t_n\}$ is defined by

$$S(f, P, E) = \sum_{i=0}^{n-1} f(s_i)\,(t_{i+1} - t_i)$$

It is pretty misleading to write the Riemann sum this way as it can make us think that the n is always the same when in fact it can change value each time we select a different P. So many of us write the definition this way instead

$$S(f, P, E) = \sum_{i \in P} f(s_i)\,(t_{i+1} - t_i)$$

and we just remember that the choice of P will determine the size of n.

Let's look at an example of all this. In Fig. 8.1, we see the graph of a typical function which is always positive on some finite interval $[a, b]$

Next, let's set the interval to be $[1, 6]$ and compute the Riemann Sum for a particular choice of Partition P and evaluation set E. This is shown in Fig. 8.2.

We can also interpret the Riemann sum as an approximation to the area under the curve as shown in Fig. 8.3.

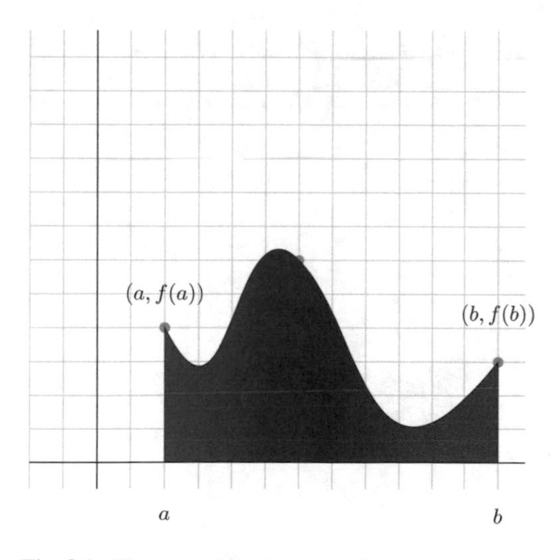

A generic curve f on the interval $[a, b]$ which is always positive. Note the area under this curve is the shaded region.

Fig. 8.1 The area under the curve f

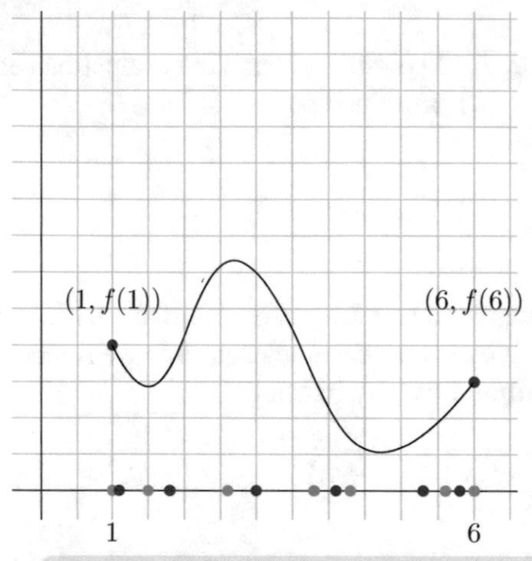

The partition is $P = \{1.0, 1.5, 2.6, 3.8, 4.3, 5.6, 6.0\}$. Hence, we have subinterval lengths of $t_1 - t_0 = 0.5$, $t_2 - t_1 = 1.1$, $t_3 - t_2 = 1.2$, $t_4 - t_3 = 0.5$, $t_5 - t_4 = 1.3$ and $t_6 - t_5 = 0.4$, giving $\| P \| = 1.3$. Thus,

$$S(f, P, E) = \sum_{i=0}^{5} f(s_i)\,(t_{i+1} - t_i)$$

For the evaluation set $E = \{1.1, 1.8, 3.0, 4.1, 5.3, 5.8\}$ shown in red in Figure 8.2, we would find the Riemann sum is

$$\begin{aligned} S(f, P, E) &= f(1.1) \times 0.5 \\ &+ f(1.8) \times 1.1 \\ &+ f(3.0) \times 1.2 \\ &+ f(4.1) \times 0.5 \\ &+ f(5.3) \times 1.3 \\ &+ f(5.8) \times 0.4 \end{aligned}$$

Of course, since our picture shows a generic f, we can't actually put in the function values $f(s_i)$!

Fig. 8.2 A simple Riemann sum

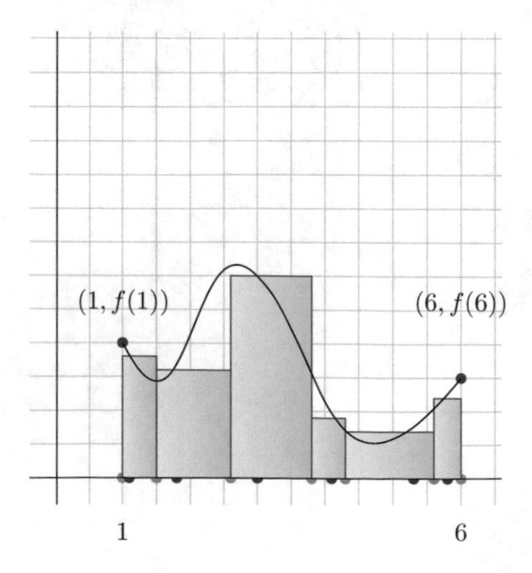

The partition is $P = \{1.0, 1.5, 2.6, 3.8, 4.3, 5.6, 6.0\}$.

Fig. 8.3 The Riemann sum as an approximate area

8.1.1 Examples

Example 8.1.1 Here's a worked out example. The only thing we won't do here is the graph. We let $f(t) = t^2 + 6t - 8$ on the interval $[2, 4]$ with P=$\{2, 2.5, 3.0, 3.7, 4.0\}$ and $E = \{2.2, 2.8, 3.3, 3.8\}$. Find the Riemann sum.

Solution *We only need to do some calculations.*

- *The partition determines subinterval lengths of* $t_1 - t_0 = 0.5$, $t_2 - t_1 = 0.5$, $t_3 - t_2 = 0.7$, *and* $t_4 - t_3 = 0.3$, *giving* $\| P \| = 0.7$.
- *For the evaluation set E the Riemann sum is*

$$\begin{aligned} S(f, P, E) &= \sum_{i=0}^{4} f(s_i)\,(t_{i+1} - t_i) \\ &= f(2.2) \times 0.5 + f(2.8) \times 0.5 \\ &\quad + f(3.3) \times 0.7 + f(3.8) \times 0.3 \\ &= 37.995 \end{aligned}$$

Example 8.1.2 Let $f(t) = 3t^2$ on the interval $[-1, 2]$ with $P = \{-1, -0.3, 0.6, 1.2, 2.0\}$ and $E = \{-0.7, 0.2, 0.9, 1.6\}$. Find the Riemann sum.

Solution *We only need to do some calculations.*

- *The partition determines subinterval lengths of* $t_1 - t_0 = 0.7$, $t_2 - t_1 = 0.9$, $t_3 - t_2 = 0.6$, *and* $t_4 - t_3 = 0.8$, *giving* $\| P \| = 0.9$.
- *For the evaluation set E the Riemann sum is*

$$\begin{aligned} S(f, P, E) &= \sum_{i=0}^{4} f(s_i)\,(t_{i+1} - t_i) \\ &= f(-0.7) \times 0.7 + f(0.2) \times 0.9 \\ &\quad + f(0.9) \times 0.6 + f(1.6) \times 0.8 \\ &= 8.739 \end{aligned}$$

8.1.2 Homework

Exercise 8.1.1 *Let* $f(t) = t^2 + 5$ *and let* $P = \{1.0, 2.0, 2.5, 4.0\}$ *be a partition of the interval* $[1.0, 4.0]$. *Let* $E = \{1.5, 2.3, 3.1\}$ *be the evaluation set. Calculate* $S(f, P, E)$ *and draw the picture that represents this Riemann sum.*

Exercise 8.1.2 *Let* $f(t) = t + 0.2t^3$ *and let* $P = \{1.0, 2.0, 2.5, 5.0\}$ *be a partition of the interval* $[1.0, 5.0]$. *Let* $E = \{1.5, 2.3, 3.9\}$ *be the evaluation set. Calculate* $S(f, P, E)$ *and draw the picture that represents this Riemann sum.*

Exercise 8.1.3 *Let* $f(t) = \sqrt{t}$ *and let* $P = \{0.0, 2.0, 6.5, 9.0\}$ *be a partition of the interval* $[0.0, 9.0]$. *Let* $E = \{1.5, 4.5, 7.1\}$ *be the evaluation set. Calculate* $S(f, P, E)$ *and draw the picture that represents this Riemann sum.*

Exercise 8.1.4 *Let* $f(t) = |t^3|$ *and let* $P = \{-1.0, 0.5, 2.0, 2.5, 4.0\}$ *be a partition of the interval* $[-1.0, 4.0]$. *Let* $E = \{-0.5, 1.3, 2.3, 2.8\}$ *be the evaluation set. Calculate* $S(f, P, E)$ *and draw the picture that represents this Riemann sum.*

Exercise 8.1.5 *Let* $f(t) = \frac{1}{t}$ *and let* $P = \{0.5, 1.0, 2.5, 4.0, 6.0\}$ *be a partition of the interval* $[0.5, 6.0]$. *Let* $E = \{0.8, 2.3, 3.1, 5.2\}$ *be the evaluation set. Calculate* $S(f, P, E)$ *and draw the picture that represents this Riemann sum.*

8.2 Riemann Sums in MatLab

Now, let's revisit Riemann sums. As you have seen, doing these by hand is tedious. Let's look at how we might do them using MatLab. Here is a typical MatLab session to do this. Let's calculate the Riemann sum for the function $f(x) = x^2$ on the interval $[1, 3]$ using the partition $P = \{1, 1.5, 2.1, 2.8, 3.0\}$ and evaluation set $E = \{1.2, 1.7, 2.5, 2.9\}$. First, set up our function like usual

Listing 8.1: Defining a Matlab Function again

```
f = @(x) (x.^2);
```

This defines the function $f(x) = x^2$. Now let's setup the partition with the command

Listing 8.2: Set up the Partition

```
x = [1;1.5;2.1;2.8;3.0]

x =

     1.0000
     1.5000
     2.1000
     2.8000
     3.0000
```

The command *diff* in MatLab is applied to a vector to create the differences we have called the Δx_i's.

Listing 8.3: Find the Delta x's

```
dx = diff(x)

dx =

     0.5000
     0.6000
     0.7000
     0.2000
```

Next, we set up the evaluation set E.

Listing 8.4: Set up the evaluation set E

```
E = [1.2;1.7;2.5;2.9]

E =

    1.2000
    1.7000
    2.5000
    2.9000
```

Now $f(s)$ will be applied to each component of **E** to create a new vector. These are the values we have called the $f(s_i)$'s. Then since dx is also a vector, we use $f(E). * dx$ to create the new vector with components $f(s_i)\Delta x_i$.

Listing 8.5: Find the vector f(E).*dx

```
g = f(E).*dx

g =

    0.7200
    1.7340
    4.3750
    1.6820
```

Finally, we add all these components together to get the Riemann sum. In MatLab, we add up the entries of a vector g with the command *sum(g)*.

Listing 8.6: Sum the vector to get the Riemann Sum

```
RS = sum(g)

RS =

    8.5110
```

Without the comments, the MatLab session is not too long.

Listing 8.7: All the Riemann Sum pieces together

```
f = @(x) (x.^2);
P = [1;1.5;2.1;2.8;3.0];
dx = diff(P);
E = [1.2;1.7;2.5;2.9];
g = f(E).*dx;
RS = sum(g);
```

8.2.1 Homework

For the given function f, partition P and evaluation set E, do the following: use Matlab to find $S(f, P, E)$ for the partition P and evaluation set E.

- Create a new **word** document for this homework.
- Do the document in single space.
- Do matlab fragments in bold font.
- The document starts with your name, Date and Homework number.

Then answer the problems like this:

1. State Problem 1.
   - insert into your doc the matlab commands you use to solve the problem. Do this in bold.
   - before each line of matlab add explanatory comments so we can check to see you know what you're doing.
2. State Problem 2.
   - same stuff

Something like this:
Your name
Today's date and HW Number,

Problem 1:
Let $f(t) = \sin(5t)$ on the interval $[1, 3]$ with $P = \{1, 1.5, 2.0, 2.5, 3.0\}$ and $E = \{1.2, 1.8, 2.3, 2.8\}$.

Listing 8.8: Sample Problem session

```
% add explanation here
f = @(x) sin(5*x);
% add explanation here
P = [1; 1.5; 2.0; 2.5; 3.0];
% add explanation here
E = [1.2; 1.8; 2.3; 2.8];
% add explanation here
dx = diff(P);
% add explanation here
g = f(E).*dx;
% add explanation here
RS = sum(g)
RS =
      0.1239
```

Exercise 8.2.1 *Let* $f(t) = t^2 + 2$ *on the interval* $[1, 3]$ *with* $P = \{1, 1.5, 2.0, 2.5, 3.0\}$ *and* $E = \{1.2, 1.8, 2.3, 2.8\}$.

Exercise 8.2.2 *Let* $f(t) = t^2 + 3$ *on the interval* $[1, 3]$ *with* $P = \{1, 1.6, 2.3, 2.8, 3.0\}$ *and* $E = \{1.2, 1.9, 2.5, 2.85\}$.

Exercise 8.2.3 *Let* $f(t) = 3t^2 + 2t$ *on the interval* $[1, 2]$ *with* $P = \{1, 1.2, 1.5, 1.8, 2.0\}$ *and* $E = \{1.1, 1.3, 1.7, 1.9\}$

Exercise 8.2.4 *Let* $f(t) = 3t^2 + t$ *on the interval* $[1, 4]$ *with* $P = \{1, 1.2, 1.5, 2.8, 4.0\}$ *and* $E = \{1.1, 1.3, 2.3, 3.2\}$

8.3 Graphing Riemann Sums

Let's recall how to graph a function f again. We did this already, but it is good to go over it again. To graph f we need to set up a variable which tells us how many data points to use in the plot. This variable is different from our partition variable. The **linspace** command below sets up a variable y to be a vector with 21 points in it. The first point is 1 and the last point is 3 and the interval $[1, 3]$ is divided into 20 equal size pieces. So this command creates y values spaced 0.1 apart:

$$\{y_1 = 1, y_2 = 1.1, y_3 = 1.2, \ldots, y_{20} = 2.9, y_{21} = 3.0\}.$$

We use the pairs $(y_i, f(y_i))$ to make a plot by connecting the dots determined by the pairs using lines. To do the plot in Matlab is easy

Listing 8.9: Setting up a function plot

```
y = linspace(1,3,21);
plot(y,f(y));
```

We can add stuff to this bare bones plot.

Listing 8.10: Adding labels to the plot

```
xlabel('x axis');
ylabel('y axis');
legend('x^2','location','best');
title('Plot of f(x) = x^2 on [1,3]');
```

where

- **xlabel** sets the name printed under the horizontal axis.
- **ylabel** sets the name printed next to the vertical axis.
- **legend** sets a blurb printed inside the graph explaining the plot. Great when you plot multiple things on the same graph.
- **title** sets the title of the graph.

The graph pops up in a separate window as you can see. Using the **file** menu, select **save as** and scroll through the choices to save the graph as a **.png** file—a Portable Network Graphics file. You'll need to give the file a name. We chose **graph1.png** and it is shown in Fig. 8.4.

But we want to do more than just graph the function. We want to graph the Riemann sums. So we need to graph those rectangles we draw by hand. To graph a rectangle, we graph 4 lines. The MatLab command

Fig. 8.4 Graph of f

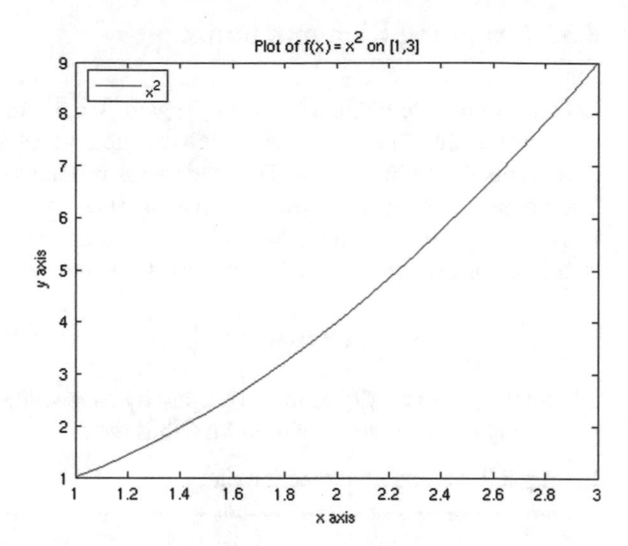

Listing 8.11: Plotting a line

```
1 plot([x1 x2], [y1 y2])
```

plots a line from the pair $(x1, y1)$ to $(x2, y2)$. So the command

Listing 8.12: Plotting a horizontal line for our rectangle

```
plot([x(i) x(i+1)],[f(s(i)) f(s(i))]);
```

plots the horizontal line segment which is the top of our rectangle and the line

Listing 8.13: Plotting a vertical line for our rectangle

```
plot([x(i) x(i)], [0 f(s(i))]);
```

plots a vertical line that starts on the x axis at x_i and ends at the function value $f(s_i)$. To plot rectangle, for the first pair of partition points, first we set the axis of our plot so we will be able to see it. We use the **axis** command in Matlab—look it up using **`help`**! If the two x points are $x1$ and $x2$ and the y value is $f(s1)$ where $s1$ is the first evaluation point, we expand the x axis to $[x1 - 1, x2 + 1]$ and expand the y axis to $[0, f(s1)]$. This allows our rectangle to be seen.

Listing 8.14: Setting the axis of our plot

```
axis([x1-1 x2+1 0 f((s1))+1]);
```

Putting this all together, we plot the first rectangle like this:

Listing 8.15: Plotting the rectangle

```
hold on
% set axis so we can see rectangle
axis([P(1)-1 P(2)+1 0 f(E(1))+1])
% plot top, LHS, RHS and bottom of rectangle
plot([P(1) P(2)],[f(E(1)) f(E(1))]);
plot([P(1) P(1)], [0 f(E(1))]);
plot([P(2) P(2)], [0 f(E(1))]);
plot([P(1) P(2)],[0 0]);
hold off
```

We use **hold on** and **hold off** to do this graph. We start with **hold on** and then all plots are kept until the **hold off** is used. This generates the rectangle in Fig, 8.5. To show the Riemann sum approximation as rectangles, we use a **for** loop in MatLab

Listing 8.16: The loop to plot all the rectangles

```
for i = 1:4
      .. do stuff for each choice of i
   end
```

To put this all together, we have to force Matlab to plot repeatedly without erasing the previous plot. Again, we use **hold on** and **hold off** to do this. We start with **hold on** and then all plots are kept until the **hold off** is used. We still think f is always positive so the **bottom** is 0 and the **top** is the $f(E(i))$ value.

Fig. 8.5 Simple rectangle

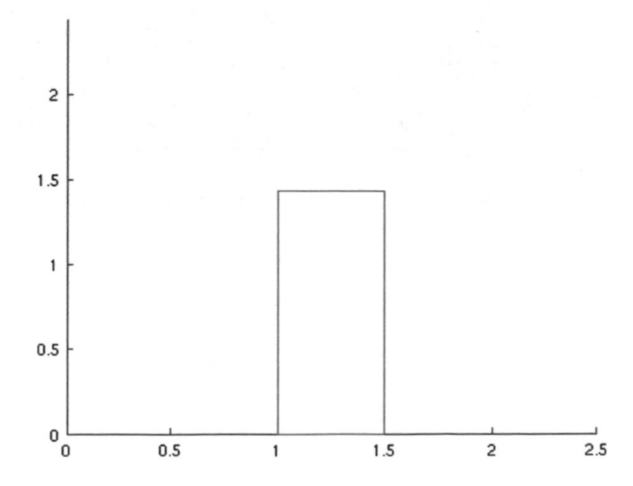

Listing 8.17: The details of the rectangle for loop

```
hold on % set hold to on
for i = 1:4  % graph rectangles
  bottom = 0;
  top = f(E(i));
  plot([P(i) P(i+1)],[f(E(i)) f(E(i))]);
  plot([P(i) P(i)], [bottom top]);
  plot([E(i) E(i)], [bottom top],'r');
  plot([P(i+1) P(i+1)], [bottom top]);
  plot([P(i) P(i+1)],[0 0]);
end
hold off % set hold off
```

Of course, we don't know if f can be negative, so we need to adjust our thinking as some of the rectangles might need to point down. We do that by setting the **bottom** and **top** of the rectangles using an **if** test.

Listing 8.18: Handling f being possibly negative

```
  bottom = 0;
  top = f(E(i));
  if f(E(i)) < 0
    top = 0;
    bottom = f(E(i));
  end
```

All together, we have

Listing 8.19: The full plot

```
hold on % set hold to on
[sizeP ,m] = size(P);
for i = 1:sizeP-1  % graph all the rectangles
  bottom = 0;
  top = f(E(i));
  if f(E(i)) < 0
    top = 0;
    bottom = f(E(i));
  end
  plot([P(i) P(i+1)],[f(E(i)) f(E(i))]);
  plot([P(i) P(i)], [bottom top]);
  plot([E(i) E(i)], [bottom top],'r');
  plot([P(i+1) P(i+1)], [bottom top]);
  plot([P(i) P(i+1)],[0 0]);
end
hold off;
```

We also want to place the plot of f over these rectangles. We plot f like so:

Listing 8.20: Adding the plot of f over the rectangles

```
y = linspace(P(1),P(sizeP), 101); % overlay the function graph
plot(y,f(y));
xlabel('x axis');
ylabel('y axis');
title('Riemann Sum overlayed on the function graph');
```

Together, we get the whole plotting package.

Listing 8.21: Plotting our Riemann Sum

```
  bottom = 0;
  top = f(E(i));
  if f(E(i)) < 0
    top = 0;
    bottom = f(E(i));
  end
```

All together, we have

Listing 8.22: The complete code

```
hold on % set hold to on
[sizeP,m] = size(P);
for i = 1:sizeP-1 % graph all the rectangles
  bottom = 0;
  top = f(E(i));
  if f(E(i)) < 0
    top = 0;
    bottom = f(E(i));
  end
  plot([P(i) P(i+1)],[f(E(i)) f(E(i))]);
  plot([P(i) P(i)], [bottom top]);
  plot([E(i) E(i)], [bottom top],'r');
  plot([P(i+1) P(i+1)], [bottom top]);
  plot([P(i) P(i+1)],[0 0]);
end
y = linspace(P(1),P(sizeP), 101); % overlay the function graph
plot(y,f(y));
xlabel('x axis');
ylabel('y axis');
title('Riemann Sum overlayed on the function graph');
hold off;
```

We can see the plot in Fig. 8.6.

8.3.1 Automating Riemann Sums

To save typing, let's learn to use a Matlab function.

- In Matlab's file menu, choose create a new Matlab function which gives

Listing 8.23: General Matlab Function

```
function [value1, value2,...] = MyFunction(arg1, arg2,...)
% stuff in here
end
```

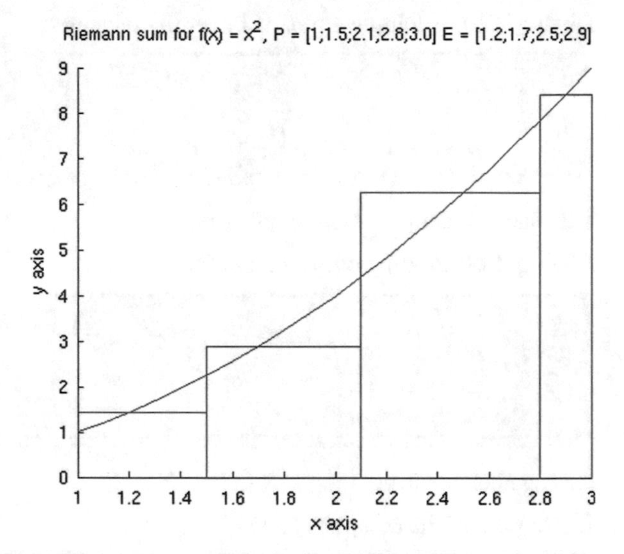

Fig. 8.6 Riemann sum for $f(x) = x^2$ for partition $\{1, 1.5, 2.1, 2.8, 3.0\}$ evaluation set $= \{1.2, 1.7, 2.5, 2.9\}$

- **[value1, value2,...]** are **returned values** the function calculates that we want to save.
- **(arg1, arg2,...)** are things the function needs to do the calculations. They are called the **arguments** to the function.
- **MyFunction** is the name of the function. This function must be stored in the file **MyFunction.m**.

Our function returns the Riemann sum, RS, and use the arguments: our function f, the partition P and the Evaluation set E. Since only one value returned **[RS] can be RS**.

Listing 8.24: Designing the Riemann Sum function

```
function RS = RiemannSum(f,P,E)
% comments alway begin with a %
matlab lines here
end
```

The name for the function **RiemannSum** must be used as the file name: i.e. we must use **RiemannSum.m** as the file name. We will build the plot of the Riemann sum into the **RiemannSum** function as follows.

Listing 8.25: Our Riemann Sum Function

```
function RS = RiemannSum(f,P,E)
% find Riemann sum
dx = diff(P);
RS = sum(f(E).*dx);
[sizeP,m] = size(P); %get size of Partition
clf; % clear the old graph
hold on % set hold to on
for i = 1:size(P)-1 % graph all the rectangles
  bottom = 0;
  top = f(E(i));
  if f(E(i)) < 0
    top = 0;
    bottom = f(E(i));
  end
  plot([P(i) P(i+1)],[f(E(i)) f(E(i))]);
  plot([P(i) P(i)], [bottom top]);
  plot([E(i) E(i)], [bottom top],'r');
  plot([P(i+1) P(i+1)], [bottom top]);
  plot([P(i) P(i+1)],[0 0]);
end
y = linspace(P(1),P(sizeP), 101); % overlay the function graph
plot(y,f(y));
xlabel('x axis');
ylabel('y axis');
title('Riemann Sum overlayed on the function graph');
hold off;
end
```

Example 8.3.1 Find the Riemann sum for the function $f(x) = sin(3x)$, partition $P = [1, 1.5, 2.1, 2.8, 3.0]$ and Evaluation set $E = [1.2, 1.7, 2.5, 2.9]$.

Solution *In Matlab, the solution just takes a few line.*

Listing 8.26: The Riemann sum

```
f = @(x) sin(3*x);
P = [1;1.5;2.1;2.8;3.0];
E = [1.2;1.7;2.5;2.9];
RS = RiemannSum(f,P,E);
```

This generates the pop up figure in Fig. 8.7 which we can save to a file.

8.3.1.1 Homework

For the given function f, partition P and evaluation set E, do the following: use Matlab to find $S(f, P, E)$ for the partition P and evaluation set E.

- Create a new **word** document for this homework.
- Do the document in single space.
- Do matlab fragments in bold font.
- The document starts with your name, Date and Homework number.

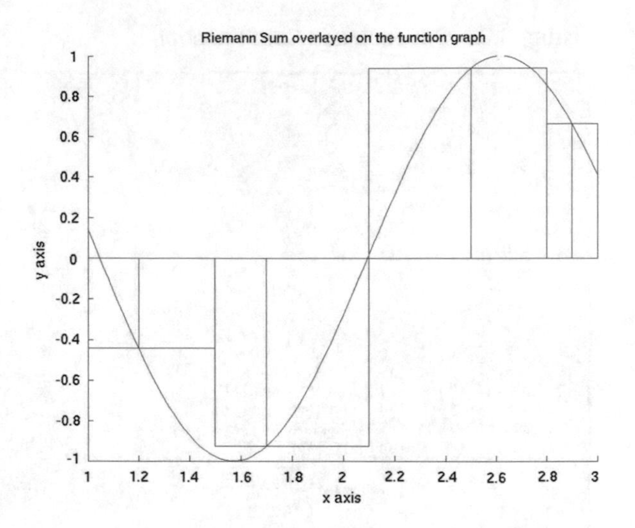

Fig. 8.7 A typical Riemann sum graph

Then answer the problems like this:

1. State Problem 1.
   - insert into your doc the matlab commands you use to solve the problem. Do this in bold.
   - before each line of matlab add explanatory comments so we can check to see you know what you're doing.
   - save the generated file to disk and insert the plot into your document nicely.
2. State Problem 2.
   - same stuff

Something like this:
Your Name
Today's date and HW Number,

Problem 1:
Let $f(t) = \sin(5t)$ on the interval $[1, 3]$ with $P = \{1, 1.5, 2.1, 2.8, 3.0\}$ and $E = \{1.2, 1.7, 2.5, 2.9\}$.

Listing 8.27: Sample Problem Session

```
% add explanation here
f = @(x) sin(5*x);
% add explanation here
P = [1;1.5;2.1;2.8;3.0];
% add explanation here
E = [1.2;1.7;2.5;2.9];
% add explanation here
RS = RiemannSum(f,P,E);
```

Now insert plot.

Exercise 8.3.1 *Let* $f(t) = 4t^2 + 2$ *on the interval* $[1, 3]$ *with* $P = \{1, 1.5, 2.0, 2.5, 3.0\}$ *and* $E = \{1.2, 1.8, 2.3, 2.8\}$.

Exercise 8.3.2 *Let* $f(t) = t^2 + 13t + 2$ *on the interval* $[1, 3]$ *with* $P = \{1, 1.6, 2.3, 2.8, 3.0\}$ *and* $E = \{1.2, 1.9, 2.5, 2.85\}$.

Exercise 8.3.3 *Let* $f(t) = 3t^2 + 2t^4$ *on the interval* $[1, 2]$ *with* $P = \{1, 1.2, 1.5, 1.8, 2.0\}$ *and* $E = \{1.1, 1.3, 1.7, 1.9\}$

Exercise 8.3.4 *Let* $f(t) = \sin(4t)$ *on the interval* $[1, 4]$ *with* $P = \{1, 1.2, 1.5, 2.8, 4.0\}$ *and* $E = \{1.1, 1.3, 2.3, 3.2\}$.

8.3.2 Uniform Partition Riemann Sums

To see graphically how the Riemann sums converge to the Riemann integral, let's write a new function: Riemann sums using uniform partitions and midpoint evaluation sets. Here is the new function.

Listing 8.28: The Uniform Partition Riemann Sum Function

```
function RS = RiemannUniformSum(f,a,b,n)
% set up a uniform partition with n+1 points
deltax = (b-a)/n;
P = [a:deltax:b]; % makes a row vector
for i=1:n
  start = a+(i-1)*deltax;
  stop = a+i*deltax;
  E(i) = 0.5*(start+stop);
end
% send in transpose of P and E so we use column vectors
% because original RiemannSum function uses columns
RS = RiemannSum(f,P',E');
end
```

We can then generate a sequence of Riemann sums for different values of n.

Here is a typical session:

Listing 8.29: Typical Session

```
f = @(x) sin(3*x);
RS = RiemannSumTwo(f,-1,4,10);
RS= RiemannSumTwo(f,-1,4,20);
RS = RiemannSumTwo(f,-1,4,30);
RS= RiemannSumTwo(f,-1,4,40);
```

These commands generate the plots of Figs. 8.8, 8.9, 8.10 and 8.11.

The actual value is $\int_{-1}^{4} \sin(3x)dx = -0.611282$. The $n = 80$ case is quite close! The experiment we just did should help you understand better what we will mean by the **Riemann Integral**. What we have shown is

$$\lim_{n\to\infty} S(f, P_n, E_n) = -0.0611282\ldots$$

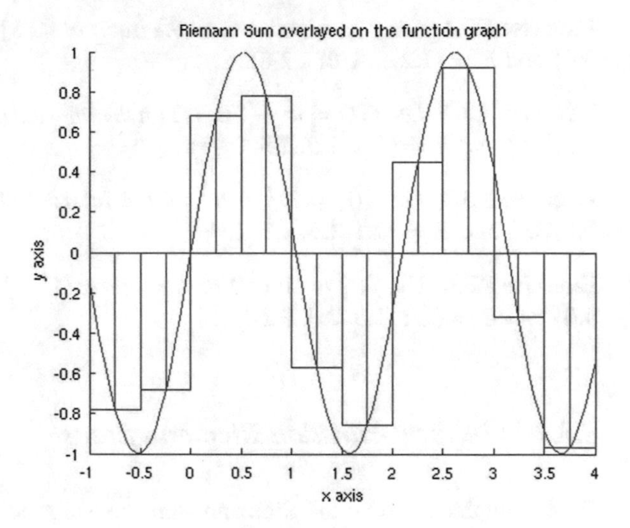

Fig. 8.8 The Riemann sum with a uniform partition P_{10} of $[-1, 4]$ for $n = 10$. The function is $\sin(3x)$ and the Riemann sum is -0.6726

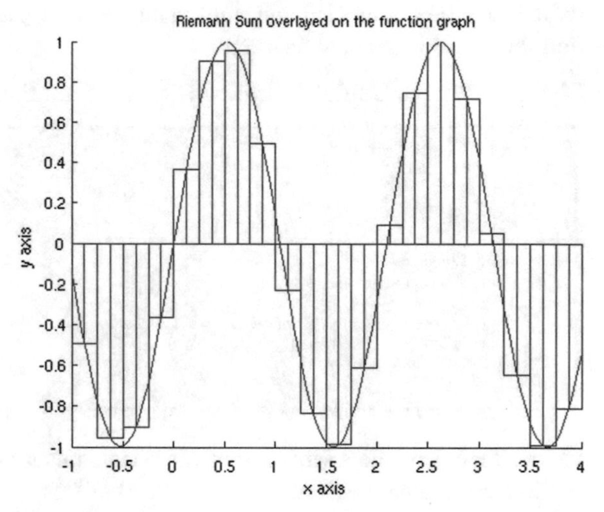

Fig. 8.9 Riemann sum with a uniform partition P_{20} of $[-1, 4]$ for $n = 20$. The function is $\sin(3x)$ and the Riemann sum is -0.6258

for the particular sequence of uniform partitions P_n with the particular choice of the evaluation sets E_n being the midpoints of each of the subintervals determined by the partition. Note the $||P_n|| = 5/n$ in each case. We will find that what we mean by the Riemann integral existing is that we get this value no matter what sequence of partitions we choose with associated evaluation sets as long as the norm of the partitions goes to 0.

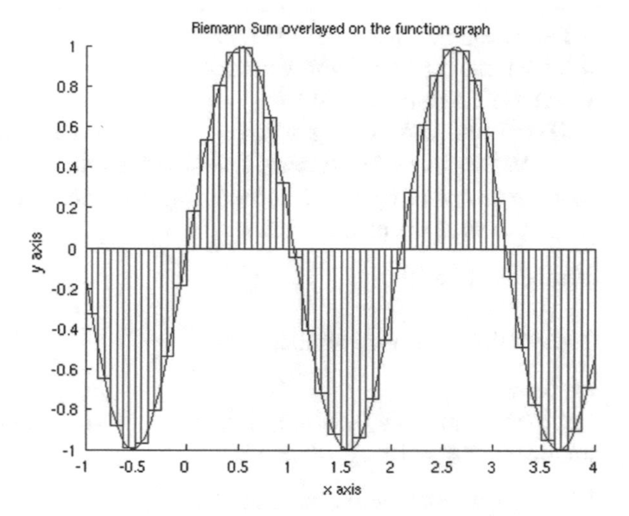

Fig. 8.10 Riemann sum with a uniform partition P_{40} of $[-1, 4]$ for $n = 40$. The function is $\sin(3x)$ and the Riemann sum is -0.6149

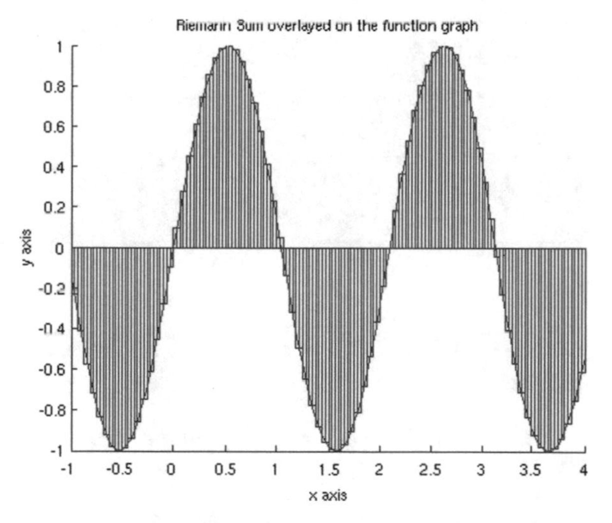

Fig. 8.11 Riemann sum with a uniform partition P_{80} of $[-1, 4]$ for $n = 80$. The function is $\sin(3x)$ and the Riemann sum is -0.6122

8.3.2.1 Homework

For the given function f, interval $[a, b]$ and choice of n, you'll calculate the corresponding uniform partition Riemann sum using the functions **RiemannSum** in file **RiemannSum.m** and **RiemannUniformSum** in file **RiemannUniformSum.m**. You can download these functions as files from the book's web site. Save them in your personal working directory.

- Create a new **word** document in single space with matlab fragments in bold font.
- The document starts with your name, Date and HW number.
- Create a new **word** document for this homework.
- Do the document in single space.

- Do matlab fragments in bold font.
- The document starts with your name, Date and Homework number.
- For each value of n, do a **save as** and save the figure with a filename like **HW#Problem#a[ ].png** where [ ] is where you put the number of the graph. Something like **HW17a.png, HW#Problem#b.png** etc.
- Insert this picture into the doc resizing as needed to make it look good. Explain in the doc what the picture shows.

Something like this:
Your Name
Today's date and HW Number,

Problem 1:
Let $f(t) = \sin(5t)$ on the interval $[-1, 4]$. Find the Riemann Uniform sum approximations for $n = 10, 20, 40$ and 80.

Listing 8.30: Sample Problem session

```
% add explanation here
f = @(x) sin(5*x);
% add explanation here
RS  = RiemannUniformSum(f,-1,4,10)
% add explanation here
RS  = RiemannUniformSum(f,-1,4,20)
% add explanation here
RS  = RiemannUniformSum(f,-1,4,40)
% add explanation here
RS  = RiemannUniformSum(f,-1,4,80)
```

Exercise 8.3.5 *Let* $f(t) = t^2 - 2t + 3$ *on the interval* $[-2, 3]$ *with* $n = 8, 16, 32$ *and* 48.

Exercise 8.3.6 *Let* $f(t) = \sin(2t)$ *on the interval* $[-1, 5]$ *with* $n = 10, 40, 60$ *and* 80.

Exercise 8.3.7 *Let* $f(t) = -t^2 + 8t + 5$ *on the interval* $[-2, 3]$ *with* $n = 4, 12, 30$ *and* 50.

8.4 Riemann Integrals

We can construct many different Riemann Sums for a given function f. If we let the norm of the partitions we use go to zero, the resulting Riemann Sums often converge to a fixed value. This fixed value is called the Riemann integral and in this section, we will make this notion more precise.

8.4.1 The Riemann Integral as a Limit

To define the Riemann Integral of f, we only need a few more things:

1. Each partition P has a maximum subinterval length $\| P \|$.
2. Each partition P and evaluation set E determines the number $S(f, P, E)$ by a simple calculation.
3. So if we took a collection of partitions P_1, P_2 and so on with associated evaluation sets E_1, E_2 etc., we would construct a sequence of real numbers $\{S(f, P_1, E_1), S(f, P_2, E_2), \ldots, S(f, P_n, E_n), \ldots\}$. Let's assume the norm of the partition P_n gets smaller all the time; i.e. $\lim_{n \to \infty} \| P_n \| = 0$. We could then ask if this sequence of numbers converges to something.

What if the sequence of Riemann sums we construct above converged to the same number I no matter what sequence of partitions whose norm goes to zero and associated evaluation sets we chose? Then, we would have that the value of this limit is *independent* of the choices above. This is indeed what we mean by the **Riemann Integral** of f on the interval $[a, b]$.

Definition 8.4.1 (*Riemann Integrability Of A Bounded Function*)
Let f be a bounded function on the finite interval $[a, b]$. If there is a number I so that

$$\lim_{n \to \infty} S(f, P_n, E_n) = I$$

no matter what sequence of partitions $\{P_n\}$ with associated sequence of evaluation sets $\{E_n\}$ we choose as long as $\lim_{n \to \infty} \| P_n \| = 0$, we will say that the Riemann Integral of f on $[a, b]$ exists and equals the value I.

The value I is dependent on the choice of f and interval $[a, b]$. So we often denote this value by $I(f, [a, b])$ or more simply as, $I(f, a, b)$. Historically, the idea of the Riemann integral was developed using area approximation as an application, so the summing nature of the Riemann Sum was denoted by the 16th century *letter* S which resembled an elongated or stretched letter S which looked like what we call the integral sign $\int$. Hence, the common notation for the Riemann Integral of f on $[a, b]$, when this value exists, is $\int_a^b f$. We usually want to remember what the independent variable of f is also and we want to remind ourselves that this value is obtained as we let the norm of the partitions go to zero. The symbol dt for the independent variable t is used as a reminder that $t_{i+1} - t_i$ is going to zero as the norm of the partitions goes to zero. So it has been very convenient to add to the symbol $\int_a^b f$ this information and use the augmented symbol $\int_a^b f(t)\,dt$ instead. Hence, if the independent variable was x instead of t, we would use $\int_a^b f(x)\,dx$. Since for a function f, the name we give to the independent variable is a matter of personal choice, we see that the choice of variable name we use in the symbol $\int_a^b f(t)\,dt$ is very arbitrary. Hence, it is common to refer to the independent variable we use in the symbol $\int_a^b f(t)\,dt$ as the dummy variable of integration.

8.4.2 Properties

If you think about it a bit, it is pretty easy to see that we can split up Riemann integrals in obvious ways. For example, **the integral of a sum should be the sum of the integrals**; i.e.

$$\int_a^b (f(x) + g(x))\, dx = \int_a^b f(x)\, dx + \int_a^b g(x)\, dx$$

and we should be able to pull out constants like so

$$\int_a^b (c\, f(x))\, dx = c \int_a^b f(x)\, dx$$

This is because in the Riemann sum, partitions of pieces like that can be broken apart and then the limits we want to do can be taken separately. Look at a typical piece of a Riemann sum which for a sum of functions would look $(f((s)i) + g(s_i))\, \Delta x_i$. We can surely split this apart to $f((s)i)\, \Delta x_i + g(s_i)\, \Delta x_i$ and then add up the pieces like usual. So we will get $RS(f + g, P, E) = RS(f, P, E) + RS(g, P, E)$ for any partition and evaluation set. Now take the limit and we get the result! To see you can pull out constants, we do the same argument. Each piece in the Riemann sum has the form $(cf(s_i)\ Deltax_i$ and it is easy to see we can whisk that constant c outside of the sum to find $RS(cf, P, E) = c\ RS(f, P, E)$. Then we take the limit and voíla!

To make it easy to see, what we are saying is this:

$$\int_1^2 (3 + 5x + 7x^2)\, dx = 3 \int_1^2 1\, dx + 5 \int_1^2 x\, dx + 7 \int_1^2 x^2)\, dx$$

Finally, the way we have setup the Riemann integral also makes it easy to see that if we do a Riemann integral over an interval of no length, the value should be 0 as all the Δx_i's are zero so the Riemann sums are 0 and hence the integral is zero.

$$\int_1^1 f(x)\, dx = 0.$$

Now look at the way the Riemann sum is pictured. If we set up Riemann sums on the interval [1, 5], say, it is pretty obvious that we could break these sums apart into Riemann sums over [1, 3] and [3, 5] for example. Then we could take limits as usual and see

$$\int_1^5 f(x)\, dx = \int_1^3 f(x)\, dx + \int_3^5 f(x)\, dx.$$

The argument is a bit more subtle than this, but now is not the time to get bogged down in those details. Subtle or not, the argument works out nicely. And we can split the interval up in any way we want. So we can say

$$\int_a^b f(x)\,dx = \int_a^c f(x)\,dx + \int_c^b f(x)\,dx.$$

for any choice of intermediate c we want.

One last thing. If we made the integration order go backwards, i.e. doing our Riemann sums from 3 to 1 instead of 1 to 3, all the Δx_i's would be flipped. So the Riemann sum would be the reverse of what it should be and the limiting value would be the negative of what we would normally have. We can say things like

$$\int_1^4 f(x)\,dx = -\int_4^1 f(x)\,dx$$

and similar things for other intervals, of course. Now you go and play around with these rules a bit and make sure you are comfortable with them!

We need a few more facts. It can be proved in more advanced courses that the following things are true about the Riemann Integral of a bounded function. First, we know when a bounded function actually has a Riemann integral from Theorem 8.4.1.

Theorem 8.4.1 (Existence Of The Riemann Integral)
Let f be a bounded function on the finite interval $[a, b]$. Then the Riemann integral of f on $[a, b]$, $\int_a^b f(t)dt$ exists if

1. *f is continuous on $[a, b]$*
2. *f is continuous except at a finite number of points on $[a, b]$.*

Further, if f and g are both Riemann integrable on $[a, b]$ and they match at all but a finite number of points, then their Riemann integrals match; i.e. $\int_a^b f(t)dt$ equals $\int_a^b g(t)dt$.

The function given by Eq. 3.1 is continuous nowhere on $[-1, 1]$ and it is indeed possible to prove it does not have a Riemann integral on that interval. However, most of our functions do have a lot of *smoothness*, i.e. continuity and even differentiability on the intervals we are interested in. Hence, Theorem 8.4.1 will apply. Here are some examples:

1. If $f(t)$ is t^2 on the interval $[-2, 4]$, then $\int_{-2}^4 f(t)dt$ does exist as f is continuous on this interval.
2. If g was defined by

$$g(t) = \begin{cases} t^2 & -2 \le t < 1 \text{ and } 1 < t \le 4 \\ 5 & t = 1 \end{cases}$$

we see g is not continuous at only one point and so it is Riemann integrable on $[-2, 4]$. Moreover, since f and g are both integrable and match at all but one point, their Riemann integrals are equal.

8.5 The Fundamental Theorem of Calculus

There is a big connection between the idea of the *antiderivative* of a function f and its Riemann integral. For a positive function f on the finite interval $[a, b]$, we can construct the area under the curve function $F(x) = \int_a^x f(t)\,dt$ where for convenience we choose an x in the open interval (a, b). We show $F(x)$ and $F(x + h)$ for a small positive h in Fig. 8.12. Let's look at the difference in these areas:

$$\begin{aligned} F(x+h) - F(x) &= \int_a^{x+h} f(t)\,dt - \int_a^x f(t)\,dt \\ &= \int_a^x f(t)\,dt + \int_x^{x+h} f(t)\,dt - \int_a^x f(t)\,dt \\ &= \int_x^{x+h} f(t)\,dt \end{aligned}$$

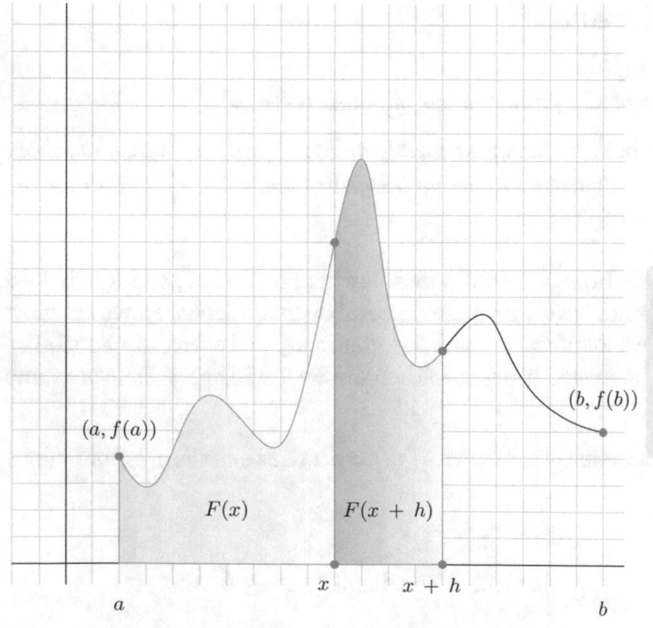

A generic curve f on the interval $[a, b]$ which is always positive. We let $F(x)$ be the area under this curve from a to x. This is indicated by the shaded region.

Fig. 8.12 The function $F(x)$

where we have used standard properties of the Riemann integral to write the first integral as two pieces and then do a subtraction. Now divide this difference by the change in x which is h. We find

$$\frac{F(x+h) - F(x)}{h} = \frac{1}{h} \int_x^{x+h} f(t)\, dt \tag{8.1}$$

The difference in area, $\int_x^{x+h} f(t)\, dt$, is the second shaded area in Fig. 8.12. If t is any number between x and $x + h$, we can see that the area of the rectangle with base h and height $f(t)$ is $f(t) \times h$ which is closely related to the area difference. In fact, the difference between this area and $F(x + h) - F(x)$ is really small when h is small. Now, of course, we are reasoning intuitively, but it is possible to make this sort of estimate very precise using the ϵ - δ language of continuity. But that is for another course! We know that f is bounded on $[a, b]$ You can easily see that f has a maximum value for the particular f we draw in Fig. 8.12. Of course, this graph is not what all such bounded functions f look like, but you should be able to get the idea that there is a number B so that $0 < f(t) \leq B$ for all t in $[a, b]$. Thus, we see

$$F(x+h) - F(x) \leq \int_x^{x+h} B\, dt = B\, h \tag{8.2}$$

From this, it follows that

$$\begin{aligned} \lim_{h \to 0} \, (F(x+h) - F(x)) &\leq \lim_{h \to 0} \, B\, h \\ &= 0 \end{aligned}$$

We conclude that F is continuous at each x in $[a, b]$ as

$$\lim_{h \to 0} \, (F(x+h) - F(x)) = 0$$

It seems that the new function F we construct by integrating the function f in this manner, always builds a new function that is continuous.

Is F differentiable at x? Recall, Eq. 8.1 can easily be estimated using areas as follows:

$$\min_{x \leq t \leq x+h} f(t)\, h \leq \int_x^{x+h} f(t) dt \leq \max_{x \leq t \leq x+h} f(t)\, h$$

Thus, we have the estimate

$$\min_{x \leq t \leq x+h} f(t) \leq \frac{F(x+h) - F(x)}{h} \leq \max_{x \leq t \leq x+h} f(t)$$

If f was continuous at x, then we must have

$$\lim_{h \to 0} \min_{x \leq t \leq x+h} f(t) = f(x)$$

and

$$\lim_{h \to 0} \max_{x \leq t \leq x+h} f(t) = f(x)$$

Note the f we draw in Fig. 8.12 is continuous all the time, but the argument we use here only needs continuity at the point x. Thus, we can infer for positive h,

$$\lim_{h \to 0^+} \frac{F(x+h) - F(x)}{h} = f(x)$$

You should be able to believe that a similar argument would work for negative values of h: i.e.

$$\lim_{h \to 0^-} \frac{F(x+h) - F(x)}{h} = \lim_{h \to 0^-} f(t) = f(x)$$

This tells us that $F'(p)$ exists and equals $f(x)$ as long as f is continuous at x as

$$F'(x^+) = \lim_{h \to 0^+} \frac{F(x+h) - F(x)}{h} = f(x)$$
$$F'(x^-) = \lim_{h \to 0^-} \frac{F(x+h) - F(x)}{h} = f(x)$$

This relationship is called the **Fundamental Theorem of Calculus**. The same sort of argument works for x equals a or b but we only need to look at the derivative from one side. We can actually prove this using fairly relaxed assumptions on f for the interval $[a, b]$. In general, f need only be Riemann Integrable on $[a, b]$ which allows for jumps in the function.

Theorem 8.5.1 (Fundamental Theorem Of Calculus)
Let f be Riemann Integrable on $[a, b]$. Then the function F defined on $[a, b]$ by $F(x) = \int_a^x f(t)\,dt$ satisfies

1. *F is continuous on all of $[a, b]$*
2. *F is differentiable at each point x in $[a, b]$ where f is continuous and $F'(x) = f(x)$.*

Proof The argument has already been made! ■

Using the same f as before, suppose G was defined on $[a, b]$ as follows

$$G(x) = \int_x^b f(t)\, dt.$$

Note that

$$\begin{aligned} F(x) + G(x) &= \int_a^x f(t)\, dt + \int_x^b f(t)\, dt \\ &= \int_a^b f(t)\, dt. \end{aligned}$$

Since the Fundamental Theorem of Calculus tells us F is differentiable, we see $G(x) = \int_a^b f(t)\, dt - F(x)$ must also be differentiable. It follows that

$$G'(x) = -F'(x) = -f(x).$$

Let's state this as a variant of the Fundamental Theorem of Calculus, the *Reversed Fundamental Theorem of Calculus* so to speak.

Theorem 8.5.2 (Fundamental Theorem Of Calculus Reversed)
Let f be Riemann Integrable on $[a, b]$. Then the function F defined on $[a, b]$ by $F(x) = \int_x^b f(t)\, dt$ satisfies

1. *F is continuous on all of $[a, b]$*
2. *F is differentiable at each point x in $[a, b]$ where f is continuous and $F'(x) = -f(x)$.*

8.6 The Cauchy Fundamental Theorem of Calculus

We can use the Fundamental Theorem of Calculus to learn how to evaluate many Riemann integrals. Let G be an antiderivative of the function f on $[a, b]$. Then, by definition, $G'(x) = f(x)$ and so we know G is continuous at each x. If we now assume f *is* continuous and define F on $[a, b]$ by

$$F(x) = f(a) + \int_a^x f(t)\, dt,$$

the Fundamental Theorem of Calculus, Theorem 8.5.1, is applicable. Thus, $F'(x) = f(x)$ at each point. But that means $F' = G' = f$ at each point. Functions whose derivatives are the same must differ by a constant. Call this constant C. We thus have $F(x) = G(x) + C$. So, we have

$$F(b) = f(a) + \int_a^b f(t)dt = G(b) + C$$
$$F(a) = f(a) + \int_a^a f(t)dt = G(a) + C$$

But $\int_a^a f(t)\, dt$ is zero, so we conclude after some rewriting

$$G(b) + C = f(a) + \int_a^b f(t)dt$$
$$G(a) + C = f(a)$$

And after subtracting, we find the important result

$$G(b) - G(a) = \int_a^b f(t)dt$$

This is huge! This is what tells us how to integrate many functions. For example, if $f(t) = t^3$, we can guess the antiderivatives have the form $t^4/4 + C$ for an arbitrary constant C. Thus, since $f(t) = t^3$ is continuous, the result above applies. We can therefore calculate Riemann integrals like these:

1.

$$\begin{aligned}\int_1^3 t^3\, dt &= \left.\frac{t^4}{4}\right|_1^3 \\ &= \frac{3^4}{4} - \frac{1^4}{4} \\ &= \frac{80}{4}\end{aligned}$$

2.

$$\begin{aligned}\int_{-2}^4 t^3\, dt &= \left.\frac{t^4}{4}\right|_{-2}^4 \\ &= \frac{4^4}{4} - \frac{(-2)^4}{4} \\ &= \frac{256}{4} - \frac{16}{4} \\ &= \frac{240}{4}\end{aligned}$$

Let's formalize this as a theorem called the *Cauchy Fundamental Theorem of Calculus*. All we really need to prove this result is that f is Riemann integrable on $[a, b]$, which for us is usually true as our functions f are continuous in general.

Theorem 8.6.1 (Cauchy Fundamental Theorem Of Calculus)
Let G be any antiderivative of the Riemann integrable function f on the interval $[a, b]$. Then $G(b) - G(a) = \int_a^b f(t)\, dt$.

With the Cauchy Fundamental Theorem of Calculus under our belt, we can use this theorem to evaluate many Riemann integrals. Let's get started.

1. The Riemann integral of the function $f(t) = t^p$ for any fraction not equal to -1 on $[a, b]$ can also be easily computed. We state this as Theorem 8.6.2.

Theorem 8.6.2 (Definite Integrals Of Simple Powers)
If p is any fractional power other than -1, then the definite integral of $f(t) = t^p$ on $[a, b]$ is $\int_a^b t^p\, dt = t^{p+1}/(p+1)\Big|_a^b$

2. The simple trigonometric functions $\sin(t)$ and $\cos(t)$ also have straightforward antiderivatives as was shown in Theorem 6.3.1.
3. The definite integrals of the sin and cos functions are then:

Theorem 8.6.3 (Definite Integrals Of Simple Trigonometric Functions)

(a) $\int_a^b \sin(t)\, dt$ *equals* $-\cos(t)\Big|_a^b$

(b) $\int_a^b \cos(t)$ *equals* $\sin(t)\Big|_a^b$

8.6.1 Examples

Example 8.6.1 Find $\int_1^2 t^2\, dt$.

Solution

$$\begin{aligned}\int_1^2 t^2\, dt &= (1/3)\, t^3\Big|_1^2\\ &= (1/3)\, \left((2)^3 - (1)^3\right)\\ &= 7/3.\end{aligned}$$

Example 8.6.2 Find $\int_{-1}^3 (2t^4 - 6)\, dt$.

Solution

$$\begin{aligned}\int_{-1}^3 (2t^4 - 6)\, dt &= \left(2t^5/5 - 6t\right)\Big|_{-1}^3\\ &= \left((2/5)(3)^5 - 6(3)\right) - \left((2/5)(-1)^5 - 6(-1)\right)\end{aligned}$$

Example 8.6.3 Find $\int_{-0.2}^{1.3} \sin(t)\, dt$.

Solution

$$\begin{aligned}\int_{-0.2}^{1.3} \sin(t)\, dt &= -\cos(t)\Big|_{-0.2}^{1.3} \\ &= -\left(\cos(1.3) - \cos(-0.2)\right).\end{aligned}$$

8.6.2 Homework

Exercise 8.6.1 *Find* $\int_1^2 (2t+5)\, dt$.

Exercise 8.6.2 *Find* $\int_0^3 (t^3 - 1)\, dt$.

Exercise 8.6.3 *Find* $\int_{-2}^{2} (2t^3 - 5t + 9)\, dt$.

Exercise 8.6.4 *Find* $\int_1^3 2\,(-t^2 + 3t + 2)\, dt$.

Exercise 8.6.5 *Find* $\int_0^{\pi/4} \cos(t)\, dt$.

Exercise 8.6.6 *Find* $\int_{-\pi/8}^{\pi/6} \sin(t)\, dt$.

8.7 Riemann Integration with Substitution

Now let's do definite integral using substitution.

Example 8.7.1 Compute $\int_1^5 (t^2+2t+1)^2\,(t+1)\, dt$.

Solution *When you look at this integral, again you should train yourself to see the simpler integral* $1/2\int u^2\, du$ *where* $u(t) = t^2 + 2t + 1$. *Here are the steps: we know* $du = (2t+2)dt$. *Thus*

$$\int_1^5 (t^2+2t+1)^2\,(t+1)\, dt = \frac{1}{2}\int_{t=1}^{t=5} u^2\, du$$

where we label the bottom and top limit of the integral in terms of the t variable to remind ourselves that the original integration was respect to t. *Then,*

$$\frac{1}{2}\int_{t=1}^{t=5} u^2\, du = \frac{1}{2}\,\frac{u^3}{3}\Big|_{t=1}^{t=5}$$

$$= \frac{1}{2}\,\frac{1}{3}(t^2 + 2t + 1)^3\Big|_1^5$$
$$= \frac{1}{6}\,\left((36)^3 - 4^3\right)$$

Example 8.7.2 Find $\int_1^2 (t^2 + 4)^4\, 8t\, dt$.

Solution • *Let* $u(t) = t^2 + 4$. *But to save time just write* $u = t^2 + 4$.
• *The differential is* $du = 2tdt$ *so write*

$$1/2du = t\, dt$$

• *Hence*

$$\int_1^2 (t^2 + 4)^4\, 8t\, dt = \int_{t=1}^{t=2} u^4\, 8/2\, du$$
$$= 8/2 \int_{t=1}^{t=2} u^4 du$$
$$= \frac{8}{2}\,\frac{1}{5}\,u^5\Big|_{t=1}^{t+2}$$
$$= 4/5\,(t^2 + 4)^5\Big|_1^2$$
$$= (4/5)(4 + 4)^5 - (4/5)(1 + 4)^5.$$

8.7.1 Homework

Write down all the steps.

- What is u?
- What is du?
- What is the new integral?
- What is the answer in u?
- What is the answer in t?
- What is the final value for the Riemann integral?

Exercise 8.7.1 $\int_1^2 (t^4 + t^2)^2\, (4t^3 + 2t)\, dt$.

Exercise 8.7.2 $\int_1^2 (t^4 + t^2)^4\, (t^3 + \frac{1}{2}t)\, dt$.

Exercise 8.7.3 $\int_0^5 \cos(2t + 25)\, 6\, dt$.

Exercise 8.7.4 $\int_{-\pi/4}^{\pi/2} \cos(3t)\, dt$.

Exercise 8.7.5 $\int_1^2 (t^5 + t^3)^{5/2}\,(5t^4 + 3t^2)\,dt$.

Exercise 8.7.6 $\int_1^2 (t^5 + t^3)^{5/2}\,(10t^4 + 6t^2)\,dt$.

Exercise 8.7.7 $\int_1^2 \sin(4t + 1)\,dt$.

Exercise 8.7.8 $\int_0^1 \sin(4t^2 + 1)\,3t\,dt$.

8.8 Integration with Jumps

Now let's look at the Riemann integral of functions which have points of discontinuity. You can think of this as enrichment. Our focus is on lots of other things and adding these jumps into the integration mix is definitely intense! But we wanted you to see it because it does come up and now you'll know you can come back to this book to answer those questions! Plus, there is always that other issue: we like abstraction and intense intellectual experiences! We are going to find the F's in the FTOC for various f's with discontinuities.

8.8.1 Removable Discontinuity

Consider the function f defined on $[-2, 5]$ by

$$f(t) = \begin{cases} 2t & -2 \le t < 0 \\ 1 & t = 0 \\ (1/5)t^2 & 0 < t \le 5 \end{cases}$$

Let's calculate $F(t) = \int_{-2}^{t} f(s)\,ds$. This will have to be done in several parts because of the way f is defined.

1. On the interval $[-2, 0]$, note that f is continuous except at one point, $t = 0$. Hence, f is Riemann integrable by Theorem 8.4.1. Also, the function $2t$ is continuous on this interval and hence is also Riemann integrable. Then since f on $[-2, 0]$ and $2t$ match at all but one point on $[-2, 0]$, their Riemann integrals must match. Hence, if t is in $[-2, 0]$, we compute F as follows:

$$\begin{aligned} F(t) &= \int_{-2}^{t} f(s)\,ds \\ &= \int_{-2}^{t} 2s\,ds \end{aligned}$$

$$= s^2 \Big|_{-2}^{t}$$
$$= t^2 - (-2)^2 = t^2 - 4$$

2. On the interval [0, 5], note that f is continuous except at one point, $t = 0$. Hence, f is Riemann integrable by Theorem 8.4.1. Also, the function $(1/5)t^2$ is continuous on this interval and so is also Riemann integrable. Then since f on [0, 5] and $(1/5)t^2$ match at all but one point on [0, 5], their Riemann integrals must match. Hence, if t is in [0, 5], we compute F as follows:

$$\begin{aligned} F(t) &= \int_{-2}^{t} f(s)\, ds \\ &= \int_{-2}^{0} f(s)\, ds + \int_{0}^{t} f(s)\, ds \\ &= \int_{-2}^{0} 2s\, ds + \int_{0}^{t} (1/5)s^2\, ds \\ &= s^2 \Big|_{-2}^{0} + (1/15)s^3 \Big|_{0}^{t} \\ &= -4 + t^3/15 \end{aligned}$$

Thus, we have found that

$$F(t) = \begin{cases} t^2 - 4 & -2 \le t < 0 \\ t^3/15 - 4 & 0 < t \le 5 \end{cases}$$

Note, we didn't define F at $t = 0$ yet. Since f is Riemann Integrable on $[-2, 5]$, we know from the Fundamental Theorem of Calculus, Theorem 8.5.1, that F must be continuous. Let's check. F is clearly continuous on either side of 0 and we note that $\lim_{t \to 0^-} F(t)$ which is $F(0^-)$ is -4 which is exactly the value of $F(0^+)$. Hence, F is indeed continuous at 0 and we can write

$$F(t) = \begin{cases} t^2 - 4 & -2 \le t \le 0 \\ t^3/15 - 4 & 0 \le t \le 5 \end{cases}$$

What about the differentiability of F? The Fundamental Theorem of Calculus guarantees that F has a derivative at each point where f is continuous and at those points $F'(t) = f(t)$. Hence, we know this is true at all t except perhaps at 0. Note at those t, we find

$$F'(t) = \begin{cases} 2t & -2 \le t < 0 \\ (1/5)t^2 & 0 < t \le 5 \end{cases}$$

which is exactly what we expect. Also, note $F'(0^-) = 0$ and $F'(0^+) = 0$ as well. Hence, since the right and left hand derivatives match, we see $F'(0)$ does exist and has the value 0. But this is not the same as $f(0) = 1$. Note, F is **not** the antiderivative of f on $[-2, 5]$ because of this mismatch.

8.8.2 Jump Discontinuity

Now consider the function f defined on $[-2, 5]$ by

$$f(t) = \begin{cases} 2t & -2 \le t < 0 \\ 1 & t = 0 \\ 2 + (1/5)t^2 & 0 < t \le 5 \end{cases}$$

Let's calculate $F(t) = \int_{-2}^{t} f(s)\, ds$. Again, this will have to be done in several parts because of the way f is defined.

1. On the interval $[-2, 0]$, note that f is continuous except at one point, $t = 0$. Hence, f is Riemann integrable by Theorem 8.4.1. Also, the function $2t$ is continuous on this interval and therefore is also Riemann integrable. Then since f on $[-2, 0]$ and $2t$ match at all but one point on $[-2, 0]$, their Riemann integrals must match. Hence, if t is in $[-2, 0]$, we compute F as follows:

$$\begin{aligned} F(t) &= \int_{-2}^{t} f(s)\, ds \\ &= \int_{-2}^{t} 2s\, ds \\ &= s^2 \Big|_{-2}^{t} \\ &= t^2 - (-2)^2 = t^2 - 4 \end{aligned}$$

2. On the interval $[0, 5]$, note that f is continuous except at one point, $t = 0$. Hence, f is Riemann integrable by Theorem 8.4.1. Also, the function $2 + (1/5)t^2$ is continuous on this interval and so is also Riemann integrable. Then since f on $[0, 5]$ and $2 + (1/5)t^2$ match at all but one point on $[0, 5]$, their Riemann integrals must match. Hence, if t is in $[0, 5]$, we compute F as follows:

$$\begin{aligned} F(t) &= \int_{-2}^{t} f(s)\, ds \\ &= \int_{-2}^{0} f(s)\, ds + \int_{0}^{t} f(s)\, ds \end{aligned}$$

$$
\begin{aligned}
&= \int_{-2}^{0} 2s\, ds + \int_{0}^{t} (2 + (1/5)s^2)\, ds \\
&= s^2 \Big|_{-2}^{0} + (2s + (1/15)s^3) \Big|_{0}^{t} \\
&= -4 + 2t + t^3/15
\end{aligned}
$$

Thus, we have found that

$$
F(t) = \begin{cases} t^2 - 4 & -2 \leq t < 0 \\ -4 + 2t + t^3/15 & 0 < t \leq 5 \end{cases}
$$

As before, we didn't define F at $t = 0$ yet. Since f is Riemann Integrable on $[-2, 5]$, we know from the Fundamental Theorem of Calculus, Theorem 8.5.1, that F must be continuous. F is clearly continuous on either side of 0 and we note that $\lim_{t \to 0^-} F(t)$ which is $F(0^-)$ is -4 which is exactly the value of $F(0^+)$. Hence, F is indeed continuous at 0 and we can write

$$
F(t) = \begin{cases} t^2 - 4 & -2 \leq t \leq 0 \\ -4 + 2t + t^3/15 & 0 \leq t \leq 5 \end{cases}
$$

What about the differentiability of F? The Fundamental Theorem of Calculus guarantees that F has a derivative at each point where f is continuous and at those points $F'(t) = f(t)$. Hence, we know this is true at all t except 0. Note at those t, we find

$$
F'(t) = \begin{cases} 2t & -2 \leq t < 0 \\ 2 + (1/5)t^2 & 0 < t \leq 5 \end{cases}
$$

which is exactly what we expect. However, when we look at the one sided derivatives, we find $F'(0^-) = 0$ and $F'(0^+) = 2$. Hence, since the right and left hand derivatives do not match, we see $F'(0)$ does not exist. Finally, note F is **not** the antiderivative of f on $[-2, 5]$ because of this mismatch.

8.8.3 Homework

Exercise 8.8.1 *Compute* $F(t) = \int_{-3}^{t} f(s)\, ds$ *for*

$$
f(t) = \begin{cases} 3t & -3 \leq t < 0 \\ 6 & t = 0 \\ (1/6)t^2 & 0 < t \leq 6 \end{cases}
$$

1. *Graph* f *and* F *carefully labeling all interesting points.*
2. *Verify that* F *is continuous and differentiable at all points but* $F'(0)$ *does not match* $f(0)$ *and so* F *is not the antiderivative of* f *on* $[-3, 6]$

Exercise 8.8.2 *Compute* $F(t) = \int_2^t f(s)\,ds$ *for*

$$f(t) = \begin{cases} -2t & 2 \le t < 5 \\ 12 & t = 5 \\ 3t - 25 & 5 < t \le 10 \end{cases}$$

1. *Graph* f *and* F *carefully labeling all interesting points.*
2. *Verify that* F *is continuous and differentiable at all points but* $F'(5)$ *does not match* $f(5)$ *and so* F *is not the antiderivative of* f *on* $[2, 10]$

Exercise 8.8.3 *Compute* $F(t) = \int_{-3}^t f(s)\,ds$ *for*

$$f(t) = \begin{cases} 3t & -3 \le t < 0 \\ 6 & t = 0 \\ (1/6)t^2 + 2 & 0 < t \le 6 \end{cases}$$

1. *Graph* f *and* F *carefully labeling all interesting points.*
2. *Verify that* F *is continuous and differentiable at all points except* 0 *and so* F *is not the antiderivative of* f *on* $[-3, 6]$

Exercise 8.8.4 *Compute* $F(t) = \int_2^t f(s)\,ds$ *for*

$$f(t) = \begin{cases} -2t & 2 \le t < 5 \\ 12 & t = 5 \\ 3t & 5 < t \le 10 \end{cases}$$

1. *Graph* f *and* F *carefully labeling all interesting points.*
2. *Verify that* F *is continuous and differentiable at all points except* 5 *and so* F *is not the antiderivative of* f *on* $[2, 10]$

Chapter 9
The Logarithm and Its Inverse

Consider the graph of the function $\frac{1}{t}$ on $(0, \infty)$. Since this function grows arbitrarily large as t get close to 0^+, we will only display the graph on the interval $[0.1, 10.0]$. This will give you the idea of what the graph looks like. This graph is shown in Fig. 9.1.

Now let's look at this more closely as shown in Fig. 9.2.

We can therefore say that e is the unique real number which satisfies Eq. 9.1.

$$\int_1^e \frac{1}{t}\, dt = 1.0 \tag{9.1}$$

9.1 The Natural Logarithm Function

Now let's do something useful with all this complicated mathematics we have been learning. Look again at the continuous function $f(t) = 1/t$ for all $t \geq 1$. We'll do this in a step by step fashion explaining all the things we know and what they imply. A bit tedious but informative!! Let $F(x) = \int_1^x 1/t\, dt$ for any $0 < x$.

$F(x)$ is smooth for $x \geq 1$

We focus on the continuous function $f(t) = 1/t$ for all $1 \geq t$.

- Pick any $x \geq 1$.
- Pick any $L > x$.
- Look at $F(x) = \int_1^x 1/t\, dt$ on the interval $[1, L]$.
- The Fundamental Theorem of Calculus applies and so we know F is continuous on $[1, L]$—in particular F is continuous at x.
- Since the integrand $1/t$ is continuous here, the Fundamental Theorem of Calculus also tells us $F'(x) = f(x) = 1/x$.
- We can do this argument for any $x \geq 1$. So we know our F is continuous at x and $F'(x) = 1/x$ for all $x \geq 1$.

J.K. Peterson, *Calculus for Cognitive Scientists*, Cognitive Science and Technology, DOI 10.1007/978-981-287-874-8_9

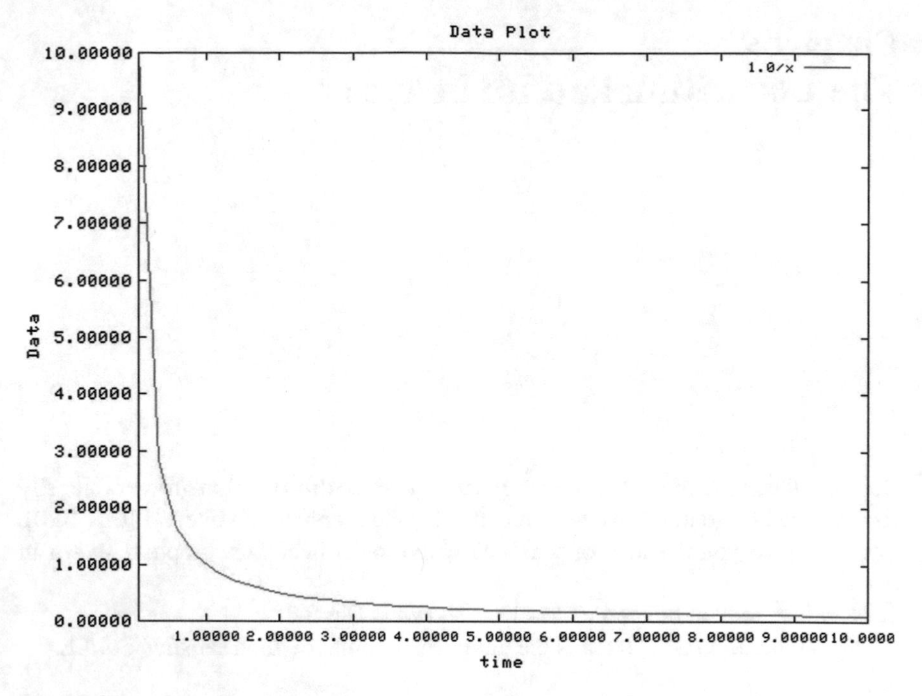

Fig. 9.1 The graph of the function 1/t

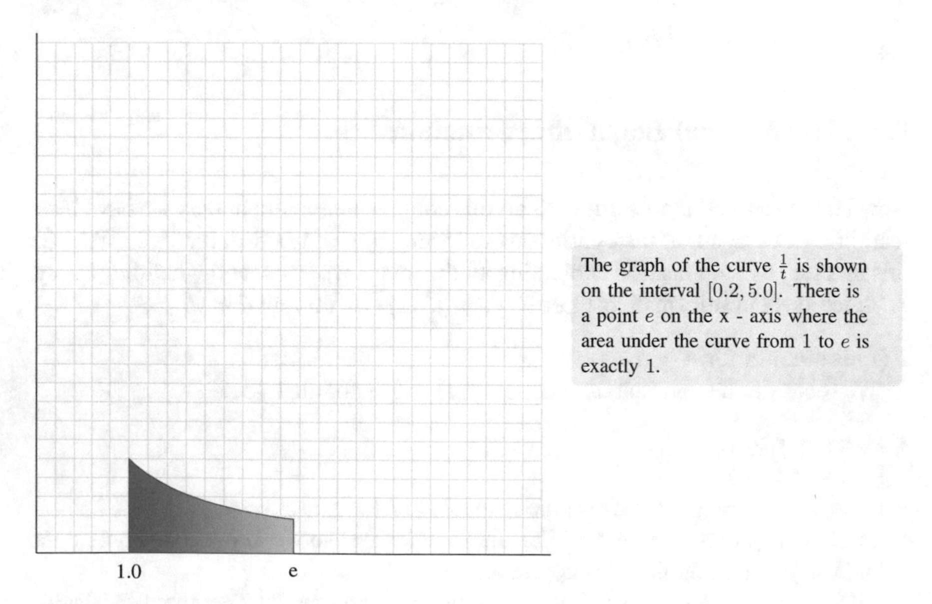

Fig. 9.2 There is a value *e* where the area under the *curve* is 1

$F(x)$ is smooth for $0 < x < 1$

We now look at the continuous function $f(t) = 1/t$ for all $0 < t \leq 1$.

- Pick any x with $0 < x \leq 1$.
- Pick any L so that $0 < L < x \leq 1$.
- Let $G(x) = \int_x^1 1/t \, dt$ on the interval $[L, 1]$.
- The Fundamental Theorem of Calculus applies and we know G is continuous on $[L, 1]$—in particular G is continuous at x.
- Since the integrand $1/t$ is continuous here, Reversed Fundamental Theorem of Calculus tells us $G'(x) = -f(x) = -1/x$.
- We can do this argument for any such x. So we know our G is continuous at x and $G'(x) = -1/x$ for all $0 < x \geq 1$.
- But for $0 < x < 1$, $G(x) = -F(x)$ and so $G'(x) = -F'(x)$. So we now know $-F'(x) = -(1/x)$ for $0 < x < 1$. Now just multiply through by -1 and we have the result we wanted: $F'(x) = 1/x$ on $0 < x < 1$.

So $F(x) = \int_1^x 1/t \, dt$ is continuous for $x > 0$ and $F'(x) = 1/x$ for $x > 0$. Very nice. This F is so special and useful it is given a special name: the **natural logarithm**. We denote this function by the symbol $\ln(x)$ and it is defined as follows:

$$\ln(x) = \int_1^x 1/t \, dt$$

As we have seen, this function is defined in two pieces:

$$\ln(x) = \begin{cases} F(x) = \int_1^x 1/t \, dt, & x \geq 1 \\ \int_1^x 1/t \, dt = -\int_x^1 1/t \, dt = -G(x), & 0 < x \leq 1. \end{cases}$$

We see $(\ln(x))' = 1/x$ for all positive x. Also, we now see that the number e is the unique number where $\ln(e) = 1$.

9.2 Logarithm Functions

We know that

$$\frac{d}{dx}(\ln(x)) = \frac{1}{x}$$

Let's start with a simple function, $u(x) = \mid x \mid$. Then $\ln(\mid x \mid)$ is nicely defined at all x not zero. Note we have

$$\ln(\mid x \mid) = \begin{cases} \ln(x) & \text{if } x > 0 \\ \ln(-x) & \text{if } x < 0 \end{cases}$$

and so since by the chain rule, if x is negative, using the chain rule with $u(x) = -x$, we have

$$\begin{aligned}\frac{d}{dx}(\ln(-x)) &= \frac{1}{-x}\frac{d}{dx}(-x)\\ &= \frac{1}{-x}(-1)\\ &= \frac{1}{x}\end{aligned}$$

Thus,

$$\begin{aligned}\frac{d}{dx}(\ln(|\,x\,|)) &= \begin{cases}\frac{d}{dx}(\ln(x)) & \text{if } x > 0\\ \frac{d}{dx}(\ln(-x)) & \text{if } x < 0\end{cases} = \begin{cases}\frac{1}{x} & \text{if } x > 0\\ \frac{1}{x} & \text{if } x < 0\end{cases}\\ &= \frac{1}{x} \text{ if } x \text{ is not } 0\end{aligned}$$

We conclude that

$$\frac{d}{dx}(\ln(|\,x\,|)) = \frac{1}{x}, \text{ if } x \neq 0$$

It then follows for a more general $u(x)$, using the chain rule that

$$\frac{d}{dx}(\ln(|\,u(x)\,|)) = \frac{1}{u(x)}\frac{du}{dx}, \text{ if } x \neq 0 \tag{9.2}$$

So the chain rule allows us to compute more interesting derivatives.

9.2.1 Worked Out Examples: Derivatives

Example 9.2.1 Find the derivative of $\ln(x^2 + x + 1)$.

Solution

$$\frac{d}{dx}\left(\ln(x^2 + x + 1)\right) = \frac{1}{x^2 + x + 1}(2x + 1)$$

Example 9.2.2 Find the derivative of $\ln(\tan(x))$.

Solution

$$\frac{d}{dx}(\ln(\tan(x))) = \frac{1}{\tan(x)}\sec^2(x)$$

Now the natural logarithm here is only defined when $\tan(x)$ *is positive. We usually just assume that we are only interested in evaluating the expression* $\ln(\tan(x))$ *for such* x*; i.e. by writing* $\ln(\tan(x))$*, we are tacitly assuming that* $\tan(x)$ *is positive. Another way of looking at this, is that the domain of the function* $\ln(\tan(x))$ *is the set of* x *where* $\tan(x)$ *is positive. However, since it is best to train ourselves to think about the restrictions on the domain without being so explicit!*

Example 9.2.3 Find the derivative of $\ln(x^6 + 3x^5 + 20)$.

Solution

$$\frac{d}{dx}\left(\ln(x^6 + 3x^5 + 20)\right) = \frac{1}{(x^6 + 3x^5 + 20)}\,(6x^5 + 15x^4)$$

Example 9.2.4 Find the derivative of $\ln(-3x)$.

Solution

$$\frac{d}{dx}(\ln(-3x)) = \frac{1}{-3x}(-3) = \frac{1}{x}$$

Here, the implied domain of $\ln(-3x)$ *is all negative* x.

Example 9.2.5 Find the derivative of $\ln(5x)$.

Solution

$$\frac{d}{dx}(\ln(5x)) = \frac{1}{5x}\,5 = \frac{1}{x}$$

Here the implied domain of $\ln(5x)$ *is all positive* x.

9.2.2 Homework: Derivatives

Exercise 9.2.1 *Find the derivative of* $\ln(t^2 + 1)$.

Exercise 9.2.2 *Find the derivative of* $\ln(t^3 + t^2 + 5)$.

Exercise 9.2.3 *Find the derivative of* $\ln(\sin(t))$.

Exercise 9.2.4 *Find the derivative of* $\ln(\sin(t^2))$.

Exercise 9.2.5 *Find the derivative of* $t\ \ln(t) - t$.

Exercise 9.2.6 *Find the derivative of* $s^2\ \ln(s^2 + 1)$.

Exercise 9.2.7 *Find the derivative of* $\ln(\sqrt{t} + t)$.

Exercise 9.2.8 *Find the derivative of* $\ln(z + 1/z)$.

9.2.3 Worked Out Examples: Integrals

We can also now find antiderivatives for a new class of functions. The chain rule for the derivatives of the logarithm function given in Eq. 9.2 implies

$$\int \frac{1}{u(x)}\, u'(x)\, dx = \ln(\mid u(x) \mid) + C$$

Another way of saying this is that $\int du/u = \ln(|u|) + C$! Let's work some examples:

Example 9.2.6 Find the integral $\int \frac{1}{x^2+x+1}\,(2x+1)dx$.

Solution

$$\begin{aligned} \int \frac{1}{x^2+x+1}\,(2x+1)dx &= \int \frac{1}{u}\, du, \textit{ use substitution } u = x^2+x+1 \\ &= \ln(\mid u \mid) + C \\ &= \ln(\mid x^2+x+1 \mid) + C \end{aligned}$$

Example 9.2.7 Find the integral $\int \frac{1}{\tan(x)}\, \sec^2(x)\, dx$.

Solution

$$\begin{aligned} \int \frac{1}{\tan(x)}\, \sec^2(x)\, dx &= \int \frac{1}{u}\, du, \textit{ use substitution } u = \tan(x)\ ;\ du = \sec^2(x)dx \\ &= \ln(\mid u \mid) + C \\ &= \ln(\mid \tan(x) \mid) + C \end{aligned}$$

Example 9.2.8 Find the integral $\int \frac{2x}{4+x^2}\, dx$.

Solution

$$\begin{aligned} \int \frac{2x}{4+x^2}\, dx &= \int \frac{1}{u}\, du, \textit{ use substitution } u = 4+x^2;\ du = 2x dx \\ &= \ln(\mid u \mid) + C \\ &= \ln(\mid 4+x^2 \mid) + C, \textit{ but } 4+x^2 \textit{ is always positive} \\ &= \ln(4+x^2) + C \end{aligned}$$

Example 9.2.9 Find the integral $\int \tan(w)dw$.

Solution

$$\begin{aligned} \int \tan(w)dw &= \int \frac{\sin(w)}{\cos(w)}\, dw \\ &= \int \frac{1}{u}\,(-du), \textit{ use substitution } u = \cos(w);\ du = -\sin(w)dw \end{aligned}$$

$$= -\ln(\mid u \mid) + C$$
$$= -\ln(\mid \cos(w) \mid) + C$$

9.2.4 Homework: Integrals

Exercise 9.2.9 *Find the integral* $\int \ln(t)/t \, dt$.

Exercise 9.2.10 *Find the integral* $\int 2t/(t^2 + 4) \, dt$.

Exercise 9.2.11 *Find the integral* $\int 2s^2/(s^3 + 9) \, ds$.

Exercise 9.2.12 *Find the integral* $\int (8z^2 + 3)/(8z^3 + 9z + 18) \, dz$.

Exercise 9.2.13 *Find the integral* $\int (40z^4 + 27z^2 + 36z)/(8z^5 + 9z^3 + 18z^2 + 20) \, dz$.

Exercise 9.2.14 *Find the integral* $\int \sin(w)/\cos(w) \, dw$.

Exercise 9.2.15 *Find the integral* $\int \cos(5t)/\sin(5t) \, dt$.

Exercise 9.2.16 *Find the integral* $\int_0^2 t/(4t^2 + 3) \, dt$.

9.3 The Exponential Function

Let's backup and talk about the idea of an inverse function. Say we have a function $y = f(x)$ like $y = x^3$. Take the cube root of each side to get $x = y^{1/3}$. Just for fun, let $g(x) = x^{1/3}$; i.e., we switched the role of x and y in the equation $x = y^{1/3}$. Now note some interesting things:

$$f(g(x)) = f(x^{1/3}) = \left(x^{1/3}\right)^3 = x$$
$$g(f(x)) = g(x^3) = \left(x^3\right)^{1/3} = x.$$

Now the function $I(x) = x$ is called the identity because it takes an x as an input and does nothing to it. The output value is still x. So we have $f(g) = I$ and $g(f) = I$. When this happens, the function g is called the **inverse** of f and is denoted by the special symbol f^{-1}. Of course, the same is true going the other way: f is the inverse of g and could be denoted by g^{-1}. Another more abstract way of saying this is

$$f^{-1}(x) = y \Leftrightarrow f(y) = x.$$

Now look at Fig. 9.3. We draw the function x^3 and its inverse $x^{1/3}$ in the unit square. We also draw the identity there which is just the graph of $y = x$; i.e. a line with

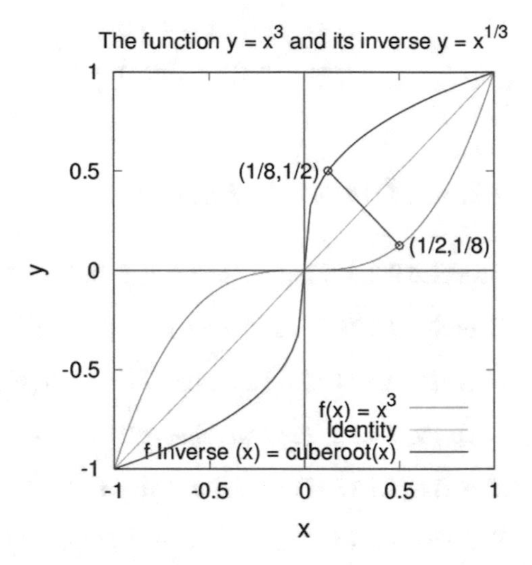

Fig. 9.3 The function x^3 and its inverse $x^{1/3}$ plotted in the unit square

slope 1. If you take the point (1/2.1/8) on the graph of x^3 and draw a line from it to the inverse point $(1/8, 1/2)$ you'll note that this line is perpendicular to the line of the identity. This will always be true with a graph of a function and its inverse. Now also note that x^3 has a positive derivative always and so is always increasing. It seems reasonable that if we had a function whose derivative was positive all the time, we could do this same thing. We could take a point on that function's graph, say (c, d), reverse the coordinates to (d, c) and the line connecting those two pairs would be perpendicular to the identity line just like in our figure. So we have a geometric procedure to define the inverse of any function that is always increasing.

What about $\ln(x)$? It has derivative $1/x$ for all positive x, so it must be always increasing. So it has an inverse which we can call $\ln^{-1}(x)$ which is called the **exponential function** which is denoted by **exp**(x). The inverse is defined by the if and only relationship

$$(\ln)^{-1}(x) = y \Leftrightarrow \ln(y) = x$$

or, using the exp notation

$$\exp(x) = y \Leftrightarrow \ln(y) = x.$$

A little thought tells us the range of $\ln(x)$ is all real numbers as for $x > 1$, $\ln(x)$ gets as large as we want and for $0 < x < 1$, as x gets closer to zero, the negative area $-\int_x^1 1/t dt$ approaches $-\infty$. By definition then

- $\ln(\exp(x)) = x$ for $-\infty < x < \infty$; i.e. for all x.
- $\exp(\ln(x)) = x$ for all $x > 0$.

Further since $\ln(\exp(x)) = x$, we can take the derivative of both sides:

$$\left(\ln(\exp(x))\right)' = \left(x\right)' = 1$$

Using the chain rule, for any function $u(x)$,

$$\left(\ln(u(x))\right)' = \frac{1}{u(x)}\, u'(x).$$

So

$$\left(\ln((\exp(x))\right)' = \frac{1}{\exp(x)} \left(\exp(x)\right)'.$$

Using this, we see $\frac{1}{\exp(x)} \left(\exp(x)\right)' = 1$ and so

$$\left(\exp(x)\right)' = \exp(x).$$

This is the only function whose derivative is itself!

9.4 Exponential Functions

Since we know now that

$$\frac{d}{dx}\,(\exp(x)) = \exp(x)$$

we can use the chain rule to compute more interesting derivatives. If u is a function that is differentiable at x then we have

$$\frac{d}{dx}\,(\exp(u(x))) = \exp(u(x))\,\frac{du}{dx}$$

9.4.1 Worked Out Examples: Derivatives

Example 9.4.1 Find the derivative of $\exp(x^2 + x + 1)$.

Solution

$$\frac{d}{dx}\,(\exp(x^2 + x + 1)) = \exp(x^2 + x + 1)\,(2x + 1)$$

Example 9.4.2 Find the derivative of $\exp(\tan(x))$.

Solution

$$\frac{d}{dx}(\exp(\tan(x))) = \exp(\tan(x))\ \sec^2(x)$$

Example 9.4.3 Find the derivative of $\exp(\exp(x^2))$.

Solution

$$\frac{d}{dx}\big(\exp(\exp(x^2))\big) = \exp(\exp(x^2))\ \exp(x^2)\ 2x$$

Example 9.4.4 Find the derivative of $\exp(-3x)$.

Solution

$$\frac{d}{dx}(\exp(-3x)) = \exp(-3x)\ (-3)$$

Example 9.4.5 Find the derivative of $\exp(5x)$.

Solution

$$\frac{d}{dx}(\exp(5x)) = \exp(5x)\ 5$$

Example 9.4.6 Find the integral $\int \frac{\exp(t)}{1+\exp(t)}\, dt$.

Solution

$$\begin{aligned}\int \frac{\exp(t)}{1+\exp(t)}\, dt &= \int \frac{1}{u}\, du, \text{ use substitution } u = 1+\exp(t);\ du = \exp(t)dt\\ &= \ln(|\ u\ |) + C\\ &= \ln(|\ 1 + |:\exp(t)\ |) + C, \text{ but } 1+\exp(t) \text{ is always positive}\\ &= \ln(1 + |:\exp(t)) + C\end{aligned}$$

9.4.2 Homework: Exponential Derivatives

Exercise 9.4.1 *Find the derivative of* $\exp(-3t)$.

Exercise 9.4.2 *Find the derivative of* $\exp(t^2 + 2t - 4)$.

Exercise 9.4.3 *Find the derivative of* $\exp(1/t)$.

Exercise 9.4.4 *Find the derivative of* $\exp(t)/(\exp(t) + \exp(-t))$.

Exercise 9.4.5 *Find the derivative of* $\cos(\exp(t^2))$.

Exercise 9.4.6 *Find the derivative of* $\sin(t\ \exp(t))$.

Exercise 9.4.7 *Find the derivative of* $(t^2 + 2t - 8)\ \exp(5t)$.

Exercise 9.4.8 *Find the derivative of* $\exp(4t)$.

Exercise 9.4.9 *Find the derivative of* $\exp(-3t)$.

Exercise 9.4.10 *Find the derivative of* $\exp(18t)$.

9.4.3 Worked Out Examples: Exponential Integrals

Further, we can now find many antiderivatives:

$$\int \exp(u(x))\ u'(x)\ dx = \exp(u(x)) + C$$

Example 9.4.7 Find the integral of $\int \exp(x^2 + x + 1)\ (2x + 1)dx$.

Solution

$$\begin{aligned}\int \exp(x^2 + x + 1)\ (2x + 1)dx &= \int \exp(u)\ du, \textit{ use substitution } u = x^2 + x + 1\\ &= \exp(u) + C\\ &= \exp(x^2 + x + 1) + C\end{aligned}$$

Example 9.4.8 Find the integral of $\int \exp(\tan(x))\ \sec^2(x)\ dx$.

Solution

$$\begin{aligned}\int \exp(\tan(x))\ \sec^2(x)\ dx &= \int \exp(u)\ du, \textit{ use substitution } u = \tan(x);\ du = \sec^2(x)dx\\ &= \exp(u) + C\\ &= \exp(\tan(x)) + C\end{aligned}$$

Example 9.4.9 Find the integral of $\int \exp(x^2)\ 3x\ dx$.

Solution

$$\begin{aligned}\int \exp(x^2)\ 3x\ dx &= \int \exp(u)3(du/2), \textit{ use substitution } u = x^2;\ du = 2xdx\\ &= (3/2) \int \exp(u)\ du\\ &= (3/2)\ \exp(u) + C\\ &= (3/2)\ \exp(x^2) + C\end{aligned}$$

Example 9.4.10 Find the integral of $\int \exp(-3x)\, dx$.

Solution

$$\begin{aligned}\int \exp(-3x)\, dx &= \int \exp(u)\, du/(-3), \textit{ use substitution } u = -3x;\, du = -3dx\\ &= \frac{1}{-3}\int \exp(u)\, du\\ &= -\frac{1}{3}\exp(u) + C\\ &= -\frac{1}{3}\exp(-3x) + C\end{aligned}$$

Example 9.4.11 Find the integral of $\int \exp(5x)\, dx$.

Solution

$$\begin{aligned}\int \exp(5x)\, dx &= \int \exp(u)\, du/(5), \textit{ use substitution } u = 5x;\, du = 5dx\\ &= \frac{1}{5}\int \exp(u)\, du\\ &= \frac{1}{5}\exp(u) + C\\ &= \frac{1}{5}\exp(5x) + C\end{aligned}$$

9.4.4 Homework: Exponential Integrals

Exercise 9.4.11 *Find the integral* $\int \exp(-3t)\, dt$.

Exercise 9.4.12 *Find the integral* $\int \exp(t^2 + 2t - 4)\, (2t + 2)dt$.

Exercise 9.4.13 *Find the integral* $\int \exp(1/t)\, (1/t^2)\, dt$.

Exercise 9.4.14 *Find the integral* $\int \exp(t^3)\, t^2\, dt$.

Exercise 9.4.15 *Find the integral* $\int 4\, \cos(\exp(t^2))\, t\, \exp(t^2)\, dt$.

Exercise 9.4.16 *Find the integral* $\int \sin(\exp(t))\, \exp(t)\, dt$.

Exercise 9.4.17 *Find the integral* $\int t\, \exp(5t^2)\, dt$.

Exercise 9.4.18 *Find the integral* $\int \exp(14t)\, dt$.

Exercise 9.4.19 *Find the integral* $\int_{-1}^{5} \exp(-33t)\, dt$.

Exercise 9.4.20 *Find the integral* $\int_0^2 \exp(4t)\, dt$.

9.5 Our Antiderivatives So Far

We now know some additional antiderivatives we can use.

- $\int (1/u)\, du = \ln(|u|) + C$.
- $\int e^u\, du = e^u + C$.

These are powerful new rules we will use a lot.

Chapter 10
Exponential and Logarithm Function Properties

Now that we know a lot about both the natural logarithm function ln and its inverse exp, let's back up and learn about their properties.

10.1 Positive Integer Powers of e

Now what is the real number c which satisfies $\ln(c) = 2$? This means

$$\ln(c) = \int_1^c \frac{1}{t}\, dt = 2.0?$$

We can see this graphically in Fig. 10.1.

Since the area under the curve $\frac{1}{t}$ is increasing as t increases, we can immediately see that c must be larger than e. Thus, using properties of the Riemann Integral, we have

$$\begin{aligned} 2 &= \int_1^c \frac{1}{t}\, dt \\ &= \int_1^e \frac{1}{t}\, dt + \int_e^c \frac{1}{t}\, dt \\ &= 1.0 + \int_e^c \frac{1}{t}\, dt \end{aligned}$$

Subtracting terms, we find

$$1 = \int_e^c \frac{1}{t}\, dt$$

J.K. Peterson, *Calculus for Cognitive Scientists*, Cognitive Science and Technology, DOI 10.1007/978-981-287-874-8_10

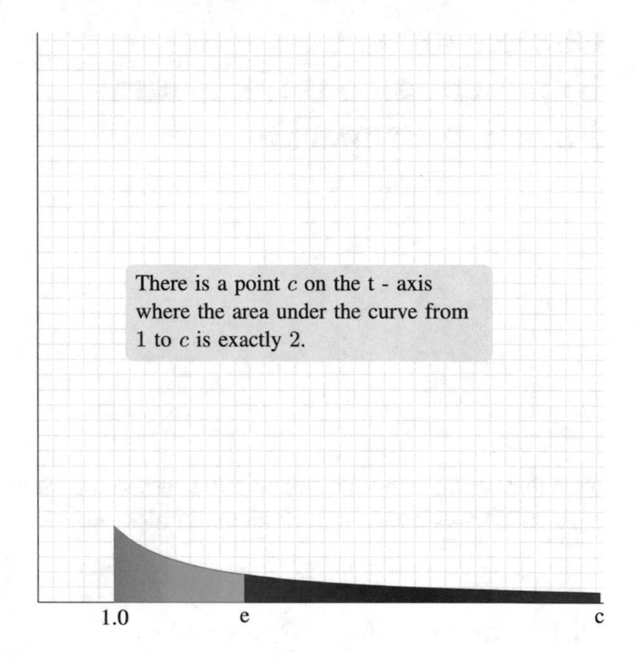

Fig. 10.1 There is a value c where the area under the *curve* from 1 to c is 2

Now, make the change of variable $u = t/e$. Then

$$\int_e^c \frac{1}{t}\, dt = \int_1^{c/e} \frac{1}{u}\, du$$

Combining, we have

$$1 = \int_1^{c/e} \frac{1}{u}\, du$$

Since the letter we use as the variable of integration is arbitrary, this can be rewritten as

$$1 = \int_1^{c/e} \frac{1}{t}\, dt$$

However, e is the unique number which marks the point on the t-axis where the area under this curve is 1. Thus, we can conclude that $c/e = e$ or $c = e^2$. A similar argument shows all of the following also to be true:

$$3 = \ln(e^3) = \int_1^{e^3} \frac{1}{t}\, dt$$

$$4 = \ln(e^4) = \int_1^{e^4} \frac{1}{t}\, dt$$

$$6 = \ln(e^6) = \int_1^{e^6} \frac{1}{t}\, dt.$$

We can thus interpret the number $\ln(e^p)$ for a positive integer power of p as an appropriate area under the graph of $\frac{1}{t}$ starting at 1 and moving to the right. That is e^p is the number that satisfies

$$p = \ln(e^p) = \int_1^{e^p} \frac{1}{t}\, dt.$$

10.1.1 Homework

Exercise 10.1.1 *Show*

$$7 = \ln(e^7) = \int_1^{e^7} \frac{1}{t}\, dt$$

Note the appropriate visual is shown in Fig. 10.2.

Exercise 10.1.2 *Show*

$$9 = \ln(e^9) = \int_1^{e^9} \frac{1}{t}\, dt$$

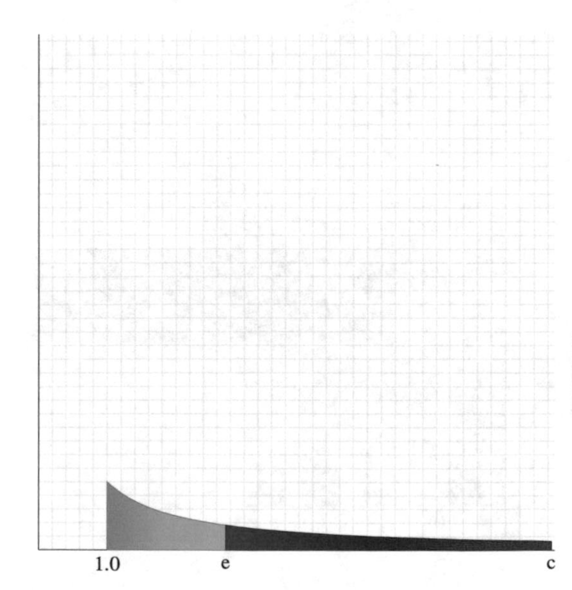

Fig. 10.2 There is a value c where the area under the *curve* from e to c is 7

10.2 Negative Integer Powers of e

We were able to interpret the number e^p for a positive integer power of p as follows:

$$p = \ln(e^p) = \int_1^{e^p} \frac{1}{t}\, dt$$

How can we interpret terms like $\frac{1}{e}$ or $\frac{1}{e^2}$? Let's start with e^{-1}. We can use a graphical analysis just like we did to define the number e. Look at Fig. 10.3.

Since the area under the curve $\frac{1}{t}$ is increasing as t increases, we can immediately see that c must be less than 1. Further, we know that

$$1 = \int_c^1 \frac{1}{t}\, dt$$

Now, make the change of variable $u = t/c$. Then

$$\int_c^1 \frac{1}{t}\, dt = \int_1^{1/c} \frac{1}{u}\, du$$

Combining, we have

$$1 = \int_1^{1/c} \frac{1}{u}\, du$$

Since the letter we use as the variable of integration is arbitrary, this can be rewritten as

$$1 = \int_1^{1/c} \frac{1}{t}\, dt$$

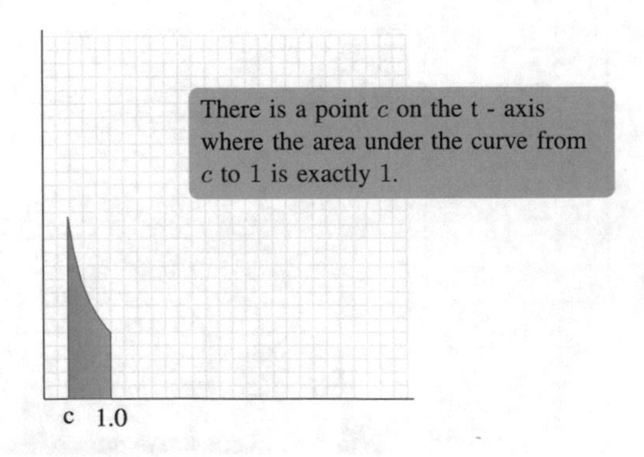

Fig. 10.3 There is a value c where the area under the *curve* which ends at 1 is 1

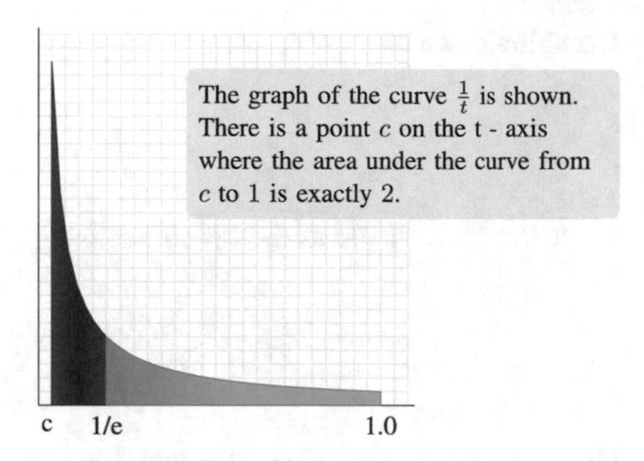

Fig. 10.4 There is a value c where the area under the *curve* which ends at 1 is 2

However, e is the unique number which marks the point on the x-axis where the area under this curve is 1. Thus, we can conclude that $1/c = e$. This implies that $c = 1/e$. Hence, we have shown

$$-1 = \ln(e^{-1}).$$

What about the interpretation of the value e^{-2}? Consider Fig. 10.4. Since the area under the curve $\frac{1}{t}$ is increasing as t increases, we can immediately see that c must be less than $1/e$ which is less than 1. Thus, using properties of the Riemann Integral, and the fact that $\int_{1/e}^{1} \frac{1}{t}\, dt = 1$, we see

$$\begin{aligned} 2 &= \int_c^1 \frac{1}{t}\, dt \\ &= \int_c^{1/e} \frac{1}{t}\, dt + \int_{1/e}^{1} \frac{1}{t}\, dt \\ &= \int_c^{1/e} \frac{1}{t}\, dt + 1.0 \end{aligned}$$

Subtracting terms, we find

$$1 = \int_c^{\frac{1}{e}} \frac{1}{t}\, dt.$$

Now, make the change of variable $u = t/c$. Then

$$\int_c^{1/e} \frac{1}{t}\, dt = \int_1^{1/(ec)} \frac{1}{u}\, du$$

Combining, we have

$$1 = \int_1^{1/(ec)} \frac{1}{u}\, du$$

Since the letter we use as the variable of integration is arbitrary, this can be rewritten as

$$1 = \int_1^{1/(ec)} \frac{1}{t}\, dt$$

However, e is the unique number which marks the point on the t-axis where the area under this curve is 1. Thus, we can conclude that $\frac{1}{ec} = e$ or $c = e^{-2}$. Hence, we have shown

$$-2 = \ln(e^{-2}).$$

A similar argument shows all of the following also to be true:

$$-3 = \ln(e^{-3}) = \int_1^{e^{-3}} \frac{1}{t}\, dt$$
$$-4 = \ln(e^{-4}) = \int_1^{e^{-4}} \frac{1}{t}\, dt$$
$$-6 = \ln(e^{-6}) = \int_1^{e^{-6}} \frac{1}{t}\, dt.$$

We can thus interpret the number $\ln(e^{-p})$ for a positive integer power of p as an appropriate area under the graph of $\frac{1}{t}$ starting at e^{-p} and moving to the 1. That is e^{-p} is the number that satisfies

$$-p = \ln(e^{-p}) = \int_1^{e^{-p}} \frac{1}{t}\, dt.$$

10.2.1 Homework

Exercise 10.2.1 *Show*

$$-3 = \ln(e^{-3}) = \int_1^{e^{-3}} \frac{1}{t}\, dt$$

Note the appropriate visual is shown in Fig. 10.5.

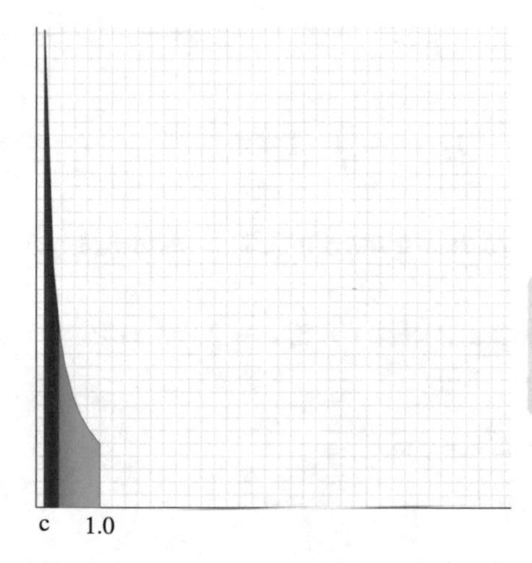

Fig. 10.5 There is a value c where the area from the point c to 1 is -3

Exercise 10.2.2 *Show*

$$-4 = \ln(e^{-4}) = \int_1^{e^{-4}} \frac{1}{t} \, dt$$

10.3 Adding Natural Logarithms

What happens if we want to add $\ln(a) + \ln(b)$ for positive numbers a and b? There are several cases. We can

- We have a is less than one but b is bigger than one: for example, $a = 0.3$ and $b = 7$.
- We have a is less than one and b is less than one also: for example, $a = 0.3$ and $b = 0.7$.
- We have a is bigger than one and b is bigger than one: for example, $a = 4$ and $b = 8$.

We can work out how to do this in general but it is easier to work through some specific examples and extrapolate from them.

10.3.1 Adding Logarithms: Both Logarithms are Bigger Than 1

Example 10.3.1 Show $\ln(3) + \ln(5) = \ln(3 \times 5) = \ln(15)$.

Solution • *First, rewrite using the definition.*

$$\ln(3) + \ln(5) = \int_1^3 1/t\, dt + \int_1^5 1/t\, dt.$$

- *Now rewrite the second integral* $\int_1^5 1/t\, dt$ *so it starts at* 3. *This requires a change of variable.*
  - *Let* $u = 3t$. *Then when* $t = 1$, $u = 3$ *and when* $t = 5$, $u = 3 \times 5 = 15$.
  - *So* $u = 3t$ *and* $du = 3dt$ *or* $1/3\, du = dt$
  - *Further,* $u = 3t$ *means* $1/3\, u = t$
  - *Make the substitutions in the second integral.*

$$\int_1^5 1/t\, dt = \int_{t=1}^{t=5} \frac{1}{1/3\, u} \frac{1}{3}\, du = \int_{t=1}^{t=5} \frac{3}{u} \frac{1}{3}\, du$$
$$= \int_{t=1}^{t=5} \frac{1}{u}\, du$$

- *Now switch the lower and upper limits of integration to u values. We didn't do this before although we could have. This gives*

$$\int_1^5 1/t\, dt = \int_{t=1}^{t=5} \frac{1}{u}\, du = \int_{u=3}^{u=15} \frac{1}{u}\, du.$$

- *Now, as we have said, the choice of letter for the name of variable in the Riemann integral does not matter. The integral above is the* **area under the curve** $f(u) = 1/u$ **between** 3 **and** 15 *which is exactly the same as* **area under the curve** $f(y) = 1/y$ **between** 3 **and** 15 *and indeed the same as* **area under the curve** $f(t) = 1/t$ **between** 3 **and** 15. *So we have*

$$\int_1^5 1/t\, dt = \int_3^{15} \frac{1}{u}\, du = \int_3^{15} \frac{1}{t}\, dt.$$

- *Now plug this new version of the second integral back into the original sum.*

$$\ln(3) + \ln(5) = \int_1^3 1/t\, dt + \int_1^5 1/t\, dt$$
$$= \int_1^3 1/t\, dt + \int_3^{15} 1/u\, du$$
$$= \int_1^3 1/t\, dt + \int_3^{15} 1/t\, dt$$

- *But* $\int_1^{15} 1/t\, dt = \ln(15)$ **so we have shown** $\ln(3) + \ln(5) = \ln(3 \times 5) = \ln(15)$.

Example 10.3.2 Show $\ln(4) + \ln(12) = \ln(4 \times 12) = \ln(48)$.

Solution • *First, rewrite using the definition.*

$$\ln(4) + \ln(12) = \int_1^4 1/t\, dt + \int_1^{12} 1/t\, dt.$$

• *Now rewrite the second integral* $\int_1^{12} 1/t\, dt$ *so it starts at* 4. *This requires a change of variable.*

 – *Let* $u = 4t$. *Then when* $t = 1$, $u = 4$ *and when* $t = 12$, $u = 4 \times 12 = 48$.
 – *So* $u = 4t$ *and* $du = 4dt$ *or* $1/4\, du = dt$
 – *Further,* $u = 4t$ *means* $1/4\, u = t$
 – *Make the substitutions in the second integral.*

$$\int_1^{12} 1/t\, dt = \int_{t=1}^{t=12} \frac{1}{1/4\, u} \frac{1}{4}\, du = \int_{t=1}^{t=12} \frac{4}{u} \frac{1}{4}\, du$$
$$= \int_{t=1}^{t=12} \frac{1}{u}\, du$$

• *Now switch the lower and upper limits of integration to u values. We didn't do this before although we could have. This gives*

$$\int_1^{12} 1/t\, dt = \int_{t=1}^{t=12} \frac{1}{u}\, du = \int_{u=4}^{u=48} \frac{1}{u}\, du.$$

• *Now, as we have said, the choice of letter for the name of variable in the Riemann integral does not matter. The integral above is the* **area under the curve** $f(u) = 1/u$ **between** 4 **and** 48 *which is exactly the same as* **area under the curve** $f(y) = 1/y$ **between** 4 **and** 48 *and indeed the same as* **area under the curve** $f(t) = 1/t$ **between** 4 **and** 48. *So we have*

$$\int_1^{12} 1/t\, dt = \int_4^{48} \frac{1}{u}\, du = \int_4^{48} \frac{1}{t}\, dt.$$

• *Now plug this new version of the second integral back into the original sum.*

$$\ln(4) + \ln(12) = \int_1^4 1/t\, dt + \int_1^{12} 1/t\, dt$$
$$= \int_1^4 1/t\, dt + \int_4^{48} 1/u\, du$$
$$= \int_1^4 1/t\, dt + \int_4^{48} 1/t\, dt$$

- *But* $\int_1^{48} 1/t\,dt = \ln(48)$ **so we have shown** $\ln(4)+\ln(12) = \ln(4 \times 12) = \ln(48)$.

10.3.1.1 Homework

Follow the steps above to show

Exercise 10.3.1

$\ln(2) + \ln(7) = \ln(14)$

Exercise 10.3.2

$\ln(3) + \ln(9) = \ln(27)$

Exercise 10.3.3

$\ln(8) + \ln(10) = \ln(80)$

10.3.2 Adding Logarithms: One Logarithm Less Than 1 and One Bigger Than 1

Example 10.3.3 Show $\ln(1/4) + \ln(5) = \ln((1/4) \times 5) = \ln((5/4))$.

Solution • *First, rewrite using the definition.*

$$\ln(1/4) + \ln(5) = \int_1^{1/4} 1/t\;dt + \int_1^{5} 1/t\;dt.$$

- *Now rewrite the second integral* $\int_1^5 1/t\;dt$ *so it starts at* $1/4$. *This requires a change of variable.*
  - *Let* $u = 1/4\,t$. *Then when* $t = 1$, $u = 1/4$ *and when* $t = 5$, $u = 1/4 \times 5 = 5/4$.
  - *So* $u = 1/4t$ *and* $du = 1/4dt$ *or* $4\;du = dt$
  - *Further,* $u = 1/4\;t$ *means* $4\;u = t$
  - *Make the substitutions in the second integral.*

$$\begin{aligned}\int_1^5 1/t\;dt &= \int_{t=1}^{t=5} \frac{1}{4\,u}\,4\,du\\ &= \int_{t=1}^{t=5} \frac{1}{u}\,du\end{aligned}$$

- *Now switch the lower and upper limits of integration to u values. We didn't do this before although we could have. This gives*

$$\int_1^5 1/t\,dt = \int_{t=1}^{t=5} \frac{1}{u}\,du = \int_{u=1/4}^{u=5/4} \frac{1}{u}\,du.$$

- *Now, as we have said, the choice of letter for the name of variable in the Riemann integral does not matter. The integral above is the* **area under the curve** $f(u) = 1/u$ **between** $1/4$ **and** $5/4$ *which is exactly the same as* **area under the curve** $f(y) = 1/y$ **between** $1/4$ **and** $5/4$ *and indeed the same as* **area under the curve** $f(t) = 1/t$ **between** $1/4$ **and** $5/4$*. So we have*

$$\int_1^5 1/t\,dt = \int_{1/4}^{5/4} \frac{1}{u}\,du = \int_{1/4}^{5/4} \frac{1}{t}\,dt.$$

- *Now plug this new version of the second integral back into the original sum.*

$$\begin{aligned}
\ln(1/4) + \ln(5) &= \int_1^{1/4} 1/t\,dt + \int_1^5 1/t\,dt \\
&= \int_1^{1/4} 1/t\,dt + \int_{1/4}^{5/4} 1/u\,du \\
&= -\int_{1/4}^{1} 1/t\,dt + \int_{1/4}^{5/4} 1/t\,dt \\
&= -\int_{1/4}^{1} 1/t\,dt + \int_{1/4}^{1} 1/t\,dt + \int_1^{5/4} 1/t\,dt \\
&= \int_1^{5/4} 1/t\,dt
\end{aligned}$$

- *But* $\int_1^{5/4} 1/t\,dt = \ln(5/4)$ **so we have shown** $\ln(1/4) + \ln(5) = \ln(1/4 \times 5) = \ln(5/4)$.

10.3.2.1 Homework

Follow the steps above to show

Exercise 10.3.4

$\ln(0.5) + \ln(7) = \ln(3.5)$

Exercise 10.3.5

$\ln(0.2) + \ln(3) = \ln(0.6)$

Exercise 10.3.6

$\ln(0.6) + \ln(9) = \ln(5.4)$

10.3.3 Generalizing These Results

Consider $\ln(1/7 \times 7)$. Since $1/7 < 1$ and $7 > 1$, this is the case above.

- $\ln(1/7) + \ln(7) = \ln(1/7 \times 7) = \ln(1) = 0$.
- So $\ln(1/7) = -\ln(7)$.
- In general, if $a > 0$, then $\ln(1/a) = -\ln(a)$.
- So we can now handle the sum of two logarithms of numbers less than one. $\ln(1/5) + \ln(1/7) = -\ln(5) - \ln(7) = -(\ln(5) + \ln(7)) = -ln(35) = ln(1/35)$!

We can combine all these results into one general rule: for any $a > 0$ and $b > 0$,

$$\ln(a) + \ln(b) = \ln(a \times b).$$

Hence, now we can do subtractions:

- $\ln(7) - \ln(5) = \ln(7) + \ln(1/5) = \ln(7/5)$
- $\ln(1/9) - \ln(1/2) = -\ln(9) - \ln(1/2) = -\ln(9/2) = ln(2/9)$.

So in general, for any $a > 0$ and $b > 0$:

$$\ln(a) - \ln(b) = \ln(a/b).$$

We can summarize all these results as follows:

Proposition 10.3.1 (Adding Natural Logarithms)
If a and b are two positive numbers, then

$$\begin{aligned} \ln(a) + \ln(b) &= \int_1^a \frac{1}{t}\, dt + \int_1^b \frac{1}{t}\, dt \\ &= \int_1^{ab} \frac{1}{t}\, dt = \ln(ab). \end{aligned}$$

10.3.4 Doing Subtracts in General

What happens if we subtract areas? We handle this like before. For any $a > 0$, we know $0 = \ln(1) = \ln(a\ (1/a)) = \ln(a) + \ln((1/a))$. Hence, $\ln((1/a)) = -\ln(a)$. From this, for any a and b which are positive, we have $\ln(a) - \ln(b) = \ln(a) + \ln((1/b)) = \ln((a/b))$. This is our general subtraction law.

Proposition 10.3.2 (Subtracting Natural Logarithms)
If a and b are two positive numbers, then

$$\ln(a) - \ln(b) = \int_1^a \frac{1}{t}\,dt - \int_1^b \frac{1}{t}\,dt$$
$$= \int_1^{a/b} \frac{1}{t}\,dt = \ln\left(\frac{a}{b}\right).$$

10.4 Fractional Powers

We now have two basic rules: for any a and b which are positive, we can say $\ln(a) + \ln(b) = \ln(ab)$ and $\ln(a) - \ln(b) = \ln((a/b))$. These two simple facts give rise to many interesting consequences. We also know that $\ln((1/a)) = -\ln(a)$. What about fractions like $\ln(a^{p/q})$ for positive integers p and q?

Let's do this for a specific fraction like 7/8. Using Proposition 10.3.1, we see

$$\ln(a^2) = \ln(a \times a) = \ln(a) + \ln(a) = 2\ln(a)$$
$$\ln(a^3) = \ln(a^2 \times a) = \ln(a^2) + \ln(a) = 2\ln(a).$$

If we keep doing this sort of expansion, we find

$$\ln(a^7) = 7\ln(a).$$

Next, note that

$$\ln(a) = \ln\left(\left(a^{\frac{1}{8}}\right)^8\right)$$
$$= 8\ln\left(a^{\frac{1}{8}}\right)$$

Dividing the last equation by 8, we find

$$\ln\left(a^{\frac{1}{8}}\right) = \frac{1}{8}\ln(a).$$

Combining these expressions, we have found that

$$\ln(a^{7/8}) = 7\ln(a^{1/8})$$
$$= \frac{7}{8}\ln(a).$$

Now the same thing can be done for the general number p/q. After similar arguments, we find

$$\ln(a^{p/q}) = \frac{p}{q}\ln(a)$$

The discussions above have led us to the proposition that if a is a positive number and $r = p/q$ is any positive rational number, then

$$\ln(a^r) = r \ln(a).$$

If r is any real number power, we know we can find a sequence of rational numbers which converges to r. We can use this to see $\ln(a^r) = r \ln(a)$. For example, consider $r = \sqrt{2}$. Let's consider how to figure out $\ln(4^{\sqrt{2}})$. We know $\sqrt{2} = 1.414214....$ So the sequence of fractions that converge to $\sqrt{2}$ is $\{p_1/q_1 = 14/10,\ p_2/q_2 = 141/100,\ p_3/q_3 = 1414/1000,\ p_4/q_4 = 14142/10000,\ p_5/q_5 = 141421/100000,\ p_6/q_6 = 1414214/1000000, \ldots\}$. So

$$\lim \ln(4^{p_n/q_n}) = \lim\ (p_n/q_n)\ \ln(4) \longrightarrow \sqrt{2}\ \ln(4).$$

Now define $4^x = \exp(\ln(4)\, x)$. Since $\exp(x)$ is continuous, so is $\exp(\ln(4)x)$. So 4^x is continuous and $\lim 4^{p_n/q_n} \longrightarrow 4^{\sqrt{2}}$. Thus, since ln is continuous at $4^{\sqrt{2}}$, we have

$$\lim \ln(4^{p_n/q_n}) = \ln(\, 4^{\sqrt{2}}\,) \longrightarrow \ln(\, 4^{\sqrt{2}}\,) = \sqrt{2}\ \ln(4).$$

This suggests that $\ln(4^{\sqrt{2}}) = \sqrt{2}\ \ln(4)$. This can be shown to be true in general even if the power is negative.

Proposition 10.4.1 (Powers)
If a is a positive number and r is any real number, then

$$\ln(a^r) = r \ln(a).$$

As a matter of fact, buried in our argument here is an idea that helps us with an old model. Do you remember the viability selection model? We had

$$\boldsymbol{P_A}(t) = \frac{1}{1 + \frac{N_B(0)}{N_A(0)} \left(\frac{V_B}{V_A}\right)^t}.$$

Now we originally thought of t as being measured in generations but in our general discussions on smoothness, we notices that really the generation time could be pretty arbitrary. So we were led to thinking of t as being measured in any time units at all. Hence, t could be any real number that represents time and it doesn't have to be thought of as a finite number of generation times like 5 or 20 h where h is the generation time unit. We now know how to interpret a function like c^t where c is some constant. Following what we did in explaining why the power law for logarithms works, we can say $c^t = \exp(t \ln(c))$. Hence we can interpret $\left(\frac{V_B}{V_A}\right)^t$ properly as $\exp\left(t \ln\left(\frac{V_B}{V_A}\right)\right)$!

10.5 The Logarithm Function Properties

From our logarithm addition laws, we now know that For any positive integer p, we also see $\ln(e^p)$ is p. Thus, since e^p approaches ∞, we see that in the limit as p goes to infinity, we have $\lim_{x \to \infty} \ln(x) = -\infty$. Further, For any positive integer p, we also see $\ln(e^{-p})$ is $-p$. Thus, since e^{-p} approaches 0 in the limit as p goes to infinity, we infer $\lim_{x \to 0^+} \ln(x) = -\infty$. We all of these properties together for convenience.

Proposition 10.5.1 (Properties Of The Natural Logarithm)
The natural logarithm of the real number x satisfies

- $\ln$ *is a continuous function of x for positive x,*
- $\lim_{x \to \infty} \ln(x) = \infty$,
- $\lim_{x \to 0^+} \ln(x) = -\infty$,
- $\ln(1) = 0$,
- $\ln(e) = 1$,
- $(\ln(x))' = \frac{1}{x}$,
- *If x and y are positive numbers then* $\ln(xy) = \ln(x) + \ln(y)$.
- *If x and y are positive numbers then* $\ln\left(\frac{x}{y}\right) = \ln(x) - \ln(y)$.
- *If x is a positive number and y is any real number then* $\ln\left(x^y\right) = y \ \ln(x)$.

It thus easy to generate a table of values for the ln function to see that the graph of $\ln(x)$ has the appearance shown in Fig. 10.6.

We can also graph different versions of ln together on the same graph. Consider graphing $\ln(2x)$ and $\ln(x)$ on the same graph. The important thing is to find where

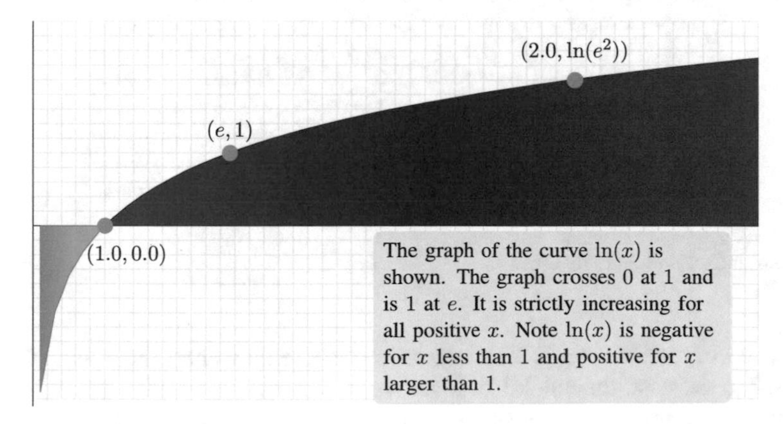

Fig. 10.6 The natural logarithm of x

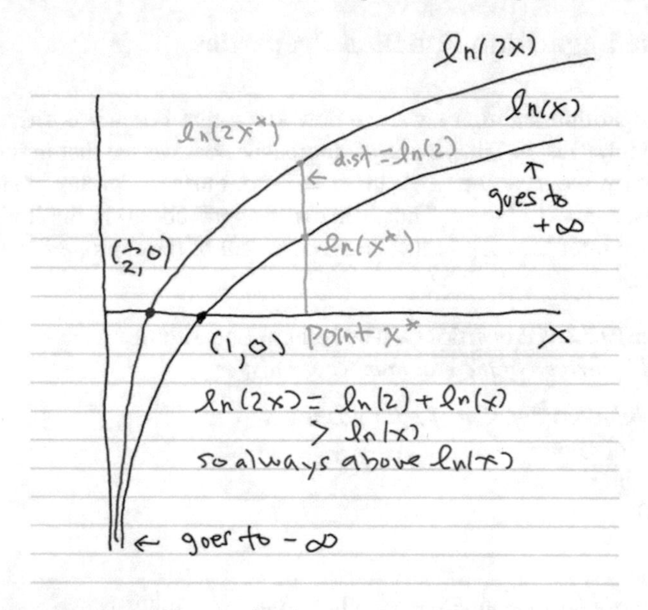

Fig. 10.7 ln(2x) and ln(3x) together

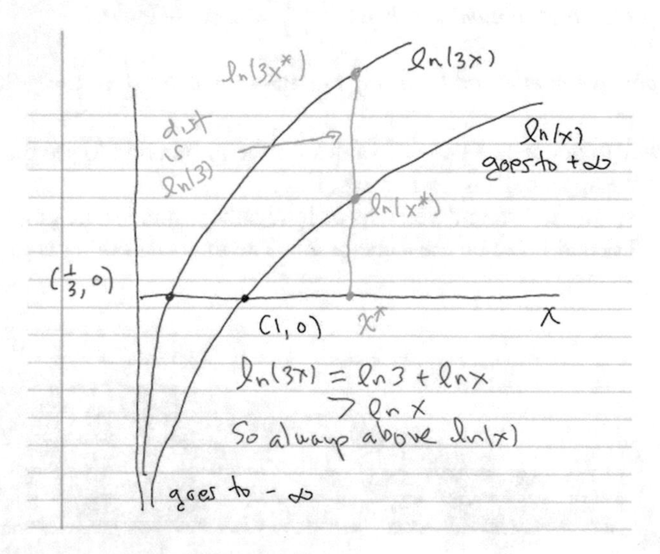

Fig. 10.8 ln(3x) and ln(x) together

these two ln functions are 0. That determines where they cross the x axis. Look at Fig. 10.7 to see what they look like together.

And to see another one, $\ln(3x)$ and $\ln(x)$ together, look at Fig. 10.8.

Now here is some homework.

10.5.1 Homework

Exercise 10.5.1 *Graph* $\ln(4x)$ *and* $\ln(3x)$ *on the same graph.*

Exercise 10.5.2 *Graph* $\ln(0.5x)$ *and* $\ln(1.5x)$ *on the same graph.*

Exercise 10.5.3 *Graph* $\ln(x)$ *and* $\ln(3x)$ *on the same graph.*

10.6 The Exponential Function Properties

We already know that the inverse of the natural logarithm function, $\exp(x) = (\ln)^{-1}(x)$, is defined by the equation

$$\exp(x) = y \Leftrightarrow \ln(y) = x.$$

So if we wanted to graph the inverse function, exp, we can just take the table of values we compute for ln and switch the roles of the x and y variables.

Listing 10.1: Natural Logarithm And Inverse Natural Logarithm Values

```
Natural Logarithm Values          Inverse Natural Logarithm Values
    x        y                          x          y
0.10000  -2.30259                  -2.30259    0.10000
0.30204  -1.19719                  -1.19719    0.30204
0.90816  -0.09633                  -0.09633    0.90816
1.11020  0.10454                   0.10454     1.11020
2.12041  0.75161                   0.75161     2.12041
...                                ...
6.16122  1.81828                   1.81828     6.16122
6.36327  1.85054                   1.85054     6.36327
...
7.77755  2.05124                   2.05124     7.77755
7.97959  2.07689                   2.07689     7.97959
8.38367  2.12629                   2.12629     8.38367
...
9.59592  2.26134                   2.26134     9.59592
10.00000  2.30259                  2.30259    10.0000
```

Thus the graph of $\exp(x)$ is shown in Fig. 10.9.

We note that we can easily graph the function $\exp(x) = \ln^{-1}(-x)$ as well. The definition of the inverse function gives

$$\begin{aligned} \exp(-x) = y &\Leftrightarrow \ln(y) = -x \\ &\Leftrightarrow -\ln(y) = x \\ &\Leftrightarrow \ln\left(\frac{1}{y}\right) = x. \end{aligned}$$

But this says

$$\exp(-x) = y \Leftrightarrow \frac{1}{y} = \exp(x).$$

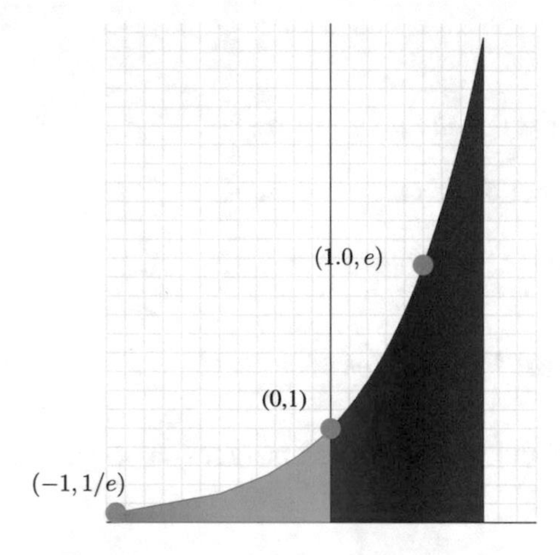

The graph of the curve $(\ln)^{-1}(x)$ is shown. The graph is asymptotic to $y = 0$ as x approaches $-\infty$ and is 1 at $x = 0$. It is strictly increasing for all x. Note $(\ln)^{-1}$ is less than 1 for negative x and larger than 1 for positive x.

Fig. 10.9 The inverse of the natural logarithm of x, $\exp(x)$

So we can see that $\exp(-x) = 1/\exp(x)$ always. So, to graph $\exp(-x)$, we take the values from the table of natural logarithm values, invert the y values keeping the x values the same, and then switch columns. This gives a table with values like this:

Listing 10.2: e^{-x} Values

```
      x        y
  -1.08974 2.97351
  -0.96154 2.61572
  -0.83333 2.30098
  -0.32051 1.37783
  ...
  -0.06410 1.06620
  0.06410 0.93791
  0.19231 0.82505
  ...
  0.83333 0.43460
  1.08974 0.33630
  1.21795 0.29584
  ...
  1.34615 0.26024
  1.60256 0.20138
  ...
  2.24359 0.10608
  2.50000 0.08208
```

We can then graph these values to see that $\exp(-x)$ looks like Fig. 10.10.

It is also traditional to call the inverse function $\exp(x)$ by the function name e^x. Note from our power rule for logarithms, for the number e and any power x, we have $\ln(e^x) = x\ln(e) = x$ as $\ln(e) = 1$. Thus, since exp is the inverse of ln, we have $\exp(\ln(e^x)) = e^x$. But since $\ln(e^x) = x$, this is the same as saying $\exp(x) = e^x$ for any x. Hence, we just call the inverse of the natural logarithm function e^x. This is read as "the exponential function of x" or "e to the x". Since the natural logarithm function has the properties listed in Proposition 10.5.1, we will also be able to derive interesting properties for e^x as well. From the graphs Fig. 10.9, we can see that

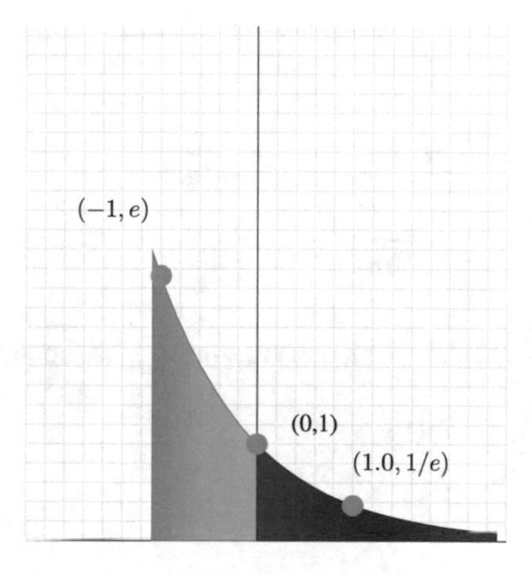

The graph of the curve $(\ln)^{-1}(-x)$ is shown. The graph is asymptotic to $y = 0$ as x approaches ∞ and is 1 at $x = 0$. It is strictly decreasing for all x. Note $(\ln)^{-1}$ is less than 1 for positive x and larger than 1 for x negative x.

Fig. 10.10 The inverse of the natural logarithm of $-x$, $\exp(-x)$

$$\lim_{x \to \infty} \exp(x) = \infty$$

It is also clear that as the argument x becomes more negative, the value of $\exp(x)$ approaches 0 from below; that is, $\exp(x)$ has a horizontal asymptote of 0.

It was easy to see that $(e^x)' = e^x$ using the properties of inverse functions. If you are skeptical of that approach, here is a traditional limit argument. It is a lot uglier!

If exp is differentiable at the point x, then the limit below must exist.

$$\lim_{h \to 0} \frac{\ln^{-1}(x+h) - \ln^{-1}(x)}{h}$$

By the definition of $\ln^{-1}(x + h)$, there is a number y_h so that $\ln(y_h) = x + h$. Similarly, there is a number y so that $\ln(y) = x$. Thus,

$$\lim_{h \to 0} \frac{\ln^{-1}(x+h) - \ln^{-1}(x)}{h} = \lim_{h \to 0} \frac{y_h - y}{\ln(y_h) - \ln(y)}$$

We also know that $\ln^{-1}$ is continuous at x. Thus,

$$\lim_{h \to 0} \ln^{-1}(x+h) = \ln^{-1}(x)$$

This can then be rewritten as

$$\lim_{h \to 0} y_h = y$$

Combining these pieces, we see

$$\lim_{h \to 0} \frac{\ln^{-1}(x+h) - \ln^{-1}(x)}{h} = \lim_{y_h \to y} \frac{y_h - y}{\ln(y_h) - \ln(y)}$$
$$= \frac{1}{\lim_{y_h \to y} \frac{\ln(y_h) - \ln(y)}{y_h - y}}$$

But we know the bottom limit since $\ln'(x)$ exists at each x and equals $\frac{1}{x}$. We conclude

$$\lim_{y_h \to y} \frac{\ln(y_h) - \ln(y)}{y_h - y} = \frac{1}{y}$$

Thus,

$$\lim_{h \to 0} \frac{\ln^{-1}(x+h) - \ln^{-1}(x)}{h} = \frac{1}{\lim_{y_h \to y} \frac{\ln(y_h) - \ln(y)}{y_h - y}}$$
$$= 1/(/1y) = y.$$

But we know $y = \exp(x)$! Hence, we have established that $\exp(x)$ has a derivative at each x and

$$(\exp)'(x) = \exp(x) \tag{10.1}$$

That was a lot harder I think than the original argument!

10.6.1 Properties of the Exponential Function

Finally, consider the number $\exp(x + y)$. Now let $\exp(x) = a$, $\exp(y) = b$ and $\exp(x + y) = c$. Then, by the definition of the inverse function $\ln^{-1} = \exp$, we must have

$$\exp(x) = a \Leftrightarrow x = \ln(a)$$
$$\exp(y) = b \Leftrightarrow y = \ln(b)$$
$$\exp(x + y) = c \Leftrightarrow x + y = \ln(c)$$

Thus, using the properties of the natural logarithm,

$$x + y = \ln(a) + \ln(b)$$
$$= \ln(a\, b)$$
$$= \ln(c)$$

We conclude the c and ab must be the same. Thus,

$$\exp(x + y) = c = ab = \exp(x)\ \exp(y)$$

In a similar way, letting $\exp(x - y) = d$

$$\begin{aligned} x - y &= \ln(a) - \ln(b) \\ &= \ln\left(\frac{a}{b}\right) \\ &= \ln(d) \end{aligned}$$

and so d and $\frac{a}{b}$ must be the same. Thus,

$$\exp(x - y) = d = \frac{a}{b} = \frac{\exp(x)}{\exp(y)}$$

From this it immediately follows that

$$\exp(-x) = d = \frac{1}{\exp(x)}$$

Finally, we need to look at $(\exp(x))^y$ for any power y. Let $\exp(x) = a$. Then, as before,

$$\exp(x) = a \Leftrightarrow x = \ln(a)$$

Thus, $x = \ln(a)$ implies

$$\begin{aligned} xy &= y \ln(a) \\ &= \ln(a^y) \end{aligned}$$

This immediately tells us that

$$\exp(xy) = a^y$$

Combining, we have

$$(\exp(x))^y = a^y = \exp(xy)$$

We can summarize what we have just done in Proposition 10.6.1.

Proposition 10.6.1 (Properties Of The Exponential Function)
The exponential function of the real number x, $\exp(x)$, *satisfies*

- $\exp$ *is a continuous function of x for all x,*
- $\lim_{x \to \infty} \exp(x) = \infty$,
- $\lim_{x \to -\infty} \exp(x) = 0$,
- $\exp(0) = 1$,
- $(\exp(x))' = \exp(x)$,
- *If x and y are real numbers then* $\exp(x + y) = \exp(x)\ \exp(y)$.
- *If x and y are real numbers then* $\exp(x - y) = \exp(x)\ \exp(-y)$ *or* $\exp(x - y) = \frac{\exp(x)}{\exp(y)}$.
- *If x and y are real numbers then* $\left(\exp(x)\right)^y = \exp(xy)$.

Before we go any further, you need to know that historically, many people got tired of writing exp all the time. They shortened the notation as follows:

$$\exp(u(x)) \equiv e^{u(x)}$$

Why is this? Well, we now have a rule that $\left(\exp(x)\right)^y = \exp(xy)$. So in particular, we have $\left(\exp(1)\right)^x = \exp(x)$. But $\exp(1) = e$ and so we know $\exp(x) = e^x$! So we can identify the two if it is convenient!! Indeed, sometimes it is easier to read a complicated exponential expression using the $e^{u(x)}$ notation and sometimes not. Judge for yourselves: the derivative problems above are now written in terms of the new notation.

1.
$$\frac{d}{dx}\left(e^{x^2+x+1}\right) = e^{x^2+x+1}\ (2x + 1)$$

2.
$$\frac{d}{dx}\left(e^{\tan(x)}\right) = e^{\tan(x)}\ \sec^2(x)$$

3.
$$\begin{aligned}\frac{d}{dx}\left(e^{e^{x^2}}\right) &= e^{e^{x^2}}\ e^{x^2}\ 2x \\ &= \text{This one is getting hard to read!}\end{aligned}$$

4.
$$\frac{d}{dx}\left(e^{-3x}\right) = e^{-3x}\ (-3)$$

5.

$$\frac{d}{dx}\left(e^{5x}\right) = e^{5x}\, 5$$

You can now see integration problems using this other notion also. Here are a few:

1.

$$\begin{aligned}\int e^{x^2+x+1}\,(2x+1) &= \int e^u\, du, \text{ use substitution } u = x^2+x+1\\ &= e^u + C\\ &= e^{x^2+x+1} + C\end{aligned}$$

2.

$$\begin{aligned}\int e^{\tan(x)}\,\sec^2(x)\,dx &= \int e^u\, du, \text{ use substitution } u = \tan(x);\, du{=}\sec^2(x)dx\\ &= e^u + C\\ &= e^{\tan(x)} + C\end{aligned}$$

3.

$$\begin{aligned}\int e^{x^2}\,3x\,dx &= \int e^u\, 3(du/2), \text{ use substitution } u = x^2;\, du = 2xdx\\ &= (3/2)\int e^u\, du\\ &= (3/2)\, e^u + C\\ &= (3/2)\, e^{x^2} + C\end{aligned}$$

4.

$$\begin{aligned}\int e^{-3x}\,dx &= \int e^u\, du/(-3), \text{ use substitution } u = -3x;\, du = -3dx\\ &= \frac{1}{-3}\int e^u\, du\\ &= -\frac{1}{3}\, e^u + C\\ &= -\frac{1}{3}\, e^{-3x} + C\end{aligned}$$

5.

$$
\begin{aligned}
\int e^{5x}\, dx &= \int e^u\, du/(5), \text{ use substitution } u = 5x;\, du = 5dx \\
&= \frac{1}{5} \int e^u\, du \\
&= \frac{1}{5}\, e^u + C \\
&= \frac{1}{5}\, e^{5x} + C
\end{aligned}
$$

Let's graph the natural logarithm function, $\ln(x)$ and its inverse $exp(x) = e^x$ on the same graph as we show in Fig. 10.11 so we can put all this together.

We can also graph different versions of the exponential function side by side. The important thing here is that all of them go through the point (0, 1). Let's graph e^{-2t} and e^{-t} on the same graph. We see this in Fig. 10.12.

We can also show two growth curves simultaneously as in Fig. 10.13.

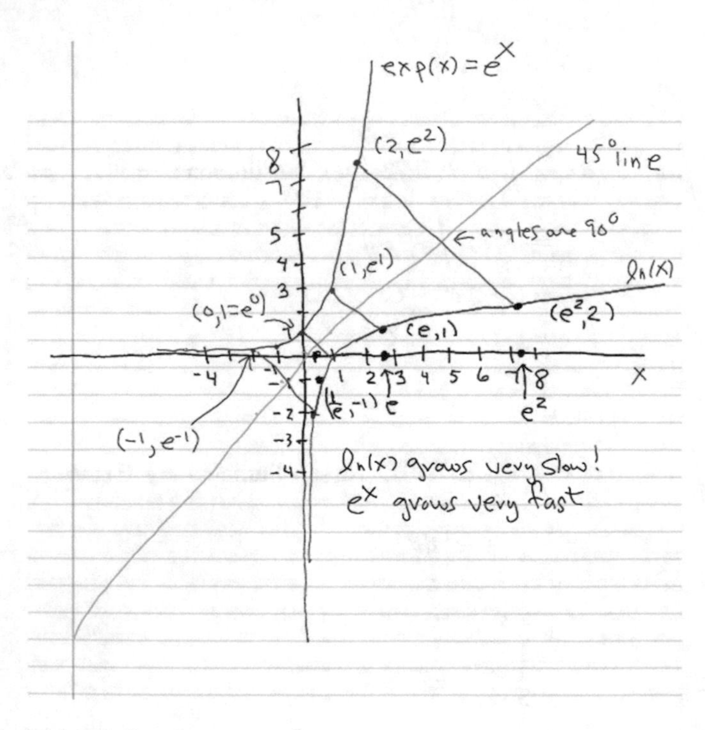

Fig. 10.11 $\ln(x)$ and e^x on the same graph

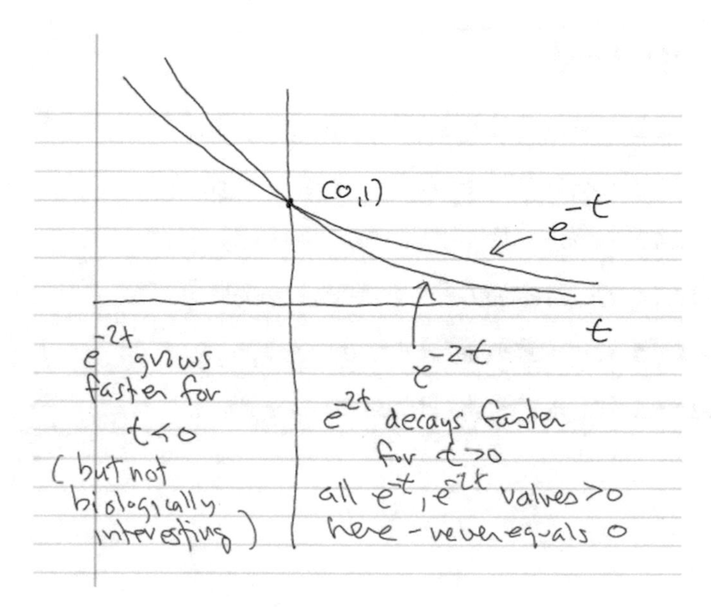

Fig. 10.12 e^{-2t} and e^{-t} on the same graph

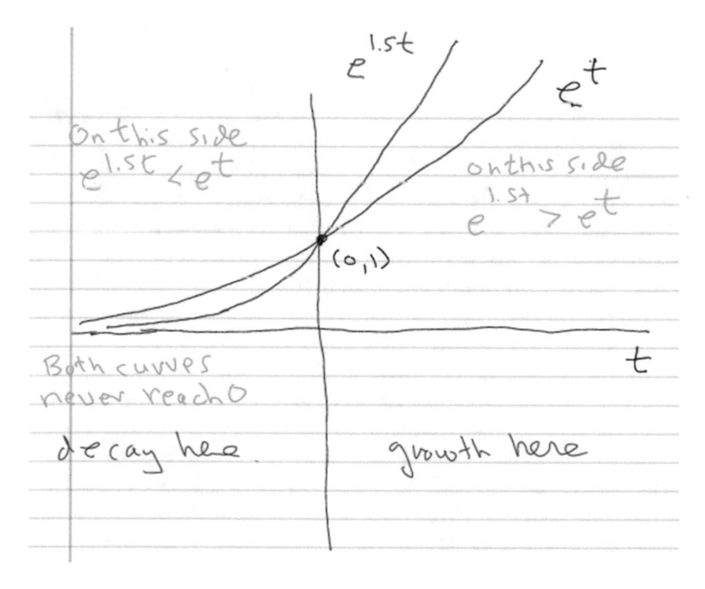

Fig. 10.13 e^{t} and $e^{1.5t}$ on the same graph

10.6.2 Homework

Exercise 10.6.1 *Graph* e^{-t} *and* $e^{-1.4t}$ *on the same graph.*

Exercise 10.6.2 *Graph* $e^{-2.3t}$ *and* $e^{-1.5t}$ *on the same graph.*

Exercise 10.6.3 *Graph* $e^{-0.8t}$ *and* $e^{-1.2t}$ *on the same graph.*

Exercise 10.6.4 *Graph* e^{t} *and* $e^{1.4t}$ *on the same graph.*

Exercise 10.6.5 *Graph* $e^{2.3t}$ *and* $e^{1.5t}$ *on the same graph.*

Exercise 10.6.6 *Graph* $e^{0.8t}$ *and* $e^{1.2t}$ *on the same graph.*

Chapter 11
Simple Rate Equations

We can take the ideas from Chaps. 9 and 10 and use them to explore what are called *simple rate problems*. A key concept in biological modeling called *exponential growth*. Suppose you had a biological process you were monitoring which depended on a variable x whose growth satisfied the following property: the rate at which this variable grew was proportional to its current level. Mathematically, we could express this by the law.

$$\frac{dx}{dt} \propto x(t) \tag{11.1}$$

where the symbol $\frac{dx}{dt}$ denotes the usual derivative of the variable x with respect to t. There is a lot going on here behind the scenes. First, we are assuming the variable of interest to us, the x, can take on any real number value. For example, if x was the population of a bacteria at a given time t, we know this population is always some integer value as we have no way to interpret a *fractional* or *irrational* value for population size. However, when we are counting the bacterial population in a petri dish, it is very convenient to think of the population as a real valued function since the number of cells we see is so large. Thus, the simple statement that $x(t)$ represents the population of the bacteria at time t represents a first choice of *model*. If $x(t)$ can only be an integer, the ratios we use to define the derivative of x with respect to t do not have well-defined limits at all values of time because x is a function that *jumps* to a new integer value whenever a bacteria in our population divides. On the other hand, if we assume $x(t)$ is real-valued and hence can take on population values between the integers, we are made another choice of *model*. The first model can be thought of as a *discrete* model as the bacterial population can only take on integer values and the second model is a *continuous* model because we assume values between integers are permitted. This discussion is similar to the one we had in Chap. 3 on how the size of the generation is built into our biological models and we make the explicit choice to move towards an abstraction that allows the time to be any number. This abstraction allows us to apply the rules we have been learning, the stuff we call **calculus**! So if we assume a continuous model for the population x, we see there are no jumps in

J.K. Peterson, *Calculus for Cognitive Scientists*, Cognitive Science and Technology, DOI 10.1007/978-981-287-874-8_11

the population size. If we further assume that the population increases smoothly, this allows us to build a model which is *differentiable*. We are therefore now assuming that the limit.

$$\lim_{\delta \to 0} \frac{x(t+\delta) - x(t)}{\delta}$$

always exists as a finite number. So we see anytime we try to take a quantity which can physically be measured and model it as a mathematical variable, we will inevitably be forced to make some assumptions about its nature. Now let's go back to the *exponential growth law* given by Eq. 11.1. For example, suppose we have collected the data given in

Listing 11.1: Collected Data

```
        x                y
  0.00000          0.05000
  0.16667          0.05907
  0.33333          0.06978
  0.50000          0.08244
  0.66667          0.09739
  0.83333          0.11505
  1.00000          0.13591
  1.16667          0.16056
  1.33333          0.18968
  1.50000          0.22408
  1.66667          0.26472
  1.83333          0.31274
  2.00000          0.36945
  2.16667          0.43646
  2.33333          0.51561
  2.50000          0.60912
  2.66667          0.71960
  2.83333          0.85010
  3.00000          1.00428
  3.16667          1.18641
  3.33333          1.40158
  3.50000          1.65577
  3.66667          1.95606
  3.83333          2.31082
  4.00000          2.72991
```

We can plot these points and as seen in Fig. 11.1.

Now consider a power function such as 2^t. We can generate a nice table of values for integer times easily.

Listing 11.2: Power Function

```
        x              y
   0.10000      10.00000
   2.08000        0.48077
   4.06000        0.24631
   6.04000        0.16556
   8.02000        0.12469
  10.00000       0.10000
```

which has corresponding graph shown in Fig. 11.2.

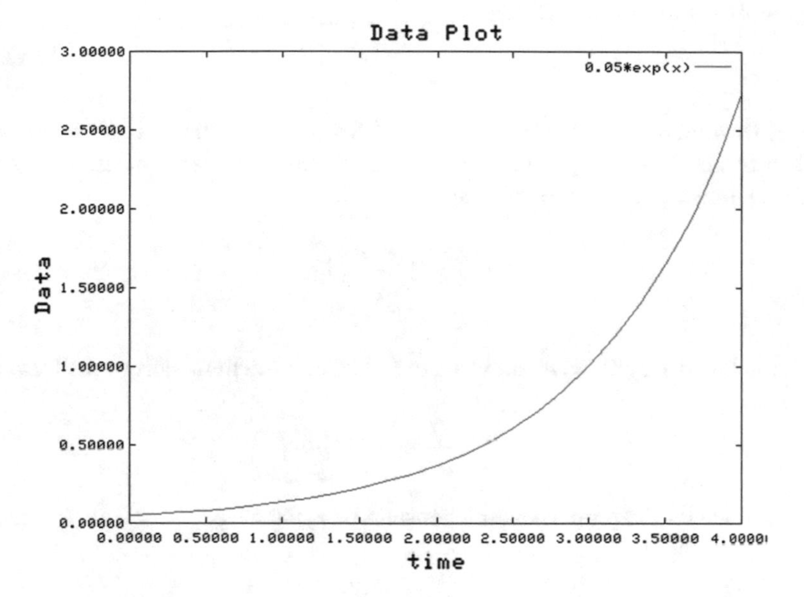

Fig. 11.1 Exponential data

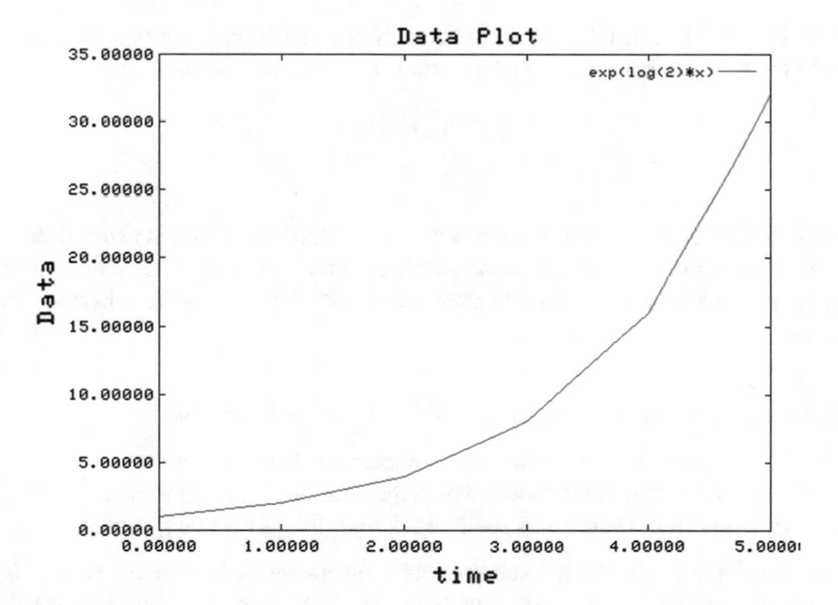

Fig. 11.2 Power data

If we calculated the derivative of the power function $y(t) = 2^t = e^{t\ln(2)}$, we would find that $y'(t) \propto y(t)$ also. We can draw similar graphs for any power function c^t for a positive constant c. All would satisfy the exponential growth law Eq. 11.1. From our earlier work in Chap. 10, we know that for any real number r.

$$\frac{de^{rt}}{dt} = re^{rt} \tag{11.2}$$

Hence, the functions of the form $x(t) = e^{rt}$ for positive time t are all appropriate models of this kind of growth. This type of function can also model things like *doubling growth*, i.e., $x(t) = 2^t$ because

$$\begin{aligned} e^{\ln(2)t} &= \left(e^{\ln(2)}\right)^t \\ &= 2^t \end{aligned}$$

Thus, the functions $x(t) = e^{rt}$ allow us to model all growth models of the form

$$\frac{dx}{dt} \propto x(t)$$

because if we let the proportionality constant be r, we have

$$\frac{dx}{dt} = rx(t)$$

Since the variable x usually starts at some initial value, a typical modeling problem would be to find a function x which satisfies these two conditions:

$$\begin{aligned} x'(t) &= rx(t) \\ x(0) &= x_0 \end{aligned}$$

Also note that we often use the *prime* notation for derivatives as it is really simple to write. So, with that said, let's do some problems. So in summary, exponential growth and decay models look like this. Mathematically, if $x(t)$ denotes the substance level at time t, then

- $x'(t) \propto x(t)$.
- Let the proportionality constant be r. Then there are three cases:
  - $x'(t) = rx(t)$, where r is a positive number: this is the growth model.
  - $x'(t) = 0x(t)$: this is the model where the substance stays constant.
  - $x'(t) = -rx(t)$, where r is a positive number; this is the decay model.

Let's do an example. What functions x solve the model $x'(t) = -3x(t)$ with $x(0) = 5$? Let's do this step by step. First, note the right hand side of the x' equation is called the **dynamics**, the condition $x(0) = 5$ is called the **Initial Condition** and together, the two are called an **Initial Value Problem** or **IVP**. Now for our **guessed** solution.

- We know that if $x(t) = Ae^{-3t}$, then $x'(t) = 3Ae^{-3t}$. This can be reorganized as $x'(t) = -3x(t)$ using the definition of $x(t)$.
- Hence, for any A at all, $x(t) = Ae^{-3t}$ is a solution to the **dynamics** $x' = -3x$ where for convenience we have stopped explicitly writing the (t).

- But we also have the **IC** $x(0) = 5$. So $5 = x(0) = Ae^{5\times 0} = A$.
- We conclude there is only one function which solves the **IVP**: $x(t) = 5e^{-3t}$.

Here are some more examples of guessing.
More examples of **guessing**.

Example 11.0.1 Solve $x' = 2.5x, x(0) = 10$

Solution *We can guess that the general solution is* $x(t) = Ae^{2.5t}$. *Since the IC is* $x(0) = 10$, *we see* $10 = x(0) = A$. *So the solution is* $x(t) = 10e^{2.5t}$.

Example 11.0.2 Solve $x' = -0.75x, x(0) = 9$

Solution *We can guess that the general solution is* $x(t) = Ae^{-0.75t}$. *Since the IC is* $x(0) = 9$, *we see* $9 = x(0) = A$. *So the solution is* $x(t) = 9e^{-0.75t}$.

However, we can't guess solutions to many of our models, so we need to learn how to use the ideas from Calculus we have introduced to solve these models. We will go over three methods: the **Antiderivative Method**, the **Riemann Integral Method** and **The Integrating Factor Method**.

11.1 Solving a Simple Rate Problem: Indefinite Approach

We are now going to solve a rate problem by integration. We also call this the **Antiderivative Method**. We are going to do this for an arbitrary value of r so that you can see how doing it for an unspecified value of r has its advantages. The other way is to solve the problem for a few specific values of r, like $r = 2$ or $r = -3$ and then just jump to the general way to doing it. We are great at building patterns in our heads and most of us can jump from what we do for $r = 2$ and $r = -3$ do how to handle a general value of r. But you should also be trained in developing the answer in general. So we will start with the general r first and do specific examples later.

So to solve the problem

$$x'(t) = rx(t) \tag{11.3}$$
$$x(0) = x_0 \tag{11.4}$$

start with the condition

$$x'(t) = rx(t)$$

This can be rewritten as

$$\frac{x'(t)}{x(t)} = rdt$$

Let's make the substitution $u = x(t)$ as we have done before. Then $du = x'(t)dt$ and so if we integrate both sides, we find

$$\int \frac{x'(t)}{x(t)} dt = \int r dt, \text{ becomes}$$
$$\int \frac{du}{u} = rt + C$$
$$\ln(\mid u \mid) = rt + C$$

Since this is an equality, we then must have that

$$\exp\left(\ln(\mid u \mid)\right) = \exp\left(rt + C\right)$$

Now, we have properties of the exponential and logarithm functions which allow us to simplify both sides; we have

$$\exp\left(\ln(\mid u \mid)\right) = \mid u \mid$$
$$\exp\left(rt + C\right) = \exp(rt)\exp(C)$$

Thus, we obtain

$$\mid u \mid = e^{rt} e^{C}$$

But we can now substitute for u to get

$$\mid x(t) \mid = e^{C} e^{rt}$$

Finally, we know

$$x(0) = x_0$$

Substituting, we see

$$\mid x(0) \mid = e^{C} e^{r0}$$
$$= e^{C}$$

Combining all of this, we see the function x that satisfies both Eqs. 11.3 and 11.4 is

$$\mid x(t) \mid = \mid x_0 \mid e^{rt} \tag{11.5}$$

11.1.1 Resolution of the Absolute Value

We can get then rid of the absolute value as follows. We will do this as a general argument which we can then just apply to our specific problems to save time. So pay close attention to how we do this. If there was a time when the solution changed algebraic sign, then there would be a time, t^*, so that $x(t^*) = 0$. This gives

$$0 = \mid x(t^*) \mid = \mid x_0 \mid e^{rt^*}.$$

But, we know that the exponential term e^{rt^*} is always positive. Thus, the equation above tells us 0 equals a nonzero number $\mid x_0 \mid e^{rt^*}$. This is impossible and so our assumption that the solution changes sign must be wrong. We know then that the solution is always positive or always negative. So the algebraic sign of the solution is completely determined by the algebraic sign of x_0. If x_0 is negative, we have the solution starts negative and so $x(t)$ is always negative. This gives

$$\begin{aligned} \mid x(t) \mid &= -x(t) \\ &= \mid x_0 \mid e^{rt} \\ &= -x_0 e^{rt} \end{aligned}$$

This implies $x(t) = x_0 e^{rt}$. In the case where x_0 is positive or zero, we see $x(t)$ is also always non-negative. Thus, in this case,

$$\begin{aligned} \mid x(t) \mid &= x(t) \\ &= \mid x_0 \mid e^{rt} \\ &= x_0 e^{rt} \end{aligned}$$

We can put these two cases together as

$$x(t) = x_0 e^{rt} \tag{11.6}$$

11.1.2 Examples

Example 11.1.1 Solve the following problem:

$$\begin{aligned} \frac{dx}{dt} &= 3x(t) \\ x(0) &= 20 \end{aligned}$$

Solution *We rewrite the problem and solve as follows:*

$$\int \frac{\frac{dx}{dt}}{x(t)} dt = \int 3dt$$
$$\ln(| x(t) |) = 3t + C$$

Since this is an equality, we then must have

$$e^{\ln(|x(t)|)} = e^{3t+C} = e^C e^{3t}$$

or letting $A = e^C$,

$$| x(t) | = Ae^{3t}$$

Thus, using the initial condition, we have $A = |x(0)| = 20$ *and so*

$$| x(t) | = 20e^{3t}$$

To resolve the absolute value, we note if there was a time when the solution changed algebraic sign, then there would be a time, t^*, *so that* $x(t^*) = 0$. *This gives*

$$0 = | x(t^*) | = 20e^{3t^*}.$$

But, we know that the exponential term e^{3t^*} *is always positive. Thus, the equation above tells us* 0 *equals a nonzero number* $20e^{3t^*}$. *This is impossible and so our assumption that the solution changes sign must be wrong. We know then that the solution is always positive or always negative. Since* $x(0) = 20$ *initially, we conclude* x *must be always positive. So* $|x(t)| = x(t)$ *and we conclude the solution is*

$$x(t) = 20e^{3t}$$

Example 11.1.2

$$\frac{d}{dt}(x) = -4x(t)$$
$$x(0) = -30$$

Solution *We rewrite the problem and solve as follows:*

$$\int \frac{\frac{dx}{dt}}{x(t)} dt = \int -4dt$$
$$\ln(| x(t) |) = -4t + C$$

We then have

$$e^{\ln(|x(t)|)} = e^{-4t+C} = e^C e^{-4t}$$

or letting $A = e^C$,

$$| x(t) | = Ae^{-4t}$$

The initial condition is $x(0) = -30$, *so*

$$| x(t) | = | -30 | e^{-4t} = 30e^{-4t}$$

To resolve the absolute value, we note if there was a time when the solution changed algebraic sign, then there would be a time, t^*, *so that* $x(t^*) = 0$. *This gives*

$$0 =| x(t^*) | = 30e^{-4t^*}.$$

But, we know that the exponential term e^{-4t^*} *is always positive. Thus, the equation above tells us* 0 *equals a nonzero number* $30e^{-4t^*}$. *This is impossible and so our assumption that the solution changes sign must be wrong. We know then that the solution is always positive or always negative. Since* $x(0) = -30$ *initially, we conclude* x *must be always negative. Thus,* $|x(t)| = -x(t)$ *and conclude the solution is*

$$x(t) = -30e^{-4t}$$

11.1.3 Homework

Solve these problems using the indefinite method.

Exercise 11.1.1

$$x'(t) = 1.8x(t)$$
$$x(0) = 12.4$$

Exercise 11.1.2

$$x'(t) = -2.5x(t)$$
$$x(0) = 250$$

Exercise 11.1.3

$$\begin{aligned} x'(t) &= 0.6x(t) \\ x(0) &= 20.2 \end{aligned}$$

Exercise 11.1.4

$$\begin{aligned} x'(t) &= -8x(t) \\ x(0) &= 200 \end{aligned}$$

Exercise 11.1.5

$$\begin{aligned} x'(t) &= 1.9x(t) \\ x(0) &= -2000 \end{aligned}$$

11.2 Solving a Simple Rate Problem: Definite Approach

We also call this the **Riemann Integral Method**. We again consider the problem

$$\begin{aligned} x'(t) &= rx(t) \\ x(0) &= x_0 \end{aligned}$$

Let's integrate both sides from the initial time 0 to a new time t. We also change the variable of integration to s to avoid confusion with our upper limit t. We find

$$\int_0^t \frac{\frac{dx}{ds}}{x(s)} ds = \int_0^t r ds,$$

becomes

$$\begin{aligned} \ln(|\ x(t)\ |) - \ln(|\ x(0)\ |) &= rt \\ \ln\left(\left|\frac{x(t)}{x(0)}\right|\right) &= rt \end{aligned}$$

We assume at this point that x_0 is not zero as otherwise the ln is not defined. Since this is an equality, we then must have

$$exp\left(\ln\left(\left|\frac{x(t)}{x(0)}\right|\right)\right) = e^{rt}$$

Now, we have properties of the exponential and logarithm functions which allow us to simplify; we obtain

$$\exp\left(\ln\left(\left|\frac{x(t)}{x(0)}\right|\right)\right) = \left|\frac{x(t)}{x(0)}\right|$$

Thus, we obtain

$$\left|\frac{x(t)}{x(0)}\right| = e^{rt}$$

Combining all of this, we see the function x that satisfies both Eqs. 11.3 and 11.4 is

$$| \, x(t) \, | = | \, x_0 \, | \, e^{rt}$$

We then apply the same algebraic sign analysis we did in the *indefinite case* to see the solution to the growth problem is given by Eq. 11.6.

11.2.1 Examples

Example 11.2.1

$$\begin{aligned} x'(t) &= 0.3x(t) \\ x(0) &= 1000 \end{aligned}$$

Solution *We rewrite the problem and solve as follows:*

$$\begin{aligned} \int_0^t \frac{\frac{dx}{ds}}{x(s)} ds &= \int_0^t 0.3 ds \\ \ln(| \, x(t) \, |) - \ln(| \, x(0) \, |) &= 0.3t \\ \ln\left(\left|\frac{x(t)}{x(0)}\right|\right) &= 0.3t \end{aligned}$$

Since this is an equality, we then must have

$$e^{\ln(|\frac{x(t)}{x(0)}|)} = e^{0.3t}$$

or

$$\left|\frac{x(t)}{x(0)}\right| = e^{0.3t}$$

Thus,

$$| x(t) | = | x(0) | e^{0.3t}$$

Now the initial condition is $x(0) = 1000$*; thus*

$$|x(t)| = 1000e^{0.3t}$$

To resolve the absolute value, we note if there was a time when the solution changed algebraic sign, then there would be a time, t^**, so that* $x(t^*) = 0$*. This gives*

$$0 =| x(t^*) | = 1000e^{0.3t^*}.$$

But, we know that the exponential term $e^{0.3t^*}$ *is always positive. Thus, the equation above tells us* 0 *equals a nonzero number* $1000e^{-4t^*}$*. This is impossible and so our assumption that the solution changes sign must be wrong. We know then that the solution is always positive or always negative. Since the solution starts positive, we have the solution is always positive. We conclude* $|x(t)| = x(t)$ *and*

$$x(t) = 1000e^{0.3t}$$

Example 11.2.2

$$\begin{aligned} x'(t) &= -0.5x(t) \\ x(1) &= -200 \end{aligned}$$

Solution *We rewrite the problem and solve as follows:*

$$\begin{aligned} \int_1^t \frac{\frac{dx}{ds}}{x(s)} &= \int_1^t -0.5ds \\ \ln(| x(t) |) - \ln(| x(1) |) &= -0.5(t-1) \\ \ln\left(\left|\frac{x(t)}{x(1)}\right|\right) &= -0.5(t-1) \end{aligned}$$

Since this is an equality, we then must have

$$e^{\ln(|\frac{x(t)}{x(1)}|)} = e^{-0.5(t-1)}$$

or

$$\left|\frac{x(t)}{x(1)}\right| = e^{-0.5(t-1)}$$

Thus,

$$| x(t) | = | x(1) | e^{-0.5(t-1)}$$

Now the initial condition is $x(1) = -200$ *and so*

$$| x(t) | = -200 | e^{-0.5(t-1)} = 200e^{-0.5(t-1)}.$$

To resolve the absolute value, we note if there was a time after 1 *when the solution changed algebraic sign, then there would be a time,* t^**, so that* $x(t^*) = 0$*. This gives*

$$0 = | x(t^*) | = 200e^{-0.5(t^*-1)}.$$

But, we know that the exponential term $e^{-0.5(t^*-1)}$ *is always positive because* $t^* \geq 1$*. Thus, the equation above tells us* 0 *equals a nonzero number. This is impossible and so our assumption that the solution changes sign must be wrong. We know then that the solution is always positive or always negative. Since the solution starts negative, we have the solution is always negative. We conclude* $|x(t)| = -x(t)$ *and*

$$x(t) = -200e^{-0.5(t^*-1)}$$

11.2.2 Homework

Solve these problems using the definite method.

Exercise 11.2.1

$$x'(t) = 2.8x(t)$$
$$x(0) = 6$$

Exercise 11.2.2

$$x'(t) = -4x(t)$$
$$x(0) = 25$$

Exercise 11.2.3

$$x'(t) = 0.007x(t)$$
$$x(1) = 2000$$

Exercise 11.2.4

$$x'(t) = -1.5x(t)$$
$$x(1) = 10$$

Exercise 11.2.5

$$x'(t) = 2.6x(t)$$
$$x(0) = -10$$

11.3 The Half Life in Exponential Decay Problems

Let's go back and examine the exponential growth problem from a new perspective.

$$x'(t) = -rx(t)$$
$$x(0) = x_0$$

The constant r is assumed to be positive and since there is a minus sign in front of it, we call this an problem of *exponential decay*. Since we start with an initial value of x_0, an obvious question is how long will it take for this initial amount to reduce by 50%. We now know the solution to this problem is

$$x(t) = x_0 e^{-rt}$$

Let $t_{\frac{1}{2}}$ denote the time at which x has decayed to one half of x_0. Then, we must solve

$$\frac{1}{2}x_0 = x_0 e^{-rt_{\frac{1}{2}}}$$

Solving, we have

$$\begin{aligned} \frac{1}{2} &= e^{-rt_{\frac{1}{2}}} \\ \ln\left(\frac{1}{2}\right) &= -rt_{\frac{1}{2}} \\ -\ln(2) &= -rt_{\frac{1}{2}} \\ t_{\frac{1}{2}} &= \frac{\ln(2)}{r} \end{aligned}$$

This value of time is called the *half life* of the substance x which is undergoing exponential decay. Many substances exist in radioactive forms and we are pretty sure how long it will take for some initial amount to decay to half the original size.

Thus, if we are told the half-life of a substance is 10,000, we can calculate the decay constant r as follows:

$$10^4 = \frac{\ln(2)}{r}$$
$$r = \frac{\ln(2)}{10^4}$$

Note for an exponential growth problem, we can do a similar analysis and arrive at the doubling time, $t_d = \ln(2)/r$. The argument is virtually the same. We start with an exponential growth model, $x_0 e^{rt}$ for some positive r. Then we ask how long does it take for the model to double? Thus, we want

$$2x_0 = x_0 e^{rt_d}$$

Solving, we have

$$2 = e^{rt_d}$$
$$\ln(2) = rt_d$$
$$\ln(2)/r = t_d$$

To make this a bit clearer, let's do some examples.

Example 11.3.1 Consider the function $x(t) = 100e^{-2t}$. How long does it take for this function to decay to the value 50? Since it has dropped to 50 % of its original value, this is the **half-life** of the function as we have discussed earlier.

- Let $t_{\frac{1}{2}}$ denote the time at which x has decayed to 50. Then, we want $50 = 100e^{-2t_{\frac{1}{2}}}$
- Solving, we have

$$\frac{1}{2} = e^{-2t_{\frac{1}{2}}}$$
$$\ln\left(\frac{1}{2}\right) = -2t_{\frac{1}{2}}$$
$$-\ln(2) = -2t_{\frac{1}{2}}$$
$$t_{\frac{1}{2}} = \frac{\ln(2)}{2}$$

Example 11.3.2 Consider the function $x(t) = 400e^{-5.4t}$. How long does it take for this function to decay to the value 200?

- Then, we want $200 = 400e^{-5.4t_{\frac{1}{2}}}$
- Solving, we have

$$\begin{aligned}\frac{1}{2} &= e^{-5.4t_{\frac{1}{2}}}\\ \ln\left(\frac{1}{2}\right) &= -5.4t_{\frac{1}{2}}\\ -\ln(2) &= -5.4t_{\frac{1}{2}}\\ t_{\frac{1}{2}} &= \frac{\ln(2)}{5.4}\end{aligned}$$

Consider the function $x(t) = 400e^{-5.4t}$ again. How long does it take for this function to decay to the value 100? Call this time $t_{\frac{1}{4}}$.

- Then, we want $100 = 400e^{-5.4t_{\frac{1}{4}}}$
- Solving, we have

$$\begin{aligned}\frac{1}{4} &= e^{-5.4t_{\frac{1}{4}}}\\ \ln\left(\frac{1}{4}\right) &= -5.4t_{\frac{1}{4}}\\ -2\ln(2) &= -5.4t_{\frac{1}{4}}\\ t_{\frac{1}{2}} &= 2\frac{\ln(2)}{5.4} = 2t_{\frac{1}{2}}\end{aligned}$$

From these sample calculations, we see for $x(t) = Ae^{-rt}$ for a positive r, we have

- At $t_{\frac{1}{2}}$ $\frac{\ln(2)}{r}$, x has decayed to 50 % of its original value.
- At $2t_{\frac{1}{2}}$ $\frac{\ln(2)}{r}$, x has decayed to 25 % of its original value.
- At $3t_{\frac{1}{2}}$ $\frac{\ln(2)}{r}$, x has decayed to 12.5 % of its original value.

and in a similar fashion, we can conclude for exponential growth, we have a similar concept: the doubling time, t_d. If $x(t) = Ae^{rt}$ for a positive r, we have

- At t_d, x has increased by a factor of to 2 times its original value.
- At $2t_d$, x has increased by a factor of to 4 times its original value.
- At $3t_d$, x has increased to 8 times its original value.

Here are two sample graphs where we use these ideas to graph two exponential functions carefully. We show $e^{-0.5t}$ and e^{-t} on the same graph using half lives as our time units in Fig. 11.3.

We graph e^t and e^{2t} on the same graph using doubling times as our time units. We show this in Fig. 11.4.

Finally, let's look at e^t and $e^{1.5t}$ as shown in Fig. 11.5.

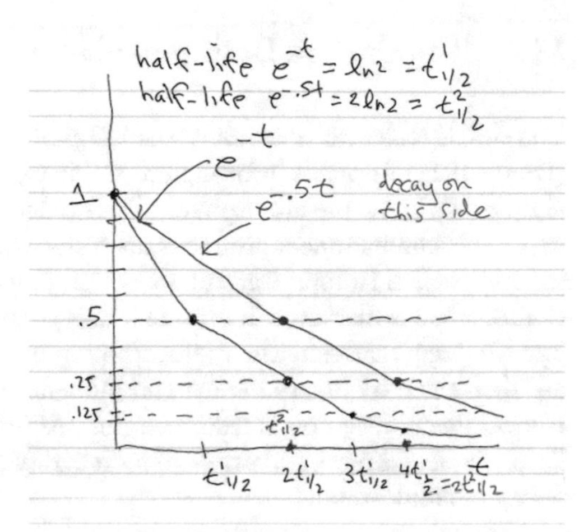

Fig. 11.3 Graph of $e^{-0.5t}$ and e^{-t}

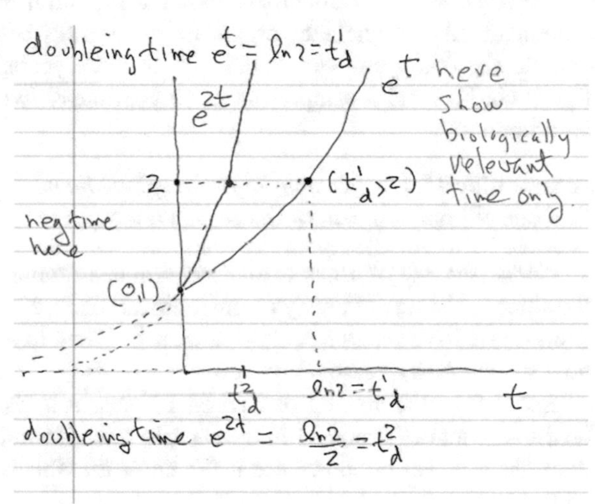

Fig. 11.4 Graph of e^t and e^{2t}

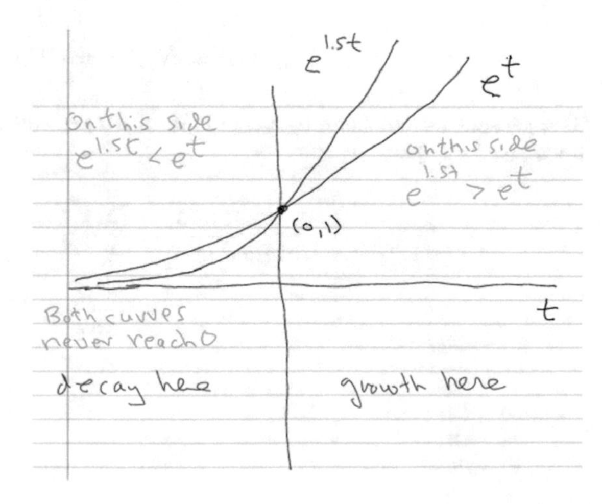

Fig. 11.5 Graph of e^t and $e^{1.5t}$

11.4 The Carbon Dating Problem

You should look at the marvelous book *Differential Equations and Their Applications* (Braun 1978) for more information of this topic. We will keep our discussion very simple. We know the atmosphere of our planet is constantly bombarded by cosmic rays. This bombardment produces neutrons which collide with Nitrogen to produce the radioactive isotope Carbon 14 which is usually called *radiocarbon* for short. Radiocarbon is then used in normal living processes and ends up being incorporated into all living material. The radiocarbon in the atmosphere is used by plants. Plants are then eaten by other animals and humans eat both plants and possibly animals. Hence, the radiocarbon is also incorporated into human material such as bones. In living tissue, the rate of *ingestion* of radiocarbon is exactly balanced by the radioactive decay of radiocarbon.

However, when an organism dies, the ingestion of radiocarbon stops and the amount of radiocarbon in the organism immediately begins to decay via the standard exponential decay law with a half life of 5,568 years. In order to use this idea to figure out how old a sample based on previously living tissue is, we make some assumptions:

Assumption 11.4.1 (*Cosmic ray atmosphere bombardment is a constant*)
The rate of bombardment of the earth's atmosphere by cosmic rays is a constant.

Hence, the rate of decay of radiocarbon in a sample measure today is considered the same as the rate of decay of the radiocarbon in the sample when it died. So let's suppose we have found a sample of charred wood in an archaeological dig. Here is how we can date it reasonably well.

Let $N(t)$ be the amount of Carbon 14 present in the sample at time t and let N_0 be the amount of Carbon 14 in the sample when it was created (i.e. when the animal died, the tree was cut down etc.). We know that this is standard exponential decay and so

$$N(t) = N_0 e^{-\lambda t} = e^{-\frac{\ln(2)}{5568}t}$$

The present rate of disintegration is then $R(t) = \lambda N(t)$ with the original rate given by $R_0 = \lambda N_0$. Thus,

$$\begin{aligned}\frac{\text{present rate}}{\text{original rate}} &= \frac{R(t)}{R_0} \\ &= \frac{\lambda N(t)}{\lambda N_0} \\ &= \frac{\lambda N_0 e^{-\lambda t}}{\lambda N_0} = e^{-\lambda t}\end{aligned}$$

This gives us

$$t = \frac{1}{\lambda} \ln \left(\frac{R_0}{R(t)} \right)$$

We measure $R(t)$ from the sample to find its current disintegration rate. Then by Assumption 11.4.1, we know the disintegration rate of the sample when it was created is the same as the disintegration in a sample of living tissue from anywhere on our earth.

11.4.1 A Simple Dating Problem

Here is a sample dating problem. Charcoal from the occupation level of the cave in France where the Cro-Magnon cave paintings have been discovered gave an average count in 1950 of 0.97 dpm/g. Living wood gave 6.68 dpm/g. What is the probable age of the sample?

Solution

$$\begin{aligned} t &= \frac{1}{\lambda} \ln \left(\frac{R_0}{R(t)} \right) \\ &= \frac{5568}{\ln(2)} \ln \left(\frac{6.68}{0.97} \right) \\ &= 15500 \end{aligned}$$

Thus, the sample was created about 13,550 BP (Before Present).

11.4.2 Homework

Exercise 11.4.1 *In a* 1950 *excavation of the Babylonian city of Nippur, charcoal from a roof beam gave a disintegration count of* 0.97 *dpm/g. Living wood gives* 6.68 *dpm/g. What is the probable age of the sample?*

Exercise 11.4.2 *Let's assume a way has been found to explore out galaxy. You are part of an archaeological team that is excavating ruins on a planet designated PX-3905 or* Bib-Shimur. *What do you need to know to do radiocarbon dating tests on organic samples from the excavations?*

Exercise 11.4.3 *Since a number of countries did nuclear weapon testing in the 1950s and 1960s, do you think these tests could have an impact on the radiocarbon dating algorithm? Think about this before you read the next section!*

11.4.3 Further Caveats

Knowing with a good degree of accuracy the actual ages of the many artifacts that have been discovered over the years in archaeological site investigations is very important. As mentioned above, an object must contain carbon, which means it must have been part of something that once was alive. A good discussion of this background material and how helpful it has been in teasing out how modern humans lived throughout the Mediterranean basin after the last ice age can be found in Mithen (2004). So read that book if you can! Mithen mentions a number of important caveats.

- Radiocarbon dates are never exact. Instead it is best to think of them as an average value with some standard deviation. So the sample exercise we did above which gave an age of 13,550 BP is misleading. If in the exercise we had been more careful with error analysis, we might have found the age to be 13,550 $\pm$ 300 years. Also, archaeologists use the term BP always using as the *present* the year 1950 even though that is over 50 years in the past now!
- Radiocarbon years are not the same length as calendar years and the length of a radiocarbon year can change over time. Thus, the radiocarbon date from our sample problem which gave 13,500 BP needs to be adjusted. It turns out that the amount of Carbon 14 in the atmosphere is not actually a constant. This amount has decreased over time which has the effect of making years seem longer. We can fix this problem by looking at the rings of trees. We can count individual calendar years quite a bit back. By using different trees, scientists have been able to find a continuous sequence of tree rings going back 11,000 years. And it turns out there are other ways to establish the calibration of calendar years to radiocarbon years for even earlier dates. Anyway, suffice it to say that the raw radiocarbon date must be adjusted to an accurate calendar date. The details are for another story.

11.5 Simple Rate Problems with Jumps

This discussion is meant to help you make the jump to more interesting models. Clearly, the biological model could switch at various time points. We will find we can still handle the model solution using the tools we already have in our tool kit. Consider the function f defined on $[0, \infty)$ by

$$f(t) = \begin{cases} -2, & 0 \leq t < 10 \\ 1, & t = 10 \\ -3, & 10 < t \end{cases}$$

We know this function is Riemann integrable on any interval of the form $[0, t]$ for a finite t as it continuous at all but a finite number of points. We even know how to integrate it as we have discussed in Sect. 8.8. Now how would we interpret the differential equation

$$\begin{aligned} x'(t) &= f(t)x(t) \\ x(0) &= 20? \end{aligned}$$

We rewrite using the definition of f as follows:

$$\begin{aligned} x'(t) &= \begin{cases} -2x(t), & t < 10 \\ 1x(t), & t = 10 \\ -3x(t), & t > 10 \end{cases} \\ x(0) &= 20 \end{aligned}$$

Thus, solving as we would any other exponential rate problem, we find

$$\begin{aligned} \frac{x'(t)}{x(t)} &= f(t) \\ \int_0^t \frac{x'(s)}{x(s)} ds &= \int_0^t f(s) ds \\ &= F(t) \end{aligned}$$

We calculate $F(t) = \int_0^t f(s)ds$ as we have done in Sect. 8.8. Again, this will have to be done in several parts because of the way f is defined.

1. On the interval $[0, 10]$, note that f is continuous except at one point, $t = 10$. Hence, f is Riemann integrable by Theorem 8.4.1. Also, the function -2 is continuous on this interval and so is also Riemann integrable. Then since f on $[0, 10]$ and -2 match at all but one point on $[0, 10]$, their Riemann integrals must match. Hence, if t is in $[0, 10]$, we compute F as follows:

$$\begin{aligned} F(t) &= \int_0^t f(s) ds \\ &= \int_0^t -2 ds \\ &= (-2s)\Big|_0^t \\ &= -2t \end{aligned}$$

2. On the interval $[10, t]$, note that f is continuous except at one point, $t = 10$. Hence, f is Riemann integrable by Theorem 8.4.1. Also, the function -3 is continuous on this interval and so is also Riemann integrable. Then since f on $[10, t]$ and -3 match at all but one point on $[10, t]$, their Riemann integrals must match. Hence, we compute F as follows:

$$
\begin{aligned}
F(t) &= \int_0^t f(s)ds \\
&= \int_0^{10} f(s)ds + \int_{10}^t f(s)ds \\
&= \int_0^{10} -2ds + \int_{10}^t -3ds \\
&= (-2s)\Big|_0^{10} + (-3s)\Big|_{10}^t \\
&= -20 - 3t + 30 \\
&= 10 - 3t
\end{aligned}
$$

Thus, we have found that

$$
F(t) = \begin{cases} -2t, & 0 \leq t \leq 10 \\ 10 - 3t, & 10 \leq t \end{cases}
$$

We know F is indeed continuous at 10 and we know $F'(10)$ does not exist. Thus, our integration of the original rate differential equation has given

$$
\begin{aligned}
\int_0^t \frac{x'(s)}{x(s)} ds &= \int_0^t f(s)ds \\
\ln \mid x(s) \mid_0^t &= \int_0^t f(s)ds \\
&= F(t) - F(0) \\
&= \begin{cases} -2t, & 0 \leq t \leq 10 \\ 10 - 3, t & 10 \leq t \end{cases} \\
\ln\left|\frac{x(t)}{x(0)}\right| &= \begin{cases} -2t, & 0 \leq t \leq 10 \\ 10 - 3t, & 10 \leq t \end{cases}
\end{aligned}
$$

The initial condition is $x(0) = 20$, which means the solution will always be positive and so the absolute values are not needed. Hence, we can calculate the exp of both sides to find

$$
\frac{x(t)}{20} = e^{F(t)}
$$

or

$$
\begin{aligned}
x(t) &= 20e^{F(t)} \\
&= \begin{cases} 20e^{-2t}, & 0 \leq t \leq 10 \\ 20e^{10-3t}, & 10 \leq t \end{cases}
\end{aligned}
$$

Note, our solution to this rate differential equation is continuous for all t and is differential for all t except at 10. Indeed, $x'(10^-)$ is $-40e^{-20} = -2x(10)$ and $x'(10^+)$ is $-60e^{-20} = -3x(10)$. So, since $x'(10)$ does not exist, we clearly can not satisfy the requirement that

$$x'(10) = 1x(10)$$

However, we can think of this function x as the solution to this *generalized* rate differential equation at all but the one point. Note if you graphed this solution, you would see a corner in x at 10 because the tangent lines at 10 don't match.

11.5.1 Homework

Exercise 11.5.1 *Solve*

$$x'(t) = \begin{cases} -4, x(t) & t < 6 \\ 3, x(t) & t = 6 \\ -5, x(t) & t > 6 \end{cases}$$
$$x(0) = 200$$

Exercise 11.5.2 *Solve*

$$x'(t) = \begin{cases} 2, x(t) & t < 6 \\ 15, x(t) & t = 6 \\ -3, x(t) & t > 6 \end{cases}$$
$$x(0) = 80$$

Exercise 11.5.3 *Solve*

$$x'(t) = \begin{cases} -2, x(t) & t < 6 \\ 30, x(t) & t = 6 \\ 3, x(t) & t > 6 \end{cases}$$
$$x(0) = 80$$

References

M. Braun, *Differential Equations and Their Applications* (Springer, New York, 1978)
S. Mithen, *After The Ice: A Global Human History, 20,000–50,000 BC* (Harvard University Press, Cambridge, 2004)

Chapter 12
Simple Protein Models

We will now discuss how to develop a simple model of protein synthesis. We will clearly be hiding a lot of detail but we want you to see how much insight we can nevertheless! But first, we have to start with a new tool. This is called the **integrating factor method**.

12.1 The Integrating Factor Approach

We consider the problem

$$x'(t) = r\,x(t)$$
$$x(0) = x_0$$

We rewrite the first equation as

$$x'(t) - r\,x(t) = 0 \tag{12.1}$$

Now from the product rule, we know that

$$\left(x(t)\,e^{-rt}\right)' = x'(t)\,e^{-rt} - r\,x(t)\,e^{-rt}$$

This suggest we multiply both sides of Eq. 12.1 by e^{-rt} to obtain

$$x'(t)\,e^{-rt} - r\,x(t)\,e^{-rt} = 0.$$

Then, using the product rule in reverse so to speak, we have

$$\left(x(t)\,e^{-rt}\right)' = 0.$$

J.K. Peterson, *Calculus for Cognitive Scientists*, Cognitive Science and Technology, DOI 10.1007/978-981-287-874-8_12

This implies that

$$x(t)\, e^{-rt} = C$$

for some constant C. Next, we apply the initial condition to see

$$x(0) \;=\; x_0 = C\, e^0 \;=\; C$$

We see we get the same solution no matter which of the three techniques we choose to use. This last one is based on finding the right function $f(t)$, so that

$$\left(x'(t) - r\, x(t)\right)\; f(t) = (x(t)\; f(t))'\,.$$

Hence, we are trying to find a function f to make the original equation into the derivative of a product. Once this is done, we can solve the problem by integrating both sides. So the f we find allows us to *integrate* the problem. For this reason, the f we find is called an *integrating factor* and this is called the *integrating factor method*. In our example, the integrating factor is $f(t) \;=\; e^{-rt}$.

12.1.1 Examples

Example 12.1.1 Solve the following problem:

$$\begin{aligned} \frac{dx}{dt} &= 3\, x(t) \\ x(0) &= 20 \end{aligned}$$

Solution *We rewrite the problem as this*

$$x'(t) - 3x(t) = 0.$$

The integrating factor here is e^{-3t}. Multiplying both sides by the integrating factor, we have

$$e^{-3t}\left(x'(t) - 3x(t)\right) = 0.$$

But as we have shown, this is the same the the derivative of a product:

$$\left(e^{-3t}x(t)\right)' = 0.$$

Hence, there is a constant C so that

$$e^{-3t} x(t) = C.$$

or

$$x(t) = Ce^{3t}$$

Applying the initial condition, we have

$$x(t) = 20e^{3t}$$

And we did not have to resolve the absolute value like we did in the other methods!

Example 12.1.2

$$\frac{d}{dt}(x) = -4\,x(t)$$
$$x(0) = -30$$

Solution *We rewrite the problem as this*

$$x'(t) + 4x(t) = 0.$$

The integrating factor here is e^{4t}. Multiplying both sides by the integrating factor, we have

$$e^{4t}\left(x'(t) + 4x(t)\right) = 0.$$

But as we have shown, this is the same the the derivative of a product:

$$\left(e^{4t} x(t)\right)' = 0.$$

Hence, there is a constant C so that

$$e^{4t} x(t) = C.$$

or

$$x(t) = Ce^{-4t}$$

Applying the initial condition, we have

$$x(t) = -30e^{-4t}.$$

12.1.2 Homework

Solve these problems using the integrating factor method.

Exercise 12.1.1

$$\begin{aligned} x'(t) &= 2\,x(t) \\ x(0) &= 3 \end{aligned}$$

Exercise 12.1.2

$$\begin{aligned} x'(t) &= -2\,x(t) \\ x(0) &= 25 \end{aligned}$$

Exercise 12.1.3

$$\begin{aligned} x'(t) &= 0.4\,x(t) \\ x(0) &= 18 \end{aligned}$$

Exercise 12.1.4

$$\begin{aligned} x'(t) &= -3\,x(t) \\ x(0) &= 35 \end{aligned}$$

Exercise 12.1.5

$$\begin{aligned} x'(t) &= 1.2\,x(t) \\ x(0) &= -45 \end{aligned}$$

12.2 The Integrating Factor Approach with a Constant on the Right

We consider now the problem with a constant right hand side.

$$\begin{aligned} x'(t) &= 2.6\,x(t) + 10 \\ x(\,0\,) &= 25 \end{aligned}$$

Let's attack this in steps like usual.

- Rewrite the first equation as

$$x'(t) - 2.6\,x(t) = 10$$

- From the product rule, we have

$$\left(x(t)\,e^{-2.6t}\right)' = x'(t)\,e^{-2.6t} - 2.6\,x(t)\,e^{-2.6t}$$

- Multiply both sides of our rewritten model by the IF $e^{-2.6t}$ to get

$$e^{-2.6t}\left(x'(t) - 2.6\,x(t)\right) = e^{-2.6t} \times 10 \;=\; 10\,e^{-2.6t}.$$

- Now recognize the left hand side is a product rule and write

$$\left(e^{-2.6t}x(t)\right)' = 10\,e^{-2.6t}.$$

- Take the Riemann integral of both sides to get

$$e^{-2.6t}x(t) - x(0) = -\frac{10}{2.6}\left(e^{-2.6t} - 1\right).$$

- Use the IC to find $e^0 x(0) = 25$. Hence

$$\begin{aligned} e^{-2.6t}x(t) &= 25 - \frac{10}{2.6}\left(e^{-2.6t} - 1\right) \\ e^{2.6t}e^{-2.6t}x(t) &= e^{2.6t}\left(25 - \frac{10}{2.6}\left(e^{-2.6t} - 1\right)\right) \\ x(t) &= \left(25 + \frac{10}{2.6}\right)e^{2.6t} - \frac{10}{2.6}. \end{aligned}$$

Example 12.2.1 Solve the following problem:

$$\frac{dx}{dt} = 3.2\,x(t) + 12.0; \quad x(0) = -3.7$$

Solution Step 1: *Rewrite as* $x'(t) - 3.2x(t) = 12.0$.
Step 2a: *Multiply both sides by the integrating factor* $e^{-3.2t}$.

$$e^{-3.2t}\left(x'(t) - 3.2x(t)\right) = 12e^{-3.2t}.$$

Step 2b: *Recognize this is a product rule and rewrite as* $\left(e^{-3.2t}x(t)\right)' = 12e^{-3.2t}$.
Step 3: *Get the antiderivative of each side to find*

$$e^{-3.2t}x(t) = -\frac{12}{3.2}e^{-3.2t} + C$$

Step 4: *Use the IC $x(0) = -3.7$.*

$$e^0 x(0) = -\frac{12}{3.2} + C \Longrightarrow C = -3.7 + \frac{12}{3.2}.$$

Step 5: *Solve for $x(t)$ to obtain*

$$e^{-3.2t} x(t) = -\frac{12}{3.2} e^{-3.2t} - 3.7 + \frac{12}{3.2}$$
$$e^{3.2t}\, e^{-3.2t}\, x(t) = e^{3.2t}\left(-\frac{12}{3.2} e^{-3.2t} - 3.7 + \frac{12}{3.2}\right)$$
$$x(t) = (-3.7 + 12/3.2)e^{3.2t} - \frac{12}{3.2}.$$

Example 12.2.2 Solve the following problem: $\frac{dx}{dt} = -0.11\, x(t) + 20;\ \ x(0) = 5.4$

Solution Step 1: *Rewrite as*

$$x'(t) + 0.11x(t) = 20.$$

Step 2a: *Multiply both sides by the integrating factor $e^{0.11t}$.*

$$e^{0.11t}\left(x'(t) + 0.11x(t)\right) = 20.$$

Step 2b: *Recognize this is a product rule and rewrite as* $\left(e^{0.11t} x(t)\right)' = 20e^{0.11t}$.

Step 3: *Get the Riemann integral of each side to find*

$$e^{0.11t} x(t) - x(0) = \frac{20}{0.11}\left(e^{0.11t} - 1\right).$$

Step 4: *Use the IC $x(0) = 5.4$.*

$$e^{0.11t} x(t) = 5.4 + \frac{20}{0.11}\left(e^{0.11t} - 1\right).$$

Step 5: *Solve for $x(t)$ to obtain*

$$e^{-0.11t}\, e^{0.11t}\, x(t) = 5.4 + \frac{20}{0.11}\left(e^{0.11t} - 1\right)$$
$$x(t) = \left(5.4 - \frac{20}{0.11}\right) + \frac{20}{0.11}$$

12.2.1 Homework

Solve these problems using the integrating factor method.

Exercise 12.2.1

$$x'(t) = 2\,x(t) + 12,\ x(0) = 3$$

Exercise 12.2.2

$$x'(t) = -2\,x(t) + 30,\ x(0) = 25$$

Exercise 12.2.3

$$x'(t) = .4\,x(t) + 15,\ x(0) = 65$$

Exercise 12.2.4

$$x'(t) = -3.3\,x(t) + 100,\ x(0) = 12.7$$

Exercise 12.2.5

$$x'(t) = 1.2\,x(t) + 35,\ x(0) = -45$$

12.3 Protein Synthesis

We now consider the problem

$$\begin{aligned} x'(t) &= -\alpha\,x(t) + \beta \\ x(\,0\,) &= x_0 \end{aligned}$$

for some nonzero choice of α and β. We will use the integrating factor method here as this is like the last section: β is a constant. Let's do this one in general. We rewrite the first equation as

$$x'(t) + \alpha\,x(t) = \beta \tag{12.2}$$

We multiply both sides of Eq. 12.2 by the integrating factor $e^{\alpha t}$ to obtain

$$x'(t)\,e^{\alpha t} + \alpha\,x(t)\,e^{\alpha t} = \beta e^{\alpha t}.$$

Then, we have

$$\left(x(t)\, e^{\alpha t}\right)' = \beta e^{\alpha t}.$$

Using the definite method, we then have

$$\int_0^t \left(x(s)\, e^{\alpha s}\right)' ds = \int_0^t \beta e^{\alpha s}\, ds$$

Integrating, we find

$$x(t)\, e^{\alpha t} - x(0) = \frac{\beta}{\alpha}\left(e^{\alpha t} - 1\right)$$

Next, we apply the initial condition and manipulate a bit to get

$$x(t)\, e^{\alpha t} = \left(x_0 - \frac{\beta}{\alpha}\right) + \frac{\beta}{\alpha} e^{\alpha t}$$

Now multiply both sides by $e^{-\alpha t}$ to obtain

$$x(t) = \left(x_0 - \frac{\beta}{\alpha}\right) e^{-\alpha t} + \frac{\beta}{\alpha}.$$

Note as $t \to \infty$, the value of x asymptotically approaches $\frac{\beta}{\alpha}$. We call $\frac{\beta}{\alpha}$ the **steady state value** of the solution which is often just denoted by **SS**.

The problem we have just solved

$$\begin{aligned} Y'(t) &= -\alpha\, Y(t) + \beta \\ Y(0) &= Y_0 \end{aligned}$$

can be interpreted in terms of protein synthesis. A given protein Y begins to be created once the proper signal S binds to the promoter region of its corresponding gene. Once the signal is bound, the gene begins to create its corresponding protein via the usual transcription process. The time it takes to reach a constant output is usually quite brief. Hence, for most purposes, we can assume the protein Y is being made at the constant rate β. However, there are other mechanisms inside a cell that degrade the protein and disassemble it into component parts for later reuse. We can typically model this degradation process as exponential decay. Hence, the protein Y is lost using the model

$$\frac{dY_{loss}}{dt} = -\alpha Y(t)$$

The net rate of change of the concentration of protein Y is the sum of the growth process and the decay or loss process. Hence, we arrive at the model

$$Y'(t) = -\alpha\, Y(t) + \beta$$
$$Y(0) = Y_0$$

From what we have already learned, we know the steady state value of the protein Y is thus $\frac{\beta}{\alpha}$.

12.3.1 The Underlying Biology

So we have a model of protein transcription, but we have only mentioned in passing the molecular machinery that is involved. It's time to look at the underlying biology more closely so you can have a better idea what is going on. Note how much of this detail we actually do not use when we build the model, yet the model still gives a lot of insight. Let's examine how a single gene $\boldsymbol{Y}$ is activated to produce its product which is a protein. We will discuss this using simple *cartoons* to represent a lot of complicated biology. We intend to show you how the mathematics we are learning has a lot of explanatory power. So remember, Protein Modeling here, is an approximation of the underlying complex biological processes which we are making in order to gain insight.

The gene $\boldsymbol{Y}$ is a string of nucleotides (**A**, **C**, **T** and **G**) with a special starting string in front of it called the **promoter**. We will draw this as shown in Fig. 12.1.

The nucleotides in the gene $\boldsymbol{Y}$ are *read* three at a time to create the amino acids which form the protein $\boldsymbol{Y^*}$ corresponding to the gene. The process is this: a special **RNA** polymerase, **RNAp**, which is a complex of several proteins, binds to the promoter region as shown in Fig. 12.2.

Once **RNAp** binds to the promoter, messenger **RNA**, **mRNA**, is synthesized that corresponds to the specific nucleotide triplets in the gene $\boldsymbol{Y}$. The process of forming this **mRNA** is called **transcription**. Once the **mRNA** is formed, the protein $\boldsymbol{Y^*}$ is then made.

The protein creation process is typically regulated. A single **regulator** works like this. An activator called $\boldsymbol{X}$ is a protein which increases the rate of **mRNA** creation

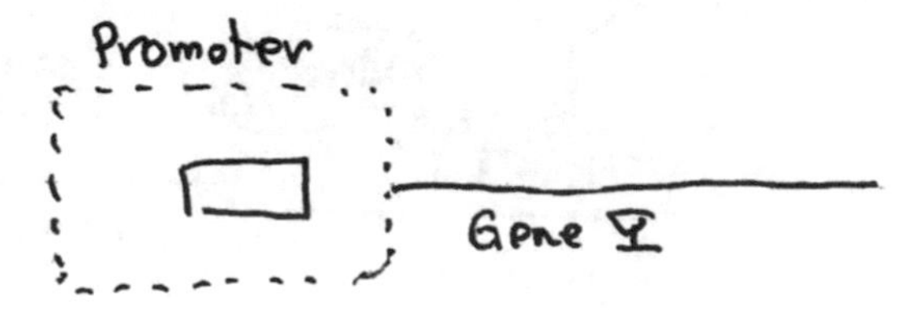

Fig. 12.1 Promoter and gene

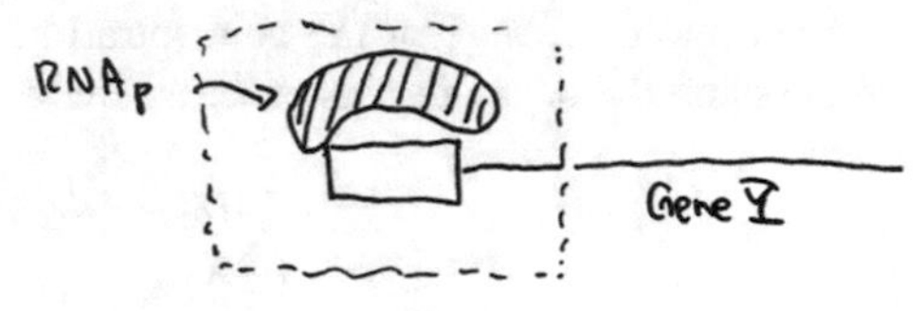

Fig. 12.2 Promoter binding

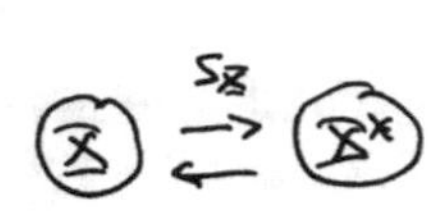

Fig. 12.3 Activator and inactivator switching

when ti binds to the promoter. The activator X switches between and active and inactive version due to a signal S_X. We let the active form be denoted by X^*. We show this in Fig. 12.3 which indicates X moves to state X^* and back.

If X^* binds in front of the promoter, **mRNA** creation increases implying an increase in the creation of the protein Y^* also. This is shown in Fig. 12.4.

We indicate all of this by the simple interaction arrow

$$X \longrightarrow Y.$$

Once the signal S_X appears, X rapidly transitions to its state X^*, binds with the front of the promoter and protein Y^* begins to accumulate. We let β denote the rate of protein accumulation which is **constant** once the signal S_X begins. However, proteins also degrade due to two processes:

- proteins are destroyed by other proteins in the cell. Call this rate of destruction α_{des}.
- the concentration of protein in the cell goes down because the cell grows and therefore its volume increases. Protein is usually measured as a concentration and the concentration goes down as the volume goes up. Call this rate αdil—the *dil* is for *dilation*.

The net or total *loss of protein* is called α and hence

$$\alpha = \alpha_{des} + \alpha_{dil}$$

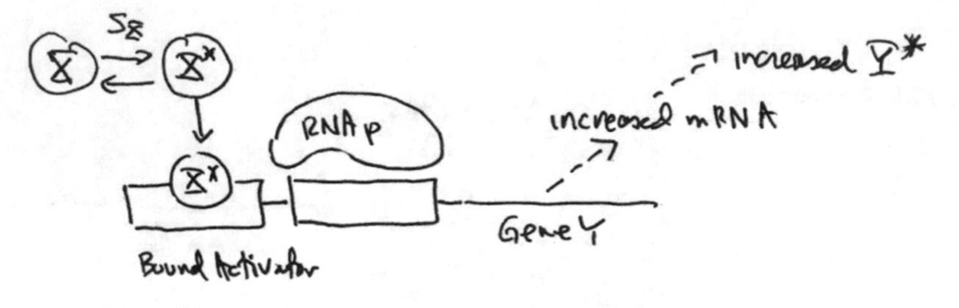

Fig. 12.4 Signal binding and protein transcription

The net rate of change of the protein concentration is then our familiar model

$$\frac{d\boldsymbol{Y^*}}{dt} = \underbrace{\boldsymbol{\beta}}_{\text{constant growth}} - \underbrace{\alpha\, \boldsymbol{Y^*}}_{\text{loss term}}$$

We usually do not make a distinction between the gene $\boldsymbol{Y}$ and its transcribed protein $\boldsymbol{Y^*}$. We usually treat the letters $\boldsymbol{Y}$ and $\boldsymbol{Y^*}$ as the same even though it is not completely correct. Hence, we just write as our model

$$\begin{aligned} Y' &= \beta - \alpha\, Y \\ Y(0) &= Y_0 \end{aligned}$$

and then solve it using the integrating factor method even though, strictly speaking, Y is the gene!

This model is easy to solve. The integrating factor here is $e^{\alpha t}$ and we solve the model as follows.

$$\begin{aligned} Y' + \alpha Y &= \beta \\ \left(Y(t)\, e^{\alpha t}\right)' &= \beta\, e^{\alpha t} \\ Y(t)\, e^{\alpha t} &= Y_0 + \frac{\beta}{\alpha}\,(e^{\alpha t} - 1) \\ Y(t) &= \left(Y_0 - \frac{\beta}{\alpha}\right) e^{-\alpha t} + \frac{\beta}{\alpha}. \end{aligned}$$

and the solution is easy to see graphically as we show in Fig. 12.5. As we know, the horizontal asymptote is called the *steady state*. With the signal active, the protein level will reach the steady state fairly rapidly. Once the signal stops, we switch from β to a constant growth rate of 0 and from that time on, the protein experiences exponential decay.

$$Y' = -\alpha\, Y$$

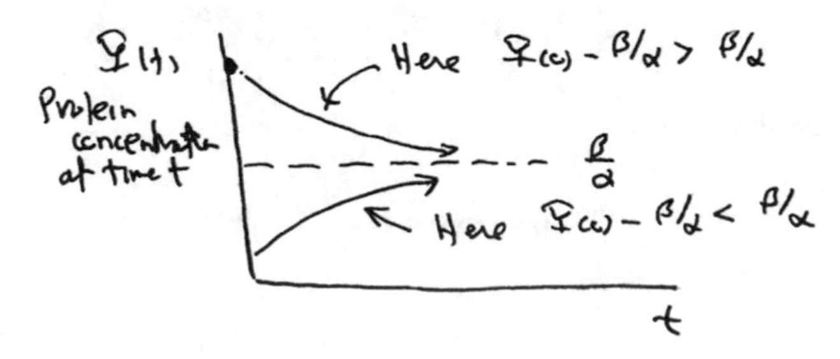

Fig. 12.5 Protein production versus time

starting from the Y value we stopped with when the signal was active. The time it takes the protein to move from its starting position to half way toward the steady state is called the response time, t_r, and we will discuss this later in more detail.

12.3.2 Worked Out Examples

Example 12.3.1 Solve the following problem:

$$x'(t) = -0.02\, x(t) + 5; \;\; x(0) = 75$$

Solution

Step 1: *Rewrite as* $x'(t) + 0.02x(t) = 5$.

Step 2a: *Multiply both sides by the integrating factor* $e^{0.02t}$.

$$e^{0.02t}\left(x'(t) + 0.02x(t)\right) = 5\, e^{0.02t}.$$

Step 2b: *Recognize this is a product rule and rewrite as* $\left(e^{0.02t}x(t)\right)' = 5\, e^{0.02t}$.

Step 3: *Integrate both sides from* 0 *to* t:

$$\int_0^t \left(e^{0.02s}x(s)\right)' ds = \int_0^t 5\, e^{0.02s}\, ds.$$

Step 4: *This gives*

$$\left(e^{0.02s}x(s)\right)\Big|_0^t = \frac{5}{0.02}\, e^{0.02s}\Big|_0^t$$

Step 5: *This gives*

$$e^{0.02t}x(t) - x(0) = \frac{5}{0.02}\left(e^{0.02t} - 1\right).$$

Step 6: *Simplify:*

$$\begin{aligned} e^{0.02t}x(t) &= x(0) + \frac{5}{0.02}\left(e^{0.02t} - 1\right) \\ &= \left(x(0) - \frac{5}{0.02}\right) + \frac{5}{0.02}e^{0.02t}. \end{aligned}$$

Step 7: *Solve for $x(t)$: multiply both sides by $e^{-0.02t}$.*

$$x(t) = e^{-0.02t}\left(\left(x(0) - \frac{5}{0.02}\right) + \frac{5}{0.02}e^{0.02t}\right)$$
$$x(t) = \left(x(0) - \frac{5}{0.02}\right)e^{-0.02t} + \frac{5}{0.02}.$$

Step 8: *Use the IC:*

$$x(t) = \left(75 - 250\right)e^{-0.02t} + 250$$
$$= -175\,e^{-0.02t} + 250.$$

Again, this is the interpretation. As $t \longrightarrow \infty$, $e^{-0.02t} \longrightarrow 0$ and so $x(t) \longrightarrow \frac{5}{0.02} = 250$. This is a horizontal asymptote and we call the fraction $\frac{\beta}{\alpha} = \frac{5}{0.02} = 250$ the **steady state value** *or* **SS***. The IC is 75 which is below the* **SS** *so $x(t)$ approaches the* **SS** *from below. The solution looks like the picture shown in Fig. 12.6. We also show what would have happened if the initial condition had been 375.*

Example 12.3.2

$$x'(t) = -3.0\,x(t) + 5.0$$
$$x(0) = 0$$

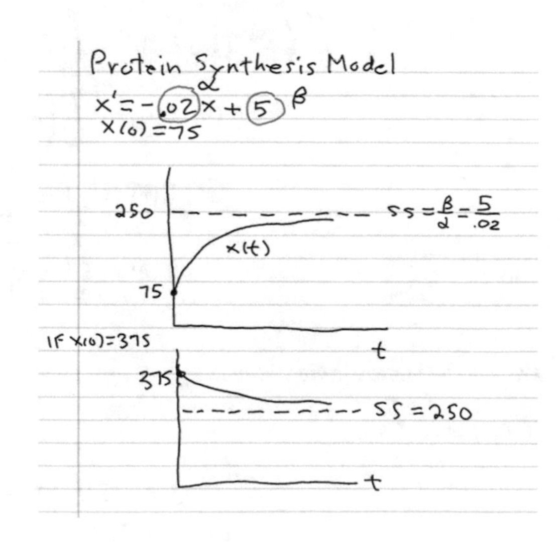

Fig. 12.6 Solution to $x'(t) = -0.02\,x(t) + 5;\ x(0) = 75$

Solution *We have, using $x(0) = 0$, that*

$$
\begin{aligned}
x'(t) + 3x(t) &= 5 \\
\left(e^{3t} x(t)\right)' &= 5e^{3t} \\
\int_0^t \left(e^{3s} x(s)\right)' ds &= \int_0^t 5e^{3s} \, ds \\
e^{3t} x(t) - x(0) &= \frac{5}{3}\left(e^{3t} - 1\right) \\
x(t) &= \frac{5}{3} - \frac{5}{3} e^{-3t}
\end{aligned}
$$

We see the steady state value is $\frac{5}{3}$.

Example 12.3.3

$$
\begin{aligned}
T'(t) &= -16.0\, T(t) + 7.0 \\
T(0) &= 1.0
\end{aligned}
$$

Solution *We have, using $T(0) = 1.0$, that*

$$
\begin{aligned}
T'(t) + 16x(t) &= 7 \\
\left(e^{16t} T(t)\right)' &= 7e^{16t} \\
\int_0^t \left(e^{16s} T(s)\right)' ds &= \int_0^t 7e^{16s} \, ds \\
e^{16t} T(t) - T(0) &= \frac{7}{16}\left(e^{16t} - 1\right) \\
e^{16t} T(t) &= \left(1 - \frac{7}{16}\right) + \frac{7}{16} e^{16t} \\
T(t) &= \left(1 - \frac{7}{16}\right) e^{-16t} + \frac{7}{16}
\end{aligned}
$$

We see the steady state value is $\frac{7}{16}$.

Example 12.3.4

$$
\begin{aligned}
T'(t) &= -2.0\, T(t) + 0.5 \\
T(0) &= 25
\end{aligned}
$$

Solution *We have, using $T(0) = 25$, that*

$$
\begin{aligned}
T'(t) + 2x(t) &= 0.5 \\
\left(e^{2t} T(t)\right)' &= 0.5e^{2t} \\
\int_0^t \left(e^{2s} T(s)\right)' ds &= \int_0^t 0.5e^{2s}\, ds \\
e^{2t} T(t) - T(0) &= \frac{0.5}{2}\left(e^{2t} - 1\right) \\
e^{2t} T(t) &= \left(25 - \frac{0.5}{2}\right) + \frac{0.5}{2} e^{2t} \\
T(t) &= \left(25 - \frac{0.5}{2}\right) e^{-2t} + \frac{0.5}{2}
\end{aligned}
$$

We see the steady state value is $\frac{0.5}{2}$.

12.3.3 Homework

For these problems, solve using the integrating factor method, state the steady state value, sketch the solution and explain the model's interpretation as a protein synthesis problem.

Exercise 12.3.1

$$
\begin{aligned}
T'(t) &= -8\, T(t) + 6 \\
T(0) &= 8
\end{aligned}
$$

Exercise 12.3.2

$$
\begin{aligned}
x'(t) &= -7\, x(t) + 3 \\
x(0) &= 1
\end{aligned}
$$

Exercise 12.3.3

$$
\begin{aligned}
y'(t) &= -45\, y(t) + 11 \\
y(0) &= 0
\end{aligned}
$$

Exercise 12.3.4

$$\Phi'(t) = -2\,\Phi(t) + 0.003$$
$$\Phi(0) = 4$$

Exercise 12.3.5

$$\Psi'(t) = -9\,\Psi(t) + 19$$
$$\Psi(0) = 6$$

12.4 The Response Time in Protein Synthesis Problems

In a protein synthesis model, the solution to

$$y'(t) = -\alpha\, y(t) + \beta$$
$$y(0) = y_0$$

is given by

$$y(t) = \left(y_0 - \frac{\beta}{\alpha}\right)e^{-\alpha t} + \frac{\beta}{\alpha}.$$

The steady state value is $\frac{\beta}{\alpha}$. How long does it take for y to increase to half way between y_0 and $\frac{\beta}{\alpha}$? This is an easy calculation. We solve for time in the following equation:

$$\frac{y_0 + \frac{\beta}{\alpha}}{2} = \left(y_0 - \frac{\beta}{\alpha}\right)e^{-\alpha t} + \frac{\beta}{\alpha}.$$

Simplifying, we find

$$\frac{y_0 - \frac{\beta}{\alpha}}{2} = \left(y_0 - \frac{\beta}{\alpha}\right)e^{-\alpha t}$$

We can cancel the $y_0 - \frac{\beta}{\alpha}$ term to obtain $\frac{1}{2} = e^{-\alpha t}$. Solving for time, we find $t = \frac{\ln(2)}{\alpha}$. This is the amount of time it takes for the protein to increase from its current level halfway up to the steady state value. We call this the Response Time of the protein. We often denote it by the symbol t_r. Note t_r has the same mathematical formula as the half—life $t_{1/2}$.

Let's look at an old example again in terms of response times.

Example 12.4.1 Solve the following problem:

$$x'(t) = -0.02\,x(t) + 5; \quad x(0) = 75$$

Solution

$$
\begin{aligned}
x'(t) + 0.02x(t) &= 5 \\
e^{0.02t}\left(x'(t) + 0.02x(t)\right) &= 5\,e^{0.02t} \\
\left(e^{0.02t}x(t)\right)' &= 5\,e^{0.02t} \\
\int_0^t \left(e^{0.02s}x(s)\right)' ds &= \int_0^t 5\,e^{0.02s}\,ds \\
\left(e^{0.02s}x(s)\right)\Big|_0^t &= \frac{5}{0.02}\,e^{0.02s}\Big|_0^t \\
e^{0.02t}x(t) - x(0) &= \frac{5}{0.02}\left(e^{0.02t} - 1\right) \\
e^{0.02t}x(t) &= x(0) + \frac{5}{0.02}\left(e^{0.02t} - 1\right) \\
&= \left(x(0) - \frac{5}{0.02}\right) + \frac{5}{0.02}e^{0.02t} \\
x(t) &= e^{\;0.02t}\left(\left(x(0) - \frac{5}{0.02}\right) + \frac{5}{0.02}e^{0.02t}\right) \\
x(t) &= \left(x(0) - \frac{5}{0.02}\right)e^{-0.02t} + \frac{5}{0.02} \\
x(t) &= \left(75 - 250\right)e^{-0.02t} + 250 \\
&= -175\,e^{-0.02t} + 250.
\end{aligned}
$$

How long does it take $x(t)$ to rise from its IC of 75 to half way to the SS of 250? This time is like a half life but in this context it is called the **response time**. *Denote it by t_r. Half way to 250 from 75 is the average $(75 + 250)/2$. This is 325/2. We find*

$$
\begin{aligned}
\frac{325}{2} &= -175\,e^{-0.02t_r} + 250 \\
-\frac{175}{2} &= -175\,e^{-0.02t_r} \\
\frac{1}{2} &= e^{-0.02t_r}
\end{aligned}
$$

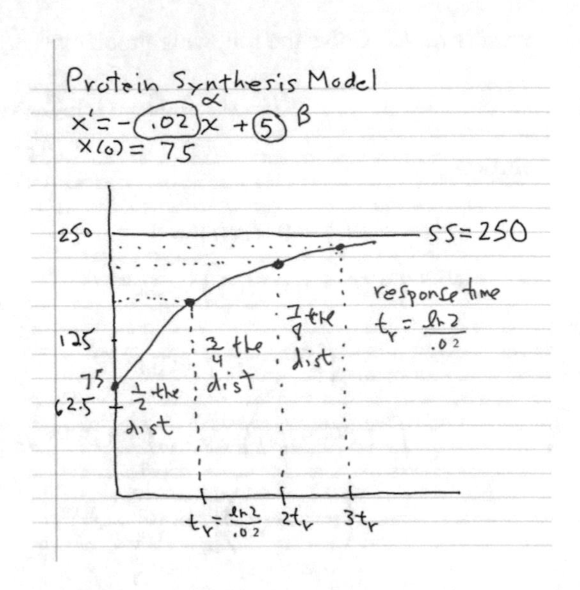

Fig. 12.7 Solution to $x'(t) = -0.02\, x(t) + 5;\ x(0) = 75$

$$
\begin{aligned}
-\ln(2) &= -0.02 t_r \\
t_r &= \frac{\ln(2)}{0.02}
\end{aligned}
$$

This is the same formula we had for the half life in exponential decay, of course. We can draw using the response time as shown in Fig. 12.7.

12.4.1 Homework

Prove the response time for these problems is given by $\ln(2)/\alpha$ for the appropriate α. Then graph the solution using the response time.

Exercise 12.4.1

$$x'(t) = -2\, x(t) + 12,\ x(0) = 3$$

Exercise 12.4.2

$$x'(t) = -1.3\, x(t) + 30,\ x(0) = 25$$

Exercise 12.4.3

$$x'(t) = -0.005\, x(t) + 15,\ x(0) = 65$$

Exercise 12.4.4

$$x'(t) = -0.0246\,x(t) + 100, \;\; x(0) = 12.7$$

Exercise 12.4.5

$$x'(t) = -0.035\,x(t) + 35, \;\; x(0) = -45$$

12.5 Signal On and Off Scenarios

Now let's look at some problems where the signal to turn on and off protein production follows a given schedule.

Consider the model

$$\begin{aligned} P' &= -0.04P + 8 \\ P(0) &= 0 \end{aligned}$$

Let's draw a careful graph of the solution to this model using response time as the time unit. The production schedule is

- Production is on for 2 response times
- Production is off for 4 response times
- Production is on for 2 response time
- Production is off from that point on.

SS is $\frac{8}{0.04} = 200$. When the signal is off the model reduces to simple exponential decay with half life $t_{1/2} = \ln(2)/0.04$ and when the signal is on the model grows using the response time $t_r = \ln(2)/0.04$. So the same time scale works. A careful graph using this time unit is shown in Fig. 12.8.

Here is another one. Consider the model

$$\begin{aligned} P' &= -0.05P + 11 \\ P(0) &= 100 \end{aligned}$$

with production schedule

- Production is on for 2 response times
- Production is off for 3 response times
- Production is on for 2 response time
- Production is off from that point on.

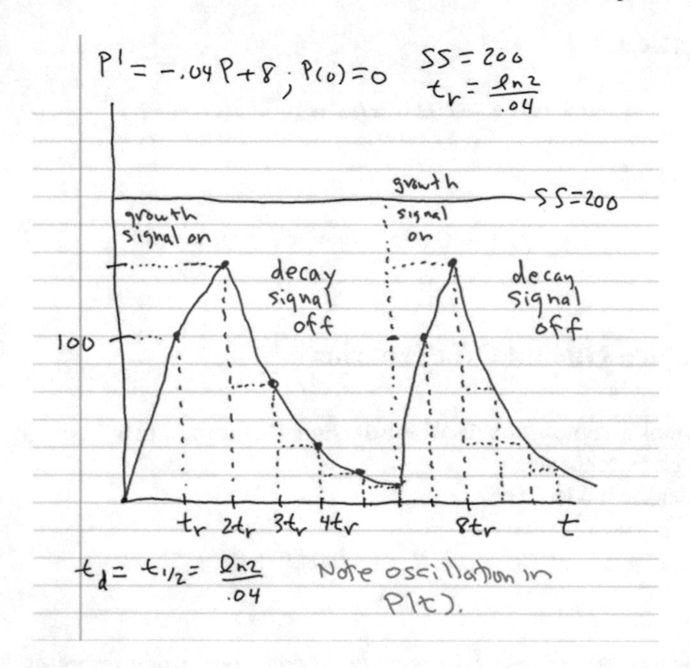

Fig. 12.8 Solution to $P' = -0.04P + 8$; $P(0) = 0$ with the scenario production is on for 2 response times, production is off for 4 response times, production is on for 2 response time and production is off from that point on

SS is $\frac{11}{0.05} = 220$. When the signal is off the model reduces to simple exponential decay with half life $t_{1/2} = \ln(2)/0.05$ and when the signal is on the model grows using the response time $t_r = \ln(2)/0.05$. The graph is shown in Fig. 12.9.

12.5.1 Homework

Exercise 12.5.1 *A protein x is modeled by the dynamics* $x' = -0.3x + 10$, $x(0) = 0$. *Draw a careful graph of how the protein's concentration changes with time for the following scenario:*

- *the signal to create is on from* $t = 0$ *to* $t = 2$ *response times.*
- *the signal is then shut off for* 3 *response times.*
- *the signal is then back on from that point on.*

Exercise 12.5.2 *A protein P is modeled by the dynamics* $P' = -0.03P + 60$, $P(0) = 0$. *Draw a careful graph of how the protein's concentration changes with time for the following scenario:*

- *the signal to create is on from* $t = 0$ *to* $t = 1$ *response times.*
- *the signal is then shut off for* 2 *response times.*

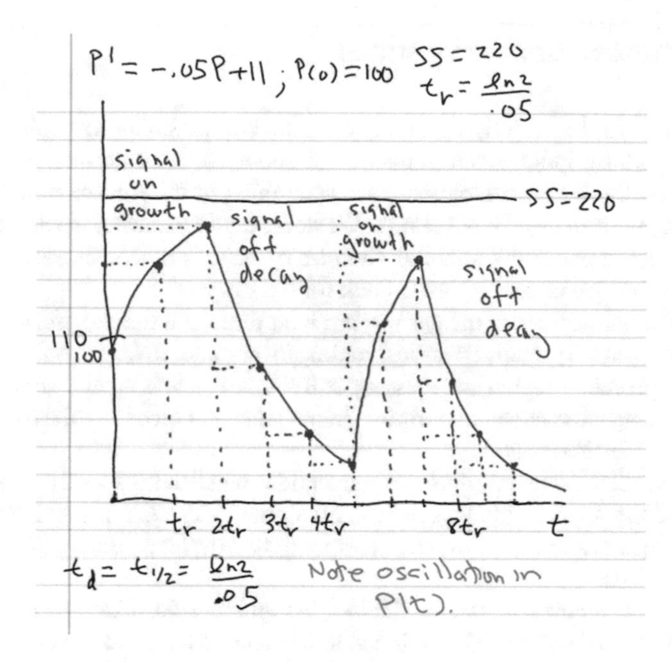

Fig. 12.9 Solution to $P' = -0.05P + 11$; $P(0) = 100$ with the scenario production is on for 2 response times, production is off for 3 response times, production is on for 2 response time and production is off from that point on

- *the signal is then back on for 2 response time.*
- *the signal is then off from that point on.*

Exercise 12.5.3 *The model is* $Q' = -0.6Q + 300$; $Q(0) = 10$. *Draw a careful graph of how the protein's concentration changes with time for the following scenario:*

- *Production is on for 2 response times*
- *Production is off for 1 response times*
- *Production is on for 3 response time*
- *Production if off from that point on.*

Exercise 12.5.4 *The model is* $\Theta' = -0.8\Theta + 80$; $\Theta(0) = 20$. *Draw a careful graph of how the protein's concentration changes with time for the following scenario:*

- *Production is on for 1 response times*
- *Production is off for 3 response times*
- *Production is on for 3 response time*
- *Production is off from that point on.*

12.6 Transcription Error Rates

Let's go back to the underlying biology again. We are going to explain why the error rate in transcription has a frequency of about 10^{-4}. To do this, we are going to look carefully at how the proteins are assembled in the ribosome and use a lot of reasoning with rates of interactions. We don't really use calculus and first order differential equations at all, although the idea of *rates* is indeed closely related. So grab a cup of coffee or tea and settle back for the story.

Recall the nucleotides in the gene $\boldsymbol{Y}$ are *read* three at a time to create the amino acids which form the protein $\boldsymbol{Y}^*$ corresponding to the gene. **RNA** polymerase, **RNAp**, binds to the promoter region and messenger **RNA**, **mRNA**, is synthesized that corresponds to the specific nucleotide triplets in the gene $\boldsymbol{Y}$. Once the **mRNA** is formed, the protein $\boldsymbol{Y}^*$ is then made.

Now each **mRNA** carries triplets of nucleotides which correspond to amino acids. Recall there are 64 possible triplets and 20 amino acids. A transfer **RNA**, **tRNA**, protein carries an amino acid which corresponds to a triplet on the **mRNA**. We show this in Fig. 12.10.

When the combined structure, **tRNA** plus amino acid, binds to **mRNA**, it is inserted into the **ribosome**. The amino acid it carries is added to the chain of amino acids already obtained and the **tRNA** is ejected. The next (**tRNA** plus amino acid) complex bound to the **mRNA** is then *read* and the process repeats. This is drawn in Fig. 12.11.

There are other (**tRNA** plus amino acid) complexes in the *soup* of components, proteins and other things that is what the inside of the cell looks like. The closest **incorrect** (**tRNA** plus amino acid) complex breaks apart or **disassociates** at the rate k_d. The rate that the correct (**tRNA** plus amino acid) complex is k_c and since we know the correct one should be read instead of the incorrect one, we must have $k_d > k_c$. We know that errors in reading are nevertheless made and experimentally we know it occurs at the frequency of 10^{-4}. Let's see if we can explain this.

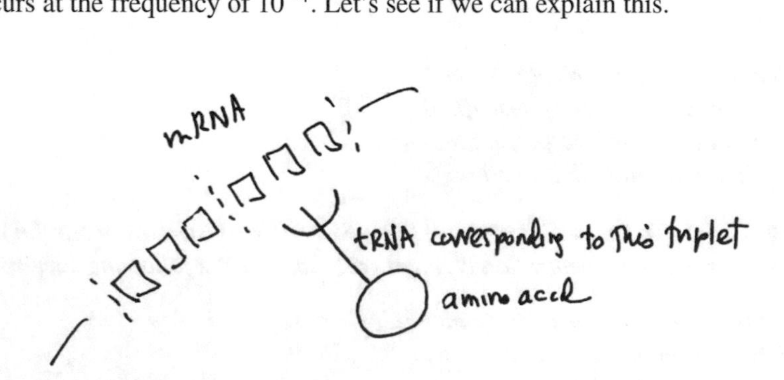

Fig. 12.10 MRNA and tRNA in more detail

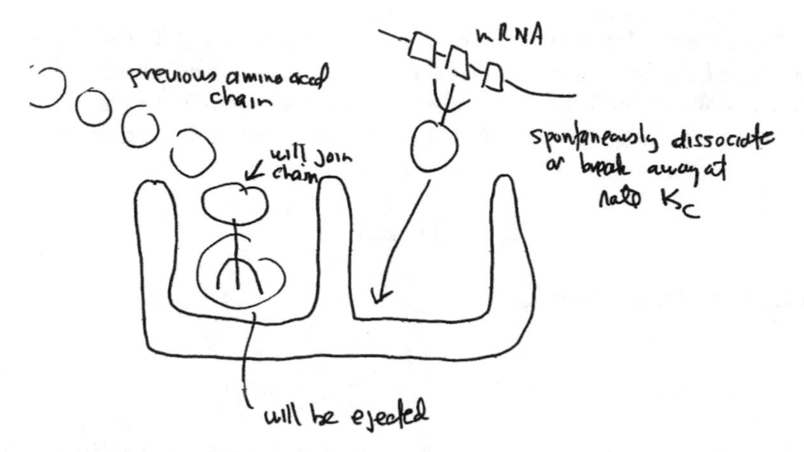

Fig. 12.11 Ribosome processing

12.6.1 A First Attempt to Explain the Error Rate

Let the correct **tRNA** complex be denoted by $\boldsymbol{c}$ and the correct **mRNA** triplet it corresponds to be $\boldsymbol{C}$. The two bind to to create another complex we will denote by $\boldsymbol{cC}$. Once the **tRNA** is created, there is a probability p per unit time the amino acid attached to the **tRNA** will be linked to the growing amino acid chain. If this happens, the freed **tRNA** unbinds from the chain and the $\boldsymbol{C}$ shifts to the next (**tRNA** plus amino acid) complex. We can analyze this with a bit of mathematical modeling like this.

$$\begin{array}{ccccc} \textbf{tRNA} + \textbf{mRNA} & k_c & \text{bound complex} & p & \\ [\boldsymbol{c}] + [\boldsymbol{C}] & \longrightarrow & [\boldsymbol{cC}] & \longrightarrow & \text{correct amino acid.} \end{array}$$

where k_c is the rate at which $\boldsymbol{c}$ and $\boldsymbol{C}$ combine to form the complex $\boldsymbol{cC}$. The complex also breaks apart at the rate k'_c which we denote in equation form as

$$[\boldsymbol{cC}] \xrightarrow{k'_c} [\boldsymbol{c}] + [\boldsymbol{C}].$$

Combining, we have the model

$$[\boldsymbol{c}] + [\boldsymbol{C}] \underset{k'_c}{\overset{k_c}{\longleftrightarrow}} [\boldsymbol{cC}] \underset{p}{\longrightarrow} \text{correct amino acid}$$

At equilibrium, the rate at which $\boldsymbol{cC}$ forms must equal the rate at which $\boldsymbol{cC}$ breaks apart. The concentration of $\boldsymbol{cC}$ is written as $[\boldsymbol{cC}]$. The concentrations of $\boldsymbol{c}$ and $\boldsymbol{C}$ are $[\boldsymbol{c}]$ and $[\boldsymbol{C}]$. The amount of $\boldsymbol{cC}$ depends on how much of the needed *recipe* ingredients are available. Hence the amount made is k_c $[\boldsymbol{c}]$ $[\boldsymbol{C}]$. The amount of $\boldsymbol{c}$ and $\boldsymbol{C}$ made because $\boldsymbol{cC}$ breaks apart depends on how much $\boldsymbol{cC}$ is available. So this amount must be k'_c $[\boldsymbol{cC}]$. This kind of balancing is a standard thing you learn later in other courses. So at equilibrium, because the formation rate and disassociation rate are the same that we must have a balance

$$k'_c\ [\boldsymbol{cC}] = k_c\ [\boldsymbol{c}]\ [\boldsymbol{C}]$$

Solving we find the relationship

$$[\boldsymbol{cC}] = \frac{k_c}{k'_c}\ [\boldsymbol{c}]\ [\boldsymbol{C}]$$

which is a really common equation to come up with in this kind of analysis. We call the fraction $\frac{k'_c}{k_c}$ the **disassociation constant** K_c and so we can write $[\boldsymbol{cC}] = \frac{1}{K_c}\ [\boldsymbol{c}]\ [\boldsymbol{C}]$ at equilibrium.

The incorporation rate of the correct amino acid is then

$$\begin{aligned}\boldsymbol{R}_{correct} &= (\text{concentration of the bound complex } \boldsymbol{cC}) \\ &\quad \times (\text{probability amino acid is linked to the protein chain}) \\ &= [\boldsymbol{cC}]\ p \\ &= \frac{p}{K_c}\ [\boldsymbol{c}]\ [\boldsymbol{C}].\end{aligned}$$

The incorrect **tRNA** will be represent by $\boldsymbol{d}$ and it binds in a similar way to form the complex $\boldsymbol{dC}$. The rates of combination and breaking apart are now given by k_d and k'_d. The same reasoning as before gives us the model

$$[\boldsymbol{d}] + [\boldsymbol{C}] \underset{k'_d}{\overset{k_d}{\longleftrightarrow}} [\boldsymbol{dC}] \underset{p}{\longrightarrow} \text{incorrect amino acid}$$

And at equilibrium (we use the same reasoning!), we find $[\boldsymbol{dC}] = \frac{1}{K_d}\ [\boldsymbol{d}]\ [\boldsymbol{C}]$ where $K_d = \frac{k'_d}{k_d}$. The rate of incorrect linking is then

$$\begin{aligned}\boldsymbol{R}_{incorrect} &= \frac{\text{incorrect rate}}{\text{correct rate}} \\ &= \frac{p\ [\boldsymbol{d}]\ [\boldsymbol{C}]}{K_d} \Big/ \frac{p\ [\boldsymbol{c}]\ [\boldsymbol{C}]}{K_c}.\end{aligned}$$

We can cancel the common p and $[\boldsymbol{C}]$ to get

$$\begin{aligned} \boldsymbol{R}_{incorrect} &= \frac{\text{incorrect rate}}{\text{correct rate}} \\ &= \frac{[\boldsymbol{d}]}{K_d} / \frac{[\boldsymbol{c}]}{K_c} \\ &= \frac{K_c}{K_d} \frac{[\boldsymbol{d}]}{[\boldsymbol{c}]}. \end{aligned}$$

Now in a cell, we know from experimental data that the concentrations of the incorrect and correct **tRNA**'s are about the same. Hence, $\frac{[\boldsymbol{d}]}{[\boldsymbol{c}]} \approx 1$ and we can say with a little bit of algebra that

$$\boldsymbol{R}_{incorrect} \approx \frac{K_c}{K_d} = \frac{k_c'}{k_c} \frac{k_d}{k_d'}.$$

From experimental data, we also know the *on* rate for binding of both $\boldsymbol{d}$ and $\boldsymbol{c}$ are limited by how molecules diffuse through the cell and because of that are about the same. Hence, $\frac{k_d}{k_c} \approx 1$ which leads to our final result.

$$\boldsymbol{R}_{incorrect} \approx \frac{k_c'}{k_d'}.$$

Now recall these terms k_d' and k_c' are rates of breaking apart the complexes. So since the incorrect binding is weaker, we know $k_d' > k_c'$ and so we know $\boldsymbol{R}_{incorrect} < 1$. We also can measure these rates and they have been determined to give the ratio $\frac{k_c'}{k_d'} \approx 10^{-2}$. So the error rate, determined by an equilibrium analysis is 100 times higher than the true rate 10^{-4}! What is wrong with our analysis? How do we explain this?

12.6.2 The Second Attempt: Kinetic Proofreading

The way to handle this discrepancy is to introduce the idea of **kinetic proofreading**. It has always been known that there is an extra step in the protein construction we saw in the ribosome. Expand our previous picture of ribosome processing to look like Fig. 12.12. Note there is a new intermediate state $\boldsymbol{c}^*$ which is formed by adding a hydroxyl $\boldsymbol{OH}$ to a special molecule on $\boldsymbol{c}$. This addition is **essentially irreversible**!

Our pathway for the reactions going on can now be rewritten so include this intermediate step,

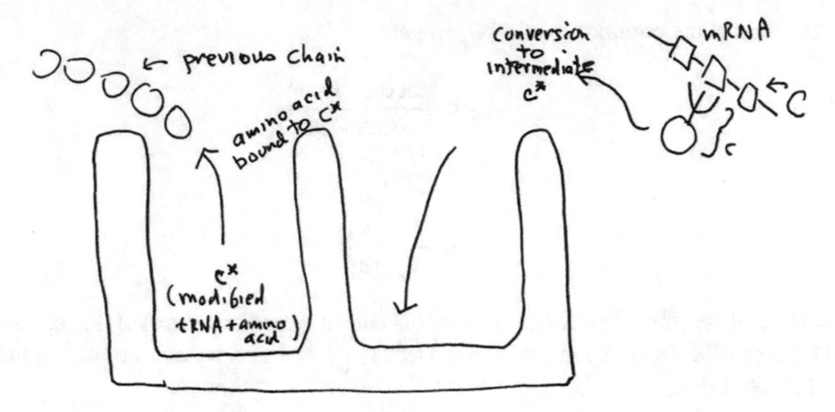

Fig. 12.12 Modified ribosome processing details

$$\begin{array}{ccccccc} & k_c & & \ell_c & & p & \\ [\boldsymbol{c}]+[\boldsymbol{C}] & \longleftrightarrow & [\boldsymbol{cC}] & \longrightarrow & [\boldsymbol{c^*C}] & \longrightarrow & \text{correct amino acid} \\ & k'_c & & & & \downarrow \ell'_c & \\ & & & & & [\boldsymbol{c}]+[\boldsymbol{C}] & \end{array}$$

where ℓ_c is the rate at which $[\boldsymbol{cC}]$ is converted to the intermediate complex $[\boldsymbol{c^*C}]$ and ℓ'_c is the rate at which $[\boldsymbol{c^*C}]$ is broken down into the components $[\boldsymbol{c}]$ plus $[\boldsymbol{C}]$. The important thing is that the modified complex, $[\boldsymbol{c^*C}]$, can *fall off* the ribosome by breaking off from the **mRNA** $\boldsymbol{C}$.

The concentration $[\boldsymbol{c^*C}]$ at equilibrium is given by the balance of the ℓ_c and ℓ'_c rates. We will ignore the pathway leading to the amino acid being linked to the previous amino acid chain as this rate is known to be much smaller that the rates given by ℓ_c and ℓ'_c. So at equilibrium, using an analysis much like we did before,

$$(\text{rate at which } \boldsymbol{c^*C} \text{forms}) = (\text{rate at which } \boldsymbol{c^*C} \text{is broken apart}).$$

Mathematically, this tells us

$$\ell_c\,[\boldsymbol{cC}] = \ell'_c\,\left[\boldsymbol{c^*C}\right] \Longrightarrow \left[\boldsymbol{c^*C}\right] = \frac{\ell_c}{\ell'_c}\,[\boldsymbol{cC}].$$

The ratio of correct incorporation is thus (we use the same arguments as before)

$$\begin{aligned} \boldsymbol{R}_{correct} &= (\text{concentration of the bound complex } \boldsymbol{cC}) \\ &\quad \times (\text{probability amino acid is linked to the protein chain}) \\ &= \left[\boldsymbol{c^*C}\right]\,p \\ &= p\,\frac{\ell_c}{\ell'_c}\,[\boldsymbol{c}]\,[\boldsymbol{C}]. \end{aligned}$$

where in the second line, we use the intermediate $[\boldsymbol{c}^*\boldsymbol{C}]$ instead of $[\boldsymbol{c}\boldsymbol{C}]$ like before. Finally, we also know that

$$[\boldsymbol{cC}] = \frac{k_c}{k'_c}\ [\boldsymbol{c}]\ [\boldsymbol{C}]$$

and so we have

$$\boldsymbol{R}_{correct} = p\ \frac{\ell_c}{\ell'_c}\ \frac{k_c}{k'_c}\ [\boldsymbol{c}]\ [\boldsymbol{C}].$$

Now do the same thing for the closest incorrect **tRNA** complex. This gives

$$\boldsymbol{R}_{incorrect} = p\ \frac{\ell_d}{\ell'_d}\ \frac{k_d}{k'_d}\ [\boldsymbol{d}]\ [\boldsymbol{C}].$$

where ℓ_d and ℓ'_d are the rates for the incorrect formation and breaking apart. Now form the ratio

$$\begin{aligned}
\boldsymbol{F} = \frac{\boldsymbol{R}_{incorrect}}{\boldsymbol{R}_{correct}} &= \frac{p\ \frac{\ell_d}{\ell'_d}\ \frac{k_c}{k'_c}\ [\boldsymbol{d}]\ [\boldsymbol{C}]}{p\ \frac{\ell_c}{\ell'_c}\ \frac{k_c}{k'_c}\ [\boldsymbol{c}]\ [\boldsymbol{C}]} \\
&= \frac{\frac{\ell_d}{\ell'_d}\ \frac{k_d}{k'_d}\ [\boldsymbol{c}]}{\frac{\ell_c}{\ell'_c}\ \frac{k_c}{k'_c}\ [\boldsymbol{c}]} \\
&= \frac{\ell_d}{\ell'_d}\ \frac{\ell'_c}{\ell_c}\ \frac{k_d}{k'_d}\ \frac{k'_c}{k_c}\ \frac{[\boldsymbol{d}]}{[\boldsymbol{c}]} \\
&= \frac{\ell_d}{\ell_c}\ \frac{\ell'_c}{\ell'_d}\ \frac{k_d}{k_c}\ \frac{k'_c}{k'_d}\ \frac{[\boldsymbol{d}]}{[\boldsymbol{c}]}
\end{aligned}$$

We know the conversion rate for $\boldsymbol{c}$ to $\boldsymbol{c}^*$ is the same as the conversion rate for $\boldsymbol{d}$ to $\boldsymbol{d}^*$ because adding the hydroxyl $\boldsymbol{OH}$ is the same not matter what **tRNA** we are working with. Hence, $\frac{\ell_d}{\ell_c} \approx 1$. Thus, we have

$$\boldsymbol{F} \approx \frac{\ell'_c}{\ell'_d}\ \frac{k_d}{k_c}\ \frac{k'_c}{k'_d}\ \frac{[\boldsymbol{d}]}{[\boldsymbol{c}]}$$

In our first argument, we reasoned that $k_c \approx k_d$. This also implies that $[\boldsymbol{d}] \approx [\boldsymbol{c}]$, so we can further simply to

$$\boldsymbol{F} \approx \frac{\ell'_c}{\ell'_d}\ \frac{k'_c}{k'_d}.$$

Finally, the *off–on* ratio for the modified **tRNA**, $\frac{\ell'_c}{\ell'_d}$ and the *off–on* ratio for the unmodified **tRNA**, $\frac{k'_c}{k'_d}$ should be the same because all are recognized by the **mRNA** triplet we call $\boldsymbol{C}$. So we can write

$$\boldsymbol{F} \approx \left(\frac{k'_c}{k'_d}\right)^2.$$

We know experimentally that $\frac{k'_c}{k'_d} \approx 10^{-2}$, so we must have $\boldsymbol{F} \approx 10^{-4}$ which is what we observe experimentally!! So **kinetic proofreading** caused by the **tRNA** modification step reduces the error rate by a factor of 100. Cool beans.

Chapter 13
Logistics Models

It is now time to look at another first order model which is of great interest to us as modelers. This model gives solutions that look like what we see in the lab when we have bacterial growth when there are limited resources. We see a rapid increase in population and then a leveling off as the resources are depleted. We are going to develop a model for this.

13.1 The Model

Another first order differential equation that is of interest is the one known as the Logistic Model or Logistics Differential Equation. This has the form

$$\begin{aligned} u'(t) &= \alpha\, u(t)\, (L - u(t)), \\ u(0) &= u_0 \end{aligned}$$

The positive parameter L is called the carrying capacity and the positive parameter α controls how quickly the solution approaches the carrying capacity L. We will see the solutions to this problem either start out above L and decay asymptotically to L from above or they begin below L and grow up to L from below. We could also solve this problem for a negative α but that is less interesting as a modeling tool. Now the case where the initial condition is $u_0 = 0$ is not very interesting either. In that case we get the initial value of the derivative is

$$\begin{aligned} u'(0) &= \alpha\, u(0)\, (L - u(0)), \\ &= \alpha\, 0\, L = 0. \end{aligned}$$

It is then easy to see the solution $u(t) = 0$ works for all time. So the zero initial condition leads to the uninteresting solution of u is always 0. A similar argument shows that is $u(0) = L$, then the solution u is the constant L. We will therefore

J.K. Peterson, *Calculus for Cognitive Scientists*, Cognitive Science and Technology, DOI 10.1007/978-981-287-874-8_13

assume that we have $0 < u_0 < L$ or $u_0 > L$. We can think of the rate of change of u as determined by the balance between a growth term and a decay or loss term.

- Let $u'_{growth} = \alpha\, L\, u$ and $u'_{decay} = -\alpha\, u^2$. Then

$$
\begin{aligned}
u' &= u'_{growth} + u'_{decay} \\
&= \alpha\, L\, u - \alpha\, u^2 \\
&= \alpha\, u\, (L - u).
\end{aligned}
$$

- The growth term could be a birth rate model where new members of the population grow following an exponential growth model.
- The decay term is an example of how to model interaction between populations. Given two populations $x(t)$ and $y(t)$, we can model their interaction with the product of the population sizes. Often the decay rate of change of a populations x and y would then have interaction components

$$
\begin{aligned}
x'_{I,\, decay} &\propto x(t)\; y(t) \\
y'_{I,\, decay} &\propto x(t)\; y(t)
\end{aligned}
$$

  You see this kind of population interaction in the Predator–Prey model and the Simple Disease model which are discussed in Peterson (2015). Here, since there is only one variable and so this is called **self interaction**. In general, a self interaction decay term would be modeled by

$$
x'_{SI,\, decay} = -a\; x(t)\; x(t) = -ax^2(t)
$$

  So for logistic models, the decay component is a **self-interaction** term.

13.1.1 An Integration Side Trip: Partial Fraction Decompositions

When we solve logistics models, we will have to integrate a function like $\frac{1}{x\,(100-x)}$. This does not fit into a simple substitution method at all. The way we do this kind of problem is to split the fraction $\frac{1}{x\,(100-x)}$ into the sum of the two simpler fractions $\frac{1}{x}$ and $\frac{1}{100-x}$. This is called the **Partial Fractions Decomposition** approach. Hence, we want to find numbers A and B so that

$$
\frac{1}{x\,(100-x)} = \frac{A}{x} + \frac{B}{100-x}
$$

If we multiply both sides of this equation by the term $x\ (100 - x)$, we get the new equation

$$1 = A\ (100 - x) + B\ x$$

Since this equation holds for $x = 0$ and $x = 100$, we can evaluate the equation twice to get

$$\begin{aligned} 1 &= \{A\ (100 - x) + B\ x\}\ |_{x=0} \\ &= 100\ A \\ 1 &= \{A\ (100 - x) + B\ x\}\ |_{x=100} \\ &= 100\ B \end{aligned}$$

Thus, we see A is 1/100 and B is 1/100. Hence,

$$\frac{1}{x\ (100 - x)} = \frac{1/100}{x} + \frac{1/100}{100 - x}$$

We could then integrate as follows

$$\begin{aligned} \int \frac{1}{x(100 - x)} dx &= \int \left(\frac{1/100}{x + 2} + \frac{1/100}{100 - x} \right) dx \\ &= \int \frac{1/100}{x}\ dx + \int \frac{1/100}{100 - x}\ dx \\ &= 1/100 \int \frac{1}{x}\ dx - 1/100 \int \frac{1}{x - 100}\ dx \\ &= 1/100\ \ln(|\ x\ |) - 1/100\ \ln(|\ x - 100\ |) + C \\ &= 1/100\ \ln \left(\frac{|\ x\ |}{|\ x - 100\ |} \right) + C \end{aligned}$$

We will use this technique when we solve the logistic models.

13.1.1.1 Homework

Exercise 13.1.1 *Evaluate* $\int \frac{1}{u\ (200-u)}\ du$.

Exercise 13.1.2 *Evaluate* $\int \frac{1}{z\ (300-z)}\ dz$.

Exercise 13.1.3 *Evaluate* $\int \frac{1}{s\ (5-s)}\ ds$.

Exercise 13.1.4 *Evaluate* $\int \frac{1}{w\ (10-w)}\ dw$.

13.1.2 Examples

Example 13.1.1 Let's solve

$$u'(t) = 3u(t)\,(100 - u(t)),$$
$$u(0) = 25$$

Solution *Note here $L = 100$ and $\alpha = 3$. The solution has quite a few steps!*

Step 1: *Divide by the two u terms to find*

$$\frac{u'(t)}{u(t)\,(100 - u(t))} = 3.$$

Step 2: *We integrate both sides with respect to s from* 0 *to t:*

$$\int_0^t \frac{u'(s)}{u(s)\,(100 - u(s))}\,ds = \int_0^t 3\,ds$$

Step 3: *Letting $w = u(s)$, a standard substitution then gives*

$$\int_{s=0}^{s=t} \frac{dw}{w\,(100 - w)} = 3t$$

Step 4: *We look for constants A and B so that*

$$\frac{1}{w\,(100 - w)} = \frac{A}{w} + \frac{B}{100 - w}, \text{ rewrite as}$$
$$1 = A(100 - w) + Bw$$

This integration trick is called Partial Fraction Decomposition *and we use it in other places too, but you can see, it is a crucial tool in solving logistic models. At $w = 0$, we find $A = 1/100$; at $w = 100$, $B = 1/100$. We will always find $A = B = 1/L$. Hence,*

$$\int_{s=0}^{s=t} \frac{dw}{w\,(100 - w)} = \frac{1}{100}\int_{s=0}^{s=t} \frac{dw}{w} + \frac{1}{100}\int_{s=0}^{s=t} \frac{dw}{100 - w}$$

Thus

$$\frac{1}{100}\int_{s=0}^{s=t}\frac{dw}{w}+\frac{1}{100}\int_{s=0}^{s=t}\frac{dw}{100-w}=3t$$

Multiply by 100.

$$\int_{s=0}^{s=t}\frac{dw}{w}+\int_{s=0}^{s=t}\frac{dw}{100-w}=300t$$

Rewrite second integral a bit

$$\int_{s=0}^{s=t}\frac{dw}{w}-\int_{s=0}^{s=t}\frac{dw}{w-100}=300t$$

Step 5: (Integration of the LHS) *Integrate:*

$$\ln\mid w\mid\Bigg|_{s=0}^{s=t}-\ln\mid w-100\mid\Bigg|_{s=0}^{s=t}=300t.$$

But $w = u(s)$ *so evaluate to get*

$$\begin{aligned}\ln|u(t)|-\ln|u(0)|\\ -\ln|u(t)-100|+\ln|u(0)-100|&=300t\\ \Big(\ln|u(t)|-\ln|u(t)-100|\Big)\\ +\Big(\ln|u(0)-100|-\ln|u(0)|\Big)&=300t.\end{aligned}$$

Use difference logarithm rule:

$$\ln\left(\frac{|u(t)|}{|u(t)-100|}\right)+\ln\left(\frac{|u(0)-100|}{|u(0)|}\right)=300t.$$

Use sum logarithm rule:

$$\ln\left(\frac{|u(t)|}{|u(t)-100||}\;\frac{|u(0)-100|}{|u(0)|}\right)=300t.$$

Now use the IC $u(0) = 25$. *This gives* $|u(0) - 100|/|u(0)| = 3$ *and so*

$$\ln\left(3\;\frac{|u(t)|}{|u(t)-100|}\right)=300t$$

Step 6: *Exponentiate both sides:*

$$\left|\frac{3|u(t)|}{|u(t) - 100|}\right| = e^{300t}$$

Rewrite as

$$|u(t)| = \frac{1}{3}\,|u(t) - 100|\,e^{300\,t}$$

Step 7: *Resolve absolute values. This argument is like the one we did for the indirect and direct method of solving a rate model when we had to resolve the absolute value there as well. But as you can see it is more complicated!*

- *If $u(t)$ changed signs, there is a time, t^*, where $u(t^*) = 0$. Using Step 6, $|u(t^*)| = \frac{1}{3}\,|u(t^*) - 100|\,e^{300\,t^*}$ or $0 = \frac{100}{3}\,e^{300\,t^*}$. But $\frac{100}{3}\,e^{300\,t^*} > 0$. So $u(t)$ can not change sign. Since $u(0) = 25 > 0$, $u(t)$ is always positive. So $|u(t)| = u(t)$.*
- *If $u(t) - 100$ changed signs, there is a time, t^*, where $u(t^*) = 100$. Using Step 6, $|u(t^*)| = \frac{1}{3}\,|u(t^*) - 100|\,e^{300\,t^*}$ or $100 = \frac{0}{3}\,e^{300\,t^*}$. But $100 = 0$ is not possible. Hence, $u(t)$ can not cross 100. Since $u(0) = 25 < 100$, we have $u(t) < 100$ always. So $|u(t) - 100| = 100 - u(t)$.*

Step 8: *From Step 7, we know $0 < u(t) < 100$ always. Hence, $|u(t)| = u(t)$ and $|u(t) - 100| = 100 - u(t)$. Using this, we find*

$$\begin{aligned}
u(t) &= \frac{1}{3}\,(100 - u(t))\,e^{300\,t} \\
e^{-300t}u(t) &= \frac{1}{3}\,(100 - u(t)) \\
\left(e^{-300t} + \frac{100}{3}\right)u(t) &= \frac{100}{3} \\
u(t) &= \frac{\frac{100}{3}}{e^{-300t} + \frac{1}{3}}.
\end{aligned}$$

Step 9: *Checks: from our solution, we see*

$$\begin{aligned}
u(0) &= \frac{\frac{100}{3}}{1 + \frac{1}{3}} = \frac{\frac{100}{3}}{\frac{4}{3}} = \frac{100}{4} = 25. \\
\lim_{t\to\infty} u(t) &= \frac{\frac{100}{3}}{\frac{1}{3}} = 100.
\end{aligned}$$

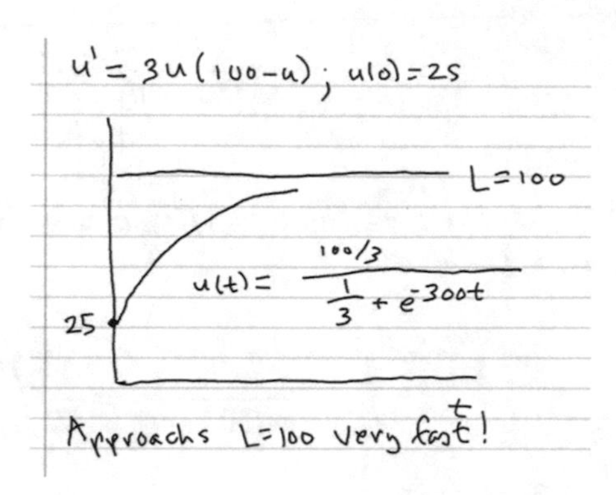

Fig. 13.1 Solution to $u'(t) = 3u(t)\,(100 - u(t))$, $u(0) = 25$

Step 10: The Solution Sketch: *A hand drawn sketch is shown in Fig. 13.1.*

Example 13.1.2 Let's solve

$$u'(t) = 2u(t)\,(50 - u(t)),$$
$$u(0) = 80$$

Solution *The solution still has quite a few steps, but it will get better. Also, these problems are a great way to use all the things we know about exponentials and logarithms!*

Step 1: *Divide by the two u terms to find*

$$\frac{u'(t)}{u(t)\,(50 - u(t))} = 2.$$

Step 2: *We integrate both sides with respect to s from* 0 *to* t:

$$\int_0^t \frac{u'(s)}{u(s)\,(50 - u(s))}\, ds = \int_0^t 2\, ds$$

Step 3: *Letting* $w = u(s)$, *a standard substitution then gives*

$$\int_{s=0}^{s=t} \frac{dw}{w\,(50 - w)} = 2t$$

We now need the partial fractions decomposition (PFD) technique again.
Step 4: *We look for constants A and B so that*

$$\frac{1}{w\,(50-w)} = \frac{A}{w} + \frac{B}{50-w}, \text{ rewrite as}$$
$$1 = A(50-w) + Bw$$

At $w = 0$, we find $A = 1/50$; at $w = 50$, $B = 1/50$. Note again that $A = B = 1/L$. Hence,

$$\int_{s=0}^{s=t} \frac{dw}{w\,(50-w)} = \frac{1}{50}\int_{s=0}^{s=t} \frac{dw}{w} + \frac{1}{50}\int_{s=0}^{s=t} \frac{dw}{50-w}$$

Thus

$$\frac{1}{50}\int_{s=0}^{s=t} \frac{dw}{w} + \frac{1}{50}\int_{s=0}^{s=t} \frac{dw}{50-w} = 2t$$

Multiply by 50.

$$\int_{s=0}^{s=t} \frac{dw}{w} + \int_{s=0}^{s=t} \frac{dw}{50-w} = 100t$$

Rewrite second integral a bit

$$\int_{s=0}^{s=t} \frac{dw}{w} - \int_{s=0}^{s=t} \frac{dw}{w-50} = 100t$$

Step 5: (Integration of the LHS) *Integrate:*

$$\ln \mid w \mid \Big|_{s=0}^{s=t} - \ln \mid w - 50 \mid \Big|_{s=0}^{s=t} = 100t.$$

But $w = u(s)$ so evaluate to get

$$\begin{aligned} &\ln|u(t)| - \ln|u(0)| \\ &- \ln|u(t)-50| + \ln|u(0)-50| = 100t \\ &\Big(\ln|u(t)| - \ln|u(t)-50|\Big) \\ &+\Big(\ln|u(0)-50| - \ln|u(0)|\Big) = 100t. \end{aligned}$$

Use difference logarithm rule:

$$\ln\left(\frac{|u(t)|}{|u(t)-50|}\right) + \ln\left(\frac{|u(0)-50|}{|u(0)|}\right) = 100t.$$

Use sum logarithm rule:

$$\ln\left(\frac{|u(t)|}{|u(t)-50|}\frac{|u(0)-50|}{|u(0)|}\right) = 100t.$$

Now use the IC $u(0) = 80$. *This gives* $|u(0) - 50|/|u(0)| = 3/8$ *and so*

$$\ln\left(\frac{3}{8}\,\frac{|u(t)|}{|u(t)-50|}\right) = 100t$$

Step 6: *Exponentiate both sides:*

$$\left.\frac{\frac{3}{8}\,|u(t)|}{|u(t)-50|}\right| = e^{100t}$$

Rewrite as

$$|u(t)| = \frac{8}{3}\,|u(t)-50|\,e^{100\,t}$$

Step 7: *Resolve absolute values.*

- *If* $u(t) - 50$ *changed signs, there is a time,* t^*, *where* $u(t^*) = 50$. *Using Step 6,* $|u(t^*)| = \frac{8}{3}\,|u(t^*) - 50|\,e^{100\,t^*}$ *or* $50 = \frac{8}{3}\,0\,e^{100\,t^*}$. *But* $50 = 0$ *is not possible. Hence,* $u(t)$ *can not cross* 50. *Since* $u(0) = 80 > 50$, *we have* $u(t) > 50$ *always. So* $|u(t) - 50| = u(t) - 50$.

Step 8: *From Step 7, we know* $u(t) > 50$ *always. Hence,* $|u(t)| = u(t)$ *and* $|u(t) - 50| = u(t) - 50$. *Using this, we find*

$$\begin{aligned}
u(t) &= \frac{8}{3}\,(u(t)-50)\,e^{100\,t}\\
e^{-100t}u(t) &= \frac{8}{3}\,(u(t)-50)\\
\left(e^{-100t}-\frac{8}{3}\right)u(t) &= -\frac{400}{3}\\
u(t) &= \frac{\frac{400}{3}}{\frac{8}{3}-e^{-100t}}.
\end{aligned}$$

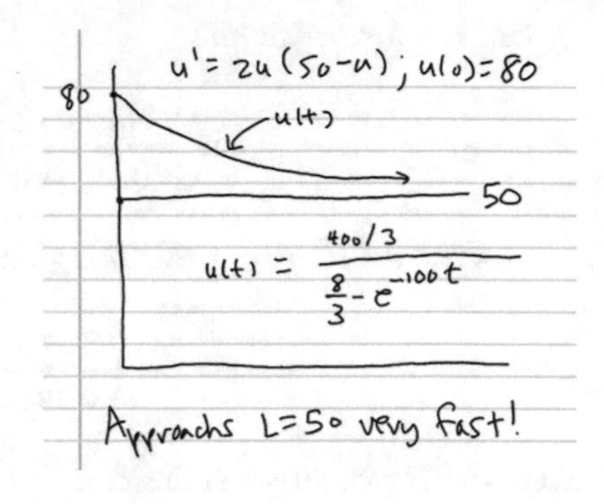

Fig. 13.2 Solution to $u'(t) = 2u(t)\,(50 - u(t))$; $u(0) = 80$

Step 9: *Checks: from our solution, we see*

$$u(0) = \frac{\frac{400}{3}}{\frac{8}{3} - 1} = \frac{\frac{400}{3}}{\frac{5}{3}} = \frac{400}{5} = 80.$$

$$\lim_{t \to \infty} u(t) = \frac{\frac{400}{3}}{\frac{8}{3}} = 50.$$

Step 10: The Solution Sketch: *A hand drawn sketch is shown in Fig. 13.2.*

13.2 The General Solution Method

Now we know how to solve the general logistics model:

$$u'(t) = \alpha\, u(t)\, (L - u(t)),$$
$$u(0) = u_0$$

We follow the steps shown in the examples.

Step 1: Divide by the two u terms to find

$$\frac{u'(t)}{u(t)\,(L - u(t))} = \alpha.$$

Step 2: We integrate both sides with respect to s from 0 to t:

$$\int_0^t \frac{u'(s)}{u(s)\,(L - u(s))}\, ds = \int_0^t \alpha\, ds$$

Step 3: Letting $w = u(s)$, a standard substitution then gives

$$\int_{s=0}^{s=t} \frac{dw}{w\,(L - w)} = \alpha t$$

The left hand side requires a partial fractions decomposition.

Step 4: We look for constants A and B so that

$$\frac{1}{w\,(L - w)} = \frac{A}{w} + \frac{B}{L - w}$$

or

$$1 = A(L - w) + Bw$$

At $w = 0$ and $w = L$, we get $A = 1/L$ and $B = 1/L$. Hence, we can rewrite the left hand side integration to find

$$\int_{s=0}^{s=t} \frac{dw}{w\,(L - w)} = \int_{s=0}^{s=t} (1/L)\frac{dw}{w} + \int_{s=0}^{s=t} (1/L)\frac{dw}{L - w}.$$

Thus, we have

$$(1/L) \int_{s=0}^{s=t} \frac{dw}{w} + (1/L) \int_{s=0}^{s=t} \frac{dw}{L - w} = \alpha t.$$

Multiply by L.

$$\int_{s=0}^{s=t} \frac{dw}{w} + \int_{s=0}^{s=t} \frac{dw}{L - w} = \alpha\, L\, t.$$

Step 5: (Integration of the LHS): Performing the integrations, we have

$$\ln \mid w \mid \Big|_{s=0}^{s=t} - \ln \mid L - w \mid \Big|_{s=0}^{s=t} = \alpha L\, t.$$

But w = u(s) so evaluate to get

$$\ln \mid u(s) \mid \Big|_{0}^{t} - \ln \mid L - u(s) \mid \Big|_{0}^{t} = \alpha L\, t.$$

or

$$\ln \mid u(t) \mid - \ln |u(0)| - \ln \mid u(t) - L \mid - \ln |u(0) - L| = \alpha L\, t.$$

Use difference logarithm rule:

$$\ln\left(\frac{|u(t)|}{|u(0)|}\right) + \ln\left(\frac{|u(0) - L|}{|u(0) - L|}\right) = \alpha\, L\, t.$$

Use the sum logarithm rule:

$$\ln\left(\left|\frac{u(t)}{L - u(t)}\right| - \ln\left|\frac{u_0}{L - u_0}\right|\right) = \ln\left(\left|\frac{u(t)}{L - u(t)}\; \frac{L - u(0)}{u(0)}\right|\right) = \alpha\, L\, t.$$

Now use the initial condition $u(0) = u_0$ and put all the function terms together and the constant terms together.

$$\ln\left(\left|\frac{u(t)}{L - u(t)}\; \frac{L - u_0}{u_0}\right|\right) = \alpha L\, t.$$

Step 6: Exponentiate both sides:

$$\left|\frac{u(t)}{L - u(t)}\; \frac{L - u_0}{u_0}\right| = e^{\alpha\, L\, t}$$

Now, multiply both sides by $|\frac{u_0}{L-u_0}|$ to obtain

$$\left|\frac{u(t)}{L - u(t)}\right| = \left|\frac{u_0}{L - u_0}\right| e^{\alpha L\, t}. \tag{13.1}$$

Step 7: Resolve absolute values. We will do this by looking at the two cases.

(i): $0 < u_0 < L$,
(ii): $u_0 > L$.

Step 7A: $0 < u_0 < L$: In case (i), $L - u_0 > 0$ and so the term $\frac{u_0}{L-u_0}$ is positive and no absolute value is needed. Thus, we have

$$\left|\frac{u(t)}{L - u(t)}\right| = \frac{u_0}{L - u_0}\, e^{\alpha L\, t}.$$

Here u starts below L so to determine the absolute values, we need to know if $u(t)$ can change sign; i.e. can $u(t)$ be zero and also if $u(t) - L$ can change sign; i.e. can $u(t) = L$. We can rewrite Eq. 13.1 as

$$|\, u(t)\, | = |\, L - u(t)\, |\, \left|\frac{u_0}{L - u_0}\right| e^{\alpha L\, t}.$$

If there is a time, t^*, where $u(t^*) = 0$, then

$$0 = \mid u(t^*) \mid = L \left| \frac{u_0}{L - u_0} \right| e^{\alpha L t^*}.$$

Now, the term $\frac{u_0}{L-u_0}\, e^{\alpha L t^*}$ is positive. So we have 0 equal to a positive number which is not possible. Hence, $u(t)$ can't become 0. Since $u(0) = u_0 > 0$, this means $u(t)$ must always be positive and so $|u(t)| = u(t)$. So, what about the second possibility? Then, we assume there is a time, t^*, so that $u(t^*) = L$. Again, evaluating at the point t^*, we get

$$\mid u(t^*) \mid = \mid L - u(t^*) \mid \; \frac{u_0}{L - u_0}\, e^{\alpha L t^*} = \mid L - L \mid \; \frac{u_0}{L - u_0}\, e^{\alpha L t^*} = 0.$$

But the left hand side is positive and the right side is 0. Hence, $u(t^*)$ can't be L either. So our assumption that there is a time t^* where $u(t) = L$ is wrong and so since $u(t)$ starts below L, it must always stay below L. We see $|u(t) - L| = L - u(t)$.

Step 7B: $u_0 > L$: In case (ii), $u_0 > L$ and so the term $\frac{u_0}{L-u_0}$ is negative, Thus, we have to switch the $|u_0 - L|$ term to $u_0 - L$ to get

$$\left| \frac{u(t)}{L - u(t)} \right| = \frac{u_0}{u_0 - L}\, e^{\alpha L t}.$$

Here u starts above L so to determine the absolute values, we need to know if $u(t)$ can change sign; i.e. can $u(t)$ be zero and also if $u(t) - L$ can change sign; i.e. can $u(t) = L$. Since $u(0)$ starts above L, we only have to check to see if $u(t)$ can cross L. If it can't cross L, $u(t)$ certainly can't go down to 0! We can rewrite Eq. 13.1 as

$$\mid u(t) \mid = \mid L - u(t) \mid \left| \frac{u_0}{u_0 - L} \right| e^{\alpha L t}.$$

Now assume there is a time, t^*, so that $u(t^*) = L$. Again, evaluating at the point t^*, we get

$$\mid u(t^*) \mid = \mid L - u(t^*) \mid \; \frac{u_0}{u_0 - L}\, e^{\alpha L t^*} = \mid L - L \mid \; \frac{u_0}{u_0 - L}\, e^{\alpha L t^*} = 0.$$

But the left hand side is positive and the right side is 0. Hence, $u(t^*)$ can't be L either. So our assumption that there is a time t^* where $u(t) = L$ is wrong and so since $u(t)$ starts above L, it must always stay above L. We see $|u(t) - L| = u(t) - L$ and since $u(t) > L$ always, we also have $|u(t)| = u(t)$.

Step 8: Solve for u**;** Once the absolute values have been resolved, we can solve for the solution $u(t)$.

Step 8A: $0 < u_0 < L$; We can thus rewrite Eq. 13.1 as

$$\frac{u(t)}{L - u(t)} = \frac{u_0}{L - u_0}\, e^{\alpha L t}.$$

Now multiply both sides by $L - u(t)$:

$$u(t) = (L - u(t))\, \frac{u_0}{L - u_0}\, e^{\alpha L t}.$$

Next, move all the u terms to the left hand side to give

$$\left(1 + \frac{u_0}{L - u_0}\, e^{\alpha L t}\right) u(t) = L\, \frac{u_0}{L - u_0}\, e^{\alpha L t}.$$

Then, upon division, we have

$$\begin{aligned} u(t) &= \frac{L\, \frac{u_0}{L - u_0}\, e^{\alpha L t}}{1 + \frac{u_0}{L - u_0}\, e^{\alpha L t}}, \\ &= \frac{L u_0\, e^{\alpha L t}}{(L - u_0) + u_0\, e^{\alpha L t}}, \\ &= \frac{L u_0}{u_0 + (L - u_0)\, e^{-\alpha L t}}. \end{aligned}$$

Hence, our solution for case (i) is

$$u(t) = \frac{L}{1 + (L/u_0 - 1)\, e^{-\alpha L t}}. \tag{13.2}$$

Step 8B: $u_0 > L$: Here $|u(t)| = u(t)$ and $|u(t) - L| = u(t) - L$. We can thus rewrite Eq. 13.1 as

$$\frac{u(t)}{u(t) - L} = \frac{u_0}{u_0 - L}\, e^{\alpha L t}.$$

Now multiply both sides by $u(t) - L$:

$$u(t) = (u(t) - L)\, \frac{u_0}{u_0 - L}\, e^{\alpha L t}.$$

Next, move all the u terms to the left hand side to give

$$\left(1 - \frac{u_0}{u_0 - L}\, e^{\alpha L t}\right) u(t) = -L\, \frac{u_0}{u_0 - L}\, e^{\alpha L t}.$$

Then, upon division, we have

$$\begin{aligned} u(t) &= \frac{-L \, \frac{u_0}{u_0-L} \, e^{\alpha L\, t}}{1 - \frac{u_0}{u_0-L} \, e^{\alpha L\, t}}, \\ &= \frac{-L u_0 \, e^{\alpha L\, t}}{(u_0 - L) - u_0 \, e^{\alpha L\, t}}, \\ &= \frac{-L u_0}{-u_0 + (u_0 - L) \, e^{-\alpha L\, t}}, \\ &= \frac{L u_0}{u_0 + (L - u_0) \, e^{-\alpha L\, t}}. \end{aligned}$$

Hence, our solution for case (ii) is exactly the same as that of case (i):

$$u(t) = \frac{L}{1 + (L/u_0 - 1) \, e^{-\alpha L\, t}}. \tag{13.3}$$

Step 9: Checks: from our solution, we see

$$\begin{aligned} u(0) &= \frac{L}{1 + (L/u_0 - 1)} = \frac{L}{L/u_0} = u_0 \\ \lim_{t \Longrightarrow \infty} u(t) &= \frac{L}{1 + (L/u_0 - 1)\,(0)} = \frac{L}{1} = L. \end{aligned}$$

Step 10: The Solution Sketch This is just like before. If $u_0 > L$, the solution decays down to L and if $0 < u_0 < L$, the solutions increases upward to L.

A few comments are in order:

1. In the partial fractions step, we also know $A = B = 1/L$ but we will still do the steps in detail so we can cement it in our minds.
2. We now know how to resolve the absolute values. There are two cases:
   - Here, the initial condition u_0 is between 0 and L our work has shown that in this case $u(t)$ can't cross L and can't go down to cross 0. So $u(t)$ is always positive and below L. Hence, $|u(t)| = u(t)$ and $|u(t) - L| = L - u(t)$.
   - Here, the initial condition u_0 is above L and so the solution $u(t)$ always stays above L. So we have $|u(t)| = u(t)$ and $|u(t) - L| = u(t) - L$.

   However, we still want you to show all the work for the resolution of the absolute values as it is good practice!

13.2.1 A Streamlined Solution

Example 13.2.1 Let's solve

$$\begin{aligned} u'(t) &= 1.2u(t)\,(150 - u(t)), \\ u(0) &= 50 \end{aligned}$$

Solution Step 1: *Divide by the two u terms to find*

$$\frac{u'(t)}{u(t)\,(150 - u(t))} = 1.2.$$

Step 2: *We integrate both sides with respect to s from* 0 *to t:*

$$\int_0^t \frac{u'(s)}{u(s)\,(150 - u(s))}\,ds = \int_0^t 1.2\,ds$$

Step 3: *Letting $w = u(s)$, a standard substitution then gives*

$$\int_{s=0}^{s=t} \frac{dw}{w\,(150 - w)} = 1.2t$$

We know the partial fractions decomposition (PFD) technique gives us $A = B = 1/L = 1/150$. Make sure you understand how we get these numbers using PFD! So we have

Step 4:

$$\int_{s=0}^{s=t} \frac{dw}{w\,(150 - w)} = \frac{1}{150}\int_{s=0}^{s=t} \frac{dw}{w} + \frac{1}{150}\int_{s=0}^{s=t} \frac{dw}{150 - w}$$

Thus

$$\frac{1}{150}\int_{s=0}^{s=t} \frac{dw}{w} + \frac{1}{150}\int_{s=0}^{s=t} \frac{dw}{150 - w} = 1.2t$$

Multiply by 150.

$$\int_{s=0}^{s=t} \frac{dw}{w} + \int_{s=0}^{s=t} \frac{dw}{150 - w} = 180t$$

Rewrite second integral a bit

$$\int_{s=0}^{s=t} \frac{dw}{w} - \int_{s=0}^{s=t} \frac{dw}{w - 150} = 180t$$

Step 5: (Integration of the LHS) *Integrate:*

$$\ln \mid w \mid \Big|_{s=0}^{s=t} - \ln \mid w - 150 \mid \Big|_{s=0}^{s=t} = 180t.$$

But $w = u(s)$ so evaluate to get

$$\begin{aligned} &\ln |u(t)| - \ln |u(0)| \\ &- \ln |u(t) - 150| + \ln |u(0) - 150| = 180t \\ &\Big(\ln |u(t)| - \ln |u(t) - 150|\Big) \\ &+ \Big(\ln |u(0) - 150| - \ln |u(0)|\Big) = 180t. \end{aligned}$$

Use difference logarithm rule:

$$\ln\left(\frac{|u(t)|}{|u(t) - 150|}\right) + \ln\left(\frac{|u(0) - 150|}{|u(0)|}\right) = 180t.$$

Use sum logarithm rule:

$$\ln\left(\frac{|u(t)|}{|u(t) - 150||} \; \frac{|u(0) - 150|}{|u(0)|}\right) = 100t.$$

Now use the IC $u(0) = 50$. This gives $|u(0) - 150|/|u(0)| = 2$ and so

$$\ln\left(2 \, \frac{|u(t)|}{|u(t) - 150|}\right) = 180t$$

Step 6: *Exponentiate both sides:*

$$\frac{2 \, |u(t)|}{|u(t) - 150|}\Bigg| = e^{180t}$$

Rewrite as

$$|u(t)| = \frac{1}{2} \, |u(t) - 150| \, e^{180\,t}$$

Step 7: *Resolve absolute values. Since the initial condition $u(0) = 50$ is between 0 and $L = 150$, we know $|u(t)| = u(t)$ and $|u(t) - 150| = 150 - u(t)$. You should know how to figure this out! So we can use this to write*

$$
\begin{aligned}
u(t) &= \frac{1}{2}\,(150 - u(t))\,e^{180\,t} \\
e^{-180t}u(t) &= \frac{1}{2}\,(150 - u(t)) \\
\left(e^{-180t} + \frac{1}{2}\right) u(t) &= -\frac{150}{2} \\
u(t) &= \frac{\frac{150}{2}}{\frac{1}{2} + e^{-180t}}.
\end{aligned}
$$

Step 9: *Checks: from our solution, we see*

$$
\begin{aligned}
u(0) &= \frac{\frac{150}{2}}{\frac{1}{2} + 1} = \frac{\frac{150}{2}}{\frac{3}{2}} = 50. \\
\lim_{t \to \infty} u(t) &= \frac{\frac{150}{2}}{\frac{1}{2}} = 150.
\end{aligned}
$$

Step 10: The Solution Sketch: *A hand drawn sketch similar to the others can then be drawn.*

13.3 Solving a Logistics Model on Paper

Here is a worked out example of how we would solve a logistics problem with pen and paper. The images of page 1–4 of our solution are shown in Figs. 13.3, 13.4, 13.5 and 13.6.

13.3.1 Homework

Solve the following logistic models. Do all the steps, do the checks and sketch the solution.

Exercise 13.3.1

$$
\begin{aligned}
u'(t) &= 0.2u(t)\,(300 - u(t)), \\
u(0) &= 350
\end{aligned}
$$

Fig. 13.3 Solution page 1

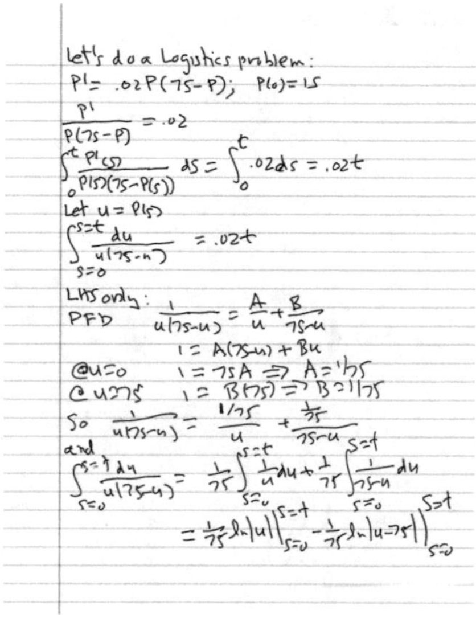

Let's do a Logistics problem:

$P' = .02P(75-P); \quad P(0) = 15$

$\frac{P'}{P(75-P)} = .02$

$\int_0^t \frac{P'(s)}{P(s)(75-P(s))}\, ds = \int_0^t .02\, ds = .02t$

Let $u = P(s)$

$\int_{s=0}^{s=t} \frac{du}{u(75-u)} = .02t$

LHS only:

PFD $\frac{1}{u(75-u)} = \frac{A}{u} + \frac{B}{75-u}$

$1 = A(75-u) + Bu$

@$u=0$ $\quad 1 = 75A \Rightarrow A = 1/75$

@$u=75$ $\quad 1 = B(75) \Rightarrow B = 1/75$

So $\frac{1}{u(75-u)} = \frac{1/75}{u} + \frac{\frac{1}{75}}{75-u}$

and $\int_{s=0}^{s=t} \frac{du}{u(75-u)} = \frac{1}{75}\int_{s=0}^{s=t} \frac{1}{u} du + \frac{1}{75}\int_{s=0}^{s=t} \frac{1}{75-u} du$

$= \frac{1}{75} \ln|u| \Big|_{s=0}^{s=t} - \frac{1}{75} \ln|u-75| \Big|_{s=0}^{s=t}$

Fig. 13.4 Solution page 2

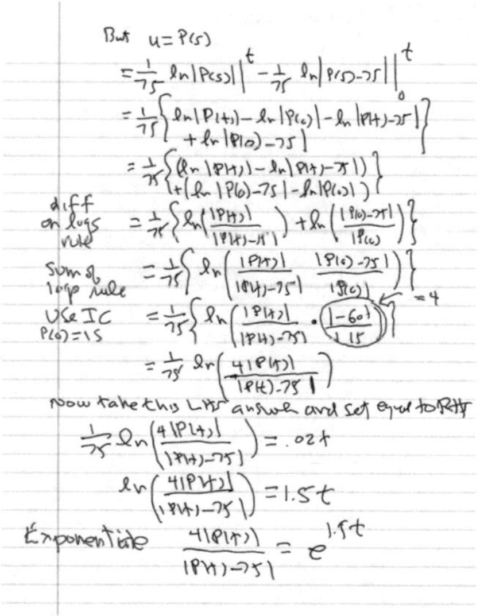

But $u = P(s)$

$= \frac{1}{75} \ln|P(s)| \Big|_0^t - \frac{1}{75} \ln|P(s)-75| \Big|_0^t$

$= \frac{1}{75}\left\{ \ln|P(t)| - \ln|P(0)| - \ln|P(t)-75| + \ln|P(0)-75| \right\}$

$= \frac{1}{75}\left\{ (\ln|P(t)| - \ln|P(t)-75|) + (\ln|P(0)-75| - \ln|P(0)|) \right\}$

diff of logs rule $= \frac{1}{75}\left\{ \ln\left(\frac{|P(t)|}{|P(t)-75|}\right) + \ln\left(\frac{|P(0)-75|}{|P(0)|}\right) \right\}$

Sum of logs rule $= \frac{1}{75}\left\{ \ln\left(\frac{|P(t)|}{|P(t)-75|} \cdot \frac{|P(0)-75|}{|P(0)|}\right) \right\}$

Use IC $P(0)=15$ $= \frac{1}{75}\left\{ \ln\left(\frac{|P(t)|}{|P(t)-75|} \cdot \frac{|-60|}{|15|}\right) \right\}$ $= 4$

$= \frac{1}{75} \ln\left(\frac{4|P(t)|}{|P(t)-75|}\right)$

Now take this LHS answer and set equal to RHS

$\frac{1}{75} \ln\left(\frac{4|P(t)|}{|P(t)-75|}\right) = .02t$

$\ln\left(\frac{4|P(t)|}{|P(t)-75|}\right) = 1.5t$

Exponentiate $\frac{4|P(t)|}{|P(t)-75|} = e^{1.5t}$

Exercise 13.3.2

$$
\begin{aligned}
u'(t) &= 1.2u(t)\,(400 - u(t)),\\
u(0) &= 150
\end{aligned}
$$

Fig. 13.5 Solution page 3

$|P(t)| = \frac{1}{4}|P(t) - 75|\, e^{1.5t}$

Resolve ||'s

$P(0) = 15$ which is < 75

So need to show $P(t)$ can't cross 0 or 75.

(i) Can $P(t)$ cross 0? If so there is t^* so that $P(t^*) = 0$

But then

$0 = |P(t^*)| = \frac{1}{4}|P(t^*) - 75|\, e^{1.5t^*}$

or

$0 = \frac{1}{4}75\, e^{1.5t^*} > 0$ always.

But 0 = pos # is not possible.

So $P(t) > 0$ always since $P(0) = 15 > 0$.

(ii) Can $P(t)$ cross 75?

If so there is t^* so that $P(t^*) = 75$.

Then

$75 = |P(t^*)| = \frac{1}{4}|P(t^*) - 75|\, e^{1.5t^*}$

or

$75 = 0$ which is not possible.

Since $P(0) = 15 < 75$ this means

$P(t) < 75$ always.

Conclude $0 < P(t) < 75$

$\Rightarrow |P(t)| = P(t)$ and $|P(t) - 75| = 75 - P(t)$

Plug this in

$P(t) = \frac{1}{4}(75 - P(t))\, e^{1.5t}$

$e^{-1.5t} P(t) = \frac{1}{4}(75 - P(t)) = \frac{75}{4} - \frac{1}{4}P(t)$

$P(t)\left(e^{-1.5t} + \frac{1}{4}\right) = 75/4$

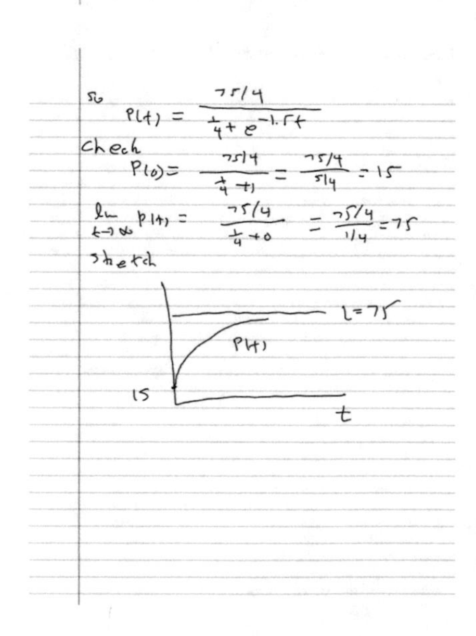

So

$$P(t) = \frac{75/4}{\frac{1}{4} + e^{-1.5t}}$$

Check

$$P(0) = \frac{75/4}{\frac{1}{4} + 1} = \frac{75/4}{5/4} = 15$$

$$\lim_{t \to \infty} P(t) = \frac{75/4}{\frac{1}{4} + 0} = \frac{75/4}{1/4} = 75$$

Sketch

Fig. 13.6 Solution page 4

Exercise 13.3.3

$$
\begin{aligned}
u'(t) &= 5u(t)\,(20 - u(t)), \\
u(0) &= 50
\end{aligned}
$$

Exercise 13.3.4

$$\begin{aligned} u'(t) &= 4u(t)\,(800 - u(t)), \\ u(0) &= 350 \end{aligned}$$

Exercise 13.3.5

$$\begin{aligned} u'(t) &= 0.004u(t)\,(1000 - u(t)), \\ u(0) &= 400 \end{aligned}$$

Reference

J. Peterson, in *Calculus for Cognitive Scientists: Higher Order Models and Their Analysis*. Springer Series on Cognitive Science and Technology, Springer Science+Business Media Singapore Pte Ltd., Singapore (2015), in press

Chapter 14
Function Approximation

In many applications, the model is too hard to deal with directly and so we replace it by a simpler one. We usually do this by using an approximation to the dynamics. So, to learn how to do that, let's discuss how to approximate functions in general.

14.1 Taylor Polynomials

We can approximate a function at a point using polynomials of various degrees. We can first find the constant that best approximates a function f at a point p. This is called the **zeroth order Taylor polynomial** and the equation we get is

$$f(x) = f(p) + E_0(x, p)$$

where $E_0(x, p)$ is the **error**. On the other hand, we could try to find the best straight line that does the job. We would find

$$f(x) = f(p) + f'(p)(x - p) + E_1(x, p)$$

where $E_1(x, p)$ is the **error** now. This straight line is the **first order Taylor polynomial** and we also know it as the **tangent line**. We can continue finding polynomials of higher and higher degree and their respective errors. In this class, our interests stop with the quadratic case. We would find

$$f(x) = f(p) + f'(p)(x - p) + \frac{1}{2} f''(p)(x - p)^2 + E_2(x, p)$$

where $E_2(x, p)$ is the **error**. This is called the **second order Taylor polynomial** or **quadratic approximation**. Now let's dig into the theory behind this so that we can better understand the error terms.

J.K. Peterson, *Calculus for Cognitive Scientists*, Cognitive Science and Technology, DOI 10.1007/978-981-287-874-8_14

14.1.1 Fundamental Tools

Let's consider a function which is defined locally at the point p. This means there is at least an interval (a, b) containing p where f is defined. Of course, this interval could be the whole x axis! Let's also assume f' exists locally at p in this same interval. There are three tools we haven't mentioned yet but it is time to talk about them. They are **Critical Points**, **Rolle's Theorem** and the **Mean Value Theorem**.

14.1.1.1 Critical Points

The first thing to think about is when a function might have a high or low place. This means that, at least locally, there is value for the function which is the highest one around or the lowest one around. Now if a function has a derivative at that high or low point—call this point p—we can say

$$f(x) = f(p) + f'(p)\,(x - p) + E(x - p).$$

Let's take the case where $f(p)$ is the highest value around. So we know $f(x)$ is smaller than $f(p)$ near p. We can then say $f(x) - f(p) < 0$ near p and this tells two things. If $x > p$, then the fraction $(f(x) - f(p))/(x - p) < 0$ as the numerator is negative and the denominator is positive. So we have

$$\frac{f(x) - f(p)}{x - p} = f'(p) + \frac{E(x - p)}{(x - p)} < 0.$$

Letting x go to p from the right hand side, since the error term goes to zero, we must have the right hand derivative $(f'(p))^{+} = f'(p) \leq 0$. On the other hand, if $x < p$, we still have $f(x) < f(p)$ and the fraction $(f(x) - f(p))/(x - p) > 0$ now as both the numerator and denominator are negative. We then have for points on the left that

$$\frac{f(x) - f(p)}{x - p} = f'(p) + \frac{E(x - p)}{(x - p)} > 0.$$

Letting x go to p from the left hand side, since the error term goes to zero, we must have the left hand derivative $(f'(p))^{-} = f'(p) \geq 0$. But we started by assuming f had a derivative at p so these right and left hand derivatives are exactly the same as $f'(p)$. So $0 \leq f'(p) \leq 0$; i.e. $f'(p) = 0$. The argument for the lowest point around is the same. We usually call the largest point locally, a **local maximum** and the smallest point locally, a **local minimum**. Our arguments have proven the following cool theorem.

Theorem 14.1.1 (Critical Points)
If f has a local maximum or local minimum at p where f has a derivative, then $f'(p) = 0$. Because of this, the places where f has a zero derivative are potentially places where f has maximum or minimum values. These are called extreme values.

Proof We just did this argument. ■

14.1.1.2 Rolle's Theorem

Rolle's Theorem is pretty easy to understand. To see this proof, stand up and take a short piece of rope in your hands and follow the reasoning!

Theorem 14.1.2 (Rolle's Theorem)
Let $f : [a, b] \rightarrow \Re$ be a function defined on the interval $[a, b]$ which is continuous on the closed interval $[a, b]$ and is at least differentiable on the open interval (a, b). If $f(a) = f(b)$, then there is at least one point c, between a and b, so that $f'(c) = 0$.

Proof Take a piece of rope in your hand and make sure you hold the right and left side at the same height. When the rope is stretched tight, it represents the graph of a constant function: at all points c, $f'(c) = 0$. If we pull up on the rope in between our hands, the graph goes up to a maximum and at that point c, again we have $f'(c) = 0$. A similar argument holds if we pull down on the rope to get a minimum. ■

14.1.1.3 The Mean Value Theorem

The proof of this is a bit technical, but not too bad. This theorem tells us that if f is smooth enough, we can find a point c so that the $f'(c)$ exactly matches the ratio $\frac{f(b)-f(a)}{b-a}$. We can use this kind of result to do all kinds of interesting things. An easy way to see this is to draw a nice smooth function on the interval of your choice. Smooth here means your function is both continuous (i.e. no jumps) and has a derivative (i.e. no corners). Take any two points on the x axis and draw the line segment connecting the pairs $(a, f(a))$ and $(b, f(b))$. Now look closely at the curve you've drawn. You'll see there is at least one place where the tangent line to the function matches the slope of the line you just drew. That is the idea of the Mean Value Theorem. Now let's do this same argument with a bit of math.

Theorem 14.1.3 (The Mean Value Theorem)
Let $f : [a, b] \rightarrow \Re$ be a function defined on the interval $[a, b]$ which is continuous on the closed interval $[a, b]$ and is at least differentiable on the open interval (a, b). Then there is at least one point c, between a and b, so that

$$\frac{f(b) - f(a)}{b - a} = f'(c).$$

Proof Define $M = (f(x) - f(a))/(x - a)$ and the function $g(t) = f(t) - f(a) - M(t - a)$. Note $g(a) = 0$ and $g(x) = f(x) - f(a) - [(f(x) - f(a))/(x - a)](x - a) = 0$. Apply **Rolle's Theorem** to g. So there is $a < c < x$ so that $g'(c) = 0$. But $g'(t) = f'(t) - M$, so $g'(c) = 0$ tells us $f'(c) = M$. We conclude there is a point c, between a and x so that $f'(c) = [f(x) - f(a)]/(x - a)]$. ■

Now let's go to town and figure out approximations!

14.2 The Zeroth Order Taylor Polynomial

Recall we have a function which is defined locally at the point p. and we assume f' exists locally at p in this same interval. Now pick any x is the interval $[p, b)$ (we can also pick a point in the left hand interval $(a, p]$ but we will leave that discussion to you!).

Our function f on the interval $[p, x]$ satisfies all the requirements of the Mean Value Theorem. So we know there is a point c_x with $p < c_x < x$ so that

$$\frac{f(x) - f(p)}{x - p} = f'(c_x).$$

This can be written as

$$f(x) = f(p) + f'(c_x)(p - a).$$

Let the constant $f(p)$, a polynomial of degree 0, be denoted by

$$P_0(p, x) = f(p).$$

We'll call this the 0th order Taylor Polynomial for f at the point p. Next, let the 0th order error term be defined by

$$E_0(x, p) = f(x) - P_0(p, x) = f(x) - f(p).$$

The error or remainder term is clearly the difference or discrepancy between the actual function value at x and the 0th order Taylor Polynomial. Since $f(x) - f(p) = f'(c_x)(x - p)$, we can write all we have above as

$$E_0(x, p) = f'(c_x)(x - p), \quad \text{some } c_x, \text{ with } p < c_x < x.$$

We can interpret what we have done by saying $f(p)$ is the best choice of 0th order polynomial or constant to approximate $f(x)$ near p. Of course, for most functions, this is a horrible approximation! So the next step is to find the best straight line that approximates f near p. Let's try our usual tangent line to f at p. We summarize this result as a theorem.

Theorem 14.2.1 (Function Approximation: Zeroth Order Taylor Polynomial)
Let $f : [a, b] \rightarrow \Re$ be continuous on $[a, b]$ and be at least differentiable on (a, b). Then for each p in $[a, b]$, there is at least one point c, between p and x, so that $f(x) = f(p) + f'(c)(x - p)$. The constant $f(p)$ is called the **zeroth order Taylor Polynomial** *for f at p and we denote it by $P_0(x; p)$. The point p is called the* **base point**. *Note we are approximating $f(x)$ by the constant $f(p)$ and the error we make is $E_0(x, p) = f'(c)(x - p)$.*

Proof We already fought the good fight for this one! ■

14.2.1 Examples

Example 14.2.1 If $f(t) = t^3$, by the theorem above, we know on the interval $[1, 3]$ that at 1 $f(t) = f(1) + f'(c)(t - 1)$ where c is some point between 1 and t. Thus, $t^3 = 1 + (3c^2)(t - 1)$ for some $1 < c < t$. So here the **zeroth order Taylor Polynomial** is $P_0(t, 1) = 1$ and the **error** is $E_0(t, 1) = (3c^2)(t - 1)$.

Example 14.2.2 If $f(t) = e^{-1.2t}$, by the theorem above, we know that at 0 $f(t) = f(0) + f'(c)(t - 0)$ where c is some point between 0 and t. Thus, $e^{-1.2t} = 1 + (-1.2)e^{-1.2c}(t - 0)$ for some $0 < c < t$ or $e^{-1.2t} = 1 - 1.2e^{-1.2c}t$. So here the **zeroth order Taylor Polynomial** is $P_0(t, 1) = 1$ and the **error** is $E_0(t, 0) = -1.2e^{-1.2c}t$.

Example 14.2.3 If $f(t) = e^{-0.00231t}$, by the theorem above, we know that at 0 $f(t) = f(0) + f'(c)(t - 0)$ where c is some point between 0 and t. Thus, $e^{-0.00231t} = 1 + (-0.00231)e^{-0.00231c}(t - 0)$ for some $0 < c < t$ or $e^{-0.00231t} = 1 - 0.00231e^{-0.00231c}t$. So here the **zeroth order Taylor Polynomial** is $P_0(t, 1) = 1$ and the **error** is $E_0(t, 0) = -0.00231e^{-0.00231c}t$.

14.2.2 Homework

For these problems,

- find the Taylor polynomial of order zero.
- state the error in terms of the first derivative.
- state the maximum error on the given interval.

Exercise 14.2.1 $f(t) = e^{-2.3 \times 10^{-4}\,t}$ *at base point* 0, *intervals* $[0, 10]$, $[0, 100]$ *and* $[0, T]$.

Exercise 14.2.2 $f(t) = e^{-6.8 \times 10^{-6}\,t}$ *at base point* 0, *intervals* $[0, 20]$, $[0, 200]$ *and* $[0, T]$.

Exercise 14.2.3 $f(t) = 2\cos(5t) + 5\sin(3t)$ *at base point* 0*, intervals* $[0, 30]$, $[0, 300]$ *and* $[0, T]$.

Exercise 14.2.4 $f(t) = -8\cos(2t) + 3\sin(7t)$ *at base point* 0*, intervals* $[0, 40]$, $[0, 400]$ *and* $[0, T]$.

14.3 The First Order Taylor Polynomial

If a function f is differentiable, we know we can approximate its value at the point p by its tangent line, T. This is an example of function approximation. We have

$$f(x) = f(p) + f'(p)\,(x - p) + E_T(x, p), \tag{14.1}$$

where the tangent line T is the function

$$T(x) = f(p) + f'(p)\,(x - p)$$

and the term $E_1(x, p)$ represents the error between the true function value $f(x)$ and the tangent line value $T(x)$. That is

$$E_1(x, p) = f(x) - T(x).$$

Another way to look at this is that the tangent line is the best straight line or *linear* approximation to f at the point p. We all know how these pictures look. If the function f is curved near p, then the tangent line is not a very good approximation to f at p unless x is very close to p. Now, let's assume f is actually two times differentiable on the local interval (a, b) also. Define the constant M by

$$M = \frac{f(x) - f(p) - f'(p)(x - p)}{(x - p)^2}.$$

In this discussion, this really is a constant value because we have fixed our value of x and p already. We can rewrite this equation as

$$f(x) = f(p) + f'(p)(x - p) + M(x - p)^2.$$

Now let's define the function g on an interval I containing p by

$$g(t) = f(t) - f(p) - f'(p)\,(t - p) - M(t - p)^2.$$

Then,

$$\begin{aligned} g'(t) &= f'(t) - f'(p) - 2M(t-p) \\ g''(t) &= f''(t) - 2M. \end{aligned}$$

Then,

$$\begin{aligned} g(x) &= f(x) - f(p) - f'(p)(x-p) - M(x-p)^2 \\ &= f(x) - f(p) - f'(p)(x-p) - f(x) + f(p) + f'(p)(x-p) \\ &= 0 \end{aligned}$$

and

$$g(p) = f(p) - f(p) - f'(p)(p-p) - M(p-p)^2 = 0.$$

We thus know $g(x) - g(p) = 0$. Also, from the Mean Value Theorem, there is a point c_x^0 between p and x so that

$$\frac{g(x) - g(p)}{p - x} = g'\left(c_x^0\right).$$

Since the numerator is $g(x) - g(p)$, we now know $g'(c_x^0) = 0$. But we also have

$$g'(p) = f'(p) - f'(p) - 2M(p-p) = 0.$$

Next, we can apply Rolle's Theorem to the function g'. This tells us there is a point c_x^1 between p and c_x^0 so that $g''(c_x^1) = 0$. Thus,

$$0 = g''\left(c_x^1\right) = f''\left(c_x^1\right) - 2M.$$

and simplifying, we have

$$M = \frac{1}{2} f''\left(c_x^1\right).$$

Remembering what the value of M was, gives us our final result

$$\frac{f(x) - f(p) - f'(p)(x-p)}{(x-p)^2} = \frac{1}{2} f''\left(c_x^1\right), \quad \text{some } c_x^1 \text{ with } p < c_x^1 < c_x^0.$$

which can be rewritten as

$$f(x) = f(p) + f'(p)(x-p) + \frac{1}{2} f''\left(c_x^1\right)(x-p)^2, \quad \text{some } c_x^1 \text{ with } p < c_x^1 < c_x^0.$$

We define the 1st order Taylor polynomial, $P_1(x, p)$ and 1st order error, $E_1(x, p)$ by

$$\begin{aligned} P_1(x, p) &= f(p) + f'(p)(x - p) \\ E_1(x, p) &= f(x) - P_1(x, p) = f(x) - f(p) - f'(p)(x - p) \\ &= \frac{1}{2} f''\left(c_x^1\right)(x - p)^2. \end{aligned}$$

Thus, we have shown, $E_1(x, p)$ satisfies

$$E_1(x, p) = f''\left(c_x^1\right) \frac{(x - p)^2}{2} \tag{14.2}$$

where c_x^1 is some point between x and p. Note the usual Tangent line is the same as the first order Taylor Polynomial, $P_1(f, p, x)$ and we have a nice representation of our error. We can state this as our next theorem:

Theorem 14.3.1 (Function Approximation: First Order Taylor Polynomial)
Let $f : [a, b] \to \Re$ be continuous on $[a, b]$ and be at least twice differentiable on (a, b). For a given p in $[a, b]$, for each x, there is at least one point c, between p and x, so that $f(x) = f(p) + f'(p)(x - p) + (1/2) f''(c)(x - p)^2$. The $f(p) + f'(p)(x - p)$ is called the **first order Taylor Polynomial** *for f at p. and we denote it by $P_1(x; p)$. The point p is again called the* **base point**. *Note we are approximating $f(x)$ by the linear function $f(p) + f'(p)(x - p)$ and the error we make is $E_1(x, p) = (1/2) f''(c)(x - p)$.*

Proof This argument was a lot more technical, but note it is a lot like the one we did for the zeroth order polynomial. This kind of reasoning sharpens the old noodle, as they say! Or drives you out for a long run to regain your zen state. ∎

14.3.1 Examples

Example 14.3.1 Let's find the tangent line approximations for a simple exponential decay function. For $f(t) = e^{-1.2t}$ on the interval $[0, 5]$ find the tangent line approximation, the error and maximum the error can be on the interval.

Solution *Using base point* 0, *we have at any t*

$$\begin{aligned} f(t) &= f(0) + f'(0)(t - 0) + (1/2) f''(c)(t - 0)^2 \\ &= 1 + (-1.2)(t - 0) + (1/2)(-1.2)^2 e^{-1.2c}(t - 0)^2 \\ &= 1 - 1.2t + (1/2)(1.2)^2 e^{-1.2c} t^2. \end{aligned}$$

where c is some point between 0 *and t. Hence, c is between* 0 *and* 5 *also. The* **first order Taylor Polynomial** *is $P_1(t, 0) = 1 - 1.2t$ which is also the tangent line to*

$e^{-1.2t}$ *at* 0. *The* **error** *is* $(1/2)(-1.2)^2 e^{-1.2c} t^2$. *Now let* **AE(t)** *denote* **actual error at t** *and* **ME** *be* **maximum error**. *The largest the error can be on* $[0, 5]$ *is when* $f''(c)$ *is the biggest it can be on the interval. Here,*

$$\textbf{\textit{AE(t)}} = (1/2)(1.2)^2 e^{-1.2c} t^2 \leq (1/2)(1.2)^2 \times 1 \times (5)^2 = (1/2)1.44 \times 25 = \textbf{\textit{ME}}.$$

Example 14.3.2 Do the previous problem on the interval [0, 10].

Solution *The approximations are the same and*

$$\textbf{\textit{AE(t)}} = (1/2)(1.2)^2 e^{-1.2c} t^2 \leq (1/2)(1.2)^2 \times 1 \times (10)^2 = (1/2)1.44 \times 100 = \textbf{\textit{ME}}.$$

Example 14.3.3 Do this same problem on the interval [0, 100].

Solution *The approximations are the same and*

$$\textbf{\textit{AE(t)}} = (1/2)(1.2)^2 e^{-1.2c}\, t^2 \leq (1/2)(1.2)^2(1)(100)^2 = (1/2)1.44 \times 10^4 = \textbf{\textit{ME}}.$$

Example 14.3.4 Do this same problem on the interval $[0, T]$.

Solution *The approximations are the same,* $0 < c < T$ *and*

$$\textbf{\textit{AE(t)}} = (1/2)(1.2)^2 e^{-1.2c} t^2 \leq (1/2)(1.2)^2(1)(T)^2 = (1/2)1.44 \times T^2 = \textbf{\textit{ME}}.$$

Example 14.3.5 Let's find the tangent line approximations for a simple exponential decay function again but let's do it a bit more generally. If $f(t) = e^{-\beta t}$, for $\beta = 1.2 \times 10^{-5}$, find the tangent line approximation, the error and the maximum error on [0, 5].

Solution *At any t*

$$\begin{aligned} f(t) &= f(0) + f'(0)(t-0) + \frac{1}{2} f''(c)(t-0)^2 \\ &= 1 + (-\beta)(t-0) + \frac{1}{2}(-\beta)^2 e^{-\beta c}(t-0)^2 \\ &= 1 - \beta t + \frac{1}{2}\beta^2 e^{-\beta c} t^2. \end{aligned}$$

where c is some point between 0 *and t which means c is between* 0 *and* 5. *The* **first order Taylor Polynomial** *is* $P_1(t, 0) = 1 - \beta t$ *which is also the tangent line to* $e^{-\beta t}$ *at* 0. *The* **error** *is* $\frac{1}{2}\beta^2 e^{-\beta c} t^2$. *The largest the error can be on* $[0, 5]$ *is when* $f''(c)$ *is the biggest it can be on the interval. Here,*

$$\begin{aligned} \textbf{\textit{AE(t)}} &= (1/2)(1.2 \times 10^{-5})^2 e^{-1.2 \times 10^{-5} c}\, t^2 \leq (1/2)(1.2 \times 10^{-5})^2(1)(5)^2 \\ &= (1/2)1.44 \times 10^{-10}(25) = \textbf{\textit{ME}} \end{aligned}$$

Example 14.3.6 Do this same problem on the interval [0, 10].

Solution *The approximation is the same, c is between* 0 *and* 10 *and*

$$\begin{aligned}\boldsymbol{AE(t)} &= (1/2)(1.2 \times 10^{-5})^2 e^{-1.2\times 10^{-5} c} t^2 \le (1/2)(1.2 \times 10^{-5})^2 (1)(10)^2 \\ &= (1/2) 1.44 \times 10^{-10} \times 100 = \boldsymbol{ME}.\end{aligned}$$

Example 14.3.7 Do this same problem on the interval [0, 100].

Solution *The approximation is the same, c is between* 0 *and* 100 *and*

$$\begin{aligned}\boldsymbol{AE(t)} &= (1/2)(1.2 \times 10^{-5})^2 e^{-1.2\times 10^{-5} c} t^2 \le (1/2)(1.2 \times 10^{-5})^2 (1)(100)^2 \\ &= (1/2) 1.44 \times 10^{-10}\ 10^4 = \boldsymbol{ME}.\end{aligned}$$

Example 14.3.8 Do this same problem on the interval $[0, T]$.

Solution *The approximation is the same, c is between* 0 *and* T *and*

$$\begin{aligned}\boldsymbol{AE(t)} &= (1/2)(1.2 \times 10^{-5})^2 e^{-1.2\times 10^{-5} c} t^2 \le (1/2)(1.2 \times 10^{-5})^2 (1)(T)^2 \\ &= (1/2) 1.44 \times 10^{-10} T^2 = \boldsymbol{ME}.\end{aligned}$$

Example 14.3.9 Let's find the tangent line approximations for a simple combination of sin's and cos's. If $f(t) = \cos(3t) + 2\sin(4t)$, then on the interval [0, 5] find the tangent line approximation, the error and the maximum error.

Solution *We have* $f(0) = 1$ *and* $f'(t) = -3\sin(3t) + 8\cos(4t)$ *so that* $f'(0) = 8$. *Further,* $f''(t) = -9\cos(3t) - 32\sin(4t)$ *so* $f''(c) = -9\cos(3c) - 32\sin(4c)$. *Thus, for some* $0 < c < 5$

$$\begin{aligned}f(t) &= f(0) + f'(0)(t-0) + \frac{1}{2} f''(c)(t-0)^2 \\ &= 1 + (8)(t-0) + \frac{1}{2}(-9\cos(3c) - 32\sin(4c))(t-0)^2,\end{aligned}$$

The **first order Taylor Polynomial** *is* $P_1(t, 0) = 1 + 8t$ *which is also the tangent line to* $\cos(3t) + 2\sin(4t)$ *at* 0. *The* **error** *is* $(1/2)(-9\cos(3c) - 32\sin(4c))(t-0)^2$. *The error is largest on* [0, 5] *when* $f''(c)$ *is the biggest it can be on* [0, 5]. *Here,*

$$\begin{aligned}\boldsymbol{AE(t)} &= (1/2)(-9\cos(3c) - 32\sin(4c))\ t^2 \\ &\le (1/2)|-9\cos(3c) - 32\sin(4c)|\ 5^2 \\ &\le (1/2)\ |9 + 32|\ 5^2 = (1/2)(41)(25) = \boldsymbol{ME}\end{aligned}$$

Example 14.3.10 Do this same problem on the interval [0, 10].

Solution *The approximations are the same,* $0 < c < 10$ *and*

$$\begin{aligned} \boldsymbol{AE(t)} &= (1/2)(-9\cos(3c) - 32\sin(4c))\, t^2 \\ &\leq (1/2)\, |-9\cos(3c) - 32\sin(4c)|\, (10)^2 \\ &= (1/2)(41)(100) = \boldsymbol{ME}. \end{aligned}$$

Example 14.3.11 Do this same problem on the interval [0, 100].

Solution *The approximations are the same,* $0 < c < 100$ *and*

$$\begin{aligned} \boldsymbol{AE(t)} &= (1/2)(-9\cos(3c) - 32\sin(4c))\, t^2 \\ &\leq (1/2)\, |-9\cos(3c) - 32\sin(4c)|\, (100)^2 \\ &= (1/2)(41)(10^4) = \boldsymbol{ME}. \end{aligned}$$

Example 14.3.12 Do this same problem on the interval $[0, T]$.

Solution *The approximations are the same,* $0 < c < T$ *and*

$$\begin{aligned} \boldsymbol{AE(t)} &= (1/2)(-9\cos(3c) - 32\sin(4c))\, t^2 \\ &\leq (1/2)\, |-9\cos(3c) - 32\sin(4c)|\, T^2 = (1/2)(41)T^2 = \boldsymbol{ME}. \end{aligned}$$

14.3.2 Homework

For these problems,

- find the Taylor polynomial of order one, i.e. the Tangent line approximation.
- state the error in terms of the second derivative.
- state the maximum error on the given interval.

Exercise 14.3.1 $f(t) = e^{-2.3\times 10^{-4}\, t}$ *at base point* 0*, intervals* [0, 10], [0, 100] *and* $[0, T]$.

Exercise 14.3.2 $f(t) = e^{-6.8\times 10^{-6}\, t}$ *at base point* 0*, intervals* [0, 20], [0, 200] *and* $[0, T]$.

Exercise 14.3.3 $f(t) = 2\cos(5t) + 5\sin(3t)$ *at base point* 0*, intervals* [0, 30], [0, 300] *and* $[0, T]$.

Exercise 14.3.4 $f(t) = -8\cos(2t) + 3\sin(7t)$ *at base point* 0*, intervals* [0, 40], [0, 400] *and* $[0, T]$.

14.4 Quadratic Approximations

We could also ask what quadratic function Q fits f best near p. Let the quadratic function Q be defined by

$$Q(x) = f(p) + f'(p)\,(x - p) + f''(p)\,\frac{(x-p)^2}{2}. \tag{14.3}$$

The new error is called $E_Q(x, p)$ and is given by

$$E_Q(x, p) = f(x) - Q(x).$$

If f is three times differentiable, we can argue like we did in the tangent line approximation (using the Mean Value Theorem and Rolle's theorem on an appropriately defined function g) to show there is a new point c_x^2 between p and c_x^1 with

$$E_Q(x, p) = f'''\left(c_x^2\right)\,\frac{(x-p)^3}{6} \tag{14.4}$$

So if f looks like a quadratic locally near p, then Q and f match nicely and the error is pretty small. On the other hand, if f is not quadratic at all near p, the error will be large. We then define the second order Taylor polynomial, $P_2(f, p, x)$ and second order error, $E_2(x, p) = E_Q(x, p)$ by

$$\begin{aligned}
P_2(x, p) &= f(p) + f'(p)(x-p) + \frac{1}{2}f''(p)(x-p)^2 \\
E_2(x, p) &= f(x) - P_2(x, p) = f(x) - f(p) - f'(p)(x-p) - \frac{1}{2}\,f''(p)(x-p)^2 \\
&= \frac{1}{6}f'''\left(c_x^2\right)\,(x-p)^3.
\end{aligned}$$

Theorem 14.4.1 (Function Approximation: Second Order Taylor Polynomial)
Let $f : [a, b] \rightarrow \Re$ be continuous on $[a, b]$ and be at least three times differentiable on (a, b). Given p in $[a, b]$, for each x, there is at least one point c, between p and x, so that $f(x) = f(p) + f'(p)(x-) + (1/2)f''(p)(x-p)^2 + (1/6)f'''(c)(x-p)^3$. The quadratic $f(p) + f'(p)(x-) + (1/2)f''(p)(x-p)^2$ is called the **second order Taylor Polynomial** *for f at p and we denote it by $P_2(x; p)$. The point p is again called the* **base point**. *Note we are approximating $f(x)$ by the quadratic $f(p) + f'(p)(x-) + (1/2)f''(p)(x-p)^2$ and the error we make is $E_2(x, p) = (1/6)f'''(c)(x-p)$.*

Proof We punted on this one! There is no need to torture you all the time, only at selected times of our choosing. So wave the flag but still absorb the result! ■

14.4.1 Examples

Let's work out some problems involving quadratic approximations.

Example 14.4.1 Look at a simple exponential decay function. If $f(t) = e^{-\beta t}$, for $\beta = 1.2 \times 10^{-5}$, find the second order approximation, the error and the maximum error on $[0, 5]$.

Solution *For each t in the interval* $[0, 5]$ *, then there is some* $0 < c << t < 5$ *so that*

$$\begin{aligned} f(t) &= f(0) + f'(0)(t-0) + (1/2)f''(0)(t-0)^2 \\ &\quad + (1/6)f'''(c)(t-0)^3 \\ &= 1 + (-\beta)(t-0) + \frac{1}{2}(-\beta)^2(t-0)^2 + (-\beta)^3 e^{\beta c}(t-0)^3 \\ &= 1 - \beta t + (1/2)\beta^2 - (1/6)\beta^3 e^{-\beta c} t^3. \end{aligned}$$

The **second order Taylor Polynomial** *is* $p_2(t, 0) = 1 - \beta t + (1/2)\beta^2\, t^2$ *which is also called the quadratic approximation to* $e^{-\beta t}$ *at* 0. *The* **error** *is* $-\frac{1}{6}\beta^3 e^{-\beta c} t^3$. *The error is largest on* $[0, 5]$ *when* $f'''(c)$ *is the biggest it can be on the interval. Here,*

$$\begin{aligned} \boldsymbol{AE(t)} &= -(1/6)(1.2 \times 10^{-5})^3 e^{-1.2\times 10^{-5} c}\, t^3 \le |-(1/6)(1.2 \times 10^{-5})^3 e^{-1.2\times 10^{-5} c}\, t^3| \\ &\le (1/6)(1.2 \times 10^{-5})^3\, (1)\, (5)^3 = (1/6)\, 1.728 \times 10^{-15}\, (125) = \boldsymbol{ME} \end{aligned}$$

Example 14.4.2 Do this same problem on the interval $[0, T]$.

Solution *The approximations are the same,* $0 < c < T$ *and*

$$\begin{aligned} \boldsymbol{AE} &= -(1/6)(1.2 \times 10^{-5})^3\, e^{-1.2\times 10^{-5} c}\, t^3 \le |-(1/6)(1.2 \times 10^{-5})^3\, e^{-1.2\times 10^{-5} c}\, t^3| \\ &\le (1/6)(1.2 \times 10^{-5})^3\, T^3 = (1/6)\, 1.728 \times 10^{-15}\, (T^3) = \boldsymbol{ME}. \end{aligned}$$

Example 14.4.3 Now a nice sin and cos problem: If $f(t) = 7\cos(3t) + 3\sin(5t)$, then on the interval $[0, 15]$ find the second order approximation, the error and the maximum error.

Solution *We have* $f(0) = 7$ *and* $f'(t) = -21\sin(3t) + 15\cos(5t)$ *so that* $f'(0) = 15$. *Further,* $f''(t) = -63\cos(3t) - 75\sin(5t)$ *so* $f''(0) = -63$ *and finally* $f'''(t) = 189\sin(3t) - 375\cos(5t)$ *and* $f'''(c) = 189\sin(3c) - 375\cos(5c)$. *Thus, for a given t, there is some* $0 < c < 15$ *so that*

$$\begin{aligned} f(t) &= f(0) + f'(0)(t-0) + (1/2)f''(0)(t-0)^2 \\ &\quad + (1/6)f'''(0)(t-0)^3 \\ &= 7 + (15)(t-0) + (1/2)(-63)(t-0)^2 \\ &\quad + (1/6)(189\sin(3c) - 375\cos(5c))(t-0)^3 \\ &= 7 + 15t - (1/2)63t^2 \\ &\quad + (1/6)(189\sin(3c) - 375\cos(5c))(t-0)^3 \end{aligned}$$

The **second order Taylor Polynomial** *is* $p_2(t, 0) = 7 + 15t - (63/2)t^2$ *which is also the quadratic approximation to* $7\cos(3t) + 3\sin(5t)$ *at* 0. *The* **error** *is* $(1/6)(189\sin(3c) - 375\cos(5c))(t - 0)^3$. *The largest the error can be on the interval* $[0, 15]$ *is then*

$$\begin{aligned} \boldsymbol{AE(t)} &= (1/6)(189\sin(3c) - 375\cos(5c))(t-0)^3 \leq (1/6)|189\sin(3c) - 375\cos(5c)t^3| \\ &\leq (1/6)(189+375)t^3 \leq (1/6)(564)(15)^3 = \boldsymbol{ME}. \end{aligned}$$

Example 14.4.4 Do this same problem on the interval $[0, T]$.

Solution *The approximations are the same,* $0 < c < T$ *and*

$$\begin{aligned} |\boldsymbol{AE(t)}| &= |(1/6)(189\sin(3c) - 375\cos(5c))(t-0)^3| \\ &\leq (1/6)(189+375)t^3 \leq (1/6)(564)T^3 = \boldsymbol{ME}. \end{aligned}$$

We can find higher order Taylor polynomials and remainders using these arguments as long as f has higher order derivatives. But, for our purposes, we can stop here.

14.4.2 Homework

It's time to use MatLab a bit. Consider this function which you can download from our web site:

Listing 14.1: Drawing the Approximations

```
function approx(f,fp,fpp,a,b,t0)
%
% don't use this function if f'(t0) = 0!
%
% t0 = base point
% f = function
% fp = f'
% fpp = f''
%
TL = @(t0,t) f(t0) + fp(t0)*(t-t0);
Q  = @(t0,t) TL(t0,t) + 0.5*fpp(t0)*(t-t0).^2;
%
% if b-a is really big, set N appropriately
h = .5;
N = ceil((b-a)/h);
% if N is too small, make it bigger so we get good plots
if N < 10
  N = 31;
end
time = linspace(a,b,N);
plot(time,f(time),time,TL(t0,time),time,Q(t0,time));
xlabel('Time');
ylabel('f and its approximations');
title('f, the tangent line to f at t0 and the quadratic approximation to f at t0
   vs. time');
legend('f','Tangent Line','Quadratic');
end
```

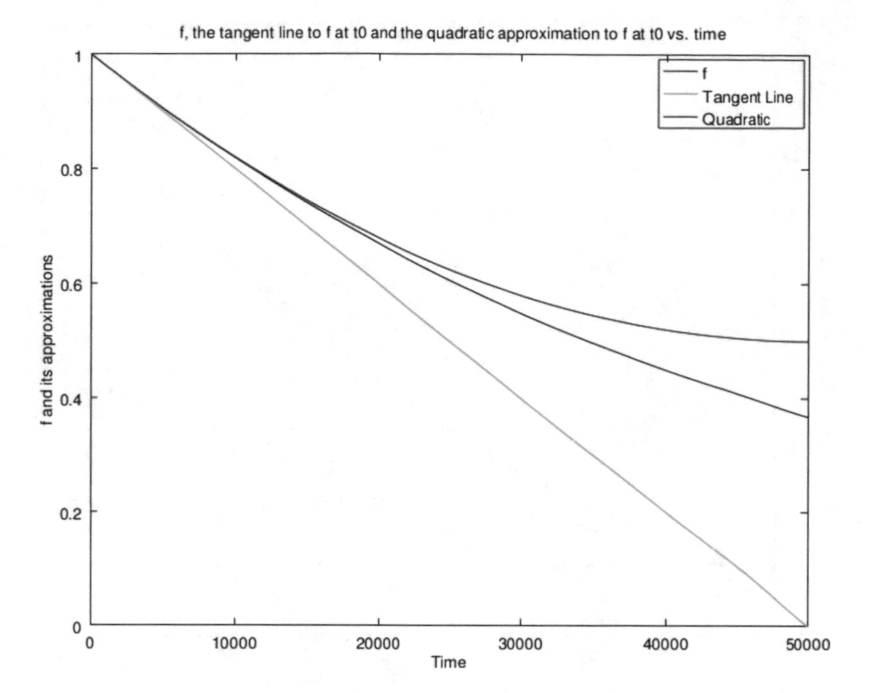

Fig. 14.1 $e^{-\beta t}$ and its tangent line and quadratic approximation

This one is easy to use: we have to define f, f' and f''. Let's do this for $f(t) = e^{-\beta t}$ for $\beta = 2.0e - 5$.

Listing 14.2: Generating approximations

```
beta = 2.0e-5;
f = @(t) e.^(-beta * t);
fp = @(t) -beta*f(t);
fpp = @(t) beta^2 * f(t);
TF=5.0e+4;
approx(f,fp,fpp,0,TF,0);
```

This generates the plot we see in Fig. 14.1.
Note, from this graph, we can see where the tangent line crosses the time axis. Here is another one. We use $f(t) = -8\cos(2t) + 3\sin(7t)$.

Listing 14.3: Generating approximations

```
f = @(t) -8*cos(2*time)+3*sin(7*t);
fp = @(t) 16*sin(2*time)+21*cos(7*t);
fpp = @(t) 32*cos(2*time)-147*sin(7*t);
TF = 2;
approx(f,fp,fpp,0,TF,0);
```

This generates the plot we see in Fig. 14.2.

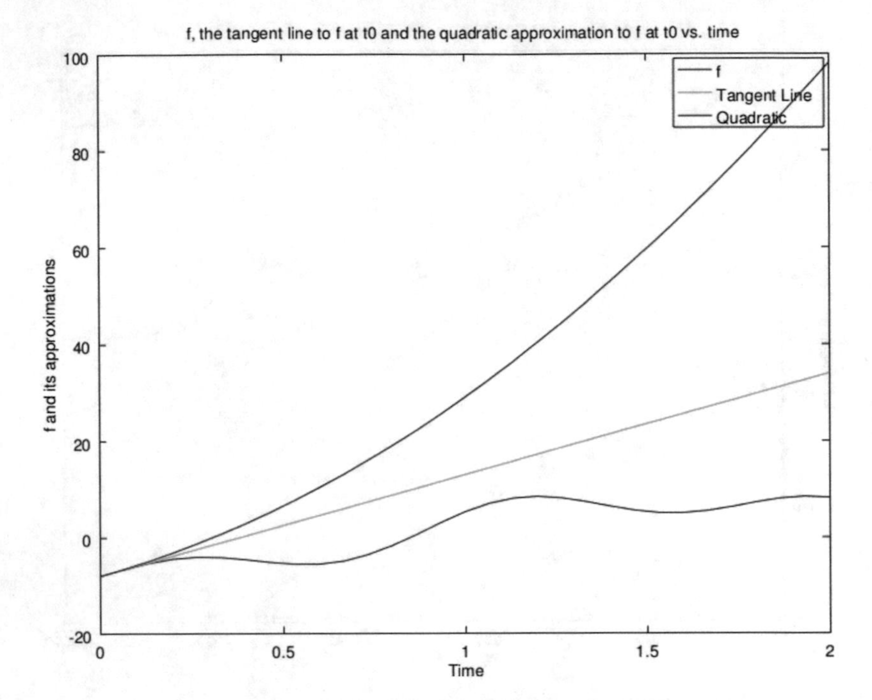

Fig. 14.2 $-8\cos(2t) + 3\sin(7t)$ and its tangent line and quadratic approximation

For these problems,

- find the Taylor polynomial of order two, i.e. the quadratic approximation.
- state the error in terms of the third derivative.
- state the maximum error on the given interval.
- Graph: use our MatLab function to get a nice plot of the function and the approximations on the interval $[0, T_F]$.

Exercise 14.4.1 $f(t) = e^{-2.3\times 10^{-4}\,t}$ *at base point* 0*, intervals* [0, 10], [0, 100] *and* $[0, T]$.

Exercise 14.4.2 $f(t) = e^{-6.8\times 10^{-6}\,t}$ *at base point* 0*, intervals* [0, 20], [0, 200] *and* $[0, T]$.

Exercise 14.4.3 $f(t) = 2\cos(5t) + 5\sin(3t)$ *at base point* 0*, intervals* [0, 30], [0, 300] *and* $[0, T]$.

Exercise 14.4.4 $f(t) = -8\cos(2t) + 3\sin(7t)$ *at base point* 0*, intervals* [0, 40], [0, 400] *and* $[0, T]$.

For these problems given the function $f(t)$ and the base point t_0:

- Tangent Line
  - Write the approximation equation $f(t) = P_1(t, t_0) + (1/2) f''(c_t)(t - t_0)^2$.
  - State the Tangent line function, $TL(t) = P_1(t, t_0)$.

- State where the point c_t is located for the tangent line approximation.
- For a given time, t, state the absolute error.
- For the given final time, T_F, determine the overestimate of the absolute error which we call **ME**, $|f(t) - TL(t)|$.

• Quadratic

- Write the approximation equation $f(t) = P_2(t, t_0) + (1/6) f'''(c_t)(1/6)(t - t_0)^3$.
- State the quadratic approximation $Q(t) = P_2(t, t_0)$.
- State where the point c_t is located for the quadratic approximation.
- For a given time, t, state the absolute error.
- For the given final time, T_F, determine the overestimate of the absolute error $|f(t) - Q(t)|$

• Graph: use our MatLab function to get a nice plot of the function and the approximations on the interval $[0, T_F]$. Note you should be able to sketch the function and its tangent line approximation by hand!

Exercise 14.4.5 *Measure t in days and use* $\beta = 0.0015$, $f(t) = e^{-\beta t}$, $t_0 = 0$ *and* $T_F = 60$ *years.*

Exercise 14.4.6 *Measure t in days and use* $\beta = 0.00025$, $f(t) = e^{-\beta t}$, $t_0 = 0$ *and* $T_F = 70$ *years.*

Exercise 14.4.7 *Measure t in days and use* $\beta = 0.0000145$, $f(t) = e^{-\beta t}$, $t_0 = 0$ *and* $T_F = 80$ *years.*

Exercise 14.4.8 *Measure t in days and use* $\beta = 0.0000015$, $f(t) = e^{-\beta t}$, $t_0 = 0$ *and* $T_F = 90$ *years.*

14.5 Exponential Approximations

Here is a nice use of these approximation ideas which we use when we study the cancer model in Chap. 19. Let's recall ideas from Sect. 14.1 and apply them to the approximation of the difference of two exponentials. Let's look at the function $f(t) = e^{-rt} - e^{-(r+a)t}$ for positive r and a. To approximate this difference, we expand each exponential function into the second order approximation plus the error as usual

$$
\begin{aligned}
e^{-rt} &= 1 - rt + r^2 \frac{t^2}{2} - r^3 e^{-rc_1} \frac{t^3}{6} \\
e^{-(r+a)t} &= 1 - (r+a)t + (r+a)^2 \frac{t^2}{2} - (r+a)^3 e^{-(r+a)c_2} \frac{t^3}{6}
\end{aligned}
$$

for some c_1 and c_2 between 0 and t. Subtracting, we have

$$\begin{aligned} e^{-rt} - e^{-(r+a)t} &= \left(1 - rt + r^2\frac{t^2}{2} - r^3 e^{-rc_1}\frac{t^3}{6}\right) \\ &\quad -\left(1 - (r+a)t + (r+a)^2\frac{t^2}{2}(r+a)^3 e^{-(r+a)c_2}\frac{t^3}{6}\right) \\ &= at - (a^2 + 2ar)\frac{t^2}{2} + \left(-r^3 e^{-rc_1} + (r+a)^3 e^{-(r+a)c_2}\right)\frac{t^3}{6} \end{aligned}$$

We conclude

$$e^{-rt} - e^{-(r+a)t} = at - (a^2 + 2ar)\frac{t^2}{2} + \left(-r^3 e^{-rc_1} + (r+a)^3 e^{-(r+a)c_2}\right)\frac{t^3}{6} \tag{14.5}$$

We can also approximate the function $g(t) = e^{-(r+a)t} - e^{-(r+b)t}$ for positive r, a and b. Using a first order tangent line type approximation, we have

$$\begin{aligned} e^{-(r+a)t} &= 1 - (r+a)t + (r+a)^2 e^{-(r+a)c_1}\frac{t^2}{2} \\ e^{-(r+b)t} &= 1 - (r+b)t + (r+b)^2 e^{-(r+a)c_2}\frac{t^2}{2} \end{aligned}$$

for some c_1 and c_2 between 0 and t. Subtracting, we find

$$\begin{aligned} e^{-(r+a)t} - e^{-(r+b)t} &= \left(1 - (r+a)t + (r+a)^2 e^{-(r+a)c_1}\frac{t^2}{2}\right) \\ &\quad -\left(1 - (r+b)t + (r+b)^2 e^{-(r+b)c_2}\frac{t^2}{2}\right) \\ &= (-a+b)t + \left((r+a)^2 e^{-(r+a)c_1} - (r+b)^2 e^{-(r+b)c_2}\right)\frac{t^2}{2} \end{aligned}$$

We conclude

$$e^{-(r+a)t} - e^{-(r+b)t} = (-a+b)t + \left((r+a)^2 e^{-(r+a)c_1} - (r+b)^2 e^{-(r+b)c_2}\right)\frac{t^2}{2} \tag{14.6}$$

14.5.1 Example

Example 14.5.1 Approximate $e^{-1.0t} - e^{-1.1t}$ using Eq. 14.5.

Solution *We know*

$$e^{-rt} - e^{-(r+a)t} = at - (a^2 + 2ar)\frac{t^2}{2} + \left(-r^3 e^{-rc_1} + (r+a)^3 e^{-(r+a)c_2}\right)\frac{t^3}{6}$$

and here $r = 1.0$ *and* $a = 0.1$. *So we have*

$$e^{-1.0t} - e^{-1.1t} \approx 0.1t - (0.01 + 0.2)\frac{t^2}{2} = 0.1t - 0.105t^2$$

and the error is on the order of t^3 *which we write as* $O(t^3)$ *where* O *stands for order.*

Example 14.5.2 Approximate $e^{-2.0t} - e^{-2.1t}$ using Eq. 14.5.

Solution *Again, we use our equation*

$$e^{-rt} - e^{-(r+a)t} = at - (a^2 + 2ar)\frac{t^2}{2} + \left(-r^3 e^{-rc_1} + (r+a)^3 e^{-(r+a)c_2}\right)\frac{t^3}{6}$$

and here $r = 2.0$ *and* $a = 0.1$. *So we have*

$$e^{-2.0t} - e^{-2.1t} \approx 0.1t - (0.01 + 2(0.1)(2))\frac{t^2}{2} = 0.1t - 0.21t^2$$

and the error is $O(t^3)$.

Example 14.5.3 Approximate $e^{-2.1t} - e^{-2.2t}$ using Eq. 14.6.

Solution *Now, we use our equation*

$$e^{-(r+a)t} - e^{-(r+b)t} = (-a + b)t + \left((r+a)^2 e^{-(r+a)c_1} - (r+b)^2 e^{-(r+b)c_2}\right)\frac{t^2}{2}$$

and here $r = 2.0$, $a = 0.1$ *and* $b = 0.2$. *So we have*

$$e^{-2.1t} - e^{-2.2t} \approx (0.2 - 0.1)t = 0.1t$$

and the error is $O(t^2)$.

Example 14.5.4 Approximate $\left(e^{-2.1t} - e^{-2.2t}\right) - \left(e^{-1.1t} - e^{-1.2t}\right)$ using Eq. 14.6.

Solution *We have*

$$\left(e^{-2.1t} - e^{-2.2t}\right) - \left(e^{-1.1t} - e^{-1.2t}\right) \approx 0.1t - 0.1t = 0$$

plus $O(t^2)$ *which is not very useful. Of course, if the numbers had been a little different, we would have not gotten* 0*. If we instead approximate using Eq. 14.5 we find*

$$\begin{aligned}\left(e^{-2.1t} - e^{-2.2t}\right) - \left(e^{-1.1t} - e^{-1.2t}\right) &\approx \left(0.1t - (0.01 + 2(0.1)(2.1))(t^2/2)\right)\\ &\quad -\left(0.1t - (0.01 + 2(0.1)(1.1))(t^2/2)\right)\\ &= -0.215t^2 + 0.230t^2 = 0.015t^2\end{aligned}$$

plus $O(t^3)$ *which is better. Note if the numbers are just right lots of stuff cancels!*

14.5.2 Homework

Exercise 14.5.1 *Approximate* $e^{-3.0t} - e^{-3.2t}$ *using Eq. 14.5; use* $r = 3.0$*,* $a = 0.2$*.*

Exercise 14.5.2 *Approximate* $e^{-0.8t} - e^{-0.85t}$ *using Eq. 14.5; use* $r = 0.8$*,* $a = 0.05$*.*

Exercise 14.5.3 *Approximate* $e^{-1.3t} - e^{-1.5t}$ *using Eq. 14.5; use* $r = 1.3$*,* $a = 0.2$*.*

Exercise 14.5.4 *Approximate* $e^{-1.1t} - e^{-1.3t}$ *using Eq. 14.6; use* $r = 1.0$*,* $a = 0.1$*,* $b = 0.3$*.*

Exercise 14.5.5 *Approximate* $e^{-0.07t} - e^{-0.09t}$ *using Eq. 14.5; use* $r = 0.06$*,* $a = 0.01$*,* $b = 0.03$*.*

Exercise 14.5.6 *Approximate* $\left(e^{-1.1t} - e^{-1.3t}\right) - \left(e^{-0.7t} - e^{-0.8t}\right)$ *using Eq. 14.6 and Eq. 14.5; for Eq. 14.5, first difference, use* $r = 1.0$*,* $a = 0.1$*,* $b = 0.3$ *second difference, use* $r = 0.6$*,* $a = 0.1$*,* $b = 0.2$*; for Eq. 14.6, first difference, use* $r = 1.1$*,* $a = 0.2$ *second difference, use* $r = 0.7$*,* $a = 0.1$*.*

Exercise 14.5.7 *Approximate* $\left(e^{-2.2t} - e^{-2.4t}\right) - \left(e^{-1.8t} - e^{-1.9t}\right)$ *using Eq. 14.6 and Eq. 14.5; for Eq. 14.5, first difference, use r = 2.0, a = 0.2, b = 0.4 second difference, use r = 1.0, a = 0.8, b = 0.9; for Eq. 14.6, first difference, use r = 2.2, a = 0.2, second difference, use r = 1.8, a = 0.1.*

Exercise 14.5.8 *Approximate* $\left(e^{-0.03t} - e^{-0.035t}\right) - \left(e^{-0.02t} - e^{-0.025t}\right)$ *using Eq. 14.6 and Eq. 14.5: for Eq. 14.5, first difference, use r = 0.02, a = 0.01, b = 0.015, second difference, use r = 0.01, a = 0.01, b = 0.015 for Eq. 14.6, first difference, use r = 0.03, a = 0.005 second difference, use r = 0.02, a = 0.005.*

Chapter 15
Extreme Values

We often want to know where a function has a minimum or maximum value. Since the functions we usually see in our models are at least differentiable, we can use ideas from the calculus of functions we have been developing to find where these extreme values occur.

15.1 Extremal Values

We know that a likely place for this is where the tangent line is flat as from simple drawings, we can see that where the tangent line is flat we often have a local minimum or local maximum of our function. We can be more precise. If p is a point where the tangent line to f is flat, then we know $f'(p) = 0$. The first order Taylor expansion is

$$f(x) = f(p) + f'(p)(x - p) + \frac{1}{2} f''(c)(x - p)^2,$$

for some c with c between p and x. Now since $f'(p) = 0$, this reduces to

$$f(x) = f(p) + \frac{1}{2} f''(c)(x - p)^2,$$

Now let's step back and talk about continuity and positive and negative values. Let's assume $f(p) > 0$. If a function is continuous at a point p, if no matter how close we got to p, we could find a q with $f(q) = 0$, we would have that the $\lim_{q \to p} f(q) = 0$. But since f is continuous at p, all such limits must have as their value $f(p)$ which is positive. So this can't happen and so there must be a value of r where in the circle about p of radius r, f is never 0. Now if inside this circle there were points q where $f(q)$ were negative no matter how close we were to p, again we could look at the limit. Let these new points were f is negative be called s. Then $\lim_{s \to p} f(s) \leq 0$

J.K. Peterson, *Calculus for Cognitive Scientists*, Cognitive Science and Technology, DOI 10.1007/978-981-287-874-8_15

and by continuity this limiting value must match $f(p)$ which is positive. So again this can't happen. We see if $f(p)$ is positive and f is continuous there, then there is a radius r where f is positive in the circle of radius r about p. We can say a similar thing if $f(p) < 0$. Let's state this more formally.

Theorem 15.1.1 (Positive and Negative Neighborhoods)
If f is continuous at p and $f(p)$ is nonzero, then there is a radius r where $f(x)$ is nonzero with the same sign on $(p - r, p + r)$.

Proof We have already sketched out the reasoning behind this result. ■

Back to our problem! So as long as f'' is continuous at p, we can say

- $f''(p) > 0$ implies $f''(c) > 0$ within some circle centered at p. This tells us $f(x) = f(p) +$ **a positive number** on this circle. Hence, $f(x) > f(p)$ locally which tells us p is a point where f has a local minimum.
- $f''(p) < 0$ implies $f''(c) < 0$ within some circle centered at p. This tells us $f(x) = f(p) -$ **a positive number** on this circle. Hence, $f(x) < f(p)$ locally which tells us p is a point where f has a local maximum.

We can state this more formally too. It is our second order test for maximum and minimum values. Note, if $f''(p) = 0$, we don't know anything. Here's why. Consider the function $f(x) = x^4$ which has $f'(0) = 0$ and $f''(0) = 0$ even though $f(0)$ is a minimum—just draw the function to see this. Then, look at $f(x) = -x^4$ which has $f'(0) = 0$ and $f''(0) = 0$ even though $f(0)$ is a maximum. Finally, note $f(x) = x^3$ doesn't have a minimum or maximum at $x = 0$ and $f''(0) = 0$. So, if $f''(p) = 0$, we could have a minimum, maximum or nothing.

Theorem 15.1.2 (Second Order Test for Extremals)
If f'' is continuous at p, $f'(p) = 0$, then $f''(p) > 0$ tells us f has a local minimum at p and $f''(p) < 0$ tells us f has a local maximum at p. If $f''(p) = 0$, we don't know anything.

Proof We have already sketched out the reasoning behind this result as well. ■

15.1.1 Example

Example 15.1.1 Show $f(x) = x^2 + 2x + 1$ has a minimum at $x = -1$.

Solution *We have $f'(x) = 2x + 2$ which is zero when $2x + 2 = 0$ or $x = -1$. We also have $f''(x) = 2 > 0$ and so we have a minimum.*

Example 15.1.2 Show $f(x) = 2x^2 + 5x + 1$ has a minimum at $x = -5/4$.

Solution *We have $f'(x) = 4x + 5$ which is zero when $4x + 5 = 0$ or $x = -5/4$. We also have $f''(x) = 4 > 0$ and so we have a minimum.*

Example 15.1.3 Show $f(x) = -2x^2 + 5x + 1$ has a maximum at $x = 5/4$.

Solution *We have* $f'(x) = -4x + 5$ *which is zero when* $-4x + 5 = 0$ *or* $x = 5/4$. *We also have* $f''(x) = -4 < 0$ *and so we have a maximum.*

Example 15.1.4 Show $f(x) = -2x^3 + 5x^2 + 1$ has a maximum at $x = 10/6$.

Solution *We have* $f'(x) = -6x^2+10x = x(-6x+10)$ *which is zero when* $x(-6x+10) = 0$ *or when* $x = 0$ *or* $x = 10/6$. *We also have* $f''(x) = -12x + 10$. *Since* $f''(0) = 10$ *we know* $x = 0$ *is a minimum. Since* $f''(10/6) = -12(10/6) + 10 < 0$, $x = 10/6$ *is a maximum.*

15.1.2 Homework

Exercise 15.1.1 *Show* $f(x) = x^2 + 4x + 1$ *has a minimum at* $x = -2$.

Exercise 15.1.2 *Show* $f(x) = 3x^2 + 8x + 1$ *has a minimum at* $x = -8/6$.

Exercise 15.1.3 *Show* $f(x) = -5x^2 + 4x + 1$ *has a maximum at* $x = 4/10$.

Exercise 15.1.4 *Show* $f(x) = 2x^3 + 9x^2 + 12x + 10$ *has a minimum at* $x = -1$ *and a maximum at* $x = -2$. *The first derivative factors nicely!*

Exercise 15.1.5 *Show* $f(x) = 2x^3 - 3x^2 - 36x + 5$ *has a minimum at* $x = 3$ *and a maximum at* $x = -2$. *The first derivative factors nicely here too!*

Exercise 15.1.6 *Show* $f(x) = \cos(x)$ *has a maximum at* $x = 0$ *and a minimum at* $x = \pi$. *Of course, there are other points too!*

Exercise 15.1.7 *Show* $f(x) = \sin(2x)$ *has a minimum at* $x = 3\pi/4$ *and a maximum at* $x = \pi/4$. *Of course, there are other points too just like you saw in the last problem.*

15.2 The Newton Cooling Project

You now know enough about solving simple differential equations to do a modeling project. Newton formulated a law of cooling by observing how the temperature of a hot object cooled. As you might expect, this is called *Newton's Law Of Cooling*. If we let $T(t)$ represent the temperature of the liquid in some container and A denote the ambient temperature of the air around the container, then Newton observed that

$$T'(t) \propto (T(t) - A).$$

We will assume that the temperature outside the container, A, is smaller than the initial temperature of the hot liquid inside. So we expect the temperature of the liquid to go down with time. We let the constant of proportionality be k. Thus, the differential equation we have to solve is

$$\begin{aligned} T'(t) &= k\,(T(t) - A), \\ T(0) &= T_0 \end{aligned}$$

where T_0 is the initial temperature of the liquid. For example, we might want to solve

$$\begin{aligned} T'(t) &= k(T(t) - 70), \\ T(0) &= 210 \end{aligned}$$

where all of our temperatures are measured in degrees Fahrenheit. We can solve using the integrating factor method. We have

$$\begin{aligned} T'(t) - kT(t) &= -70\,k, \\ \left(e^{-kt}T(t)\right)' &= -70ke^{-kt} \\ e^{-kt}T(t) - T(0) &= 70(e^{-kt} - 1) \\ T(t) &= 70 + (T(0) - 70)\,e^{kt} \end{aligned}$$

Using the initial condition, we find

$$T(t) = 70 + 140e^{kt}.$$

Now since the temperature of our liquid is going down, it is apparent that the proportionality constant k must be negative. In fact, our common sense tells us that as time increases, the temperature of the liquid approaches the ambient temperature 70 asymptotically from above. Note also, we can choose to measure time in any units we want.

Example 15.2.1 Solve $T'(t) = k(T(t) - 70)$ with $T(0) = 210$ and then use the conditions $T(10) = 140$ to find k. Here time is measured in minutes.

Solution *In this problem, the extra condition at* $t = 10$, *will enable us to find the constant of proportionality* k. *First, we solve as usual to find*

$$T(t) = 70 + 140e^{kt}.$$

Next, we know $T(10) = 140$, *so we must have*

$$T(10) = 140 = 70 + 140e^{10k}.$$

Thus,

$$
\begin{aligned}
70 &= 140e^{10k}, \\
\frac{1}{2} &= e^{10k}, \\
-\ln(2) &= 10k,
\end{aligned}
$$

and so $k = -\ln(2)/10 = -0.0693$.

15.2.1 Homework

Exercise 15.2.1 *Solve* $T'(t) = k(T(t) - 90)$ *with* $T(0) = 205$ *and then use the conditions* $T(20) = 100$ *to find k. Here time is measured in minutes.*

Exercise 15.2.2 *Solve* $T'(t) = k(T(t) - 40)$ *with* $T(0) = 195$ *and then use the conditions* $T(5) = 190$ *to find k. Here time is measured in minutes.*

Exercise 15.2.3 *Solve* $T'(t) = k(T(t) - 85)$ *with* $T(0) = 207$ *and then use the conditions* $T(9) = 182$ *to find k. Here time is measured in minutes.*

15.2.2 Your Newton Cooling Project

Your assignment will be to take the liquid of your choice and place it in a cup of some sort like a coffee cup. Measure the ambient temperature A of the room you are in. Microwave the contents until the liquid is close to boiling. Then take it out of the microwave and measure the initial temperature, T_0. Then measure the temperature every 30 s for the first five minutes and then every minute for the next 10 mins and then every five minutes thereafter until the temperature of the liquid in the cup reaches $A + 10$. Record all this data in a notebook. You will not reach room temperature in any reasonable length of time, so stopping at $A + 10$ is just fine. We will then model the decay of the temperature of your liquid using Newton's Law of Cooling. So it we let $T(t)$ be the temperature of your liquid at time t in minutes, we will assume that the cooling of the liquid satisfies Newton's Law of Cooling. Hence, we know

$$
\begin{aligned}
T'(t) &= k(T(t) - A), \\
T(0) &= T_0.
\end{aligned}
$$

This has solution

$$
T(t) = A + Be^{kt}.
$$

Applying the initial condition, we get

$$T_0 = A + B.$$

or $B = T_0 - A$. Hence, the solution is

$$T(t) = A + (T_0 - A)e^{kt}.$$

The one parameter we don't know is the constant of proportionality, k. However, we do have all the data we collected. So, rewrite our solution as

$$\frac{T(t) - A}{T_0 - A} = e^{kt}.$$

Now take the logarithm of both sides to get

$$\ln\left(\frac{T(t) - A}{T_0 - A}\right) = kt. \tag{15.1}$$

This tell us how to estimate the value of k. Let the variable $U(t)$ be defined to be

$$U(t) = \ln\left(\frac{T(t) - A}{T_0 - A}\right)$$

Then, Eq. (15.1) can be written as

$$U(t) = kt.$$

Thus, the variable U is linear in t; i.e. if we graph U versus t we should see a straight line with slope k. Our collected data consists of pairs of the form (time, temperature). Suppose we collected N such pairs. Label them as (t_i, T_i) for $1 \leq i \leq N$ (you might have 80 points say so for you $N = 80$). Compute the corresponding $U(t_i) \equiv U_i$ points

$$U_i = \ln\left(\frac{T_i - A}{T_0 - A}\right)$$

and plot them as a scatter plot in MatLab. Note the first value, $U_1 = \ln \frac{T_0 - A}{T_0 - A} = \ln(1) = 0$. So our line has U intercept 0. Hence, our line is of the form $y = mt$ and we have to find a good candidate for the slope m. This slope will then be used in our cooling model.

15.2.2.1 Estimating the Slope

From your data, you now have a collection of pairs (t_i, U_i) for $1 \leq i \leq N$. We can find a line which comes close in an optimal way to all of this data (although the line does not have to include all the data!) by using a bit of calculus. At each time, the line has the value mt_i. The discrepancy or error between the actual U value and the line value is then $U_i - mt_i$. We want the cumulative error between our line and the data, so we don't want errors to cancel. Hence, we could choose an error term like $|U_i - mt_i|$ or $(U_i - mt_i)^2$. Both of these are fine, but the squared one is differentiable which will allow us to use calculus methods. Define the *energy* function $E(m)$ where m is the slope to be

$$E(m) = \sum_{i=1}^{N} (U_i - mt_i)^2.$$

This is a nice function and so let's find where this function has a minimum by setting its derivative is zero. Setting the derivative to zero, we find

$$\frac{dE}{dm} = 2 \sum_{i=1}^{N} (U_i - mt_i)\,(-t_i) = 0.$$

Simplifying a bit, we have

$$\sum_{i=1}^{N} U_i t_i = m \left(\sum_{i=1}^{N} t_i^2 \right).$$

Solving for the critical point m, we find

$$m = \frac{\sum_{i=1}^{N} U_i t_i}{\sum_{i=1}^{N} t_i^2}$$

Finally, it is easy to see

$$\frac{d^2 E}{dm^2} = 2 \sum_{i=1}^{N} t_i^2 > 0.$$

for all choices of m. Hence, the value m which is the critical point must be the global minimum Hence, the optimal slope is

$$m^* = \frac{\sum_{i=1}^{N} U_i t_i}{\sum_{i=1}^{N} t_i^2}$$

Then we can see how good our model is by plotting the model solution $T(t) = T(t) = A + (T_0 - A)e^{m^* t}$ on the same plot as the actual data.

15.2.3 Your Report

First some notes of formatting of the report in your choice of word processor.

1. Use a one inch margin on all sides.
2. Use 11 pt font.

Now let's talk about the structure of the document. It consists of the following pieces:

- **Introduction:** Here you discuss the Newton's Law of Cooling. You should do some research on this at the library or online to find out more about this law. Your findings about this will then go in this section. This is **6 Points**.
- **The Mathematical Model:** Here you state the mathematical form of Newton's Law of Cooling. This means that you need to type the appropriate mathematical equations. Take care to do this nicely so that the report has a nice visual look. Discuss the meaning of each symbol in the mathematical equations, of course.

  – Derive the solution to the Newton cooling problem showing all the mathematics. This is to make sure you learn how to use mathematics in a word processing document.
  – Derive the optimal slope value as we have done in the text: again, this is to make sure you can type mathematics.
  – Derive the appropriate log plot that you will use to find the Newton Cooling Constant.

  This is **10 Points**.
- **The Experiment:** Here you discuss that your experiment is going to try find out if your liquid obeys Newton's Law of Cooling. You need to explicitly go over all of the information that is relevant to the experiment. This includes the choice of liquid, the cup it was placed in, an analysis of the kind of cup it was (ceramic, Styrofoam etc.), how the temperature of the liquid was measured and so forth. Be very explicit here. The idea is that the reader would have enough information to run the experiment themselves. Include a picture of the cup and your temperature measurement device in this section. Don't use *water*; that is an automatic project score of 0! Take pictures of your experimental setup and embed them in the document. This is **10 Points**.
- **The MatLab Analysis:** Here you include

  1. the MatLab code fragments that plot your original data using circles for the data points.

2. the MatLab code to create the appropriate log plot annotated carefully.
3. the plot of your log data using circles for the data points.
4. the MatLab code to find the optimal slope from log plot and your value of the slope.

This is **10 Points**.

- **The Model:** Here you discuss the model you found from your data and show how it compares to the data you collected. The MatLab code to plot the model and the data simultaneously needs to be here as well as the actual plot. Your original data is again plotted with circles for the data while the model is plotted as a smooth curve. This is **8 Points**.
- **Conclusions:** Here you discuss whether or not you feel Newton's Law of Cooling provides a good theoretical model for your collected data. Discuss also flaws you might have found in the experiment. This is **6 points**.
- **References:** Any references you used need to be placed here.

15.2.4 Some Sample Calculations

Let's see how we can manage our experimental data. Here is a typical MatLab session. Save your experimental data in a file with two columns. Column one is for time and column two is for the temperature of your liquid. For a typical liquid (orange juice, water, tomato juice etc.), we might collect the data here.

Listing 15.1: Cooling Data

```
0.0  205
0.5 201
2.0 201
5.0 190
8.0 178
10.0 175
13.0 167
15.0 161
25.0 141
30.0 135
40.0 127
50.0 117
65.0 108
80.0 100
95.0 95
110.0 89
130.0 83
165.0 80
```

We save this data in the file **`Newton.dat`** in a path of your choosing. We have the ambient or room temperature is 76° for our experiment.

Listing 15.2: Loading The Data Into Matlab

```
% load in the experimental data from the
% file "Newton.dat" as two columns
Data = load('Newton.dat');
% use the first column as time
time = Data(:,1);
% use the second column as temperature
Temperature = Data(:,2);
% generate a plot of the experimental data
plot(time,Temperature,'o');
% set xlabel
xlabel('Time in Minutes');
% set y label
ylabel('Temperature in Fahrenheit');
% set title
title('Newton Law Of Cooling Experimental Data');
```

We see the plot of the measured data in Fig. 15.1.

From the theory we have discussed, we know we should graph the transformed data given by $\log\left(\frac{T-A}{T_0-A}\right)$ where A is the room temperature, T and T_0 are the temperature and initial temperature of the liquid, respectively. Here $A = 76$ and $T_0 = 205$. Thus, $T_0 - A = 129$.

Listing 15.3: Get the Logarithms of the Temperature Data and Its Plot

```
% for this data, T_0 - A = 205 - 76 = 129
A = 76;
LogTemperature = log((Temperature-A)/(129.0));
plot(time,LogTemperature,'o');
xlabel('Time in Minutes');
ylabel('Transformed Variable ln( (T-A)/(T_0-A))');
title('Transformed Data Versus Time');
```

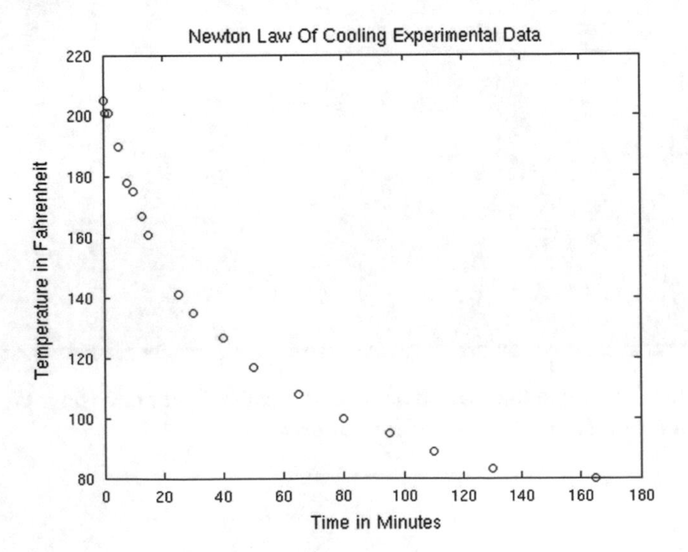

Fig. 15.1 Experimental data

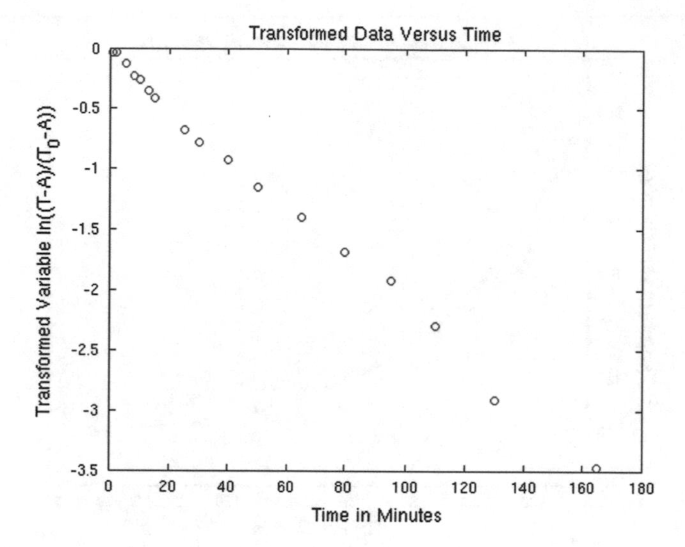

Fig. 15.2 Correct log plot to find slope

You can see this transformed data in Fig. 15.2.
Then, look for the best straight line through the plot making sure you have the line go through the point (0, 0) at the top. The optimal slope is then

$$m^* = \frac{\sum_{i=1}^N U_i t_i}{\sum_{i=1}^N t_i^2}.$$

In MatLab, we can add up all the entries in the variable **U** and **time** using the **sum** command. The optimal slope calculation in MatLab is then

Listing 15.4: Find Optimal Slope for Model

```
mstar = sum(LogTemperature.*time)/sum(time.*time)

mstar =

   -0.0215
```

where the variable **LogTemperature** plays the role of U. The syntax **LogTemperature.*time** means to multiply each component of **LogTemperature** and **time** separately to create the new column of values

$$\begin{matrix} \text{LogTemperature}(1) & \text{time}\ (1) \\ \text{LogTemperature}(2) & \text{time}(2) \\ & \vdots \\ \text{LogTemperature}(N) & \text{time}\ (N) \end{matrix}$$

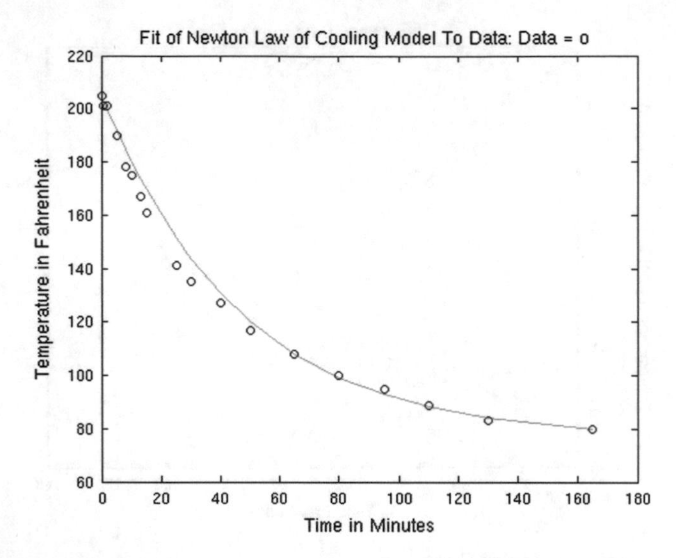

Fig. 15.3 The model versus the data

The command **sum** applied to this column then creates the sum $\sum_{i=1}^{N} U_i t_i$ like we need. A similar thing is happening with the term **sum(time.*time)**.

The Model:

The model is thus

$$u(t) = 76.0 + 129\, e^{-0.0215t}$$

which we enter in MatLab as

Listing 15.5: Build Model of Cooling Data

```
% compute the model
u2 = 76.0 + 129.0*exp(mstar*time);
```

Then we plot both the experimental data and the model on the same plot, Fig. 15.3, to see how we did.

Listing 15.6: Plot the data and model on the same graph

```
% plot the model and the measure data on the same plot
plot(time,u2,'r-',time,Temperature,'o');
xlabel('Time In Minutes');
ylabel('Temperature in Fahrenheit');
title('Fit of Newton Law of Cooling Model To Data: Data = o');
```

Looks pretty good! We could also have added a legend like **legend ('Model','Data','Location','BestOutside');** but we didn't show that. You should do that though in your report so it is really easy to see what is going on. So we conclude the Newton Cooling Model is a good model to use here.

Chapter 16
Numerical Methods Order One ODEs

When we try to solve systems like

$$\frac{dy}{dt} = f(t, y) \tag{16.1}$$

$$y(t_0) = y0 \tag{16.2}$$

where f is continuous in the variables t and y, and y_0 is some value the solution is to have at the time point t_0, we will quickly find that it is very hard in general to do this by *hand* like we have done in Chaps. 11 and 13. So it is time to begin looking at how the MatLab environment can help us. We will use MatLab to solve these differential equations with what are called *numerical* methods.

16.1 Euler's Method

The simplest way to try to approximate the solution of a differential equation like Eqs. 16.1 and 16.2 is to approximate the derivative term y' by using tangent lines or linear approximations. We know we can approximate a function g at a point x_0 using a tangent line. The tangent line to the differentiable function g at the point x_0 is given by $T(x, x_0)$ where

$$T(x, x_0) = g(x_0) + g'(x_0)\,(x - x_0)$$

and the error between the true function value $g(x)$ and tangent line value $T(x)$ is given by $E(x - x_0)$ with

$$\begin{aligned} E(x_0, x) &= g(x) - T(x, x_0) \\ &= g(x) - g(x_0) - g'(x_0)\,(x - x_0). \end{aligned}$$

J.K. Peterson, *Calculus for Cognitive Scientists*, Cognitive Science and Technology, DOI 10.1007/978-981-287-874-8_16

For example, if $g(x) = 2x^2 + 6x + 2$, then at the point $x_0 = 1$, we have the tangent line $T(x, 1)$ is

$$\begin{aligned} T(x,1) &= g(1) + g'(1)\,(x-1) \\ &= 10 + (4x+6)\Big|_1 \,(x-1) \\ &= 10 + 10\,(x-1). \end{aligned}$$

At $x = 2$, the true function value is $g(2) = 22$ and the tangent line value is $T(2, 1) = 20$. The absolute error is then

$$\begin{aligned} \mid E(2-1) \mid &= \mid g(2) - T(2,1) \mid \\ &= \mid 22 - 20 \mid = 2. \end{aligned}$$

Note we could rewrite this by thinking of the point $x_0 = 1$ as the base point and the new point $x = 2$ as the point $x_0 + h$ where $h = 1$. Then we would have the tangent line at x_0 has the value at $x_0 + h$ given by

$$T(x_0 + h, x_0) = g(x_0) + g'(x_0)\,h$$

and the error at $x_0 + h$ would be

$$\begin{aligned} E(h) &= g(x_0 + h) - T(x_0 + h, x_0) \\ &= g(x_0 + h) - g(x_0) - g'(x_0)\,h. \end{aligned}$$

Now, let's calculate both $g(1 + h)$ and $T(1 + h, 1)$ for arbitrary h. We have

$$\begin{aligned} g(1+h) &= \left(2x^2 + 6x + 2\right)\Big|_{1+h} \\ &= 2(1+h)^2 + 6(1+h) + 2 \\ &= 10 + 10h + 2h^2, \end{aligned}$$

and

$$\begin{aligned} T(1+h,1) &= \left(10 + 10\,(x-1)\right)\Big|_{1+h} \\ &= 10 + 10h. \end{aligned}$$

Then, for our sample problem we could compute the absolute error for many values of h via the equation

$$\begin{aligned} \mid E(h) \mid &= \mid g(1+h) - T(1+h,1) \mid \\ &= \left|10 + 10h + 2h^2 - 10 - 10h\right| \\ &= 2h^2. \end{aligned}$$

Letting the parameter h be called the step size, we can generate the table below:

Step size h	Absolute error $\mid E(1, 1+h) \mid$
1.000	$2(1.000)^2 = 2.000000$
0.500	$2(0.500)^2 = 0.500000$
0.250	$2(0.250)^2 = 0.125000$
0.100	$2(0.100)^2 = 0.020000$
0.005	$2(0.005)^2 = 0.000050$
0.002	$2(0.002)^2 = 0.000008$
0.001	$2(0.001)^2 = 0.000001$

We can easily see that the error drops significantly as we let the step size decrease to 0.001. Roughly speaking, we say that approximating the function $g(x)$ by its tangent line at $x = 1$ is a good idea as long as we stay close to the base point $x = 1$! We also know from our earlier discussions that the absolute error is always given by

$$\begin{aligned} E(x - x_0) &= g(x) - T(x, x_0) \\ &= g(x) - g(x_0) - g'(x_0)\,(x - x_0) \\ &= g''(c)/2\,h^2. \end{aligned}$$

where the number c is somewhere between the base point x_0 and $x_0 + h$. In general, this is not terribly helpful as we may have little practical knowledge about the second derivative of g! But here, $g''(x) = 4$ always and so we can write

$$\begin{aligned} E(x - x_0) &= g(x) - T(x, x_0) \\ &= g(x) - g(x_0) - g'(x_0)\,(x - x_0) \\ &= 4/2\,h^2 = 2h^2 \end{aligned}$$

which is exactly the results we see in our table above.

16.1.1 Approximating the Solution to First Order Differential Equations

Let's use this idea to approximate the solution to a differential equation. Let's look at the problem

$$\begin{aligned} \frac{dy}{dt} &= 3\,y(t) \\ y(0) &= 20 \end{aligned}$$

Pick a step size h. We know the solution to this problem already, so we know the solution ($20e^{3t}$) is differentiable and so has very nice tangent lines at each time t. So, using base time point $t_0 = 0$, we have

$$y(0+h) = y(0) + y'(0)\,h + y''(c_h)/2\,h^2.$$

where c_h is some time between 0 and h. Let's also assume we are interested in the solution to this differential equation on the interval $[0, 0.5]$ We know the solution $20e^{3t}$ always grows and so on the interval $[0, 0.5]$, the solution is bounded above by 90, its derivative is bounded by 270 and its second derivative by 810 (this is a simple calculator computation). Since this is a differential equation, we know $y'(0) = 3y(0)$ and so we can plug that in to get

$$y(0+h) = y(0) + 3y(0)\,h + y''(c_h)/2\,h^2.$$

But, we can estimate how big the error can be because

$$\begin{aligned}|y(0+h) - y(0) - 3y(0)\,h| &= y''(c_h)/2\,h^2 \\ &< 810/2\,h^2 = 405h^2.\end{aligned}$$

We will call the term $y(0) + 3y(0)h$ the approximate solution to the differential equation at the time $t = 0$ and label it by the symbol $\hat{y}(0+h)$. For convenience, we will set the initial approximate solution, $\hat{y}(0)$ to match the initial value $y(0)$. Also, we will call the error we make at this first step from 0 to $0+h$, E_h. Thus, using this new notation,

$$\begin{aligned}\hat{y}(0+h) &= y(0) + 3y(0)h \\ &= \hat{y}(0) + 3\hat{y}(0)h,\end{aligned}$$

and

$$\begin{aligned}\mid E_h \mid &= \mid y(0+h) - \hat{y}(0+h) \mid \\ &< 405\,h^2.\end{aligned}$$

We can do this approximation again. The tangent line approximation to $y(0+2h)$ is

$$y(0+2h) = y(0+h) + y'(0+h)\,h\;y''(c_{2h})/2\,h^2.$$

where c_{2h} is between $0+h$ and $0+2h$. But since $y'(0+h) = 3y(0+h)$, we have

$$y(0+2h) = y(0+h) + 3\,y(0+h)\,h + y''(c_{2h})/2\,h^2.$$

Let the second approximate solution value be given by

$$\hat{y}(0+2h) = \hat{y}(0+h) + 3\hat{y}(0+h)h.$$

Then, the error in replacing $y(0+2h)$ by $\hat{y}(0+2h)$ is E_{2h} or

$$\begin{aligned}
|E_{2h}| &= |y(0+2h) - \hat{y}(0+2h)| \\
&= |y(0+h) + 3\,y(0+h)\,h - \hat{y}(0+h) - 3\hat{y}(0+h)h + \left(y''(c_{2h})/2\right) h^2| \\
&= \left|(y(0+h) - \hat{y}(0+h)) + 3\,(y(0+h) - \hat{y}(0+h))\,h + \left(y''(c_{2h})/2\right) h^2\right|
\end{aligned}$$

Now in the above, we are taking the absolute value of three things added. Clearly, if we take the absolute value of each thing first and then add, we eliminate the possibility of cancellation due to negative stuff. Hence, the error is overestimated by

$$\begin{aligned}
|E_{2h}| &\leq |y(0+h) - \hat{y}(0+h)| + 3\,|y(0+h) - \hat{y}(0+h)||h| + \left(\left|y''(c_{2h})\right|/2\right) h^2 \\
&= |E_h| + 3|E_h|\,|h| + \left(|y''(c_{2h})|/2\right) h^2
\end{aligned}$$

But we know that $|E_h| < 405h^2$ and the second derivative is bounded by 810. Thus,

$$|E_{2h}| \leq 2 \times 405h^2 + 3 \times 405\,|h|^3.$$

Now it gets pretty messy, but we can do this all again to find

$$\begin{aligned}
|E_{3h}| &= |y(0+3h) - \hat{y}(0+3h)| \\
&\leq |y(0+2h) - \hat{y}(0+2h)| + 3\,|y(0+2h) - \hat{y}(0+2h)||h| + \left(|y''(c_{3h})|/2\right) h^2
\end{aligned}$$

for some number c_{3h} between $0 + 2h$ and $0 + 3h$. Then, using the same sort of inequality estimates, we find

$$\begin{aligned}
|E_{3h}| &= |y(0+3h) - \hat{y}(0+3h)| \\
&\leq 2 \times 405|h|^2 + 3 \times 405\,|h|^3 + 3\left(2 \times 405|h|^2 + 3 \times 405\,|h|^3\right)|h| + 405h^2 \\
&= 3 \times 405|h|^2 + 9 \times 405\,|h|^3 + 9 \times 405\,|h|^4.
\end{aligned}$$

Now if h is smaller than 1, h^3 and h^4 are smaller that h^2. Hence, the contribution to the error from these terms is so small it is usually termed negligible. We write

$$\begin{aligned}
|E_{3h}| &= |y(0+3h) - \hat{y}(0+3h)| \\
&\approx 3 \times 405|h|^2,
\end{aligned}$$

where the symbol $\approx$ indicates that the largest error is due to the h^2 term. So far, we have generated the following approximate solution values for the interval [0, 0.5].

True solution	Approximate solution	Absolute error	Estimate of error
$y(0)$	$\hat{y}(0)$	$\|y(0) - \hat{y}(0)\|$	0
$y(0+h)$	$\hat{y}(0+h)$	$\|y(0+h) - \hat{y}(0+h)\|$	$405h^2$
$y(0+2h)$	$\hat{y}(0+2h)$	$\|y(0+2h) - \hat{y}(0+2h)\|$	$2 \times 405h^2$
$y(0+3h)$	$\hat{y}(0+3h)$	$\|y(0+3h) - \hat{y}(0+3h)\|$	$3 \times 405h^2$

We could continue to work out these approximations. We would find that after n steps to get us from 0 to $0 + nh$, the error would be

$$\begin{aligned} |E_{nh}| &= |y(0+nh) - \hat{y}(0+nh)| \\ &\approx n \times 405|h|^2 \approx (nh)\,(405h). \end{aligned}$$

After some number of steps, say N, we will reach the end of our interval [0, 0.05]. Hence, we can say after N steps, $Nh = 0.05$ and we have a total error estimate of

$$\begin{aligned} |E_{Nh}| &= |y(0+Nh) - \hat{y}(0+Nh)| \\ &\approx N \times 405|h|^2 \approx (Nh)\,(405h) \\ &= (0.05)\,(405h). \end{aligned}$$

We could do a similar analysis on any interval such as $[0, T]$ for some positive T. Now the constants 405 and so forth would change, but the basic argument would not. After N steps, we would reach the end of the interval $[0, T]$ and our total error estimate would be

$$\begin{aligned} |E_{Nh}| &= |y(0+Nh) - \hat{y}(0+Nh)| \\ &\approx N \times C|h|^2 \approx (Nh)\,(Ch) \\ &= (T)\,(Ch). \end{aligned}$$

for some constant C. Hence, the **Euler's method** appears to generate a sequence of approximate solutions with the total error made over the interval in question proportional to h; i.e. $|E_{Nh}| \propto h$. Let's do some specific numbers. We are on the interval [0, 0.5] so for various values of the stepsize h, we can look at the overall error to go from $y(0)$ to $y(0.5)$. We summarize in the table below:
The first row in the table gives the error estimate using the first step size of $h = 0.1$. When we half the step size to $h = 0.05$, we need twice as many steps to reach the final time of 0.5 and hence we have more individual errors to add up. Note the error when we reach 0.5 is $1/2$ the first error due to $h = 0.1$. The same thing happens

True solution	Stepsize	Steps to get to 0.5	Overall absolute error
$y(0.5)$	0.1	$0.5/0.1 = 5$	$\approx 5 \times 405\ (0.1)^2 = \mathbf{E}$
$y(0.5)$	0.05	$0.5/0.05 = 10$	$\approx 10 \times 405\ (0.1/2)^2$ $\approx 5/2 \times 405\ (0.1)^2 = \mathbf{E/2}$
$y(0.5)$	0.025	$0.5/0.025 = 20$	$\approx 20 \times 405\ (0.1/4)^2$ $\approx 5/4 \times 405\ (0.1)^2 = \mathbf{E/4}$

when we compare row 2 and row 3. So, the pattern that emerges is that **halving the step size halves the total error**. Our example seems to suggest the following:

- The error made at each step is like h^2 and this is called the **local** error.
- The error made in going from the initial value to the final value of y is called the **global** error for a given step size h. We see the global error goes down linearly with h. That is, if we change from step size h to step size $r\ h$ for some r in $(0, 1)$, then the **global** error, E, goes down to $r\ E$.

The stuff we are doing above could be done for any first order differential equation

$$\frac{dy}{dt} = f(y(t))$$
$$y(0) = y_0$$

where f is some function of y. In our example above, $f(y) = 3y$.

16.1.2 Euler Approximates Again

We are going to redo our arguments more generally and switch to the variable x so that you can see the notation using a different letter. The ideas are still the same and you need to be able to move back and forth between different variable names recognizing they are simply placekeepers for the ideas. This time our arguments are more abstract so it might make you a bit uncomfortable, but it is worth it for you to try to go through our reasoning. Trust me: as they say what doesn't kill you makes you stronger! Do you hear music in the background? But we digress.

So let's try to approximate the solution to $x' = f(x)$ with $x(0) = x_0$ on the interval $[0, T]$ using these tangent line ideas. We already looked carefully at the model $y' = 3y$ with $y(0) = 20$ but now we want to examine what is happening more generally. The solution $x(t)$ can be written

$$x(t) = x(0) + x'(0)(t - 0) + x''(c_t)(t - 0)^2/2$$

where c_t is some number between 0 and t. To approximate the solution, we divide the interval $[0, T]$ into pieces of length h, the **stepsize**. If h does not evenly divide T, we just use the last subinterval even though it may be a bit short. Let N be the number of subintervals we get by doing this. For example, divide $[0, 5]$ using $h = 0.4$. Then $5/0.4 = 12.5$ so we create 13 subintervals with the last one of length 0.2 instead of 0.4. So $N = 13$. On the other hand, if we divide $[0, 10]$ using $h = 0.2$, then $10/0.2 = 50$ and we get $N = 50$. To approximate the **true** solution $x(h)$ we then have

$$\begin{aligned} x(h) &= x(0) + x'(0)(h-0) + x''(c_h)(h-0)^2/2 \\ &= x_0 + x'(0)h + x''(c_h)h^2/2, \end{aligned}$$

where c_h is between 0 and h. We can rewrite this more. Note $x' = f(x)$ tells us we can replace $x'(0)$ by $f(x(0)) = f(x_0)$. Also, since $x' = f(x)$, the chain rule tells us $x'' = (df/dx)\, x' = (df/dx)\, f$. So $x''(c_h) = (df/dx)(x(c_h))\, f(x(c_h))$. Hence,

$$\begin{aligned} x(h) &= x_0 + f(x_0)h + (df/dx)(x(c_h))\, f(x(c_h))\, h^2/2 \\ &= x_0 + f(x_0)h + f'(x(c_h))\, f(x(c_h))\, h^2/2 \end{aligned}$$

where we just write f' instead of df/dx for convenience. Now let x_1 be the true solution $x(h)$. and let $\hat{x}_0$ be the starting or zeroth Euler approximate which is defined by $\hat{x}_0 = x_0$. So we make no error at first. The first Euler approximate $\hat{x}_1$ is defined by $\hat{x}_1 = x_0 + f(x_0)\, h = \hat{x}_0 + f(\hat{x}_0)\, h$. Thus, we have

$$x_1 = \hat{x}_1 + f'(x(c_h))\, f(x(c_h))\, h^2/2.$$

The error at the first step is then $E_1 = |x_1 - \hat{x}_1|$. and

$$E_1 = |x_1 - \hat{x}_1| = |f'(x(c_h))|\, |f(x(c_h))|\, h^2/2.$$

Now let's do bounds. Since x is continuous on $[0, T]$, we know x is bounded. Hence, $|x(t)|$ is bounded on $[0, T]$ also, Let $||x||_\infty = \max_{0 \le t \le T} |x(t)|$. This is some finite number which we will call D. Thus, $x(t)$ lives in the interval $[-D, D]$ which is an interval of the x axis. We know f is continuous on $[-D, D]$, so $||f||_\infty = \max_{-D \le x \le D} |f(x)|$ is some finite number. Also, f' is continuous on $[-D, D]$, so $||f'||_\infty = \max_{-D \le x \le D} |f'(x)|$ is finite as well. The specific numbers we used for the example $y' = 3y$ are an example of these bounds. Using these bounds, we have

$$\begin{aligned} E_1 = |x_1 - \hat{x}_1| &= |f'(x(c_h))|\, |f(x(c_h))|\, h^2/2 \\ &\le ||f||_\infty\, ||f'||_\infty\, h^2/2 = B\, h^2/2. \end{aligned}$$

where we let $B = ||f||_\infty\, ||f'||_\infty$.
Now let's do the approximation for $x(2h)$. Let $x_2 = x(2h)$ and note the second Euler approximate is $\hat{x}_2 = \hat{x}_1 + f(\hat{x}_1)\, h$. The tangent line approximation to x and h gives

$$
\begin{aligned}
x(2h) &= x(h) + x'(h)\,h + x''(c_{2h})\,h^2/2 \\
&= x(h) + f(x(h))h + f'(x(c_{2h}))\,f(x(c_{2h}))\,h^2/2 \\
&= x_1 + f(x_1)h + f'(x(c_{2h}))\,f(x(c_{2h}))\,h^2/2.
\end{aligned}
$$

Now add and subtract $\hat{x}_2 = \hat{x}_1 + f(\hat{x}_1)\,h$ to this equation.

$$
\begin{aligned}
x(2h) &= x_1 + f(x_1)h + f'(x(c_{2h}))\,f(x(c_{2h}))\,h^2/2 \\
&= x_1 + f(x_1)h + \hat{x}_2 - \hat{x}_2 + f'(x(c_{2h}))\,f(x(c_{2h}))\,h^2/2 \\
&= (x_1 + f(x_1)h - \hat{x}_1 - f(\hat{x}_1)\,h) + \hat{x}_2 \\
&\quad + f'(x(c_{2h}))\,f(x(c_{2h}))\,h^2/2 \\
&= \hat{x}_2 + (x_1 - \hat{x}_1) + (f(x_1) - f(\hat{x}_1))\,h \\
&\quad + f'(x(c_{2h}))\,f(x(c_{2h}))\,h^2/2.
\end{aligned}
$$

We are almost there! We can apply the Mean Value Theorem to the difference $f(x_1) - f(\hat{x}_1)$ to find $f(x_1) - f(\hat{x}_1) = f'(x_d)(x_1 - \hat{x}_1)$ with x_d between x_1 and $\hat{x}_1$. Plugging this in, we find

$$
\begin{aligned}
x_2 &= \hat{x}_2 + (x_1 - \hat{x}_1) + (f(x_1) - f(\hat{x}_1))\,h \\
&\quad + f'(x(c_{2h}))\,f(x(c_{2h}))\,h^2/2 \\
&= \hat{x}_2 + (x_1 - \hat{x}_1) + f'(x_d)\,(x_1 - \hat{x}_1)\,h \\
&\quad + f'(x(c_{2h}))\,f(x(c_{2h}))\,h^2/2 \\
&= \hat{x}_2 + (x_1 - \hat{x}_1)\left(1 + f'(x_d)h\right) + f'(x(c_{2h}))\,f(x(c_{2h}))\,h^2/2
\end{aligned}
$$

Thus

$$
x_2 - \hat{x}_2 = (x_1 - \hat{x}_1)\,(1 + f'(x_d)h) + f'(x(c_{2h}))\,f(x(c_{2h}))\,h^2/2.
$$

Now $E_2 = |x_2 - \hat{x}_2|$, so we can overestimate

$$
\begin{aligned}
E_2 = |x_2 - \hat{x}_2| &\le |x_1 - \hat{x}_1|\,(1 + |f'(x_d)|h) \\
&\quad + |f'(x(c_{2h}))|\,|f(x(c_{2h}))|\,h^2/2 \\
&\le E_1\,(1 + ||f'||_\infty h) + ||f'||_\infty\,||f||_\infty\,h^2/2 \\
&= E_1\,(1 + Ah) + B\,h^2/2.
\end{aligned}
$$

where we let $A = ||f||_\infty$ to save typing. Since $E_1 \le Bh^2/2$, we find

$$
\begin{aligned}
E_2 &\le \left(B\,h^2/2\right)(1 + Ah) + B\,h^2/2 = B\,(2 + Ah)\,h^2/2 \\
&= 2\,Bh^2/2 + ABh^3/2.
\end{aligned}
$$

We can continue to do this and although the algebra gets pretty intense, we find that after N steps

$$E_N \leq N\ Bh^2/2 + Ch^3.$$

where C is some constant. After N steps, we reach the end of the interval T. So $Nh = T$. Rewriting, we have the absolute error we make after N Euler approximation steps is

$$E_N \leq N\ Bh^2/2 + Ch^3 = (Nh)\ Bh/2 + Ch^2\ (T/N)$$

as $h = T/N$. This simplifies to

$$E_N \leq (BT/2)\ h + (CT)/N\ h^2.$$

We conclude

- The dominant part of the error using step size h over $[0, T]$ is proportional to h. We say the **global error** is proportional to h.
- The local error at each step is proportional to h^2.
- However, these errors add up and over N steps, the error we make to find the approximate solution on $[0, T]$ is larger; indeed, it is proportional to h but that is a bit harder to see with this approach.

16.1.3 Euler Approximates the Last Time

So let's look at this approach one more time.

Again, we approximate the solution to the model $x' = f(x)$ with $x(0) = x_0$. The solution $x(t)$ can be written

$$x(t) = x(0) + x'(0)(t-0) + x''(c_t)(t-0)^2/2$$

where c_t is some number between 0 and t. To approximate the solution, we will divide the interval $[0, T]$ into pieces of length h as usual. To approximate the **true** solution $x(h)$, we have

$$x(h) = x_0 + x'(0)h + x''(c_h)h^2/2,$$

where c_h is between 0 and h. As usual, we note $x'(0)$ by $f(x(0)) = f(x_0)$ and $x''(c_h) = f'(x(c_h))\ f(x(c_h))$. Thus,

$$x(h) = x_0 + f(x_0)h + f'(x(c_h))\ f(x(c_h))\ h^2/2$$

Again, we let x_1 be the true solution $x(h)$ and $1\hat{x}_0$ be the starting or zeroth Euler approximate which is defined by $\hat{x}_0 = x_0$. Hence, we make no error at first. The first Euler approximate $\hat{x}_1$ is $\hat{x}_1 = x_0 + f(x_0)\,h = \hat{x}_0 + f(\hat{x}_0)\,h$. which is the tangent line approximation to x at the point $t = 0$. Then we have

$$x_1 = \hat{x}_1 + f'(x(c_h))\, f(x(c_h))\, h^2/2.$$

Define the error at the first step by $E_1 = |x_1 - \hat{x}_1|$. Thus,

$$E_1 = |x_1 - \hat{x}_1| = |f'(x(c_h))|\, |f(x(c_h))|\, h^2/2.$$

Since x is continuous on $[0, T]$, we know x is bounded. Hence, $||x||_\infty = \max_{0 \leq t \leq T} |x(t)|$ is some finite number. Call it D. We see $x(t)$ lives in the interval $[-D, D]$ which is on the x axis. Then

- f is continuous on $[-D, D]$, so $||f||_\infty = \max_{-D \leq x \leq D} |f(x)|$ is some finit number.
- f' is continuous on $[-D, D]$, so $||f'||_\infty = \max_{-D \leq x \leq D} |f'(x)|$ is some finite number.

Using these bounds, we have

$$\begin{aligned} E_1 &= |x_1 - \hat{x}_1| = |f'(x(c_h))|\, |f(x(c_h))|\, h^2/2 \\ &\leq ||f||_\infty\, ||f'||_\infty\, h^2/2 = B\, h^2/2 \leq C\, h^2/2. \end{aligned}$$

where we let $A = ||f||_\infty$, $B = ||f||_\infty\, ||f'||_\infty$ and C be the maximum of A and B.

Now let $x_2 = x(2h)$ and we define the second Euler approximate by $\hat{x}_2 = \hat{x}_1 + f(\hat{x}_1)\,h$. The tangent line approximation to x at h gives

$$x(2h) = x_1 + f(x_1)h + f'(x(c_{2h}))\, f(x(c_{2h}))\, h^2/2.$$

Now add and subtract $\hat{x}_2 = \hat{x}_1 + f(\hat{x}_1)\,h$ to this equation.

$$\begin{aligned} x(2h) &= x_1 + f(x_1)h + f'(x(c_{2h}))\, f(x(c_{2h}))\, h^2/2 \\ &= x_1 + f(x_1)h + \hat{x}_2 - \hat{x}_2 + f'(x(c_{2h}))\, f(x(c_{2h}))\, h^2/2 \\ &= (x_1 + f(x_1)h - \hat{x}_1 - f(\hat{x}_1)\,h) + \hat{x}_2 \\ &\quad + f'(x(c_{2h}))\, f(x(c_{2h}))\, h^2/2 \\ &= \hat{x}_2 + (x_1 - \hat{x}_1) + (f(x_1) - f(\hat{x}_1))\,h \\ &\quad + f'(x(c_{2h}))\, f(x(c_{2h}))\, h^2/2. \end{aligned}$$

Next, we can apply the Mean Value Theorem to the difference $f(x_1) - f(\hat{x}_1)$ and find $f(x_1) - f(\hat{x}_1) = f'(x_d)(x_1 - \hat{x}_1)$ with x_d between x_1 and $\hat{x}_1$. Plugging this in, we have

$$\begin{aligned} x_2 &= \hat{x}_2 + (x_1 - \hat{x}_1) + (f(x_1) - f(\hat{x}_1))\, h \\ &\quad + f'(x(c_{2h}))\, f(x(c_{2h}))\, h^2/2 \\ &= \hat{x}_2 + (x_1 - \hat{x}_1) + f'(x_d)\, (x_1 - \hat{x}_1)\, h \\ &\quad + f'(x(c_{2h}))\, f(x(c_{2h}))\, h^2/2 \\ &= \hat{x}_2 + (x_1 - \hat{x}_1)\, (1 + f'(x_d)h) + f'(x(c_{2h}))\, f(x(c_{2h}))\, h^2/2 \end{aligned}$$

Thus

$$x_2 - \hat{x}_2 = (x_1 - \hat{x}_1)\, (1 + f'(x_d)h) + f'(x(c_{2h}))\, f(x(c_{2h}))\, h^2/2.$$

Now $E_2 = |x_2 - \hat{x}_2|$, so we can overestimate

$$\begin{aligned} E_2 = |x_2 - \hat{x}_2| &\le |x_1 - \hat{x}_1|\, (1 + |f'(x_d)|h) \\ &\quad + |f'(x(c_{2h}))|\, |f(x(c_{2h}))|\, h^2/2 \\ &\le E_1\, (1 + ||f'||_\infty h) + ||f'||_\infty\, ||f||_\infty\, h^2/2 \\ &= E_1\, (1 + Ah) + B\, h^2/2 \le E_1(1 + Ch) + Ch^2/2. \end{aligned}$$

Since $E_1 \le Ch^2/2$, we find

$$E_2 \le C\, h^2/2\, (1 + Ch) + C\, h^2/2 = (C\, h^2/2) \left(1 + (1 + Ch)\right)$$

We know using our approximations

$$e^{ut} = 1 + ut + u^2 e^{uc} t^2/2$$

for some c between 0 and u. But the error term is positive so we know $e^{ut} \ge 1 + ut$. Letting $u = Ch$ and using $t = 1$, we have

$$1 + (1 + Ch) \le 1 + e^{Ch}.$$

and so

$$E_2 \le (C\, h^2/2)\, (1 + e^{Ch})$$

Now let's do the approximation for $x(3h)$. We will let $x_3 = x(3h)$ and we will define the third Euler approximate by $\hat{x}_3 = \hat{x}_2 + f(\hat{x}_2)\, h$. The tangent line approximation to x at $2h$ gives

$$
\begin{aligned}
x(3h) &= x(2h) + x'(2h)\,h + x''(c_{3h})\,h^2/2 \\
&= x(2h) + f(x(2h))h + f'(x(c_{3h}))\,f(x(c_{3h}))\,h^2/2 \\
&= x_2 + f(x_2)h + f'(x(c_{3h}))\,f(x(c_{3h}))\,h^2/2.
\end{aligned}
$$

Now add and subtract $\hat{x}_3 = \hat{x}_2 + f(\hat{x}_2)\,h$ to this equation.

$$
\begin{aligned}
x(3h) &= x_2 + f(x_2)h + f'(x(c_{3h}))\,f(x(c_{3h}))\,h^2/2 \\
&= x_2 + f(x_2)h + \hat{x}_3 - \hat{x}_3 + f'(x(c_{3h}))\,f(x(c_{3h}))\,h^2/2 \\
&= (x_2 + f(x_2)h - \hat{x}_2 - f(\hat{x}_2)\,h) + \hat{x}_3 \\
&\quad + f'(x(c_{3h}))\,f(x(c_{3h}))\,h^2/2 \\
&= \hat{x}_3 + (x_2 - \hat{x}_2) + (f(x_2) - f(\hat{x}_2))\,h \\
&\quad + f'(x(c_{3h}))\,f(x(c_{3h}))\,h^2/2.
\end{aligned}
$$

But we can apply the Mean Value Theorem to the difference $f(x_2) - f(\hat{x}_2)$. We find $f(x_2) - f(\hat{x}_2) = f'(x_u)(x_2 - \hat{x}_2)$ with x_u between x_2 and $\hat{x}_2$. Plugging this in, we find

$$
\begin{aligned}
x_3 &= \hat{x}_3 + (x_2 - \hat{x}_2) + (f(x_2) - f(\hat{x}_2))\,h \\
&\quad + f'(x(c_{3h}))\,f(x(c_{3h}))\,h^2/2 \\
&= \hat{x}_3 + (x_2 - \hat{x}_2) + f'(x_u)\,(x_2 - \hat{x}_2)\,h \\
&\quad + f'(x(c_{3h}))\,f(x(c_{3h}))\,h^2/2 \\
&= \hat{x}_3 + (x_2 - \hat{x}_2)\,(1 + f'(x_u)h) + f'(x(c_{3h}))\,f(x(c_{3h}))\,h^2/2
\end{aligned}
$$

Thus

$$
x_3 - \hat{x}_3 = (x_2 - \hat{x}_2)\,(1 + f'(x_u)h) + f'(x(c_{3h}))\,f(x(c_{3h}))\,h^2/2.
$$

Now $E_3 = |x_3 - \hat{x}_3|$, so we can overestimate

$$
\begin{aligned}
E_3 = |x_3 - \hat{x}_3| &\le |x_2 - \hat{x}_2|\,(1 + |f'(x_u)|h) \\
&\quad + |f'(x(c_{3h}))|\,|f(x(c_{3h}))|\,h^2/2 \\
&\le E_2\,(1 + ||f'||_\infty h) + ||f'||_\infty\,||f||_\infty\,h^2/2 \\
&= E_2\,(1 + Ah) + B\,h^2/2.
\end{aligned}
$$

Since $E_2 \le (C\,h^2/2)\,(2 + Ch)$, we find

$$
\begin{aligned}
E_3 &\le \big((C\,h^2/2)\,(2 + Ch)\big)\;(1 + Ah) + B\,h^2/2 \\
&\le \big((C\,h^2/2)\,(2 + Ch)\big)\;(1 + Ch) + C\,h^2/2 \\
&= C\;\big(1 + (1 + Ch) + (1 + Ch)^2\big)\;h^2/2
\end{aligned}
$$

We have already shown that $1+u \leq e^u$ for any u. It follows $(1+u)^2 \leq (e^u)^2 = e^{2u}$. So we have

$$\begin{aligned} E_3 &\leq C\,\left(1+(1+Ch)+(1+Ch)^2\right)\,h^2/2 \\ &\leq C\,\left(1+e^{Ch}+e^{2Ch}\right)\,h^2/2. \end{aligned}$$

We also know $1+u+u^2 = (u^3-1)/(u-1)$. So we have

$$E_3 \leq C\,\frac{e^{3Ch}-1}{e^{Ch}-1}\,h^2/2.$$

Continuing we find after N steps

$$E_N \leq C\,\frac{e^{NCh}-1}{e^{Ch}-1}\,h^2/2.$$

Now after N steps, we reach the end of the interval T. So $Nh = T$. Rewriting, we have the absolute error we make after N Euler approximation steps is

$$E_N \leq C\,\frac{e^{CT}-1}{e^{Ch}-1}\,h^2/2.$$

The total error is what we get when we add up $E_1 + \cdots + E_N$. Note,

$$\begin{aligned} E_1 &\leq C\,\frac{e^{Ch}-1}{e^{Ch}-1}\,h^2/2 \leq C\,\frac{e^{CT}-1}{e^{Ch}-1}\,h^2/2 \\ E_2 &\leq C\,\frac{e^{2Ch}-1}{e^{Ch}-1}\,h^2/2 \leq C\,\frac{e^{CT}-1}{e^{Ch}-1}\,h^2/2 \\ E_3 &\leq C\,\frac{e^{3Ch}-1}{e^{Ch}-1}\,h^2/2 \leq C\,\frac{e^{CT}-1}{e^{Ch}-1}\,h^2/2 \end{aligned}$$

and so on, so we have

This gives the estimate

$$\begin{aligned} E_1 + \cdots + E_N &\leq N\,C\,\frac{e^{CT}-1}{e^{Ch}-1}\,h^2/2 \\ &\leq (Nh)\,\frac{e^{CT}-1}{e^{Ch}-1}\,h \\ &= T\,\frac{e^{CT}-1}{e^{Ch}-1}\,h \end{aligned}$$

So the **local error** at each step is on the order of h^2 and the **global error** we make to go N steps to find the solution on the whole interval $[0, T]$ is of order h.

16.1.4 Euler's Algorithm

Here is **Euler's Algorithm** to approximate the solution to $x' = f(x)$ with $x(0) = x_0$ using step size h for as many steps as we want.

- $\hat{x}_0 = x_0$ so $E_0 = 0$.
- $\hat{x}_1 = \hat{x}_0 + f(\hat{x}_0)\,h$; $E_1 = |x_1 - \hat{x}_1|$.
- $\hat{x}_2 = \hat{x}_1 + f(\hat{x}_1)\,h$; $E_2 = |x_2 - \hat{x}_2|$.
- $\hat{x}_3 = \hat{x}_2 + f(\hat{x}_2)\,h$; $E_3 = |x_3 - \hat{x}_3|$.
- $\hat{x}_4 = \hat{x}_3 + f(\hat{x}_3)\,h$; $E_4 = |x_4 - \hat{x}_4|$.
- Continue as many steps as you want.

Recursively:

- $\hat{x}_0 = x_0$ so $E_0 = 0$.
- $\hat{x}_{n+1} = \hat{x}_n + f(\hat{x}_n)\,h$; $E_{n+1} = |x_{n+1} - \hat{x}_{n+1}|$ for $n = 0, 1, 2, 3, \ldots$.

The approximation scheme above is called **Euler's Method**.

Example 16.1.1 Find the first three Euler approximates for $x' = -2x$, $x(0) = 3$ using $h = 0.3$. Find the true solution values and errors also.

Solution *Here $f(x) = -2x$ and the true solution is $x(t) = 3e^{-2t}$.*

- $\hat{x}_0 = x_0 = 3$ *so* $E_0 = 0$.
-
$$\begin{aligned}
\hat{x}_1 &= \hat{x}_0 + f(\hat{x}_0)\,h = 3 + f(3)\,(0.3)\\
&= 3 + (-2(3))\,(0.3) = 3 - 6(0.3) = 3 - 1.8 = 1.2.\\
x_1 &= x(h) = 3e^{-2h} = 3e^{-0.6} = 1.646.\\
E_1 &= |x_1 - \hat{x}_1| = |1.646 - 1.2| = 0.446.
\end{aligned}$$

-
$$\begin{aligned}
\hat{x}_2 &= \hat{x}_1 + f(\hat{x}_1)\,h = 1.2 + f(1.2)\,(0.3)\\
&= 1.2 + (-2(1.2))\,(0.3) = 1.2 - 2.4(0.3) = 0.48.\\
x_2 &= x(2h) = 3e^{-2(2h)} = 3e^{-4h} = 3e^{-1.2} = 0.9036.\\
E_2 &= |x_2 - \hat{x}_2| = |0.9036 - 0.48| = 0.4236.
\end{aligned}$$

-
$$\begin{aligned}
\hat{x}_3 &= \hat{x}_2 + f(\hat{x}_2)\,h = 0.48 + f(0.48)\,(0.3)\\
&= 0.48 + (-2(0.48))\,(0.3) = 0.48 - 0.96(0.3) = 0.192.\\
x_3 &= x(3h) = 3e^{-2(3h)} = 3e^{-6h} = 3e^{-1.8} = 0.4959.\\
E_3 &= |x_3 - \hat{x}_3| = |0.4959 - 0.192| = 0.3039.
\end{aligned}$$

Example 16.1.2 Find the first three Euler approximates for $x' = 2x, x(0) = 4$ using $h = 0.2$. Find the true solution values and errors also.

Solution *Here* $f(x) = 2x$ *and the true solution is* $x(t) = 4e^{2t}$.

- $\hat{x}_0 = x_0 = 4$ *so* $E_0 = 0$.
-

$$\begin{aligned}\hat{x}_1 &= \hat{x}_0 + f(\hat{x}_0)\,h = 4 + f(4)\,(0.3)\\ &= 4 + (2(4))\,(0.2) = 4 + 8(0.2) = 5.6.\\ x_1 &= x(h) = 4e^{2h} = 4e^{0.4} = 5.9673.\\ E_1 &= |x_1 - \hat{x}_1| = |5.9673 - 5.6| = 0.3673.\end{aligned}$$

-

$$\begin{aligned}\hat{x}_2 &= \hat{x}_1 + f(\hat{x}_1)\,h = 5.6 + f(5.6)\,(0.2)\\ &= 5.6 + (2(5.6))\,(0.2) = 5.6 + 11.2(0.2) = 7.84.\\ x_2 &= x(2h) = 4e^{2(2h)} = 4e^{4h} = 4e^{0.8} = 8.9032.\\ E_2 &= |x_2 - \hat{x}_2| = |8.9032 - 7.84| = 1.0622.\end{aligned}$$

-

$$\begin{aligned}\hat{x}_3 &= \hat{x}_2 + f(\hat{x}_2)\,h = 7.84 + f(7.84)\,(0.2)\\ &= 7.84 + (2(7.84))\,(0.2) = 7.84 + 15.68(0.2) = 10.976.\\ x_3 &= x(3h) = 4e^{2(3h)} = 4e^{6h} = 4e^{1.2} = 13.2805.\\ E_3 &= |x_3 - \hat{x}_3| = |13.2805 - 10.976| = 2.3045.\end{aligned}$$

Example 16.1.3 Find the first three Euler approximates for $x' = 0.2x(100 - x)$, $x(0) = 20$ using $h = 0.05$. Find the true solution values and errors also.

Solution *Here* $f(x) = 0.2x(100 - x)$ *and the true solution is* $x(t) = 100/[1 + (100/20 - 1)e^{-20t}]$ *or* $x(t) = 100/[1 + 4e^{-20t}]$ *(see the general logistics solutions). We have* $\hat{x}_0 = x_0 = 20$ *so* $E_0 = 0$ *and*

$$\begin{aligned}\hat{x}_1 &= \hat{x}_0 + f(\hat{x}_0)\,h = 20 + f(20)\,(0.05)\\ &= 20 + \{0.2(20)(100 - 20)\}\,(0.05)\\ &= 20 + 0.2(20)(80)(0.05) = 20 + 16 = 36\\ x_1 &= x(h) = 100/[1 + 4e^{-20h}] = 100/[1 + 4e^{-20(0.05)}]\\ &= 100/[1 + 4e^{-1}] = 40.46\end{aligned}$$

And so

$$E_1 = |x_1 - \hat{x}_1| = |40.46 - 36| = 4.46.$$

Next

$$\begin{aligned}
\hat{x}_2 &= \hat{x}_1 + f(\hat{x}_1)\,h = 36 + f(36)\,(0.05) \\
&= 36 + \{0.2(36)(100 - 36)\}\,(0.05) \\
&= 36 + 0.2(36)(64)(0.05) = 36 + 23.04 = 59.04 \\
x_2 &= x(2h) = 100/[1 + 4e^{-20(2h)}] = 100/[1 + 4e^{-20(0.1)}] \\
&= 100/[1 + 4e^{-2}] = 64.8786 \\
E_2 &= |x_2 - \hat{x}_2| = |64.8786 - 59.04| = 5.8386.
\end{aligned}$$

Finally

$$\begin{aligned}
\hat{x}_3 &= \hat{x}_2 + f(\hat{x}_2)\,h = 59.04 + f(59.04)\,(0.05) \\
&= 59.04 + \{0.2(59.04)(100 = 59.04)\}\,(0.05) \\
&= 59.04 + 0.2(59.04)(40.96)(0.05) = 59.04 + 24.1828 \\
&= 83.2228 \\
x_3 &= x(3h) = 100/[1 + 4e^{-20(3h)}] = 100/[1 + 4e^{-20(0.15)}] \\
&= 100/[1 + 4e^{-3}] = 83.3925 \\
E_3 &= |x_3 - \hat{x}_3| = |83.3925 - 83.2228| = 0.1697.
\end{aligned}$$

16.1.4.1 Homework

Exercise 16.1.1 *Find the first 4 Euler approximates for* $x' = 3x$ *with* $x(0) = 2$ *for* $h = 0.4$. *Find the true values and errors also.*

Exercise 16.1.2 *Find the first 3 Euler approximates for* $x' = -3x$ *with* $x(0) = 8$ *for* $h = 0.3$. *Find the true values and errors also.*

Exercise 16.1.3 *Find the first 4 Euler approximates for* $x' = 0.3x(75 - x)$ *with* $x(0) = 20$ *for* $h = 0.02$. *Find the true values and errors also.*

Exercise 16.1.4 *Find the first 4 Euler approximates for* $x' = -0.2x + 5$ *with* $x(0) = 10$ *for* $h = 0.2$. *Find the true values and errors also.*

16.1.5 Adding Time to Euler's Method

The really neat thing about this idea of approximating the solution to the differential equation, is that we can use the technique even when we have no idea how to use calculus to find the answer. For example, the function f could also depend on t and we could ask to solve

$$\begin{aligned}\frac{dy}{dt} &= f(t, y(t))\\ y(0) &= y_0\end{aligned}$$

where f is some function of t and y. For example, we could look at the logistics model with t added into the function f as we see below.

$$\begin{aligned}\frac{dy}{dt} &= 0.02y(t)(200 - y(t))\, e^{-0.04t^2}\\ y(0) &= 150.\end{aligned}$$

This is certainly not a logistics model! **If** we had good reasons to expect the solution to this problem does exist and its second order derivative is nice enough to be bounded on our interval $[0, 0.5]$ (and this is not an easy question to answer in general!), we could use the more general Euler's scheme

$$\begin{aligned}\hat{y}_0 &= y(0) = y_0\\ \hat{y}_1 &= \hat{y}_0 + f(0, \hat{y}_0)\, h\\ \hat{y}_2 &= \hat{y}_1 + f(h, \hat{y}_1)\, h\\ \hat{y}_3 &= \hat{y}_2 + f(2h, \hat{y}_2)\, h\\ \vdots &= \vdots\\ \hat{y}_n &= \hat{y}_{n-1} + f((n-1)h, \hat{y}_{n-1})\, h\end{aligned}$$

In the example above, $f(t, y) = 0.02y(200 - y)\, e^{-0.04t^2}$.

16.1.6 Simple MatLab Experiments

Let's work our way through some examples.

Example 16.1.4 Find the Euler approximates to

$$\begin{aligned}x'(t) &= 3x(t)\\ x(0) &= 2.\end{aligned}$$

using Matlab.

Solution *We know the solution is* $x(t) = 2e^{3t}$. *Define the function* $f(x) = 3x$.

Listing 16.1: Define dynamics $f(x) = 3x$

```
f = @(x) (3*x);
```

Now, we'll do Euler's method the long way with individual variables for each step. We will let the Euler approximate values be given by **xhat0**, **xhat1** *and so forth with the true values* **x0**, **x1** *etc. Here is the session with comments.*

Listing 16.2: Euler Approximates The Long Way

```
% Set up initial value
x0 = 2;
% Setup first Euler approximate
xhat0 = 2;
% Choose a step size
h = .3;
% Find second Euler approximate
xhat1 = xhat0 + f(xhat0)*h
xhat1 =
      3.8000
% Set up the true function:
true = @(t) (2*exp(3*t));
% Get x1 = phi(h) = true(h) here
x1 = true(h)
x1 =
      4.9192

% Get the first abs error
error0 = abs(x0 - xhat0)
error0 =
       0
% Get the second abs error
error1 = abs(x1-xhat1)
error1  =
      1.1192
%Get next euler approx
xhat2 = xhat1 + f(xhat1)*h
xhat2 =
      7.2200
%Get next true      this is true(2*h) here
x2 = true(2*h)
x2 =
     12.0993
%Get next abs error
error2 = abs(x2-xhat2)
error2 =
      4.8793
And you can continue like this.
```

But it is easier to set up additional functions and use a **for i=1:N** loop. Here is another example.

Example 16.1.5 Find the first three Euler approximates for $x' = -2x$, $x(0) = 3$ using $h = 0.3$. Find the true solution values and errors also. But this time, do it in Matlab!

Solution *We use a simple for loop.*

Listing 16.3: Simple Euler Approximations In Matlab With a for loop

```
% set up the dynamics and true solution
f = @(x) -2*x;
true = @(t) 3*exp(-2*t);
% set up the euler step function
euler = @(x,h) x + f(x)*h;
% set the number of steps to take, the initial condition and the stepsize
N = 3;
x0 = 3;
h = .3;
% initialize true values x, approxis xhat and errors e
xhat(1) = x0;
x(1) = x0;
e(1) = 0;
% calculate euler steps and true solution values
for i=1:N
xhat(i+1) = euler(xhat(i),h);
x(i+1) = true(i*h);
e(i+1) = abs( x(i+1) - xhat(i+1) );
end
% display results
xhat, x, e
xhat =
      3.0000    1.2000    0.4800    0.1920
x =
      3.0000    1.6464    0.9036    0.4959
e =
           0    0.4464    0.4236    0.3039
```

Comment 16.1.1 *Numbering in Matlab is off by* 1. *So* $\hat{x}_0$ *is* **`xhat(1)`** *and so on. So you have to remember to move everything up by* 1.

16.1.6.1 Homework

Exercise 16.1.5 *Find the first 4 Euler approximates for* $x' = 1.5x$ *with* $x(0) = 6$ *for* $h = 0.3$. *Find the true values and errors also. This time do this in Matlab.*

Exercise 16.1.6 *Find the first 6 Euler approximates for* $x' = -2.3x$ *with* $x(0) = 5$ *for* $h = 0.1$. *Find the true values and errors also. This time do this in Matlab.*

Exercise 16.1.7 *Find the first 10 Euler approximates for* $x' = 0.4x(95 - x)$ *with* $x(0) = 10$ *for* $h = 0.01$. *Find the true values and errors also. This time do this in Matlab.*

Exercise 16.1.8 *Find the first 8 Euler approximates for* $x' = -0.4x + 15$ *with* $x(0) = 20$ *for* $h = 0.1$. *Find the true values and errors also. This time do this in Matlab.*

16.1.7 Matlab Euler Functions

Without the annotated comments in the code, the Matlab to do this is quick. So why not set this up as a function? Make up the function **DoEuler** and save it in the file **DoEuler.m** in your directory. Do this like before when we created and saved **RiemannSum.m** etc.

Listing 16.4: DoEuler.m

```
function [xhat,x,e] = DoEuler(f,true,N,x0,h)
% N = number of approximations to find
% f = the model dynamics function
% true = the true solution
% x0 = the initial condition
% h = the step size
%
euler = @(x,h) x + f(x)*h;
xhat(1) = x0;
% initialize x, xhat and e
x(1) = x0;
e(1) = 0;
for i=1:N
    xhat(i+1) = euler(xhat(i),h);
    x(i+1) = true(i*h);
    e(i+1) = abs( x(i+1) - xhat(i+1) );
end
end
```

Now let's do some examples using **DoEuler**.

Example 16.1.6 Find the first 6 Euler approximates for $x' = 1.9x$ with $x(0) = 10$ for $h = 0.12$. Find the true values and errors also. This time do this in Matlab using the function **DoEuler**.

Solution *Simpler now!*

Listing 16.5: Euler Approximates for $x' = 1.9x$, $x(0) = 10$

```
f = @(x) 1.9*x;
true = @(t) 10*exp(1.9*t);
[xhat,x,e] = DoEuler(f,true,6,10,.12);
xhat,x,e
xhat =
    10.0000    12.5609    15.7775    19.8179    24.8930    31.2677    39.2749
x =
    10.0000    12.2800    15.0798    18.5180    22.7402    27.9249    34.2918
e =
          0     0.2809     0.6977     1.2998     2.1528     3.3428     4.9831
```

Example 16.1.7 Find the first 5 Euler approximates for $x' = 0.5x(60 - x)$ with $x(0) = 20$ for $h = 0.01$. Find the true values and errors also. This time do this in Matlab using the function **DoEuler**.

Solution

Listing 16.6: Euler Approximates for $x' = 0.5x(60 - x)$, $x(0) = 20$

```
f = @(x) .5*x.*(60-x);
true = @(t) 60/( 1 + (60/20 - 1)*exp(-.5*60*t) ) ;
[xhat,x,e] = DoEuler(f,true,5,20,.01);
xhat,x,e
xhat =
    20.0000    24.1776    28.6038    33.0918    37.4441    41.4863
x =
    20.0000    24.0000    28.3200    32.8059    37.2665    41.5025
e =
          0     0.1776     0.2838     0.2859     0.1776     0.0162
```

Example 16.1.8 Find the first 6 Euler approximates for $x' = -0.3x + 13$ with $x(0) = 10$ for $h = 0.3$. Find the true values and errors also. This time do this in Matlab using the function **DoEuler**.

Solution

Listing 16.7: Euler Approximates for $x' = -0.3x + 13$, $x(0) = 10$

```
f = @(x) -.3*x + 13;
% see page 112 in text for true solution formula
true = @(t) (10 - 13/.3)*exp(-.3*t) + 13/.3 ;
[xhat,x,e] = DoEuler(f,true,6,10,.3);
xhat,x,e
xhat =
    10.0000    12.8690    15.4910    17.8874    20.0775    22.0791    23.9084
x =
    10.0000    12.4000    14.4400    16.1740    17.6479    18.9007    19.9656
e =
          0     0.4690     1.0510     1.7134     2.4296     3.1783     3.9428
```

16.1.8 Homework

Exercise 16.1.9 *Find the first 6 Euler approximates for* $x' = 2.5x$ *with* $x(0) = 6$ *for* $h = 0.3$. *Find the true values and errors also. This time use the function* **DoEuler**.

Exercise 16.1.10 *Find the first 6 Euler approximates for* $x' = -1.43x$ *with* $x(0) = 9$ *for* $h = 0.1$. *Find the true values and errors also. This time use the function* **DoEuler**.

Exercise 16.1.11 *Find the first 5 Euler approximates for* $x' = 0.2x(65 - x)$ *with* $x(0) = 20$ *for* $h = 0.01$. *Find the true values and errors also. This time use the function* **DoEuler**.

Exercise 16.1.12 *Find the first 6 Euler approximates for* $x' = -0.2x + 15$ *with* $x(0) = 10$ *for* $h = 0.2$. *Find the true values and errors also. This time use the function* **DoEuler**.

16.1.9 The True Versus the Euler Approximate Solution

We can also plot the true solution versus the Euler approximates so we can see how they are doing. The code is not too bad. Here it is with annotations. We are solving $x' = 0.5x(60 - x)$ with $x(0) = 20$.

Listing 16.8: True versus Euler Approximate Solutions

```
% Define the dynamics
f = @(x) .5*x.*(60-x);
% Define the true solution: note the ./ as we will
% use a vector for t
true = @(t) 60./( 1 + (60/20 - 1)*exp(-.5*60*t) ) ;
% set up final time and step size
T = .6;
h = .01;
% find number of steps
N = ceil(T/h)
N =
     60
% set up a time vector for the true solution plot
time = linspace(0,T,31);
% find the euler approximates
[xhat,x,e] = DoEuler(f,true,N,20,h);
% set up a time variable for the approximates
htime =(0:h:N*h);
% plot them both
plot(time,true(time),htime,xhat,'*');
xlabel('Time');
ylabel('x');
title('True Solution vs. Euler Approximate');
legend('True','Euler','Location','Best');
```

The line where we use **`N=ceil(T/h)`** does the division of T by h and rounds it up to the nearest integer. That way we get the right value of N even if h does not divide T evenly. We set up the time points for the Euler approximate values using the command **`htime =(0:h:N*h);`** which is a new one. It sets up time points every h units from 0 to the final one which is Nh. Sometimes it is convenient to use this command. When we run this code, we generate the plot you see in Fig. 16.1 and you can see visually that the Euler approximates are doing a good job.

It is also easy to plot just the Euler approximations alone by simply changing what we plot. For example, the Matlab session below does just that. We don't define a **`time`** variable as we are not plotting the true solution.

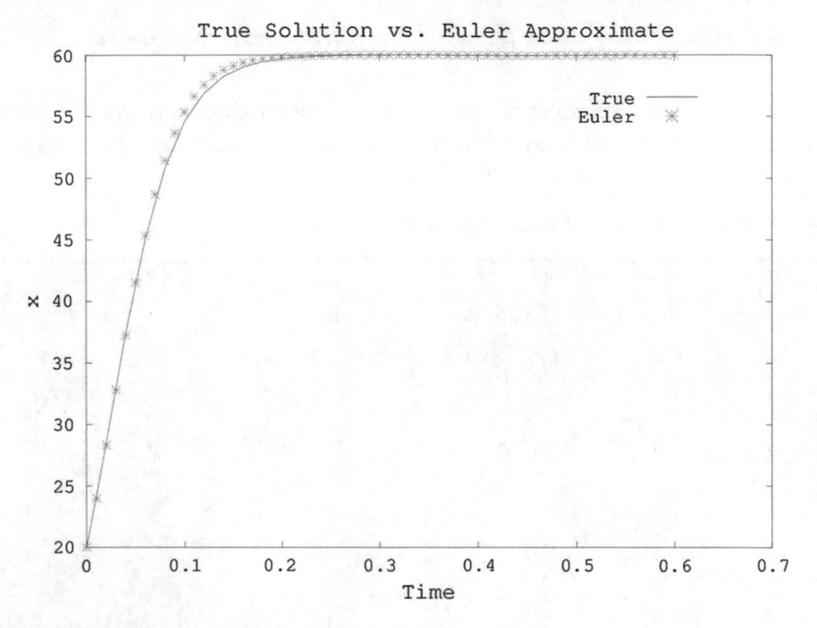

Fig. 16.1 True versus Euler for $x' = 0.5x(60 - x)$, $x(0) = 20$

Listing 16.9: Plotting Euler Approximations Only

```
f = @(x) .5*x.*(60-x);
true = @(t) 60./( 1 + (60/20 - 1)*exp(-.5*60*t) );
T = .6;
h = .01;
N = ceil(T/h);
% find the euler approximates
[xhat,x,e] = DoEuler(f,true,N,20,h);
% set up a time variable for the approximates
htime =(0:h:N*h);
% plot them both
plot(htime,xhat,'*');
xlabel('Time');
ylabel('x');
title('Euler Approximate');
```

We can also plot Euler approximates for different step sizes at the same time, with or without the true solution.

Listing 16.10: Plotting Multiple Euler Approximations Only

```
f = @(x) .5*x.*(60-x);
true = @(t) 60./( 1 + (60/20 - 1)*exp(-.5*60*t) ) ;
T = .6;
% set up a time vector for the true solution plot
time = linspace(0,T,31);
% Do step size 1
h1 = .04;
N1 = ceil(T/h1);
% find the euler approximates for h1
[xhat1,x1,e1] = DoEuler(f,true,N1,20,h1);
% set up a time variable for the approximates
htime1 =(0:h1:N1*h1);
% Do step size 2
h2 = .02;
N2 = ceil(T/h2);
% find the euler approximates for h2
[xhat2,x2,e2] = DoEuler(f,true,N2,20,h2);
% set up a time variable for the approximates
htime2 =(0:h2:N2*h2);
% plot all three
plot(time,true(time),htime1,xhat1,'*',htime2,xhat2,'o');
xlabel('Time');
ylabel('x');
title('True Solution vs. Euler Approximates');
legend('True','Euler h1=.04','Euler h2=.02','Location','Best');
```

The plot is shown in Fig. 16.2.

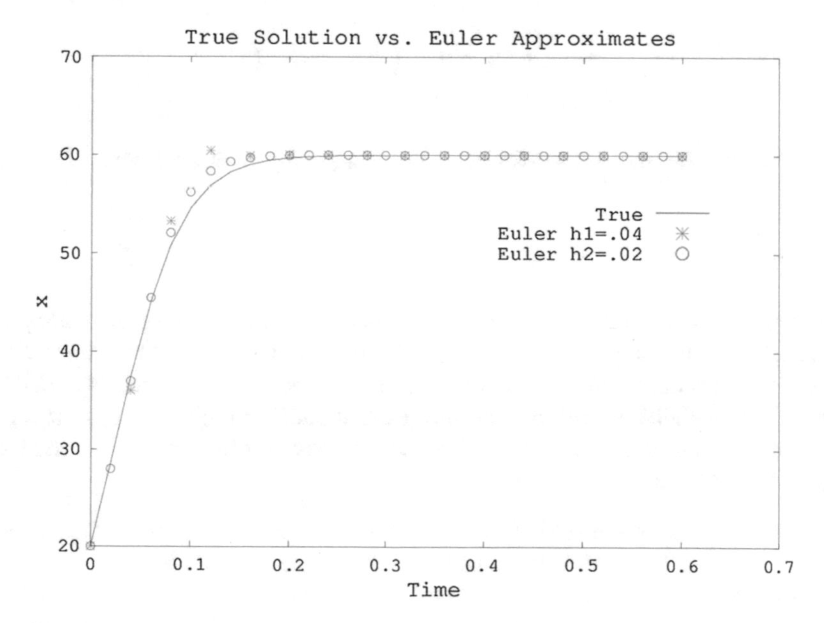

Fig. 16.2 True versus Euler for $x' = 0.5x(60 - x)$, $x(0) = 20$

16.1.10 Homework

For these problems, use the Matlab session fragments above to find the plots of the true solution versus the Euler approximates for the list of step sizes given but put all the plots in the same graph.

Exercise 16.1.13 $x' = 1.58x$ *with* $x(0) = 4$ *for* $h = 0.4$ *and* $h = 0.2$.

Exercise 16.1.14 $x' = -1.03x$ *with* $x(0) = 8$ *for* $h = 0.6$ *and* $h = 0.3$.

Exercise 16.1.15 $x' = 0.4x(105 - x)$ *with* $x(0) = 19$ *for* $h = 0.06$ *and* $h = 0.03$.

Exercise 16.1.16 $x' = -0.2x + 15$ *with* $x(0) = 10$ *for* $h = 0.5$ *and* $h = 0.25$.

16.2 Runge-Kutta Methods

These methods are based on more sophisticated ways of approximating the solution y'. These methods use multiple function evaluations at different time points around a given t^* to approximate $y(t^*)$. In more advanced discussions, we can show this technique generates a sequence $\{y_n\}$ starting at y_0 using the following recursion equation:

$$\begin{aligned} y_{n+1} &= y_n + h \times F^o(t_n, y_n, h, f) \\ y_0 &= y0 \end{aligned}$$

where h is the step size we use for our underlying partition of the time space giving

$$t_i = t_0 + i \times h$$

for appropriate indices and F^o is a fairly complicated function of the previous approximate solution, the step size and the right hand side function f. The Runge-Kutta methods are available for various choices of the superscript o which is called the order of the method. We will not discuss much about F^o in this course, as it is best served up in a more advanced text. What we can say is this: For order o, the **local** error is like h^{o+1}. So

Order One: **Local** error is h^2 and this method is the same as the Euler method. The **global** error then goes down linearly with h.

Order Two: **Local** error is h^3 and this method is better than the Euler method. If the **global** error for a given stepsize h is then E, halving the stepsize to $h/2$ gives a new global error of $E/4$. Thus, the global error goes down quadratically. This means halving the stepsize has a dramatic effect of the global error.

Order Three: **Local** error is h^4 and this method is better than the Euler method. If the **global** error for a given stepsize h is E, then halving the stepsize to $h/2$ gives a new global error of $E/8$. Thus, the global error goes down as a cubic power. This means halving the stepsize has an even more dramatic effect of the global error.

Order Four: **Local** error is h^5 and this method is better than the order three Method. If the **global** error for a given stepsize h is E, then halving the stepsize to $h/2$ gives a new global error of $E/16$. Thus, the global error goes down as a fourth power! This means halving the stepsize has huge effect of the global error.

We will now look at MatLab code that allows us to solve our differential equation problems using the Runge-Kutta method instead of the Euler method of Sect. 16.1.

16.2.1 The MatLab Implementation

The basic code to implement the Runge-Kutta methods is broken into two pieces. The first one, **`RKstep.m`** implements the evaluation of the next approximation solution at point (t_n, y_n) given the old approximation at (t_{n-1}, y_{n-1}). We then loop through all the steps to get to the chosen final time using the code in **`FixedRK.m`**. The details of these algorithms are beyond the scope of this text although we discuss this more carefully in two follow up texts (Peterson 2015a, b). In this code, we are allowing for the dynamic functions to depend on time also. Previously, we have used dynamics like **`f = @(x) 3*x`** and we expect the dynamics functions to have that form in **`DoEuler`**. However, we want to have more complicated dynamics now—at least the possibility of it!—so we will adapt what we have done before. We will now define our dynamics as if they depend on time. So from now on, we would write **`f=@(t,x) 3*x`** even though there is no time dependence. We then rewrite our **`DoEuler`** to **`DoEulerTwo`** so that we can use these more general dynamics. This code is in **`DoEulerTwo.m`** and as you can see is changed where we define **`Euler`** because of its dependence on the function **`f`**. Also note, in general, we won't be able to find the true solution. Hence, we have removed the true and error results.

Listing 16.11: DoEulerTwo.m

```
function [xhat,x,e] = DoEulerTwo(f,true,N,x0,h)
% N = number of approximations to find
% f = the model dynamics function of (t,x)
% true = the true solution
% x0 = the initial condition
% h = the step size
%
euler = @(t,x,h) x + f(t,x)*h;
xhat(1) = x0;
% initialize x, xhat and e
x(1) = x0;
e(1) = 0;
for i=1:N
   xhat(i+1) = euler(i*h,xhat(i),h);
   x(i+1) = true(i*h);
   e(i+1) = abs( x(i+1) - xhat(i+1) );
end
end
```

So the Runge-Kutta code below uses the new dynamics functions. In it you see the lines like **feval(fname,t,x)** which means take the function **fname** passed in as an argument and **evaluate** it as the pair **(t,x)**. Hence, **fname(t,x)** is the same as **f(t,x)**.

Listing 16.12: RKstep.m: Runge-Kutta Codes

```
function [tnew,ynew,fnew] = RKstep(fname,tc,yc,fc,h,k)
%
% fname     the name of the right hand side function f(t,y)
%           t is a scalar usually called time and
%           y is a vector of size d
% yc        approximate solution to y'(t) = f(t,y(t)) at t=tc
% fc        f(tc,yc)
% h         The time step
% k         The order of the Runge-Kutta Method 1<= k <= 4
%
% tnew      tc+h
% ynew      approximate solution at tnew
% fnew      f(tnew,ynew)
%
if k==1
  k1 = h*fc;
  ynew = yc+k1;
elseif k==2
  k1 = h*fc;
  k2 = h*feval(fname,tc+(h/2),yc+(k1/2));
  ynew = yc + k2;
elseif k==3
  k1 = h*fc;
  k2 = h*feval(fname,tc+(h/2),yc+(k1/2));
  k3 = h*feval(fname,tc+h,yc-k1+2*k2);
  ynew = yc+(k1+4*k2+k3)/6;
elseif k==4
  k1 = h*fc;
  k2 = h*feval(fname,tc+(h/2),yc+(k1/2));
  k3 = h*feval(fname,tc+(h/2),yc+(k2/2));
  k4 = h*feval(fname,tc+h,yc+k3);
  ynew = yc+(k1+2*k2+2*k3+k4)/6;
else
  disp(sprintf('The RK method %2d order is not allowed!',k));
end
tnew = tc+h;
fnew = feval(fname,tnew,ynew);
end
```

The code above does all the heavy lifting. It manages all of the multiple tangent line calculations that Runge-Kutta needs at each step. We loop through all the steps to get to the chosen final time using the code in **FixedRK.m** which is shown below.

Listing 16.13: FixedRK.m: The Runge-Kutta Solution

```
function [tvals,yvals,fcvals] = FixedRK(fname,t0,y0,h,k,n)
%
%        Gives approximate solution to
%          y'(t) = f(t,y(t))
%          y(t0) = y0
%        using a kth order RK method
%
% t0     initial time
% y0     initial state
% h      stepsize
% k      RK order  1<= k <= 4
% n      Number of steps to take
%
% tvals  time values of form
%        tvals(j) = t0 + (j-1)*h, 1 <= j <= n
% yvals  approximate solution
%        yvals(:j) = approximate solution at
%        tvals(j),  1 <= j <= n
%
tc = t0;
yc = y0;
tvals = tc;
yvals = yc;
fc = feval(fname,tc,yc);
for j=1:n-1
  [tc,yc,fc] = RKstep(fname,tc,yc,fc,h,k);
  yvals = [yvals yc];
  tvals = [tvals tc];
  fcvals = [fcvals fc];
end
end
```

Here is an example where we solve a specific model using all four Runge-Kutta choices and plot them all together. Note when we use **RKstep**, we only return the first two outputs; that is, for our returned variables, we write **[htime1,xhat1] = FixedRK(f,0,20,0.06,1,N1);** instead of returning the full list of outputs which includes function evaluations **[htime1,xhat1,fhat1] = FixedRK (f,0,20,0.06,1,N1);**. We can do this as it is all right to not return the third output. However, you still have to return the arguments in the order stated when the function is defined. For example, if we used the command **[htime1,fhat1] = FixedRK(f,0,20,0.06,1,N1);**, this would return the approximate values and place them in the variable **fhat1** which is not what we would want to do.

Listing 16.14: True versus All Four Runge-Kutta Approximations

```
f = @(t,x) .5*x.*(60-x);
true = @(t) 60./( 1 + (60/20 - 1)*exp(-.5*60*t) ) ;
T = .6;
time = linspace(0,T,31);
h1 = .06;
N1 = ceil(T/h1);
[htime1,xhat1] = FixedRK(f,0,20,.06,1,N1);
[htime2,xhat2] = FixedRK(f,0,20,.06,2,N1);
[htime3,xhat3] = FixedRK(f,0,20,.06,3,N1);
[htime4,xhat4] = FixedRK(f,0,20,.06,4,N1);
% the ... at the end of the line allows us to continue
% a long line to the start of the next line
plot(time,true(time),htime1,xhat1,'*',htime2,xhat2,'o',...
     htime3,xhat3,'+',htime4,xhat4,'.');
xlabel('Time');
ylabel('x');
% We want to use the derivative symbol x' so
% since Matlab treat ' as the start and stop of the label
% we write ''. That way Matlab will treat '' as a single quote
% that is our differentiation symbol
title('RK Approximations to x''=.5x(60-x), x(0) - 20');
% Notice we continue this line too
legend('True','RK 1, h=.06','RK 2, h=.06','RK 3, h = .06',...
'RK 4, h = .06','Location','Best');
```

This generates Fig. 16.3.

Note Runge-Kutta Order 4 does a great job even with a large step size.

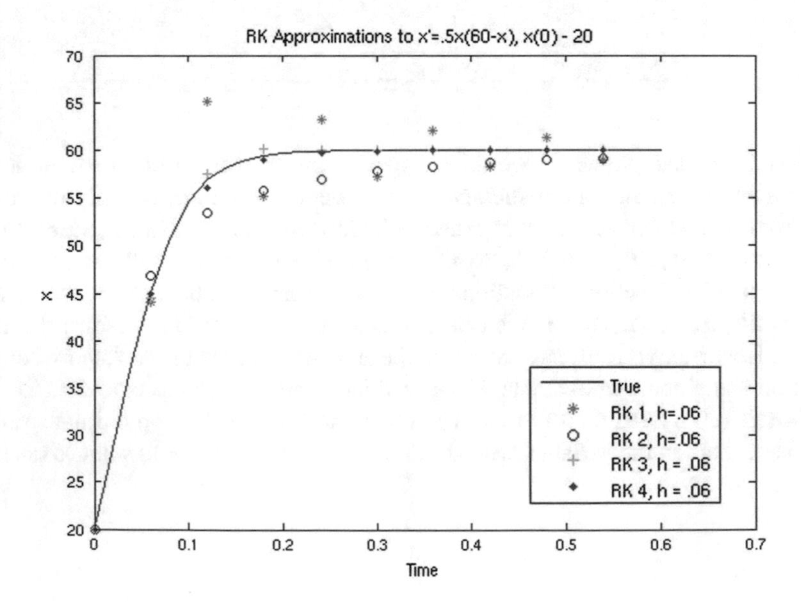

Fig. 16.3 True versus Euler for $x' = 0.5x(60 - x)$, $x(0) = 20$

16.2.2 Homework

You are now ready to solve a few on your own. For these problems,

(i): Write the dynamics function code as `f = @(t,x)`....
(ii): Write the true function code as usual `true = @(t) etc.`.
(iii): Solve the problem with all four Runge-Kutta Methods for the given step size h and appropriate time interval. Plot all four Runge-Kutta approximate solutions and the true solution on the same plot.
(iv): Do all of this in a word doc as usual.

Exercise 16.2.1 $h = 0.05$ *and the model is*

$$y' = 0.07\, y\,(75 - y), \quad y(0) = 5.$$

Exercise 16.2.2 $h = 0.06$ *and the model is*

$$u'(t) = 0.13\, u(t)\,(30 - u(t)),$$
$$u(0) = 45.$$

Exercise 16.2.3 $h = 0.4$ *and the model is*

$$x' = 1.9x, \quad x(0) = 6.$$

Exercise 16.2.4 $h = 0.5$ *and the model is*

$$x' = -1.95x + 30, \quad x(0) = 16.$$

References

J. Peterson, *Calculus for Cognitive Scientists: Higher Order Models and Their Analysis*. Springer Series on Cognitive Science and Technology (Springer Science+Business Media Singapore Pte Ltd. Singapore, 2015a)

J. Peterson, *Calculus for Cognitive Scientists: Partial Differential Equation Models*. Springer Series on Cognitive Science and Technology (Springer Science+Business Media Singapore Pte Ltd. Singapore, 2015b)

Chapter 17
Advanced Protein Models

Now that we have a bit more mathematics under our belts, let's go back and look at proteins again. A great review of this, with lots of great pictures and diagrams, is in the paper by Sneppen et al. (2010). It is well worth your time to check it out. We will do some of what is discussed in this paper and use appropriate MatLab to get some insight, but view this chapter as a pointer to more fascinating stuff! As you know from your first biology courses, the inside of the cell is quite crowded. There are protein filaments that guide the movement of other proteins, proteins that help the amino acid strands coming out of the ribosome to coil properly so that they can take their correct shape and so on. And all of this is taking place in water. Water is a very complicated substance with interesting and complex abilities to assemble into cages because its electric charge distribution is not symmetric. Oxygen and two hydrogens give a molecule with a definite minus to plus look. You should read another survey paper on this to get a better understanding. Look at Pascale Mentré's paper on water in the cell (Mentré 2012) to get a nice overview.

17.1 Binding Time Estimates

Regulation of protein transcription is very dynamic as transcription factors **TF**'s rapidly associate and disassociate to the sites on the portion of DNA that has been exposed from the genome. We looked at some of this action in Chap. 12 and we called this special site the **promoter**. Now a single transcription factor moves slowly across a cell because it slowly drifts by bouncing off other molecules. Imagine a tiny robotic submarine being driven in a pond just chock full of debris, living plants and so on. If it was small enough it would even bounce off of tadpoles and small fish! If it was smaller yet, it would bounce off of small multi cellular creatures too. You can see the progress of the tiny submarine would be slow and halting and even though the captain of the submarine is trying to move in a straight line towards her target, she can't really do that. Instead the submarine bounces around and finds the target in an indirect way. Well, the transcription factor is like the tiny submarine and we

J.K. Peterson, *Calculus for Cognitive Scientists*, Cognitive Science and Technology, DOI 10.1007/978-981-287-874-8_17

measure the speed at which the **TF** moves in the cell with what is called the **diffusion coefficient**, $\boldsymbol{D}$. This coefficient is measured in units of molecules of **TF** $\times$ $(\mu\text{m})^2/\text{s}$ where the symbol μm is called a micrometer and has the value $\mu\text{m} = 10^{-6}\,\text{m}$. Typically $\boldsymbol{D} \approx 5 - 50 \frac{\text{molecules}}{}(\mu\text{m})^2/\text{s}$. The transcription factor is trying to hit a specific promoter whose diameter, a, is on average is about 5 nm. Here, the unit nm is called a *nano meter* and it is $10^-3\,\mu$m is size. The promoter is hidden in the cell which has volume **V** So the product $\boldsymbol{D}\,\boldsymbol{a}$ has units of molecules $(\mu\text{m})^3/\text{s}$ and so represents a kind of search speed. If we divide this by the volume of the cell, we divide by the units $(\mu\text{m})^3$ and so the fraction $\frac{V}{\boldsymbol{D}\,a}$ gives an estimate of how long it would take to search through the entire cell volume to find the target promoter. A typical cell has volume 1 $(\mu\text{m})^3$. So, the time to find the target is roughly

$$\tau_{on} \propto \frac{V}{\boldsymbol{D}\,a}$$

and careful reasoning which you can find in biophysics texts tells us the proportionality constant is $1/4\pi$—not too surprising as the **TF** is moving through a sphere and the volume of a sphere is $(4/3)\pi a^3$ here. So we have

$$\tau_{on} = \frac{V}{4\pi\,\boldsymbol{D}\,a}$$

Plugging in our values, since $V = 4/3\pi(1^3)\ (\mu\text{m})^3$ or $V = 1.33\ (\mu\text{m})^3$, we have

$$\begin{aligned}\tau_{on} &= 523.64\pi\ (5\text{ to }50)\ (5 \times 10^{-3})\ \text{s}\\ &= \frac{1.33}{(0.100\text{ to }1.0)}\ \text{s}\\ &= 1.33\text{ to }13.3\text{ s.}\end{aligned}$$

Next, if you think about it, it is easy to see that it costs energy for the transcription factor **TF** to break off the promoter. We won't go through how we come with a quantitative estimate here. Instead, we will just say the the time it takes for **TF** to break off is proportional to τ_{on}. We can show

$$\tau_{off} \propto \frac{\tau_{on}}{K}$$

where **K** is a number which is related to the binding energy. The proportionality constant then turns out to be $1/(6 \times 10^8)$. For a reasonably strong binding energy of about -13 kcal/mole , where kcal is a unit of *kilo calories* which is a measure of energy used and *mole* is a term you probably saw in your chemistry class which represents 6.02×10^{23} molecules, we find $K = 10^{-9}$. This kind of detail though is not relevant for our purposes although some of you will probably take courses in the future where you will go through the reasoning needed to figure this out. So we have in general

$$\tau_{off} = \frac{\tau_{on}}{6 \times 10^8\, K}$$

And for a binding energy of 13 kcal/mole, $K = 10^{-9}$, we get

$$\tau_{off} = \frac{\tau_{on}}{0.6} = 1.67\ \tau_{on}.$$

So both τ_{on} and τ_{off} have nominal ranges of 1–23 s. Our purpose here is just to make you see the rough time order for the **TF** search and bind to the promoter and its release from the promoter. The back and forth bind and release takes place a small fraction of a minute. We also know from experiments and theoretical reasoning that protein transcription takes time on the order of minutes. So the transcription factor binding and releasing is occurring on a time scale that is significantly less than the time scale we have for full protein transcription. Signal changes to the promoter, i.e. **TF** movements due to external signals that release the molecules needed to release the **TF**, occur much faster than the time scale we see when protein production is at equilibrium.

17.2 The Bound Fraction

Let the promoter be denoted by **P** and the **TF** plus promoter complex by **TFP**. We have the following *reaction*

$$[\boldsymbol{P}] + [\boldsymbol{TF}] \overset{k}{\longrightarrow} [\boldsymbol{TFP}]$$

where k is the rate at which $\boldsymbol{P}$ and $\boldsymbol{TF}$ combine to form the complex $\boldsymbol{TFP}$. The complex also breaks apart at the rate k' which we denote in equation form as

$$[\boldsymbol{TFP}] \overset{k'}{\longrightarrow} [\boldsymbol{P}] + [\boldsymbol{TF}].$$

Combining, we have the model

$$[\boldsymbol{P}] + [\boldsymbol{TF}] \underset{k'}{\overset{k}{\longleftrightarrow}} [\boldsymbol{TFP}]$$

At equilibrium, the rate at which $\boldsymbol{TFP}$ forms must equal the rate at which $\boldsymbol{TFP}$ breaks apart. The concentration of $\boldsymbol{TFP}$ is written as $[\boldsymbol{TFP}]$. The concentrations of $\boldsymbol{P}$ and $\boldsymbol{TF}$ are then $[\boldsymbol{P}]$ and $[\boldsymbol{TF}]$. The amount of $\boldsymbol{TFP}$ depends on how much of the needed *recipe* ingredients are available. Hence the amount made is $k\ [\boldsymbol{P}]\ [\boldsymbol{TF}]$. The amount of $\boldsymbol{P}$ and $\boldsymbol{TF}$ made because $\boldsymbol{TFP}$ breaks apart depends on how much $\boldsymbol{TFP}$

is available which is k' $[\boldsymbol{TFP}]$. So at equilibrium, because the formation rate and disassociation rate are the same that we must have a balance

$$k' \ [\boldsymbol{TFP}] = k \ [\boldsymbol{P}] \ [\boldsymbol{TF}]$$

Solving we find the relationship

$$[\boldsymbol{TFP}] = \frac{k}{k'} \ [\boldsymbol{P}] \ [\boldsymbol{TF}]$$

We call the fraction $\frac{k'}{k}$ the **disassociation constant** K and this is the same variable we use earlier which had to do with the binding energy! We have $[\boldsymbol{TFP}] = \frac{1}{K} \ [\boldsymbol{P}] \ [\boldsymbol{TF}]$ at equilibrium. We see the fraction of **TF** bound to the promoter can be written as

$$\begin{aligned} \text{bound fraction} &= \frac{[\boldsymbol{TFP}]}{[\boldsymbol{P}] + [\boldsymbol{TFP}]} \\ &= \frac{\frac{1}{K} \ [\boldsymbol{P}] \ [\boldsymbol{TF}]}{[\boldsymbol{P}] + [\boldsymbol{TF}]}. \end{aligned}$$

Now simplify a bit to get

$$\begin{aligned} \text{bound fraction} &= \frac{[\boldsymbol{TFP}]}{[\boldsymbol{P}] + [\boldsymbol{TFP}]} \\ &= \frac{K \ [\boldsymbol{TFP}]}{K \ [\boldsymbol{P}] + K \ [\boldsymbol{TFP}]} \\ &= \frac{[\boldsymbol{P}] \ [\boldsymbol{TF}]}{K \ [\boldsymbol{P}] + [\boldsymbol{P}] \ [\boldsymbol{TF}]} \\ &= \frac{[\boldsymbol{TF}]}{K + [\boldsymbol{TF}]} \\ &= \frac{[\boldsymbol{TF}]/K}{1 + [\boldsymbol{TF}]/K}. \end{aligned}$$

So we express the *disassociation constant* K in two ways to get insight. The first uses energy and movement ideas to estimate how much time it takes for the transcription factor and the promoter binding to reach equilibrium—tens of seconds at most. The second uses familiar reaction ideas to estimate the fraction of transcription factor that is bound to the promoter. If we let $\boldsymbol{R}$ denote the concentration of our transcription factor and $\boldsymbol{K}$ denote its disassociation constant, we have a generic relationship.

$$\text{bound fraction} = \frac{\frac{\boldsymbol{R}}{\boldsymbol{K}}}{1 + \frac{\boldsymbol{R}}{\boldsymbol{K}}}.$$

We can do a similar bit of analysis to figure out the *unbound* fraction. We have

$$\begin{aligned}
\text{unbound fraction} &= \frac{[\boldsymbol{P}]}{[\boldsymbol{P}] + [\boldsymbol{TFP}]} \\
&= \frac{K\ [\boldsymbol{P}]}{K\ [\boldsymbol{P}] + K\ [\boldsymbol{TFP}]} \\
&= \frac{K\ [\boldsymbol{P}]}{K\ [\boldsymbol{P}] + [\boldsymbol{P}]\ [\boldsymbol{TF}]} \\
&= \frac{K}{K + [\boldsymbol{TF}]} \\
&= \frac{1}{1 + \frac{[\boldsymbol{TF}]}{K}}.
\end{aligned}$$

Hence, the unbound fraction, using $\boldsymbol{R}$ to denote the concentration of the transcription factor is

$$\text{unbound fraction} = \frac{1}{1 + \frac{\boldsymbol{R}}{\boldsymbol{K}}}$$

It turns out we can do more. Sometimes transcription factors bind cooperatively or other factors can bind to the transcription factor prior to binding to the promoter. We can model this sort of thing too, but the details are not for us here. In such cases, the bound and unbound fractions can be represented by adding a power **b** to the term $\frac{\boldsymbol{R}}{\boldsymbol{K}}$. Hence, the more general fractions are

$$\begin{aligned}
\text{bound fraction} &= \frac{\left(\frac{\boldsymbol{R}}{\boldsymbol{K}}\right)^b}{1 + \left(\frac{\boldsymbol{R}}{\boldsymbol{K}}\right)^b}. \\
\text{unbound fraction} &= \frac{1}{1 + \left(\frac{\boldsymbol{R}}{\boldsymbol{K}}\right)^b}
\end{aligned}$$

where $\mathbf{b} \geq 1$ is called a **Hill** coefficient. We think of the ratio $\frac{\boldsymbol{R}}{\boldsymbol{K}}$ in this way. If the amount of $[\boldsymbol{TF}]$ exactly matched the disassociation rate $\boldsymbol{K}$, then the ratio would be 1 and the bound and unbound fraction are both $1/2$. Just like the probability of being bound and unbound is exactly the same. However, the mixture can easily be skewed by simply letting $[\boldsymbol{TF}]$ be $r\boldsymbol{K}$ for different values of r. If $r \gg 1$, the bound fraction moves past $1/2$ and is closer to saturation: the bound fraction is 1. Similarly, if $r \ll 1$, the bound fraction is quite small with the unbound fraction close to 1.

17.2.1 Example

Example 17.2.1 Calculate the bound and unbound fractions assuming the fraction R/K is 0.2 for a Hill's coefficient $b = 1$.

Solution *We know*

$$
\begin{aligned}
\textit{bound fraction} &= \frac{\left(\frac{R}{K}\right)^b}{1 + \left(\frac{R}{K}\right)^b}. \\
&= \frac{0.2}{1 + 0.2} = 0.1667 \\
\textit{unbound fraction} &= \frac{1}{1 + \left(\frac{R}{K}\right)^b} \\
&= \frac{1}{1 + 0.2} = 0.8333
\end{aligned}
$$

That's all there is to it! So most of the transcription factor is unbound here.

17.2.2 Homework

Exercise 17.2.1 *Calculate the bound and unbound fractions assuming the fraction* R/K *is* 0.6 *for a Hill's coefficient* $b = 1$.

Exercise 17.2.2 *Calculate the bound and unbound fractions assuming the fraction* R/K *is* 1.2 *for a Hill's coefficient* $b = 2$.

Exercise 17.2.3 *Calculate the bound and unbound fractions assuming the fraction* R/K *is* 10.5 *for a Hill's coefficient* $b = 1$.

Exercise 17.2.4 *Calculate the bound and unbound fractions assuming the fraction* R/K *is* 1.6 *for a Hill's coefficient* $b = 4$.

17.3 Transcription Regulation

The transcription factor can activate or repress a promoter. If there is activation, then

$$\text{activation} \propto \text{bound fraction}$$

$$\propto \frac{\left(\frac{R}{K}\right)^b}{1+\left(\frac{R}{K}\right)^b}.$$

but if there is repression, then

$$\begin{aligned} \text{repression} &\propto \text{unbound fraction} \\ &\propto \frac{1}{1+\left(\frac{R}{K}\right)^b}. \end{aligned}$$

For convenience, let $x = \frac{R}{K}$. Let's graph the some of these activation and repression functions. We'll use MatLab. Here is a simple session.

Listing 17.1: Activation for $b = 1$

```
f = @(b,x) ( (x).^b./( 1+ (x).^b ) );
X = linspace(0,10,101);
b = 1;
plot(X,f(1,X));
xlabel('x = [TF]/K');
ylabel('activation');
title('activation with b = 1');
```

You can see the plot in Fig. 17.1. This curve will be asymptotic to 1 as $[\boldsymbol{TF}]/K$ get large. Now if you think of $[\boldsymbol{TF}]$ as fixed, varying K means we are changing the effectiveness of the binding. If K is large, the fraction is small and the activation is small. On the other hand, as K decreases, the fraction increases and the activation increases towards 1. We can make the switch to full activation happen faster by increasing b.

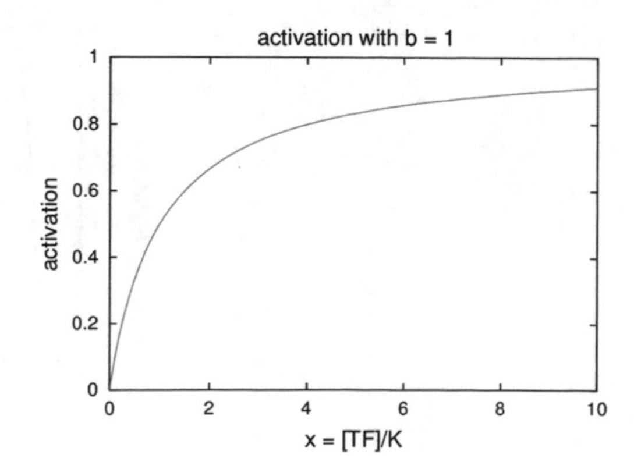

Fig. 17.1 The activation when $b = 1$

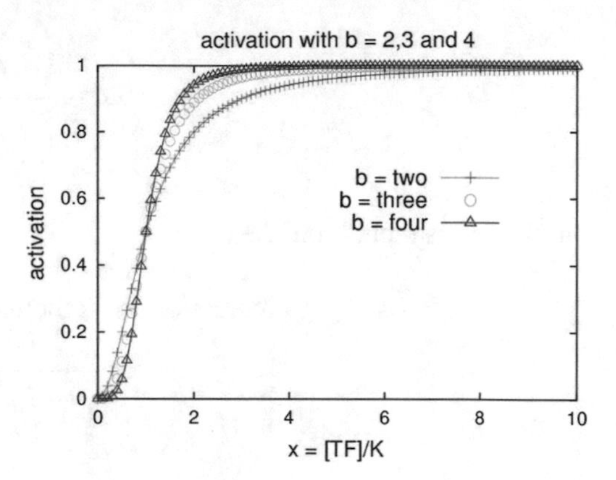

Fig. 17.2 The activation when $b = 2$, 3 and 4

Consider the activation graphs for $b = 2$, $b = 3$ and $b = 4$ on the same graph as shown in Fig. 17.2. Note that all pass through the same pair $(1, 1/2)$ but approach the saturation value of 1 in different ways. Effectively, increasing b ramps protein production up to its maximum value quicker. Another way of looking at it is that a larger Hill's coefficient b makes the activation more sensitive to variations in the $[\boldsymbol{TF}]$ level. As $[\boldsymbol{TF}]$ moves past $\boldsymbol{K}$, production ramps quickly. Indeed, if b is very large, the activation is essentially a step function which moves from 0 to 1 very fast indeed. This is shown in Fig. 17.3.

Given all we have discussed, we are now ready to show you our new protein transcription model. We start with the gene being *repressed*.

$$C' = \Lambda + \frac{\gamma}{1 + \left(\frac{R}{K}\right)^b} - \frac{1}{\tau} C \tag{17.1}$$

$$C(0) = C_0; \tag{17.2}$$

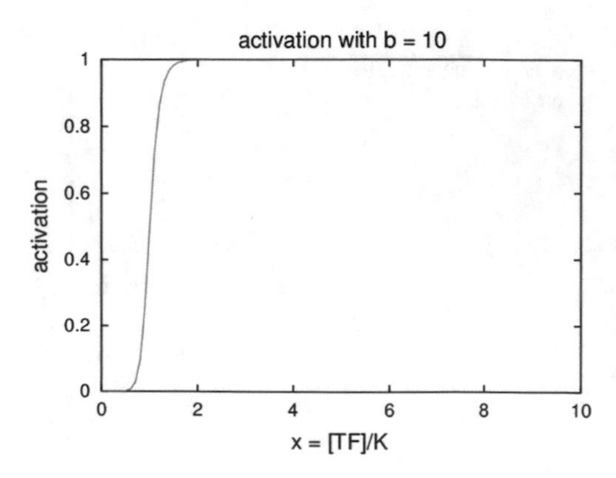

Fig. 17.3 The activation when $b = 10$

The new terms in our equation need to be defined:

- Λ is the leakage production; i.e. it is possible that the gene cannot be fully repressed and so there is some production always.
- γ is the maximum production rate over the base rate set by the leakage value Λ.
- τ is the lifetime of the protein which is essentially the term that includes all the losses that occur due to degradation, cell volume increase and so forth.

This is still a standard $x' = -\alpha x + \beta$ model though. We know the steady state $C_{SS}^{\textbf{repression}}$ is β/α. Here

$$\alpha = \frac{1}{\tau}$$
$$\beta = \Lambda + \frac{\gamma}{1 + \left(\frac{R}{K}\right)^b}$$

and so

$$C_{SS}^{\textbf{repression}} = \tau \left(\Lambda + \frac{\gamma}{1 + \left(\frac{R}{K}\right)^b} \right). \tag{17.3}$$

and the response time is therefore $t_R = \tau \ln(2)$. The production is fully repressed when $R \gg K$, say $R = 20K$ at which point the production model is $C' = \Lambda + 0 - C/\tau = \Lambda - C/\tau$. This fully repressed model has steady state $C_{SS}^{\textbf{full repression}} = \tau \Lambda$. On the other hand, if there is no repression, i.e. $R \ll K$, say $R = 0.01K$, then the minimally repressed model is $C' = \Lambda + \gamma - C/\tau$ with steady state $C_{SS}^{\textbf{no repression}} = \tau(\Lambda + \gamma)$. How quickly production switches between these extremes depends on the Hill coefficient b as can be seen in Fig. 17.4. The code to generate this figure is as follows:

Listing 17.2: The Steady State for the Repression model with $b = 2$

```
g = @(b,tau,Lambda,gamma,x) tau.*(Lambda+gamma./(1 + x.^b));
X = linspace(0,20,101);
b = 2;
Lambda = 1;
gamma = 10;
tau = 0.5;
plot(X,g(b,tau,Lambda,gamma,X));
```

We see in Fig. 17.4 that the production is quickly shut off.

We can also activate protein production by allowing the transcription factor to activate production. The model is similar.

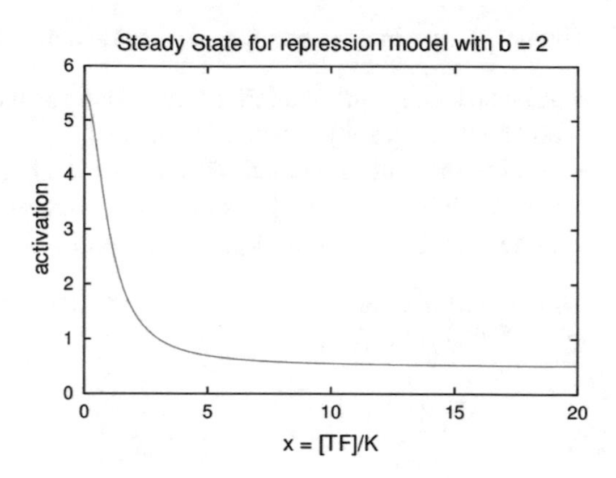

Fig. 17.4 The steady State value for the repression model when $b = 2$

$$C' = \Lambda + \gamma \frac{\left(\frac{R}{K}\right)^b}{1 + \left(\frac{R}{K}\right)^b} - \frac{1}{\tau} C \tag{17.4}$$

$$C(0) = C_0; \tag{17.5}$$

We know the steady state $C_{SS}^{\mathbf{activation}}$ is β/α. Here

$$\alpha = \frac{1}{\tau}$$

$$\beta = \Lambda + \gamma \frac{\left(\frac{R}{K}\right)^b}{1 + \left(\frac{R}{K}\right)^b}$$

and so

$$C_{SS}^{\mathbf{activation}} = \tau \left(\Lambda + \gamma \frac{\left(\frac{R}{K}\right)^b}{1 + \left(\frac{R}{K}\right)^b} \right). \tag{17.6}$$

17.3.1 Example

Example 17.3.1 Calculate $C_{SS}^{\mathbf{repression}}$ and $C_{SS}^{\mathbf{activation}}$ for $\Lambda = 4, \gamma = 2, b = 4, \tau = 1$ and the ratio $R/K = 0.1, 1, 2, 5, 20$. Comment on your results.

Solution • $R/K = 0.1$.

$$C_{SS}^{\textbf{repression}} = \tau\left(\Lambda + \frac{\gamma}{1+\left(\frac{R}{K}\right)^b}\right)$$

$$= 4 + \frac{2}{1+(0.1)^4} = 5.9998$$

$$C_{SS}^{\textbf{activation}} = \tau\left(\Lambda + \gamma\,\frac{\left(\frac{R}{K}\right)^b}{1+\left(\frac{R}{K}\right)^b}\right)$$

$$= 4 + 2\,\frac{(0.1)^4}{1+(0.1)^4} = 4.0002.$$

- $R/K = 1$.

$$C_{SS}^{\textbf{repression}} = \tau\left(\Lambda + \frac{\gamma}{1+\left(\frac{R}{K}\right)^b}\right)$$

$$= 4 + \frac{2}{1+(1)^4} = 5$$

$$C_{SS}^{\textbf{activation}} = \tau\left(\Lambda + \gamma\,\frac{\left(\frac{R}{K}\right)^b}{1+\left(\frac{R}{K}\right)^b}\right)$$

$$= 4 + 2\,\frac{(1)^4}{1+(1)^4} = 5.$$

- $R/K = 2$.

$$C_{SS}^{\textbf{repression}} = \tau\left(\Lambda + \frac{\gamma}{1+\left(\frac{R}{K}\right)^b}\right)$$

$$= 4 + \frac{2}{1+(2)^4} = 4.11765$$

$$C_{SS}^{\textbf{activation}} = \tau \left(\Lambda + \gamma \frac{\left(\frac{R}{K}\right)^b}{1 + \left(\frac{R}{K}\right)^b} \right)$$

$$= 4 + 2 \, \frac{(2)^4}{1 + (2)^4} = 5.88235.$$

- $R/K = 5.$

$$C_{SS}^{\textbf{repression}} = \tau \left(\Lambda + \frac{\gamma}{1 + \left(\frac{R}{K}\right)^b} \right)$$

$$= 4 + \frac{2}{1 + (5)^4} = 4.00001$$

$$C_{SS}^{\textbf{activation}} = \tau \left(\Lambda + \gamma \frac{\left(\frac{R}{K}\right)^b}{1 + \left(\frac{R}{K}\right)^b} \right)$$

$$= 4 + 2 \, \frac{(5)^4}{1 + (5)^4} = 5.9984.$$

- $R/K = 20.0.$

$$C_{SS}^{\textbf{repression}} = \tau \left(\Lambda + \frac{\gamma}{1 + \left(\frac{R}{K}\right)^b} \right)$$

$$= 4 + \frac{2}{1 + (20)^4} = 4.0000$$

$$C_{SS}^{\textbf{activation}} = \tau \left(\Lambda + \gamma \frac{\left(\frac{R}{K}\right)^b}{1 + \left(\frac{R}{K}\right)^b} \right)$$

$$= 4 + 2 \, \frac{(20)^4}{1 + (20)^4} = 6.000.$$

Note both the unbound and bound are in the range [4, 6].

Example 17.3.2 Plot the steady state for the repression model for $\Lambda = 0$, $\gamma = 5$, $b = 1$ and $\tau = 1$ as a function of R/K. Comment on your results. We don't show the plot here.

Solution

Listing 17.3: Problem Code

```
g = @(b,tau,Lambda,gamma,x) tau.*(Lambda+gamma./(1 + x.^b));
X = linspace(0,20,101);
b = 1.0;
Lambda = 0;
gamma = 5;
tau = 1.0;
plot(X,g(b,tau,Lambda,gamma,X));
```

17.3.2 Homework

Exercise 17.3.1 *Calculate* $C_{SS}^{\textbf{repression}}$ *and* $C_{SS}^{\textbf{activation}}$ *for* $\Lambda = 2$, $\gamma = 3$, $b = 2$, $\tau = 2$ *and the ratio* $R/K = 0.1, 1, 2, 5, 20$. *Comment on your results.*

Exercise 17.3.2 *Calculate* $C_{SS}^{\textbf{repression}}$ *and* $C_{SS}^{\textbf{activation}}$ *for* $\Lambda = 5$, $\gamma = 2$, $b = 3$, $\tau = 1$ *and the ratio* $R/K = 0.1, 1, 2, 5, 20$. *Comment on your results.*

Exercise 17.3.3 *Plot the steady state for the repression model for* $\Lambda = 2$, $\gamma = 3$, $b = 2$ *and* $\tau = 2$ *as a function of* R/K. *Comment on your results.*

Exercise 17.3.4 *Plot the steady state for the repression model for* $\Lambda = 0$, $\gamma = 1$, $b = 1$ *and* $\tau = 1$ *as a function of* R/K. *Comment on your results.*

17.4 Simple Regulations

Now for simplicity, let's set the leakage $\Lambda = 0$ and all the other constants in these two models to 1 except for the concentration of the transcription factor R and τ. Then we have two simple models we can look at.

$$C' = \frac{1}{1+R} - \frac{1}{\tau} C \tag{17.7}$$

$$C_{SS}^{\textbf{repression}} = \tau\left(\frac{1}{1+R}\right)$$

$$C' = \frac{R}{1+R} - \frac{1}{\tau} C \tag{17.8}$$

$$C_{SS}^{\textbf{activation}} = \frac{\tau R}{1+R}$$

We also a simple model where the transcription factor enhances the *loss* of the protein by selectively increasing production of the proteins that break down C. We will assume the transcription factor binds to C and degrades it through some process. We have the following reactions:

$$[\boldsymbol{C}] + [\boldsymbol{TF}] \xrightarrow{k} [\boldsymbol{TFC}]$$

where k is the rate at which $\boldsymbol{C}$ and $\boldsymbol{TF}$ combine to form the complex $\boldsymbol{TFC}$ which then immediately degrades. Hence, although we also know the complex also breaks apart at the rate k' which we denote in equation form as:

$$[\boldsymbol{TFC}] \xrightarrow{k'} [\boldsymbol{C}] + [\boldsymbol{TF}].$$

effectively, the backwards rate is $k' = 0$ and so we only have the forward association reaction. The rate of change of $[\boldsymbol{C}]$ is then

$$\begin{aligned}\frac{d\,[\boldsymbol{C}]}{dt} &= \text{amount of } [\boldsymbol{C}] \text{ lost in the reaction} \\ &= [\boldsymbol{TFC}] \\ &= k\ [\boldsymbol{C}]\ [\boldsymbol{TF}]\end{aligned}$$

using the fact the concentration of the complex is the product of the concentrations of the components used to produce it—standard chemistry! So replacing $[\boldsymbol{C}]$ by C and the transcription factor $[\boldsymbol{TF}]$ by R, we see this use of the transcription factor leads to a loss term of $C'_{loss} = -RC$. We then add that to our standard protein transcription model to obtain Eq. 17.9.

$$C' = 1 - R\,C - \frac{1}{\tau}\,C \tag{17.9}$$

$$C_{SS}^{\textbf{degradation}} = \frac{\tau}{1 + R\tau}$$

Now depending on the value of τ, we can see that a protein can be **cleared** out of the biological system more rapidly using this degradation method rather than repression. We show this in two graphs. In Fig. 17.5, we show the steady state values for the degradation and repression models as a function of R for $\tau = 0.5$. Now the response time for repression is $\tau \ln(2)$, so protein production ramps down quickly for this value of τ. So we see repression is a better strategy for clearing the protein. However, it τ is larger, degradation is a better way to go. In Fig. 17.6, we show how the steady states change with R and degradation is clearer faster.

A good way to view these plots is to look at $R = 0.5$ on the graph and see what happens to the steady states if R suddenly ramps up by a factor of 20 to $R = 10$. Note how the degradation policy clears much more protein. The models are simplified as usual so all constants are 1 except for τ and R. The code to generate these plots is shown below.

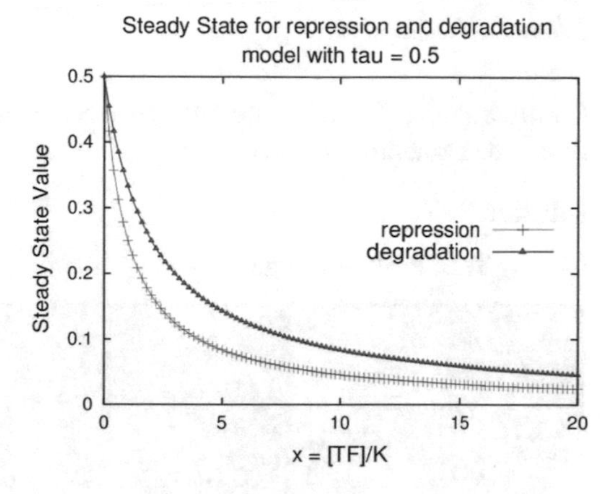

Fig. 17.5 Repression and Degradation steady state values for $\tau = 0.5$

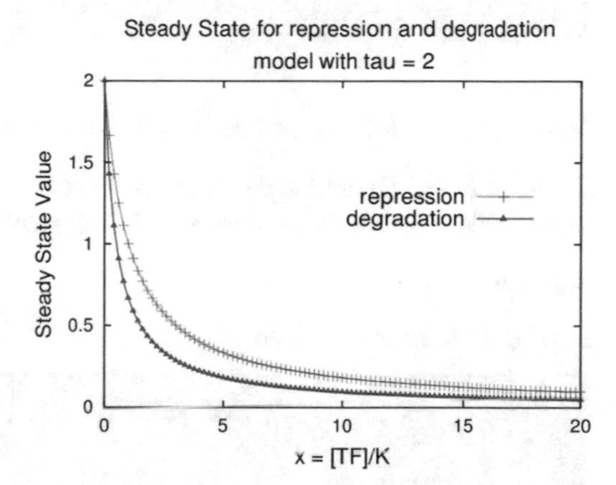

Fig. 17.6 Repression and degradation steady state values for $\tau = 2.0$

Listing 17.4: Steady State Values for Repression and Degradation Models

```
% tau = 0.5
CSSREP = @(tau,R) tau./(1+R);
CSSDEG = @(tau,R) tau./(1 + tau.*R);
tau = 0.5;
r = linspace(0,20,101);
plot(r,CSSREP(tau,r),'-+',r,CSSDEG(tau,r),'-^');
xlabel('x = [TF]/K');
ylabel('Steady State Value');
title('Steady State for repression and degradation model with tau = 0.5');
legend('repression', 'degradation','location','east');

% tau = 2.0
tau = 2.0;
plot(r,CSSREP(tau,r),'-+',r,CSSDEG(tau,r),'-^');
xlabel('x = [TF]/K');
ylabel('Steady State Value');
title('Steady State for repression and degradation model with tau = 2');
legend('repression', 'degradation','location','east');
```

17.4.1 Example

Example 17.4.1 Examine the difference between the steady state values for repression and degradation in the case $\tau = 1.0$. Comment on the results.

Solution

Listing 17.5: Problem Code

```
CSSREP = @(tau,R) tau./(1+R);
CSSDEG = @(tau,R) tau./(1 + tau.*R);
r = linspace(0,20,101);
tau = 1.0;
plot(r,CSSREP(tau,r),'-+',r,CSSDEG(tau,r),'-^');
xlabel('x = [TF]/K');
ylabel('Steady State Value');
title('Steady State for repression and degradation model with tau = 1');
legend('repression', 'degradation','location','east');
```

Looking at the plot, you can see which strategy wins: repression or degradation.

Example 17.4.2 Examine the difference between the steady state values for repression and degradation in the case $\tau = 3.0$. Comment on the results.

Solution

Listing 17.6: Problem Code

```
CSSREP = @(tau,R) tau./(1+R);
CSSDEG = @(tau,R) tau./(1 + tau.*R);
r = linspace(0,20,101);
tau = 3.0;
plot(r,CSSREP(tau,r),'-+',r,CSSDEG(tau,r),'-^');
xlabel('x = [TF]/K');
ylabel('Steady State Value');
title('Steady State for repression and degradation model with tau = 3');
legend('repression', 'degradation','location','east');
```

Looking at the plot, you can see which strategy wins: repression or degradation.

17.4.2 Homework

Exercise 17.4.1 *Examine the difference between the steady state values for repression and activation in the case* $\tau = 1.5$*. Here you would use* `CSSACT = @(tau,R) tau.*R./(1+R);` *in your code. Comment on the results.*

Exercise 17.4.2 *Examine the difference between the steady state values for repression and activation in the case* $\tau = 2.5$*. Here you would use* `CSSACT = @(tau,R) tau.*R./(1+R);` *in your code. Comment on the results.*

Exercise 17.4.3 *Examine the difference between the steady state values for degradation and activation in the case* $\tau = 4.0$. *Comment on the results.*

Exercise 17.4.4 *Examine the difference between the steady state values for repression and degradation in the case* $\tau = 6.0$. *Comment on the results.*

17.5 Feedback Loops

The final models we want to discuss involve feedback of the transcription factor signal onto itself. There are two forms: negative feedback and positive feedback and in both this can be done also with a time delay. For these equations, we are again setting the leakage $\Lambda = 0$ all the other constants to be 1 for simplicity. The feedback equations are seen in Eqs. 17.10 and 17.11.

$$R' = 1 + \frac{1}{1+R} - \frac{1}{\tau} R, \quad \text{negative feedback} \tag{17.10}$$

$$R' = 1 + \frac{R}{1+R} - \frac{1}{\tau} R, \quad \text{positive feedback} \tag{17.11}$$

These models can not be solved with our Calculus tools, so we must use numerical techniques as we have discussed. Here are some sample runs to show you how to generate interesting plots. In this code, we are using the more general model rather than the simpler one in Eq. 17.10. Hence, in the code, the C which represented the protein concentration is now being thought of as its own transcription factor. So the R is the model is variable and we get the following equations.

$$R' = \Lambda + \gamma \left(\frac{1}{1+R^b} \right) - \frac{1}{\tau} R, \quad \text{negative feedback} \tag{17.12}$$

$$R' = \Lambda + \gamma \left(\frac{R^b}{1+R^b} \right) - \frac{1}{\tau} R, \quad \text{positive feedback} \tag{17.13}$$

Listing 17.7: A Sample Negative Feedback Model

```
f = @(t,Lambda,gamma,b,tau,R) Lambda + gamma./(1+R.^b) - R/tau;
g = @(t,R) f(t,Lambda,gamma,b,tau,R);
h = .1;
tau = 1;
5 Lambda = 1;
gamma = 4;
b = 1;
FinalTime = 30.0;
RInit = 2;
10 MaxIters = ceil(FinalTime/h);
[htime,Rhat] = FixedRK(g,0,RInit,h,4,MaxIters);
plot(htime,Rhat);
```

This generates the plot we see in Fig. 17.7.

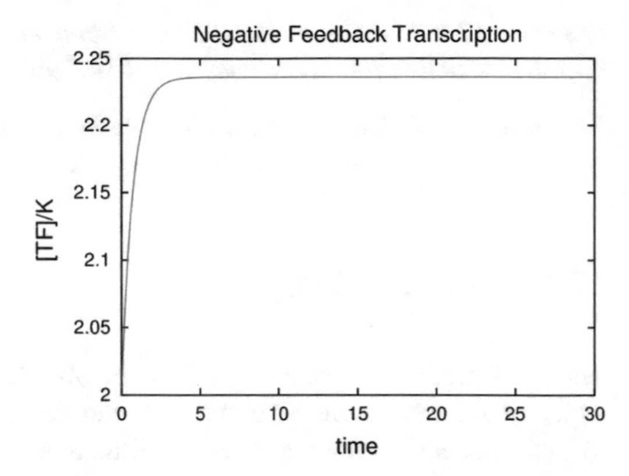

Fig. 17.7 Negative Feedback Transcription

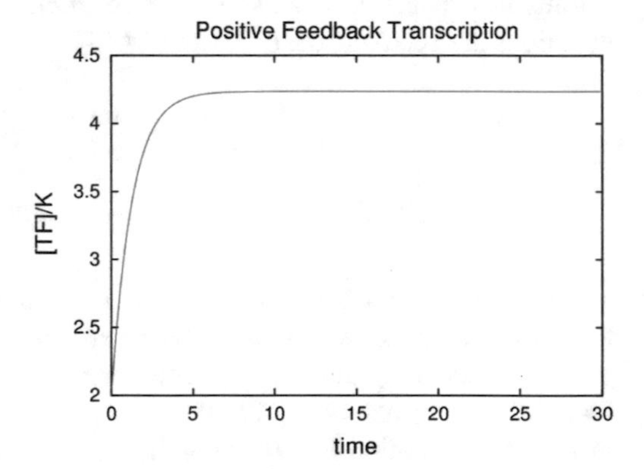

Fig. 17.8 Positive Feedback Transcription

The positive feedback case is next. We have

Listing 17.8: A Sample Positive Feedback Model

```
f = @(t,Lambda,gamma,b,tau,R) Lambda + gamma*(R.^b)./(1+R.^b) - R/tau;
g = @(t,R) f(t,Lambda,gamma,b,tau,R);
h = .1;
tau = 1;
Lambda = 1;
gamma = 4;
b = 1;
TF = 30.0;
RInit = 2;
MaxIters = ceil(TF/h);
[htime,Rhat] = FixedRK(g,0,RInit,h,4,MaxIters);
plot(htime,Rhat);
```

This generates the plot we see in Fig. 17.8.

If we add a delay of time d, the models are changed to Eqs. 17.14 and 17.15.

$$R' = \Lambda + \gamma \left(\frac{1}{1 + R(t-d)} \right) - \frac{1}{\tau} R, \text{ negative feedback} \tag{17.14}$$

$$R' = \Lambda + \gamma \left(\frac{R(t-d)}{1 + R(t-d)} \right) - \frac{1}{\tau} R, \text{ positive feedback} \tag{17.15}$$

We don't know how to solve a model with delays, so we are going to show you a quick way to an estimate of the solution. We would have to be more sophisticated if we were really serious, but this will be enough for now. We will use a modified Euler approach. Let's introduce the code **`eulerdelay`** as seen below. The basic idea is like this. If the delay is D and the step size is h, it takes $m = D/h$ steps to reach the time when the delayed information starts coming back. So remove the delay part of the model for those time steps and for the first m time steps solve

$$R' = \Lambda + \gamma - \frac{1}{\tau} R, \text{ negative feedback}$$
$$R' = \Lambda + \gamma - \frac{1}{\tau} R, \text{ positive feedback}$$

using Euler or Runge-Kutta methods. Then switch to the delayed models for the remainder of the time steps. So we need to define a function g for the model dynamics before the delay kicks in and a function f for the steps with a delay.

Listing 17.9: eulerdelay.m

```
function [htime,Rhat] = eulerdelay(Lambda,gamma,b,tau,delay,TF,RInit,h,f,g)
%
% Lambda is the leakage
% gamma is the capacity
% b is the Hill's coefficient
% tau is like the response time
% delay is the time delay
% TF is the final time
% h is the step size
% f is the dynamics after the delay time has been passed
% g is the dynamics before the delay time has been reached
%
M - delay/h;
N = ceil(TF/h);

htime = zeros(1,N);
Rhat = zeros(1,N);
Rhat(1) = RInit;
htime(1) = 0;
% solve for steps before the delay so we use the dynamics g
% which are not delayed
for i=1:M
   Rhat(i+1) = Rhat(i) + g(i*h,Lambda,gamma,b,tau,Rhat(i))*h;
   htime(i+1) = i*h;
end
% now we have reached the delay and we switch to the delay model
% with dynamics f
for i = M+1:N-1
    Rhat(i+1) = Rhat(i) + f(i*h,Lambda,gamma,b,tau,Rhat(i-M),Rhat(i))*h;
    htime(i+1) = i*h;
end

end
```

17.5.1 Examples

Example 17.5.1 Let's use this code on the negative feedback with delay model for $\tau = 1$, a delay of 1.5, $\Lambda = 0$, $\gamma = 4$ and the Hill coefficient $b = 1$. We'll use a step size $h = 0.1$ with a final time of 30.0 and initialize with $R = 0$.

Solution

Listing 17.10: A Sample Negative Feedback With Delay

```
g = @(t,Lambda,gamma,b,tau,R) Lambda + gamma - R/tau;
f = @(t,Lambda,gamma,b,tau,S,R) Lambda + gamma./(1+S.^b) - R/tau;
h = .1;
tau = 1;
delay = 1.5;
Lambda = 0;
gamma = 4;
b = 1;
TF = 30.0;
RInit = 0;
[htime,Rhat] = eulerdelay(Lambda,gamma,b,tau,delay,30,RInit,h,f,g);
plot(htime,Rhat);
```

This generates the plot we see in Fig. 17.9.

Example 17.5.2 Finally, let's use this code on the positive feedback with delay model with the same parameter values: $\tau = 1$, a delay of 1.5, $\Lambda = 0$, $\gamma = 4$ and the Hill coefficient $b = 1$. We'll use a step size $h = 0.1$ with a final time of 30.0 and initialize with $R = 0$.

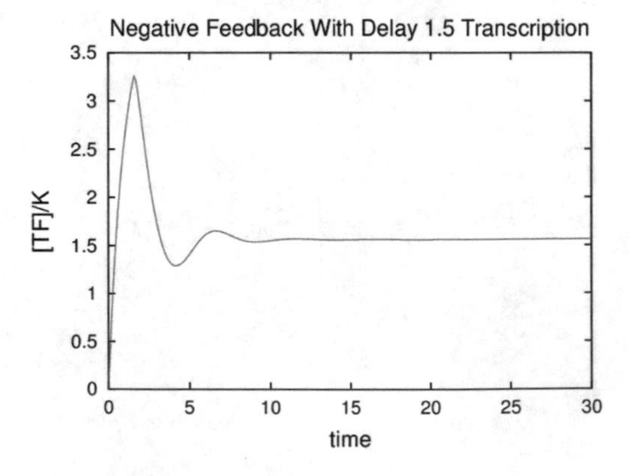

Fig. 17.9 Positive Feedback Transcription

Solution

Listing 17.11: A Sample Positive Feedback With Delay

```
v = @(t,Lambda,gamma,b,tau,R)  Lambda  - R/tau;
u = @(t,Lambda,gamma,b,tau,S,R)  Lambda + gamma*(S.^b)./(1+S.^b) - R/tau;
h = .1;
tau = 1;
delay = 1.5;
Lambda = 0;
gamma = 4;
b = 1;
TF = 30.0;
RInit = 2;
[htime,Rhat] = eulerdelay(Lambda,gamma,b,tau,delay,30,RInit,h,u,v);
plot(htime,Rhat);
```

This generates the plot we see in Fig. 17.10.

17.5.2 Homework

Exercise 17.5.1 *Use this code on the negative feedback with delay model for* $\tau = 2$, *a delay of* 3.5, $\Lambda = 1$, $\gamma = 2$ *and the Hill coefficient* $b = 2$. *Use a step size* $h = 0.05$ *with a final time of* 30.0 *and initialize with* $R = 2$.

Exercise 17.5.2 *Use this code on the positive feedback with delay model for* $\tau = 0.5$, *a delay of* 1.5, $\Lambda = 0$, $\gamma = 2$ *and the Hill coefficient* $b = 2$. *We'll use a step size* $h = 0.05$ *with a final time of* 30.0 *and initialize with* $R = 2$.

Exercise 17.5.3 *Use this code on the negative feedback with delay model for* $\tau = 4$, *a delay of* 1.5, $\Lambda = 0$, $\gamma = 3$ *and the Hill coefficient* $b = 1$. *We'll use a step size* $h = 0.1$ *with a final time of* 30.0 *and initialize with* $R = 2$.

Exercise 17.5.4 *Use this code on the positive feedback with delay model for* $\tau = 1$, *a delay of* 3.0, $\Lambda = 1$, $\gamma = 2$ *and the Hill coefficient* $b = 2$. *We'll use a step size* $h = 0.05$ *with a final time of* 30.0 *and initialize with* $R = 2$.

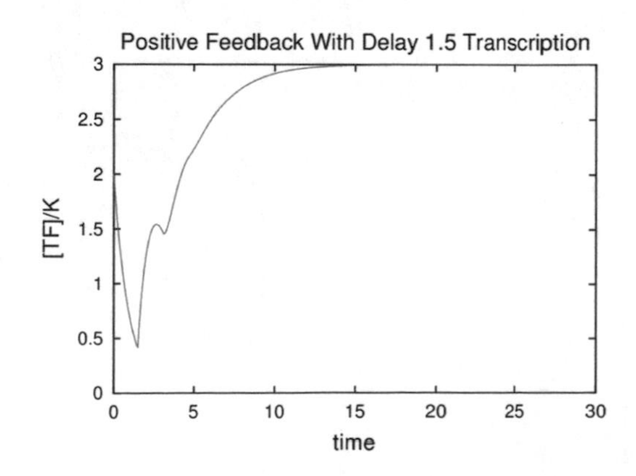

Fig. 17.10 Positive Feedback Transcription

References

P. Mentré, Water in the orchestration of the cell machinery. Some misunderstandings: a short review. J. Biol. Physics **38**, 13–26 (2012)

K. Sneppen, S. Krishna, S. Semsey, Simplified models of biological networks. Annu. Rev. Biophys. **39**, 43–59 (2010)

Part III
Using Multiple Variables

Chapter 18
Matrices and Vectors

We will need to use both *vector* and *matrix* ideas in this course. Most of you have probably seen this kind of material before in earlier classes and we certainly are using the ideas of vectors when we do our MatLab stuff, but we are going to lay it out for you here.

18.1 Matrices

A **matrix** is a rectangular collection of real numbers organized like this:

$$\begin{bmatrix} -2 & 4 & 5 & 1 \\ -12 & 14 & 15 & -6 \\ 20 & 4 & 1 & 2 \\ 8 & 14 & 5 & 11 \end{bmatrix} \tag{18.1}$$

In Eq. 18.1, we have a collection of numbers which are organized into 4 rows and 4 columns. We call this a *square* matrix because the number of rows and columns are the same. This particular matrix has only positive or negative integers in it, but of course the number 0 could be used as well as real numbers like 1.2356, π and e. It is just easier to type integers! We would call this a 4×4 matrix and read this as a 4 by 4 matrix. A matrix can also have a different number of rows and columns. Consider the matrices shown in Eq. 18.2 which is a 5×4 matrix and Eq. 18.3, which is a 4×3 matrix. We call 5×4 the **size** of the matrix in Eq. 18.2. In general, if a matrix has m rows and n columns, we say its size is $m \times n$.

$$\begin{bmatrix} -2 & 4 & 5 & 1 \\ -12 & 14 & 15 & -6 \\ 20 & 4 & 1 & 2 \\ 8 & 14 & 5 & 11 \\ -2 & -23 & 7 & -3 \end{bmatrix} \tag{18.2}$$

J.K. Peterson, *Calculus for Cognitive Scientists*, Cognitive Science and Technology, DOI 10.1007/978-981-287-874-8_18

$$\begin{bmatrix} 4 & 5 & 1 \\ 14 & 15 & -6 \\ 4 & 1 & 2 \\ 14 & 5 & 11 \\ -23 & 7 & -3 \end{bmatrix} \quad (18.3)$$

We usually denote a matrix by a capital letter such as $\boldsymbol{A}$. Hence, Eqs. 18.1–18.3 could be labeled as follows:

$$\boldsymbol{A} = \begin{bmatrix} -2 & 4 & 5 & 1 \\ -12 & 14 & 15 & -6 \\ 20 & 4 & 1 & 2 \\ 8 & 14 & 5 & 11 \end{bmatrix}, \ \boldsymbol{B} = \begin{bmatrix} -2 & 4 & 5 & 1 \\ -12 & 14 & 15 & -6 \\ 20 & 4 & 1 & 2 \\ 8 & 14 & 5 & 11 \\ -2 & -23 & 7 & -3 \end{bmatrix}, \text{ and } \boldsymbol{C} = \begin{bmatrix} 4 & 5 & 1 \\ 14 & 15 & -6 \\ 4 & 1 & 2 \\ 14 & 5 & 11 \\ -23 & 7 & -3 \end{bmatrix}$$

Each entry in a matrix can be labeled by the row and column it occurs in. Thus, the entry in row 2 and column 3 of a matrix is labeled as A_{23}. So, the matrix in Eq. 18.2 has the labels shown in Eq. 18.4

$$\boldsymbol{B} = \begin{bmatrix} A_{11} & A_{12} & A_{13} & A_{14} \\ A_{21} & A_{22} & A_{23} & A_{24} \\ A_{31} & A_{32} & A_{33} & A_{34} \\ A_{41} & A_{42} & A_{43} & A_{44} \\ A_{51} & A_{52} & A_{53} & A_{54} \end{bmatrix} = \begin{bmatrix} -2 & 4 & 5 & 1 \\ -12 & 14 & 15 & -6 \\ 20 & 4 & 1 & 2 \\ 8 & 14 & 5 & 11 \\ -2 & -23 & 7 & -3 \end{bmatrix}. \quad (18.4)$$

18.1.1 The Zero Matrices

There are some special matrices. A matrix that only has 0 as its entries is called a *zero* matrix. Now, since there are matrices of all different sizes, we can not pick just one to call the *zero* matrix. So when we are working on a problem, we just use the size of the *zero* matrix that is appropriate for the problem's context. For example, a 4×3 zero matrix would be

$$\boldsymbol{0} = \begin{bmatrix} 0 & 0 & 0 \\ 0 & 0 & 0 \\ 0 & 0 & 0 \\ 0 & 0 & 0 \end{bmatrix}$$

while a 2×2 zero matrix would be

$$\boldsymbol{0} = \begin{bmatrix} 0 & 0 \\ 0 & 0 \end{bmatrix}$$

We could denote these two matrices by say $\mathbf{0}_{4\times3}$ or $\mathbf{0}_{2\times2}$ to distinguish them. However, that is really cumbersome! So as we mentioned above, it is usually pretty easy to figure out the appropriate size of the zero matrix from context.

18.1.2 Square Matrices

Square matrices often occur in our work, i.e. matrices that have the same number of rows and columns. Consider

$$\boldsymbol{A} = \begin{bmatrix} A_{11} & A_{12} & A_{13} & A_{14} \\ A_{21} & A_{22} & A_{23} & A_{24} \\ A_{31} & A_{32} & A_{33} & A_{34} \\ A_{41} & A_{42} & A_{43} & A_{44} \end{bmatrix} = \begin{bmatrix} -2 & 4 & 5 & 1 \\ -12 & 14 & 15 & -6 \\ 20 & 4 & 1 & 2 \\ 8 & 14 & 5 & 11 \end{bmatrix} \tag{18.5}$$

A square matrix has three important parts which you are subsets of the original matrix.

1. The Lower Triangular Part of $\boldsymbol{A}$ is $\boldsymbol{L}$ given by

$$\boldsymbol{L} = \begin{bmatrix} A_{11} & 0 & 0 & 0 \\ A_{21} & A_{22} & 0 & 0 \\ A_{31} & A_{32} & A_{33} & 0 \\ A_{41} & A_{42} & A_{43} & A_{44} \end{bmatrix} = \begin{bmatrix} -2 & 0 & 0 & 0 \\ -12 & 14 & 0 & 0 \\ 20 & 4 & 1 & 0 \\ 8 & 14 & 5 & 11 \end{bmatrix} \tag{18.6}$$

2. The Upper Triangular Part of $\boldsymbol{A}$ is $\boldsymbol{U}$ given by

$$\boldsymbol{U} = \begin{bmatrix} A_{11} & A_{12} & A_{13} & A_{14} \\ 0 & A_{22} & A_{23} & A_{24} \\ 0 & 0 & A_{33} & A_{34} \\ 0 & 0 & 0 & A_{44} \end{bmatrix} = \begin{bmatrix} -2 & 4 & 5 & 1 \\ 0 & 14 & 15 & -6 \\ 0 & 0 & 1 & 2 \\ 0 & 0 & 0 & 11 \end{bmatrix} \tag{18.7}$$

3. The Diagonal Part of $\boldsymbol{A}$ is $\boldsymbol{D}$ given by

$$\boldsymbol{D} = \begin{bmatrix} A_{11} & 0 & 0 & 0 \\ 0 & A_{22} & 0 & 0 \\ 0 & 0 & A_{33} & 0 \\ 0 & 0 & 0 & A_{44} \end{bmatrix} = \begin{bmatrix} -2 & 0 & 0 & 0 \\ 0 & 14 & 0 & 0 \\ 0 & 0 & 1 & 0 \\ 0 & 0 & 0 & 11 \end{bmatrix} \tag{18.8}$$

18.1.3 The Identity Matrices

We can also define what is called the **identity** matrix. An identity matrix is a square matrix whose only nonzero entries are one's on the diagonal. For example,

$$I = \begin{bmatrix} 1\,0\,0 \\ 0\,1\,0 \\ 0\,0\,1 \end{bmatrix}$$

is a 3×3 identity matrix, while a 4×4 identity matrix would be

$$I = \begin{bmatrix} 1\,0\,0\,0 \\ 0\,1\,0\,0 \\ 0\,0\,1\,0 \\ 0\,0\,0\,1 \end{bmatrix}$$

We could denote these two matrices by say $I_{3\times3}$ or $I_{4\times4}$ to distinguish them. However, this notation is just as irritating to use as the previous one like this for the zero matrices. Hence, we usually figure out the appropriate size of from context.

18.1.4 The Transpose of a Matrix

We can illustrate this easiest with a simple example. Consider the 5×4 matrix A defined by

$$A = \begin{bmatrix} -2 & 4 & 5 & 1 \\ -12 & 14 & 15 & -6 \\ 20 & 4 & 1 & 2 \\ 8 & 14 & 5 & 11 \\ -2 & -23 & 7 & -3 \end{bmatrix}$$

The **transpose** of A is the matrix formed by switching the rows and columns of A. We denote this new matrix by A^T or sometimes A'. If the entries of A are as usual given by A_{ij} where i is the row number and j is the column number, the entries of A^T are reversed and become A_{ji}. That sounds confusing, doesn't it? Try this: if $A_{14} = \beta$, then the row 4 and column 1 entry in A^T becomes β. Note the switching of rows and columns? It is that easy to find the transpose of a matrix. So A^T here is

$$A^T = \begin{bmatrix} -2 & -12 & 20 & 8 & -2 \\ 4 & 14 & 4 & 14 & -23 \\ 5 & 15 & 1 & 5 & 7 \\ 1 & -6 & 2 & 11 & -3 \end{bmatrix}$$

So you should be able to see that the row i, column j entry of A^T is precisely the row j, column i entry of A. We can say also by stating $A^T{}_{ij} = A_{ji}$

Comment 18.1.1 *If a matrix A has size $m \times n$, then its transpose, A^T, has size $n \times m$*

Comment 18.1.2 *If a matrix* $\boldsymbol{A}$ *equals its own transpose, then first, we know* $\boldsymbol{A}$ *must be a square matrix of size* $n \times n$ *for some positive integer* n*. Thus,*

$$\left(\boldsymbol{A}^T\right)_{ij} = \boldsymbol{A}_{ij} = \boldsymbol{A}_{ji}$$

In this case, we say $\boldsymbol{A}$ *is* **symmetric***.*

Thus, the matrix $\boldsymbol{A}$ below is symmetric.

$$\boldsymbol{A} = \begin{bmatrix} -2 & -12 & 20 & 8 \\ -12 & 14 & 4 & 14 \\ 20 & 4 & 1 & 5 \\ 8 & 14 & 5 & 11 \end{bmatrix}$$

18.1.5 Homework

Exercise 18.1.1 *Find the transpose of*

$$\begin{bmatrix} 2 & 3 & 4 \\ -1 & 4 & 90 \end{bmatrix}$$

Exercise 18.1.2 *Find the transpose of*

$$\begin{bmatrix} 2 & 3 & 4 \\ -11 & 4 & 9 \\ 6 & -3 & 8 \end{bmatrix}$$

Exercise 18.1.3 *Is this matrix symmetric?*

$$\begin{bmatrix} 2 & 3 & 4 \\ 3 & 4 & -3 \\ 4 & -3 & 8 \end{bmatrix}$$

Exercise 18.1.4 *Is this matrix symmetric?*

$$\begin{bmatrix} 2 & -3 \\ 3 & 4 \end{bmatrix}$$

18.2 Operations on Matrices

We can also perform many operations on matrices. It is easiest to show these operations with examples.

1. We can add matrices of the same size by adding their components.

$$\begin{bmatrix} 1 & -2 & 3 \\ 4 & 1 & -8 \\ -7 & 6 & 12 \end{bmatrix} + \begin{bmatrix} -20 & 3 & -11 \\ 16 & 9 & 5 \\ 16 & 2 & -8 \end{bmatrix} = \begin{bmatrix} 1-20 & -2+3 & 3-11 \\ 4+16 & 1+9 & -8+5 \\ -7+16 & 6+2 & 12-8 \end{bmatrix} = \begin{bmatrix} -19 & 1 & -8 \\ 20 & 10 & -3 \\ 9 & 8 & 4 \end{bmatrix}$$

2. We can subtract matrices of the same size by subtracting their components.

$$\begin{bmatrix} 1 & -2 & 3 \\ 4 & 1 & -8 \\ -7 & 6 & 12 \end{bmatrix} - \begin{bmatrix} -20 & 3 & -11 \\ 16 & 9 & 5 \\ 16 & 2 & -8 \end{bmatrix} = \begin{bmatrix} 1--20 & -2-3 & 3--11 \\ 4-16 & 1-9 & -8-5 \\ -7-16 & 6-2 & 12--8 \end{bmatrix} = \begin{bmatrix} 21 & -5 & 14 \\ -12 & -8 & -13 \\ -23 & 4 & 20 \end{bmatrix}$$

3. We can *scale* a matrix by multiplying each component of the matrix by the same number.

$$-3 \times \begin{bmatrix} 1 & -2 & 3 \\ 4 & 1 & -8 \\ -7 & 6 & 12 \end{bmatrix} = \begin{bmatrix} -3 & 6 & -9 \\ -12 & -3 & 24 \\ 21 & -18 & -36 \end{bmatrix}$$

4. We can multiply two matrices $\boldsymbol{A}$ and $\boldsymbol{B}$ if their sizes are just right. The number of columns of $\boldsymbol{A}$ must match the number of rows of $\boldsymbol{B}$. In the example below, the number of columns of the first matrix is 3 which matches the number of rows in the second matrix. So the matrix multiplication is defined. Since the size of $\boldsymbol{A}$ is 4×3 and the size of $\boldsymbol{B}$ is 3×2, the size of the product will be 4×2. In this example, each row of the first matrix has 3 entries and each column of the second matrix has 3 rows. Look at row 1 of the first matrix and column 1 of the second matrix. We multiply row 1 and column 1 like this:

$$\begin{bmatrix} 1 & -2 & 3 \end{bmatrix} \times \begin{bmatrix} -20 \\ 16 \\ 16 \end{bmatrix} = (1)(-20) + (-2)(16) + (3)(16).$$

In general, if the entries were generic, we would have for the ith row of $\boldsymbol{A}$ and the jth column of $\boldsymbol{B}$

$$\begin{aligned} \begin{bmatrix} A_{i1} & A_{i2} & A_{i3} \end{bmatrix} \times \begin{bmatrix} B_{1j} \\ B_{2j} \\ B_{3j} \end{bmatrix} &= (A_{i1})(B_{1j}) + (A_{i2})(B_{2j}) + (A_{i3})(B_{3j}) \\ &= \sum_{k=1}^{3} A_{ik}\, B_{kj}. \end{aligned}$$

where the individual components of $\boldsymbol{A}$ are denoted by A_{ij} and those of $\boldsymbol{B}$ by B_{ij} for appropriate indices i and j. Hence, the full matrix multiplication of these two matrices is given by

$$\begin{bmatrix} 1 & -2 & 3 \\ 4 & 1 & -8 \\ -7 & 6 & 12 \\ 12 & -2 & 3 \end{bmatrix} \times \begin{bmatrix} -20 & 3 \\ 16 & 9 \\ 16 & 2 \end{bmatrix}$$

$$= \begin{bmatrix} (1)(-20) + (-2)(16) + (3)(16) & (1)(3) + (-2)(9) + (3)(2) \\ (4)(-20) + (1)(16) + (-8)(16) & (4)(3) + (1)(9) + (-8)(2) \\ (-7)(-20) + (6)(16) + (12)(16) & (-7)(3) + (6)(9) + (12)(2) \\ (12)(-20) + (-2)(16) + (3)(16) & (12)(3) + (-2)(9) + (3)(2) \end{bmatrix}$$

$$= \begin{bmatrix} -20 + -32 + 48 & 3 + -18 + 6 \\ -80 + 16 + -128 & 12 + 9 + -16 \\ 140 + 96 + 192 & -21 + 54 + 24 \\ -240 + -32 + 48 & 36 + -18 + 6 \end{bmatrix} = \begin{bmatrix} -4 & -9 \\ -192 & 5 \\ 429 & 57 \\ -224 & 24 \end{bmatrix}$$

Comment 18.2.1 *If* $\boldsymbol{A}$ *is a square matrix of size* $n \times n$*, then if* $\boldsymbol{I}$ *denotes the identity matrix of size* $n \times n$*, both multiplications* $\boldsymbol{I}\ \boldsymbol{A}$ *and* $\boldsymbol{A}\ \boldsymbol{I}$ *are possible and give the answer* $\boldsymbol{A}$*. This is why* $\boldsymbol{I}$ *is called the identity matrix!*

Comment 18.2.2 *If* $\boldsymbol{A}$ *is a matrix of any size and* $\boldsymbol{0}$ *is the appropriate zero matrix of the same size, then both* $\boldsymbol{0} + \boldsymbol{A}$ *and* $\boldsymbol{A} + \boldsymbol{0}$ *are nicely defined operations and the result is just* $\boldsymbol{A}$*.*

Comment 18.2.3 *Matrix multiplication is not commutative: i.e. for square matrices* $\boldsymbol{A}$ *and* $\boldsymbol{B}$*, the matrix product* $\boldsymbol{A}\ \boldsymbol{B}$ *is not necessarily the same as the product* $\boldsymbol{B}\ \boldsymbol{A}$*.*

18.2.1 Homework

Exercise 18.2.1 *Compute*

$$\begin{bmatrix} -1.0 & 2.5 & 6.0 \\ 8.0 & -1.0 & 2.5 \\ -3.0 & 4.2 & 12.0 \end{bmatrix} \begin{bmatrix} 2 \\ -5 \\ 7 \end{bmatrix}$$

Exercise 18.2.2 *Compute*

$$\begin{bmatrix} -1.0 & 2.5 & 6.0 & 5 \\ 8.0 & -1.0 & 2.5 & -8.2 \\ -3.0 & 4.2 & 12.0 & 6.1 \end{bmatrix} \begin{bmatrix} 2 & -5 & 7 \\ -10 & 0.3 & 8 \\ 1 & -1 & 2 \\ 6 & 16.5 & -2 \end{bmatrix}$$

Exercise 18.2.3 *Consider*

$$\boldsymbol{C} = \begin{bmatrix} -1.0 & 2.5 & 6.0 & 5 \\ 8.0 & -1.0 & 2.5 & -8.2 \\ -3.0 & 4.2 & 12.0 & 6.1 \\ 2 & -3 & 7.2 & 9.4 \end{bmatrix} \text{ and } \boldsymbol{D} = \begin{bmatrix} 2 & -5 & 7 & 9 \\ -10 & 0.3 & 8 & 10 \\ 1 & -1 & 2 & -5 \\ 6 & 16.5 & -2 & 14 \end{bmatrix}$$

1. *Compute* $\boldsymbol{C} + \boldsymbol{D}$
2. *Compute* $\boldsymbol{C} - \boldsymbol{D}$
3. *Compute* $2\boldsymbol{C} + 3\boldsymbol{D}$
4. *Compute* $-4\boldsymbol{C} + 5\boldsymbol{D}$
5. *Compute* $\boldsymbol{C}\ \boldsymbol{D} - \boldsymbol{D}\ \boldsymbol{C}$

Exercise 18.2.4 *Consider*

$$\boldsymbol{C} = \begin{bmatrix} -1.0 & 2.5 & 6.0 \\ 8.0 & -1.0 & 2.5 \\ -3.0 & 4.2 & 12.2 \end{bmatrix} \text{ and } \boldsymbol{D} = \begin{bmatrix} 12 & -5 & 7.1 \\ -10 & 8.3 & 8 \\ 10 & -1 & 2.4 \end{bmatrix}$$

1. *Compute* $\boldsymbol{C} + \boldsymbol{D}$
2. *Compute* $\boldsymbol{C} - \boldsymbol{D}$
3. *Compute* $3\boldsymbol{C} + 2\boldsymbol{D}$
4. *Compute* $-2\boldsymbol{C} + 4\boldsymbol{D}$
5. *Compute* $\boldsymbol{C}\ \boldsymbol{D} - \boldsymbol{D}\ \boldsymbol{C}$

Exercise 18.2.5 *Consider*

$$\boldsymbol{P} = \begin{bmatrix} -1.0 & 12.5 \\ 18.0 & -1.0 \end{bmatrix} \text{ and } \boldsymbol{Q} = \begin{bmatrix} 2 & -6 \\ -20 & 3.3 \end{bmatrix}$$

1. *Compute* $\boldsymbol{P} + \boldsymbol{Q}$
2. *Compute* $\boldsymbol{P} - \boldsymbol{Q}$
3. *Compute* $8\boldsymbol{P} + 2\boldsymbol{Q}$
4. *Compute* $-6\boldsymbol{P} + 5\boldsymbol{Q}$
5. *Compute* $\boldsymbol{P}\ \boldsymbol{Q} - \boldsymbol{Q}\ \boldsymbol{P}$

18.3 Vectors

A **vector** is a *matrix* which has only one row or one column. We will call *vectors* with one column, **column vectors** and those with one row as **row vectors**. Thus,

$$\boldsymbol{V} = \begin{bmatrix} 4 \\ 14 \\ 4 \\ 14 \\ -23 \end{bmatrix} \text{ and } \boldsymbol{W} = \begin{bmatrix} -2 & 4 & 5 & 1 \end{bmatrix}$$

define a 4×1 column vector $\boldsymbol{V}$ and a 1×4 row vector $\boldsymbol{W}$. The addition and subtraction of vectors follows the usual matrix type rules. However, vector–vector is not defined as the sizes do not match!

18.4 Operations on Vectors

It is easiest to show these operations with examples.

1. We can add row vectors of the same size by adding their components.

$$\begin{bmatrix}1 & -2 & 3\end{bmatrix} + \begin{bmatrix}-20 & 3 & -11\end{bmatrix} = \begin{bmatrix}1-20 & -2+3 & 3-11\end{bmatrix} = \begin{bmatrix}-19 & 1 & -8\end{bmatrix}$$

   Addition of column vectors is similar, of course.
2. We can subtract column vectors of the same size by subtracting their components. Subtraction of row vectors is similar!

$$\begin{bmatrix}1\\4\\-7\end{bmatrix} - \begin{bmatrix}-20\\16\\16\end{bmatrix} = \begin{bmatrix}1--20\\4-16\\-7-16\end{bmatrix} = \begin{bmatrix}21 & -5 & 14\\-12 & -8 & -13\\-23 & 4 & 20\end{bmatrix}$$

3. We can *scale* a vector by multiplying each component of the vector by the same number.

$$-3 \times \begin{bmatrix}1\\4\\-7\end{bmatrix} = \begin{bmatrix}-3\\-12\\21\end{bmatrix}$$

4. The transpose of a row vector is just a column vector and vice versa. If $\boldsymbol{V}$ is defined by

$$\boldsymbol{V} = \begin{bmatrix}1\\4\\-7\end{bmatrix}$$

   then the transpose of the column vector $\boldsymbol{V}$ is the row vector $\boldsymbol{V}^T$ given by

$$\boldsymbol{V}^T = \begin{bmatrix}1 & 4 & -7\end{bmatrix}$$

18.5 The Magnitude of a Vector

The magnitude of a column or a row vector is defined to be its length. We start with a vector of two components.

$$\boldsymbol{V} = \begin{bmatrix}a\\c\end{bmatrix}$$

where for convenience, we use letters to denote components of $\boldsymbol{V}$. We usually say $\boldsymbol{V}$ is a **Two Dimensional** vector because it has two components. Since it is easy to casually identify row and column vectors, we typically think of a 1×2 row vector also as **two dimensional** even though intellectually, there are actually different mathematical beasts! Let's graph this vector using its components as coordinates in the standard $x - y$ plane. So we identity $\boldsymbol{V}$ with the ordered pair (a, c). Note this ordered pair (a, c) defines a line of length $\sqrt{(a)^2 + (b)^2}$ as is shown in Fig. 18.1. Hence, the vector $\boldsymbol{V}$ has a representation with

$$
\begin{aligned}
a &= r\ \cos(\theta) \\
c &= r\ \sin(\theta)
\end{aligned}
$$

just like before. This is called the **polar coordinate representation**.

Since the vector $\boldsymbol{V}$ is graphed in the $x - y$ plane as the coordinate (a, c) as shown in Fig. 18.1, we see the line connecting $(0, 0)$ and (a, c) has equation $y = (a/c)x$. Thus, one way to visualize the vector $\boldsymbol{V}$ is as a line segment starting at $(0, 0)$ with an arrow at its head (a, c).

Example 18.5.1 For the vector

$$
\boldsymbol{V} = \begin{bmatrix} -5 \\ 3 \end{bmatrix}
$$

find its magnitude, its associated angle and graph it carefully,

Solution *The magnitude is* $||\boldsymbol{V}|| = \sqrt{(-5)^2 + (3)^2}$. *This vector is in Quadrant 2 and so the associated angle is* $\pi - \tan^{-1}\left(|\frac{3}{-5}|\right) = \pi - 0.54 = 2.60$ *radians. The graph is for you to do.*

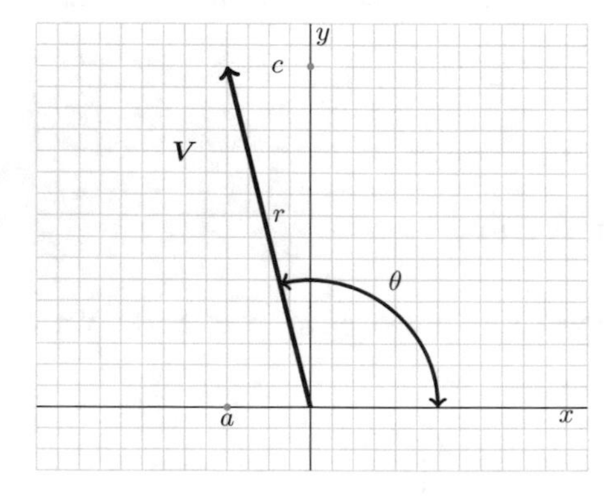

A vector $\boldsymbol{V}$ can be identified with an order pair (a, c). The components (a, c) are graphed in the usual Cartesian manner as an ordered pair in plane. The magnitude of $\boldsymbol{V}$ is $\sqrt{(a)^2 + (c)^2}$ which is shown on the graph as r. The angle associated with $\boldsymbol{V}$ is drawn as an arc of angle θ

Fig. 18.1 Graphing a two dimensional vector

Example 18.5.2 For the vector

$$V = \begin{bmatrix} -8 \\ -2 \end{bmatrix}$$

find its magnitude, its associated angle and graph it carefully,

Solution *The magnitude is* $||V|| = \sqrt{(-8)^2 + (-2)^2}$. *This vector is in Quadrant 3 and so the associated angle is* $\pi + \tan^{-1}\left(|\frac{-2}{-8}|\right) = \pi + 0.24 = 3.38$ *radians. The graph is for you to do.*

What if V was of size 3×1? We would call this column vector a **Three Dimensional** vector because it has three components. Assume V is defined by

$$V = \begin{bmatrix} a \\ c \\ e \end{bmatrix}$$

where our use of e here is not at all related to the e we use in defining the natural logarithm! Again, think **context** at all times!! We can identify this vector with a triple (a, c, e) in three dimensional space as shown in Fig. 18.2.

Now let's return to vectors of size $n \times 1$ which we will call n Dimensional Vectors. We clearly can only graph two and three dimensional vectors, but the formula we have used for the magnitude of the two and three dimensional vectors $\sqrt{(a)^2 + (c)^2}$ and $\sqrt{(a)^2 + (c)^2 + (e)^2}$, respectively, suggests the n dimensional magnitude should be defined by

Definition 18.5.1 (*The Norm Of A Vector*)
The norm of the row or column vector V with components $\{V_1, \ldots, V_n\}$ is

$$|| V || = \sqrt{V_1^2 + \ldots + V_n^2}$$

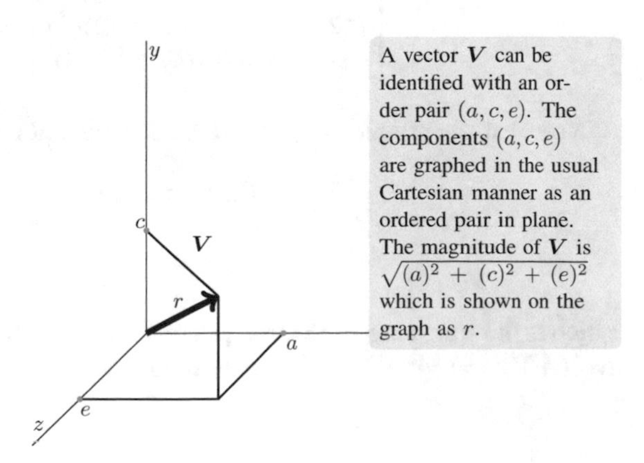

Fig. 18.2 Graphing a three dimensional vector

18.5.1 Homework

For each vector, find its magnitude, its associated angle, and graph it carefully.

Exercise 18.5.1

$$\boldsymbol{V} = \begin{bmatrix} 5 \\ 4 \end{bmatrix}$$

Exercise 18.5.2

$$\boldsymbol{V} = \begin{bmatrix} -2 \\ 4 \end{bmatrix}$$

Exercise 18.5.3

$$\boldsymbol{V} = \begin{bmatrix} -5 \\ -3 \end{bmatrix}$$

18.5.2 Some Matrix–Vector Calculations

Note that the multiplication of a matrix $\boldsymbol{A}$ and a column vector $\boldsymbol{V}$ is defined as long as the number of column of $\boldsymbol{A}$ is the same as the number of rows of $\boldsymbol{V}$. Thus, if

$$\boldsymbol{A} = \begin{bmatrix} -1 & 2 & 6 \\ 12 & -1 & 2 \\ -3 & 4 & 9 \end{bmatrix} \text{ and } \boldsymbol{V} = \begin{bmatrix} 1 \\ 4 \\ 3 \end{bmatrix}$$

then $\boldsymbol{A}\ \boldsymbol{V}$ is

$$\begin{bmatrix} (-1)(1) + (2)(4) + (6)(3) \\ (12)(1) + (-1)(4) + (2)(3) \\ (-3)(1) + (4)(4) + (9)(3) \end{bmatrix} = \begin{bmatrix} 25 \\ 14 \\ 40 \end{bmatrix}$$

Note that the product $\boldsymbol{V}^T\ \boldsymbol{A}^T$ is also defined and gives

$$\begin{bmatrix} 1 & 4 & 3 \end{bmatrix} \begin{bmatrix} -1 & 12 & -3 \\ 2 & -1 & 4 \\ 6 & 2 & 9 \end{bmatrix} = \begin{bmatrix} 25 & 14 & 40 \end{bmatrix}$$

which is the transpose of the first calculation. Note that in this case, we have found that $(\boldsymbol{A}\ \boldsymbol{V})^T = \boldsymbol{V}^T\ \boldsymbol{A}^T$. This is true in general.

18.6 The Inner Product of Two Column Vectors

If $\boldsymbol{V}$ and $\boldsymbol{W}$ are two column vectors of size $n \times 1$, then the product $\boldsymbol{V}^T \boldsymbol{W}$ is a matrix of size 1×1 which we identify with a real number. We see if

$$\boldsymbol{V} = \begin{bmatrix} V_1 \\ V_2 \\ V_3 \\ \vdots \\ V_n \end{bmatrix} \text{ and } \boldsymbol{W} = \begin{bmatrix} W_1 \\ W_2 \\ W_3 \\ \vdots \\ W_n \end{bmatrix}$$

then we define the 1×1 matrix

$$\boldsymbol{V}^T \boldsymbol{W} = \boldsymbol{W}^T \boldsymbol{V} = [V_1 W_1 + V_2 W_2 + V_3 W_3 + \cdots + V_n W_n]$$

and we identify this one by one matrix with the real number

$$V_1 W_1 + V_2 W_2 + V_3 W_3 + \cdots + V_n W_n$$

This product is so important, it is given a special name: it is the **inner product** of the two vectors $\boldsymbol{V}$ and $\boldsymbol{W}$. Let's make this formal with Definition 18.6.1.

Definition 18.6.1 (*The Inner Product Of Two Vectors*)
If $\boldsymbol{V}$ and $\boldsymbol{W}$ are two column vectors of size $n \times 1$, the inner product of these vectors is denoted by $< \boldsymbol{V}, \boldsymbol{W} >$ which is defined as the matrix product $\boldsymbol{V}^T \boldsymbol{W}$ which is equivalent to the $\boldsymbol{W}^T \boldsymbol{V}$ and we interpret this 1×1 matrix product as the real number

$$V_1 W_1 + V_2 W_2 + V_3 W_3 + \cdots + V_n W_n$$

where V_i are the components of $\boldsymbol{V}$ and W_i are the components of $\boldsymbol{W}$.

Now it is possible to show that there is a geometrical interpretation to this inner product. You can see it yourself with simple examples. Any time the dot product of two vectors is 0, the vectors are perpendicular! This is true for vectors of any dimension, although, of course, we can't see that if they are in 4D! So tuck this fact away because we will use it here and there. To help you remember, let's make it a theorem.

Theorem 18.6.1 (Perpendicular Vectors Test)
If the dot product or inner product of the two nonzero vectors $\boldsymbol{V}$ *and* $\boldsymbol{W}$ *is* 0, *then the two vectors are perpendicular.*

Proof The proof of this is not hard, but takes some time and setup to do. We do this in the next course so if you take the second Calculus for Biologist's course, a mystery will be solved! ■

Example 18.6.1 Find the dot product of the vectors $\boldsymbol{V}$ and $\boldsymbol{W}$ given below.

$$\boldsymbol{V} = \begin{bmatrix} -8 \\ 10 \end{bmatrix} \text{ and } \boldsymbol{W} = \begin{bmatrix} 3 \\ 4 \end{bmatrix}.$$

Solution

$$< \boldsymbol{V}, \boldsymbol{W} > = (-8)\,(3) + (10)\,(4) = -24 + 40 = 16.$$

18.6.1 Homework

Exercise 18.6.1 *Find the dot product of the vectors* $\boldsymbol{V}$ *and* $\boldsymbol{W}$ *given by*

$$\boldsymbol{V} = \begin{bmatrix} 6 \\ 1 \end{bmatrix} \text{ and } \boldsymbol{W} = \begin{bmatrix} 7 \\ 2 \end{bmatrix}.$$

Exercise 18.6.2 *Find the dot product of the vectors* $\boldsymbol{V}$ *and* $\boldsymbol{W}$ *given by*

$$\boldsymbol{V} = \begin{bmatrix} -6 \\ -8 \end{bmatrix} \text{ and } \boldsymbol{W} = \begin{bmatrix} 2 \\ 6 \end{bmatrix}.$$

Exercise 18.6.3 *Find the dot product of the vectors* $\boldsymbol{V}$ *and* $\boldsymbol{W}$ *given by*

$$\boldsymbol{V} = \begin{bmatrix} 10 \\ -4 \end{bmatrix} \text{ and } \boldsymbol{W} = \begin{bmatrix} 2 \\ 80 \end{bmatrix}.$$

Chapter 19
A Cancer Model

We are going to examine a relatively simple cancer model as discussed in Novak (2006). Our model is a small fraction of what Novak shows is possible, but it should help you to continue to grow in the art of modeling. This chapter is adapted from Novak's work, so you should really go and look at his full book after you work your way through this chapter. However, we are more careful in the underlying mathematical analysis than Novak and after this chapter is completed, you should have a better understanding of why the mathematics and the science need to work together. You are not ready for more sophisticated mathematical models of cancer, but it wouldn't hurt you to look at a few to get the lay of the land, so to speak. Check out Ribba et al. (2006) and Tatabe et al. (2001) when you can.

We are built of individual cells that have their own reproductive machinery. These cells can sometimes revert to uncontrolled self-replication; this is, of course, a change from their normal programming. Cancer is a disease of organisms that have more than one cell. In order for organisms to have evolved multicellularity, individual cells had to establish and maintain cooperation with each other. From one point of view then, cancer is a breakdown of cellular cooperation; cells must divide when needed by the development program, but not otherwise. Complicated genetic controls on networks of cells had to evolve to make this happen. Indeed, many genes are involved to maintain integrity of the genome, to make sure cell division does not make errors and to help establish the development program that tells cells when to divide. Some of these genes monitor the cell's progress and, if necessary, induce cell death, a process called *apoptosis*. Hence, most cells listen to many signals from other cells telling them they are ok. If the signals saying this fail to arrive, the *default* program is for the cell to commit suicide; i.e. to trigger apoptosis. Hence, apoptosis is a critical defense against cancer. Here is a rough timeline (adapted from Novak) that shows the important events that are relevant to our simple cancer modeling exercise.

J.K. Peterson, *Calculus for Cognitive Scientists*, Cognitive Science and Technology, DOI 10.1007/978-981-287-874-8_19

1890	David von Hansermann noted cancer cells have *abnormal cell division events*.
1914	Theodore Boveri sees that something is wrong in the chromosomes of cancer cells: they are *aneuploid*. That is, they do not have the normal number of chromosomes.
1916	Ernst Tyzzer first applied the term *somatic mutation* to cancer.
1927	Herman Muller discovered ionizing radiation which was known to cause cancer (i.e. was *carcinogenic*) was also able to cause genetic mutations (i.e. was *mutagenic*).
1951	Herman Muller proposed cancer requires a single cell to receive multiple mutations.
1950–1959	Mathematical modeling of cancer begins. It is based on statistics.
1971	Alfred Knudson proposes the concept of a *Tumor Suppressor Gene* or **TSG**. The idea is that it takes a two point mutation to inactivate a TSG. TSG's play a central role in regulatory networks that determine the rate of cell cycling. Their inactivation modifies regulatory networks and can lead to increased cell proliferation.
1986	A *Retinoblastoma* TSG is identified which is a gene involved in a childhood eye cancer.

Since 1986, about 30 more TSG's have been found. An important TSG is **p53**. This is mutated in more than 50 % of all human cancers. This gene is at the center of a control network that monitors genetic damage such as *double stranded breaks* (**DSB**) of DNA. In a *single stranded break* (**SSB**), at some point the double stranded helix of DNA breaks apart on one strand only. In a DSB, the DNA actually separates into different pieces giving a complete gap. DSB's are often due to ionizing radiation. If a certain amount of damage is achieved, cell division is paused and the cell is given time for repair. If there is too much damage, the cell will undergo apoptosis. In many cancer cells, **p53** is inactivated. This allows these cells to divide in the presence of substantial genetic damage. In 1976, Michael Bishop and Harold Varmus introduced the idea of *oncogenes*. These are another class of genes involved in cancer. These genes increase cell proliferation if they are mutated or inappropriately expressed. Now a given gene that occupies a certain position on a chromosome (this position is called the *locus* of the gene) can have a number of alternate forms. These alternate forms are called *alleles*. The number of alleles a gene has for an individual is called that individual's *genotype* for that gene. Note, the number of alleles a gene has is therefore the number of viable DNA codings for that gene. We see then that mutations of a TSG and an oncogene increase the net reproductive rate or *somatic fitness* of a cell. Further, mutations in genetic instability genes also increase the mutation rate.

For example, mutations in mismatch repair genes lead to 50–100 fold increases in point mutation rates. These usually occur in repetitive stretches of short sequences of DNA. Such regions are called *micro satellite regions* of the genome. These regions are used as genetic markers to track inheritance in families. They are short sequences of nucleotides (i.e. **ATCG**) which are repeated over and over. Changes can occur such as increasing or decreasing the number of repeats. This type of instability is thus called a *micro satellite* or **MIN** instability. It is known that 15 % of colon cancer cells have MIN.

Another instability is *chromosomal instability* or **CIN**. This means an increase or decrease in the rate of gaining or losing whole chromosomes or large fractions of chromosomes during cell division.

Let's look at what can happen to a typical TSG. If the first allele of a TSG is inactivated by a point mutation while the second allele is inactivated by a loss of one parent's contribution to part of the cell's genome, the second allele is inactivated because it is lost in the copying process. This is an example of CIN and in this case is called *loss of heterozygosity* or **LOH**. It is known that 85 % of colon cancer cells have CIN.

19.1 Two Allele TSG Models

Let's look now at colon cancer itself. Look at Fig. 19.1.
At the bottom of the crypt, a small number of *stem* cells slowly divide to produce *differentiated* cells. These differentiated cells divide a few times while migrating to the top of the crypt where they undergo apoptosis. This architecture means only a small subset of cells are at risk of acquiring mutations that become fixed in the permanent cell lineage. Many mutations that arise in the differentiated cells will be removed by apoptosis.

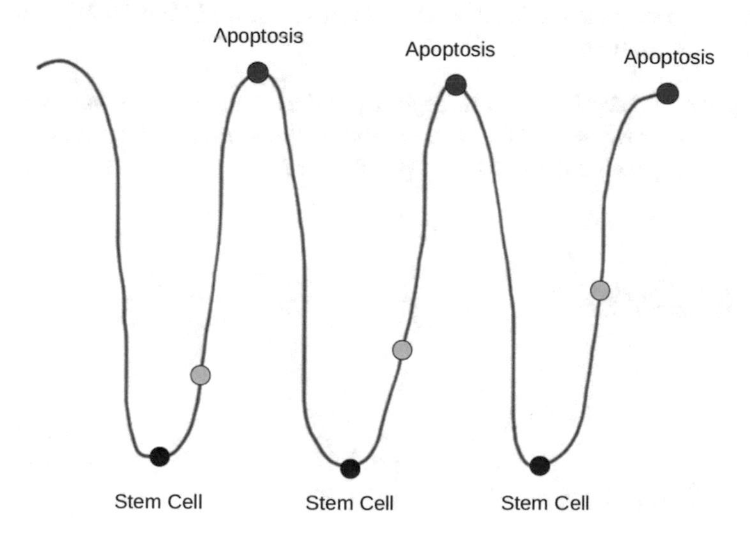

Fig. 19.1 Typical colon crypts

Colon rectal cancer is thought to arise as follows. A mutation inactivates the **Adenomatous Polyposis Coli** or **APC** TSG pathway. Ninety-five percent of colorectal cancer cells have this mutation with other mutations accounting for the other 5 %. The crypt in which the **APC** mutant cell arises becomes **dyplastic**; i.e. has abnormal growth and produces a *polyp*. Large *polyps* seem to require additional oncogene activation. Then 10–20 % of these large polyps progress to cancer.

A general model of cancer based on TSG inactivation is as follows. The tumor starts with the inactivation of a **TSG** called **A**, in a small compartment of cells. A good example is the inactivation of the **APC** gene in a colonic crypt, but it could be another gene. Initially, all cells have two active alleles of the **TSG**. We will denote this by $\boldsymbol{A}^{+/+}$ where the superscript "$+/+$" indicates both alleles are active. One of the alleles becomes inactivated at mutation rate u_1 to generate a cell type denoted by $\boldsymbol{A}^{+/-}$. The superscript $+/-$ tells us one allele is inactivated. The second allele becomes inactivated at rate $\hat{u}_2$ to become the cell type $\boldsymbol{A}^{-/-}$. In addition, $\boldsymbol{A}^{+/+}$ cells can also receive mutations that trigger **CIN**. This happens at the rate u_c resulting in the cell type $\boldsymbol{A}^{+/+\,CIN}$. This kind of a cell can inactivate the first allele of the **TSG** with normal mutation rate u_1 to produce a cell with one inactivated allele (i.e. a $+/-$) which started from a CIN state. We denote these cells as $\boldsymbol{A}^{+/-\,CIN}$. We can also get a cell of type $\boldsymbol{A}^{+/-\,CIN}$ when a cell of type $\boldsymbol{A}^{+/-}$ receives a mutation which triggers **CIN**. We will assume this happens at the same rate u_c as before. The $\boldsymbol{A}^{+/-\,CIN}$ cell then rapidly undergoes **LOH** at rate $\hat{u}_3$ to produce cells of type $\boldsymbol{A}^{-/-\,CIN}$. Finally, $\boldsymbol{A}^{-/-}$ cells can experience **CIN** at rate u_c to generate $\boldsymbol{A}^{-/-\,CIN}$ cells. We show this information in Fig. 19.2.

Let N be the population size of the compartment. For colonic crypts, the typical value of N is 1000–4000. The first allele is inactivated by a point mutation. The rate at which this occurs is modeled by the rate u_1 as shown in Fig. 19.2. We make the following assumptions:

Assumption 19.1.1 (*Mutation rates for u_1 and u_c are population independent*) The mutations governed by the rates u_1 and u_c are **neutral**. This means that these rates do not depend on the size of the population N.

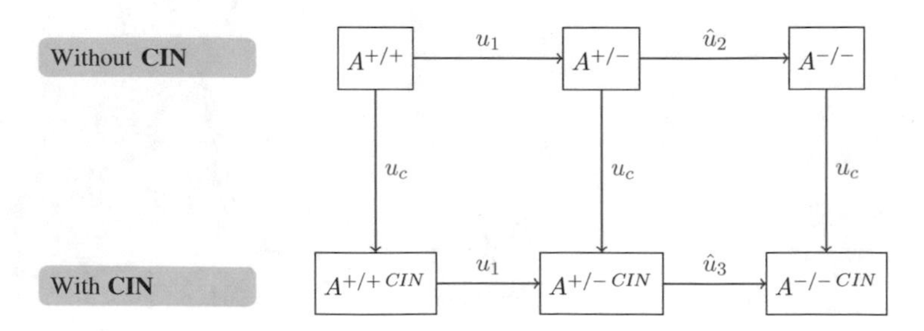

Fig. 19.2 The pathways for the **TSG** allele losses

Assumption 19.1.2 (*Mutation rates $\hat{u}_2$ and $\hat{u}_3$ give selective advantage*)
The events governed by $\hat{u}_2$ and $\hat{u}_3$ give what is called **selective advantage**. This means that the size of the population size does matter.

Using these assumptions, we will model $\hat{u}_2$ and $\hat{u}_3$ like this:

$$\hat{u}_2 = N\, u_2$$

and

$$\hat{u}_3 = N\, u_3.$$

where u_2 and u_3 are neutral rates. We can thus redraw our figure as Fig. 19.3. The mathematical model is then setup as follows. Let

$X_0(t)$ is the probability a cell in cell type $A^{+/+}$ at time t.
$X_1(t)$ is the probability a cell in cell type $A^{+/-}$ at time t.
$X_2(t)$ is the probability a cell in cell type $A^{-/-}$ at time t.
$Y_0(t)$ is the probability a cell in cell type $A^{+/+\,CIN}$ at time t.
$Y_1(t)$ is the probability a cell in cell type $A^{+/-\,CIN}$ at time t.
$Y_2(t)$ is the probability a cell in cell type $A^{-/-\,CIN}$ at time t.

Looking at Fig. 19.3, we can generate rate equations. First, let's rewrite Fig. 19.3 using our variables as Fig. 19.4.

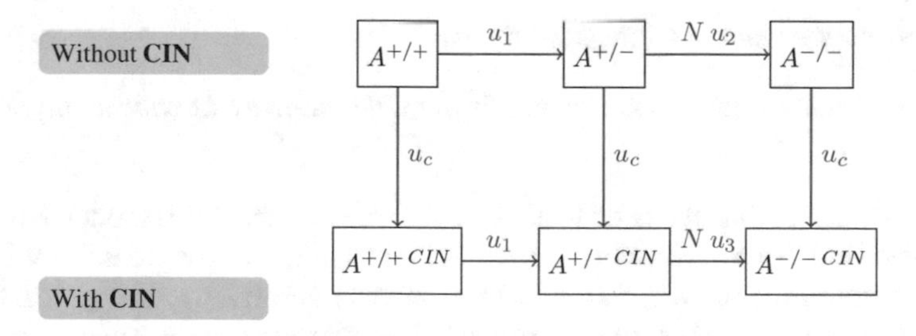

Fig. 19.3 The pathways for the **TSG** allele losses rewritten using selective advantage

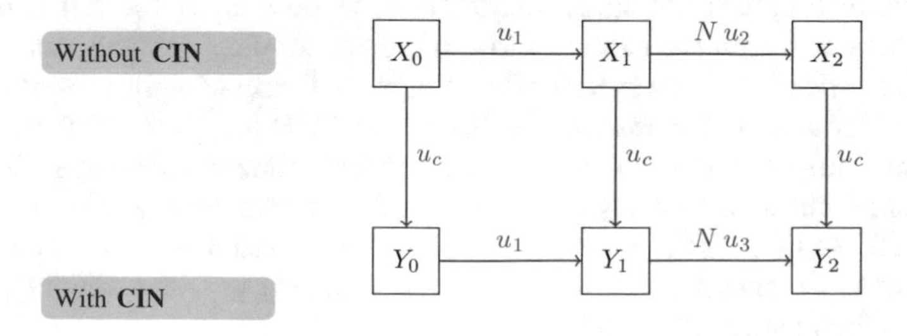

Fig. 19.4 The pathways for the **TSG** allele losses rewritten using mathematical variables

To generate the equations we need, note each box has arrows coming into it and arrows coming out of it. The **arrows in** are **growth** terms for the net change of the variable in the box and the **arrows out** are the **decay or loss** terms. We model **growth** as **exponential growth** and **loss** as **exponential decay**. So X_0 only has arrows going out which tells us it only has **loss** terms. So we would say $\left(X_0'\right)_{loss} = -u_1 X_0 - u_c X_0$ which implies $X_0' = -(u_1 + u_c)X_0$. Further, X_1 has arrows going in and out which tells us it has **growth** and **loss** terms. So we would say $\left(X_1'\right)_{loss} = -N u_2 X_1 - u_c X_1$ and $\left(X_1'\right)_{growth} = u_1 X_0$ which implies $X_1' = u_1 X_0 - (N u_2 + u_c) X1$. We can continue in this way to find all the model equations. We can then see the *Cancer Model* rate equations are

$$X_0' = -(u_1 + u_c)\, X_0 \tag{19.1}$$
$$X_1' = u_1\, X_0 - (u_c + N\, u_2)\, X_1 \tag{19.2}$$
$$X_2' = N\, u_2\, X_1 - u_c\, X_2 \tag{19.3}$$
$$Y_0' = u_c\, X_0 - u_1\, Y_0 \tag{19.4}$$
$$Y_1' = u_c\, X_1 + u_1\, Y_0 - N\, u_3\, Y_1 \tag{19.5}$$
$$Y_2' = N\, u_3\, Y_1 + u_c\, X_2 \tag{19.6}$$

Initially, at time 0, all the cells are in the state X_0, so we have

$$X_0(0) = 1, \;\; X_1(0) = 0, \;\; X_2(0) = 0 \tag{19.7}$$
$$Y_0(0) = 0, \;\; Y_1(0) = 0, \;\; Y_2(0) = 0. \tag{19.8}$$

The problem we want to address is this one:

Question *Under what circumstances is the CIN pathway to cancer the dominant one?*

In order to answer this, we need to analyze the trajectories of this model. Note, if we were interested in the asymptotic behavior of this model as t goes to infinity, then it is clear everything ends up with value 0. However, our interest is over the typical lifetime of a human being and thus, we never reach the asymptotic state. Thus, our analysis in the sections that follow is always concerned with the values of our six variables at the end of a human life span. In Sect. 19.4 we do this for the cell populations without CIN and then we can use those results to develop approximations to cell populations with CIN. However, for this text, we will stop with the top pathway and leave the CIN work to the next course. What we will do here is to develop some key approximations using our tangent line and quadratic function approximation ideas. But, make no mistake, the analysis is a tricky and requires a fair amount of intellectual effort. Still, we think at the end of it all, you will understand and appreciate how we use mathematics and science together to try to come to grips with a difficult problem. Now, let's look at the full model.

19.2 Model Assumptions

Since our interest in these variables is over the typical lifetime of a human being, we need to pick a maximum typical lifetime.

Assumption 19.2.1 (*Average human lifetime*)
The average human life span is 100 years. We also assume that cells divide once per day and so a good choice of time unit is *days*. The final time for our model will be denoted by T and hence

$$T = 3.65 \times 10^4 \text{ days.}$$

Next, recall our colonic crypt, N is from 1000 to 4000 cells. For estimation purposes, we often think of N as the upper value, $N = 4 \times 10^3$.

Assumption 19.2.2 (*Loss of allele rates for neutral mutations*)
We assume

$$u_1 \approx 10^{-7}$$
$$u_2 \approx 10^{-7}.$$

We will assume the rate $N\,u_3$ is quite rapid and so it is close to 1. We will set u_3 as follows:

Assumption 19.2.3 (*Losing the second allele due to CIN is close to probability one*)
We assume

$$N\,u_3 \approx 1 - r.$$

for small positive values of r.

Hence, once a cell reaches the Y_1 state, it will rapidly transition to the end state Y_2 if r is sufficiently small.
We are not yet sure how to set the magnitude of u_c, but it certainly is at least u_1. For convenience, we will assume

Assumption 19.2.4 (u_c *is proportional to* u_1)
We assume

$$u_c = R\,u_1.$$

where R is a number at least 1. For example, if $u_c = 10^{-5}$, this would mean $R = 100$.

19.3 Solving the Top Pathway Exactly

We will now solve the top pathway model exactly using the tools we have developed in this course.

19.3.1 The $X_0 - X_1$ Subsystem

The $X_0 - X_1$ subsystem has two models in it we can easily solve.

$$\begin{aligned} X_0' &= -(u_1 + u_c)X_0 \\ X_0(0) &= 1. \end{aligned}$$

This has the solution

$$X_0(t) = e^{-(u_1+u_c)t}.$$

The X_1' equation is

$$\begin{aligned} X_0' &= u_1 X_0 - (u_c + N u_2)X_1 \\ X_1(0) &= 0. \end{aligned}$$

This is a standard integrating factor problem. We rewrite as

$$\begin{aligned} X_0' + (u_c + N u_2)X_1 &= u_1 X_0 \\ X_1(0) &= 0. \end{aligned}$$

The integrating factor is $e^{(u_c+Nu_2)t}$ leading to

$$\begin{aligned} \left(X_0(t)e^{(u_c+Nu_2)t} \right)' &= u_1\, e^{(u_c+Nu_2)t}\, X_0(t) \\ X_1(0) &= 0. \end{aligned}$$

Integrating from 0 to t, we find

$$\begin{aligned} X_0(t)e^{(u_c+Nu_2)t} &= u_1 \int_0^t e^{(u_c+Nu_2)s}\, X_0(s)\, ds \\ X_1(0) &= 0. \end{aligned}$$

The integration is pretty standard. Plugging in for the solution $X_0(t)$, we have

$$\begin{aligned} & u_1 \int_0^t e^{(u_c+Nu_2)s}\, X_0(s)\, ds \\ &= u_1 \int_0^t e^{(u_c+Nu_2)s}\, e^{-(u_1+u_c)t}\, ds \\ &= u_1 \int_0^t e^{(Nu_2-u_1)s}\, ds \\ &= \frac{u_1}{Nu_2 - u_1} \left(e^{(Nu_2-u_1)t} - 1 \right). \end{aligned}$$

Solving for $X_1(t)$, we find

$$\begin{aligned} X_1(t) &= \frac{u_1}{Nu_2 - u_1}\, e^{-(u_c+Nu_2)t} \left(e^{(Nu_2-u_1)t} - 1 \right) \\ &= \frac{u_1}{Nu_2 - u_1} \left(e^{-(u_c+u_1)t} - e^{-(u_c+Nu_2)t} \right). \end{aligned}$$

19.3.2 Solving for X_2

Using the X_1 solution, we have to solve

$$\begin{aligned} X_2'(t) + u_c X_2(t) &= Nu_2 X_1(t) \\ &= \frac{Nu_1u_2}{Nu_2 - u_1} \left(e^{-(u_1+u_c)t} - e^{-(u_c+Nu_2)t} \right) \end{aligned}$$

This can be solved using the integrating factor $e^{u_c t}$. We find

$$\begin{aligned} \left(e^{u_c t}\, X_2(t) \right)' &= \frac{Nu_1u_2}{Nu_2 - u_1} \left(e^{-(u_1+u_c)t} - e^{-(u_c+Nu_2)t} \right) e^{u_c t} \\ &= \frac{Nu_1u_2}{Nu_2 - u_1} \left(e^{-u_1 t} - e^{-Nu_2 t} \right) \end{aligned}$$

Integrating from 0 to t and applying the initial conditions, we obtain

$$e^{u_c t}\, X_2(t) = \frac{Nu_1u_2}{Nu_2 - u_1} \left(\frac{1}{u_1}\left(1 - e^{-u_1 t}\right) - \frac{1}{Nu_2}\left(1 - e^{-Nu_2 t}\right) \right)$$

Then, multiplying through by $e^{-u_c t}$, we finally find $X_2(t)$.

$$X_2(t)\frac{Nu_1u_2}{Nu_2 - u_1}\left(\frac{1}{u_1}\left(e^{-u_ct} - e^{-(u_1+u_c)t}\right) - \frac{1}{Nu_2}\left(e^{-u_ct} - e^{-(u_c+Nu_2)t}\right)\right) \tag{19.9}$$

19.4 Approximation of the Top Pathway

These top pathway solutions are impossible to get a feel for because they are so complicated. But they are written in terms of **differences of exponentials** and **differences of differences of exponentials** and we have tools from Sect. 14.5 we can use. Yes, they are really messy, but we think you will appreciate the payoff in terms of insight! To do this, we can take advantage of some key approximations.

19.4.1 Approximating X_0

Although X_0 is a nice exponential decay, we can approximate it with a tangent line.

$$X_0(t) = 1 - (u_1 + u_c)t + (u_1 + u_c)^2 e^{-(u_c+u_1)c_1}\, \frac{t^2}{2}$$

for some c_1 between 0 and t. Hence, $X_0(t) \approx 1 - (u_1 + u_c)t$ with error $E_0(t)$

$$\begin{aligned} E_0(t) &= (u_1 + u_c)^2 e^{-(u_c+u_1)c_1}\, \frac{t^2}{2} \\ &\leq (u_1 + u_c)^2 \frac{T^2}{2}. \end{aligned}$$

Thus, the maximum error made in approximating $X_0(t)$ by $1-(u_1+u_c)t$ is E_0 given by

$$E_0 = (u_1 + u_c)^2 \frac{T^2}{2} = \left(10^{-7}(1 + R)\right)^2 6.67 \times 10^8 = 6.67(1 + R)^2\, 10^{-6}.$$

Now we want this estimate for X_0 be reasonable; i.e. first, give a positive number over the human life time range and second, have the discrepancy between the true $X_0(t)$ and this approximation is small. For example, if we want the maximum error in X_0 to be 0.05, then we want

$$E_0 \leq (1 + R)^2\, 6.67 \times 10^{-6} < 0.05.$$

This implies that

$$(1 + R)^2 < 75.0 \times 10^2$$

or

$$R < 85.7$$

Since $u_c = R\, u_1$, we see

$$u_c < 8.57\ 10^{-6}.$$

to have a good X_0 estimate.

19.4.2 Approximating X_1

Recall

$$X_1(t) = \frac{u_1}{Nu_2 - u_1}\left(e^{-(u_c+u_1)t} - e^{-(u_c+Nu_2)t}\right)$$

Since $X_1(t)$ is written as the difference of two exponentials, we can use a first order approximation as discussed in Eq. 14.6, to find

$$\begin{aligned} X_1(t) &= \frac{u_1}{Nu_2 - u_1}\left((Nu_2 - u_1)t + \left((u_c + u_1)^2 e^{-(u_c+u_1)c_1} - (u_c + Nu_2)^2 e^{-(u_c+Nu_2)c_2}\right)\frac{t^2}{2}\right) \\ &= u_1 t + \frac{u_1}{Nu_2 - u_1}\left((u_c + u_1)^2 e^{-(u_c+u_1)c_1} - (u_c + Nu_2)^2 e^{-(u_c+Nu_2)c_2}\right)\frac{t^2}{2}. \end{aligned}$$

Hence, approximating with $X_1(t) \approx u_1 t$, gives a maximum error of

$$\begin{aligned} E_1 &= \max_{0\le t\le T} \frac{u_1}{Nu_2 - u_1}\left(\left((u_c + u_1)^2 e^{-(u_c+u_1)c_1} - (u_c + Nu_2)^2 e^{-(u_c+Nu_2)c_2}\right)\frac{t^2}{2}\right) \\ &\le \frac{u_1}{Nu_2 - u_1}\left((u_c + u_1)^2 + (u_c + Nu_2)^2\right)\frac{T^2}{2}. \end{aligned}$$

We have already found the $R < 85.7$ and for N at most 4000 with $u_1 = u_2 = 10^{-7}$, we see $Nu_2 - u_1 \approx Nu_2$. Further, $u_c = Ru_1 \le 10^{-5}$ and $Nu_2 \le 4 \times 10^{-4}$ so that in the second term, Nu_2 is dominant. We therefore have

$$E_1 \le \frac{u_1}{Nu_2}\left((1 + R)^2 (u_1)^2 + (Nu_2)^2\right)\frac{T^2}{2}.$$

But the term u_1^2 is negligible in the second term, so we have

$$\begin{aligned} E_1 &\approx \frac{u_1}{Nu_2}\,(Nu_2)^2\,\frac{T^2}{2} \\ &= Nu_1 u_2\,\frac{T^2}{2} = 4000 \times 10^{-14} \times 6.67 \times 10^8 = 0.027. \end{aligned}$$

19.4.3 Approximating X_2

Now here comes the messy part. Apply the second order difference of exponentials approximation from Eq. 14.5 above to our X_2 solution. To make our notation somewhat more manageable, we will define the error term $E(r, a, t)$ by

$$E(r, a, t) = \left| -r^3 e^{-rc_1} + (r+a)^3 e^{-(r+a)c_2} \right| \frac{t^3}{6} \leq 2(r+a)^3 \frac{t^3}{6}$$

Note the maximum error over human life time is thus $E(r, a)$ which is

$$E(r, a) = 2(r+a)^3 \frac{T^3}{6}. \tag{19.10}$$

Now let's try to find an approximation for $X_2(t)$. We have

$$\begin{aligned} X_2(t) &= \frac{Nu_1u_2}{Nu_2 - u_1} \left(\frac{1}{u_1} \left(e^{-u_c t} - e^{-(u_1+u_c)t} \right) - \frac{1}{Nu_2} \left(e^{-u_c t} - e^{-(u_c+Nu_2)t} \right) \right) \\ &= \frac{Nu_1u_2}{Nu_2 - u_1} \left(\frac{1}{u_1} \left(u_1 t - (u_1^2 + 2u_1u_c) \frac{t^2}{2} + E(u_c, u_1, t) \right) \right) \\ &\quad - \frac{Nu_1u_2}{Nu_2 - u_1} \left(\frac{1}{Nu_2} \left(Nu_2 t - ((Nu_2)^2 + 2Nu_2u_c) \frac{t^2}{2} + E(u_c, Nu_2, t) \right) \right) \end{aligned}$$

Now do the divisions above to obtain

$$\begin{aligned} X_2(t) = & \frac{Nu_1u_2}{Nu_2 - u_1} \left(\left(t - (u_1 + 2u_c) \frac{t^2}{2} + \frac{E(u_c, u_1, t)}{u_1} \right) \right) \\ & - \frac{Nu_1u_2}{Nu_2 - u_1} \left(\left(t - (Nu_2 + 2u_c) \frac{t^2}{2} + \frac{E(u_c, Nu_2, t)}{Nu_2} \right) \right) \end{aligned}$$

We can then simplify a bit to get

$$\begin{aligned} X_2(t) &= \frac{Nu_1u_2}{Nu_2 - u_1} \left(\left(t - (u_1 + 2u_c) \frac{t^2}{2} + \frac{E(u_c, u_1, t)}{u_1} - t + (Nu_2 + 2u_c) \frac{t^2}{2} \right. \right. \\ &\qquad \left. \left. - \frac{E(u_c, Nu_2, t)}{Nu_2} \right) \right) \\ &= \frac{Nu_1u_2}{Nu_2 - u_1} \left(\left((Nu_2 - u_1) \frac{t^2}{2} + \frac{E(u_c, u_1, t)}{u_1} - \frac{E(u_c, Nu_2, t)}{Nu_2} \right) \right) \\ &= Nu_1u_2 \frac{t^2}{2} + \frac{Nu_2}{Nu_2 - u_1} E(u_c, u_1, t) - \frac{u_1}{Nu_2 - u_1} E(u_c, Nu_2, t). \end{aligned}$$

Hence, we see $X_2(t) \approx N u_1 u_2 \frac{t^2}{2}$ with maximum error E_2 over human life time T given by

$$\begin{aligned} E_2 &= \max_{0 \le t \le T} \left(\frac{N u_2}{N u_2 - u_1} E(u_c, u_1, t) - \frac{u_1}{N u_2 - u_1} E(u_c, N u_2, t). \right) \\ &\le \frac{N u_2}{N u_2 - u_1} E(u_c, u_1) + \frac{u_1}{N u_2 - u_1} E(u_c, N u_2) \\ &= \left(\frac{N u_2}{N u_2 - u_1} 2(u_1 + u_c)^3 + \frac{u_1}{N u_2 - u_1} 2(u_c + N u_2)^3 \right) \frac{T^3}{6} \end{aligned}$$

From our model assumptions, an upper bound on N is 4000. Since $u_1 = u_2 = 10^{-7}$, we see the term $N u_2 - u_1 \approx N u_2 = 4 \times 10^{-4}$. Also, $u_1 + u_c$ is dominated by u_c. Thus,

$$\begin{aligned} E_2 &\approx \left(2u_c^3 + \frac{u_1}{N u_2 - u_1} (N u_2)^3 \right) \frac{T^3}{6} \\ &\approx \left(2u_c^3 + \frac{u_1}{N u_2} (N u_2)^3 \right) \frac{T^3}{6} \\ &\approx \left(2u_c^3 + u_1 (N u_2)^2 \right) \frac{T^3}{6} \end{aligned}$$

But the term $u_1 (N u_2)^2$ is very small ($\approx 1.6e - 14$) and can also be neglected. So, we have

$$E_2 \approx 2u_c^3 \frac{T^3}{6}$$

Thus, if $R \approx 90$, we have $R^3 = 7.3 \times 10^2$ and

$$\begin{aligned} E_2 &\approx 2 \times (7.3 \times 10^{-16})\, 8.1\; 10^{12} \\ &= 118.3 \times 10^{-4} = 0.0018. \end{aligned}$$

We summarize our approximation results for the top pathway in Table 19.1.

Table 19.1 The Non CIN Pathway Approximations with error estimates

Approximation	Maximum error
$X_0(t) \approx 1 - (u_1 + u_c)\, t$	$(u_1 + u_c)^2 \frac{T^2}{2}$
$X_1(t) \approx u_1\, t$	$\frac{u_1}{N u_2 - u_1} \left((u_c + u_1)^2 + (u_c + N u_2)^2 \right) \frac{T^2}{2}$
$X_2(t) \approx N\, u_1\, u_2\, \frac{t^2}{2}$	$2u_c^3 \frac{T^3}{6}$

So $X_2 \propto t^2$ and $X_1 \propto t$! Who would have thunk it? The approximations, while yucky and intense, have a great payoff in terms of insight. This is a good lesson to remember. **Insight is only achieved with great effort**! In the next course, if you take it (and you should), we go back to the Cancer Model and handle the bottom pathway and figure out which pathway to cancer is the dominant one: top or bottom. But that is another story. As a final note, although we did not solve for Y_2, note $Y_2(T)$ is the fraction of cells that have lost both TSG alleles and hence is a measurement of cancer for the bottom pathway. Also $X_2(T)$ is the fraction of cells that have lost both TSG alleles and measures cancer in the top pathway. It turns out we can approximate Y_2 by $u_1 u_c T^2$ over human lifetime. So over human life time, the ratio of Y_2 to X_2 can be approximated by

$$\frac{Y_2(T)}{X_2(T)} \approx \frac{u_1 u_c T^2}{N u_1 u_2 \frac{T^2}{2}} = \frac{2 u_1 u_c}{N u_1 u_2} = \frac{2 u_c}{N u_2}.$$

Since $u_1 = u_2$ and $u_c = R u_1$, so

$$\frac{Y_2(T)}{X_2(T)} \approx \frac{2 R u_1}{N u_1} = \frac{2R}{N}$$

So the **TOP** pathway to cancer is the dominant one if this fraction is less than one. That is $2R/N < 1$ implies the **TOP** is the dominant pathway. The careful analysis in the next course tells us that our approximations have reasonable error as long as $R < 85$ for our range of $N = 1000$ to $N = 4000$. So as long as $u_c \approx 8.5 \times 10^{-6}$ or less, the **TOP** pathway is dominant.

References

M. Novak, *Evolutionary Dynamics: Exploring the Equations of Life* (Belknap Press, Cambridge, 2006)

B. Ribba, T. Colin, S. Schnell, A multiscale mathematical model of cancer, and its use in analyzing irradiation therapies. Theor. Biol. Med. Model. **3**, 1–19 (2006)

Y. Tatabe, S. Tavare, D. Shibata, Investigating stem cells in human colon using methylation patterns. Proc. Natl Acad. Sci. **98**, 10839–10844 (2001)

Chapter 20
First Order Multivariable Calculus

We will now talk about functions of more than one variable. This is a very complicated topic, so we will start slowly and add a little at a time. We begin by using MatLab to show you how to visualize functions of two variables as surfaces. This approach also helps you learn a bit more MatLab which is part of the master plan!

20.1 Functions of Two Variables

Let's start by looking at the x–y plane as a collection of two dimensional vectors. Each vector is rooted at the origin and the head of the vector corresponds to our usual coordinate pair (x, y). The set of all such x and y determines the x–y plane which we will also call $\Re^2$. The superscript two is used because we are now explicitly acknowledging that we can think of these ordered pairs as vectors also with just a slight identification on our part. Since we know about vectors, note if we have a vector we can rewrite it, using our standard rules for vector arithmetic and scaling of vectors as

$$\begin{bmatrix} 6 \\ 7 \end{bmatrix} = 6 \begin{bmatrix} 1 \\ 0 \end{bmatrix} + 7 \begin{bmatrix} 0 \\ 1 \end{bmatrix}$$

A little thought will let you see we can do this for any vector and so we define special vectors $\boldsymbol{i} = \boldsymbol{e}_1$ and $\boldsymbol{j} = \boldsymbol{e}_2$ as follows:

$$\boldsymbol{i} = \boldsymbol{e}_1 = \begin{bmatrix} 1 \\ 0 \end{bmatrix} \text{ and } \boldsymbol{j} = \boldsymbol{e}_2 = \begin{bmatrix} 0 \\ 1 \end{bmatrix}$$

J.K. Peterson, *Calculus for Cognitive Scientists*, Cognitive Science and Technology, DOI 10.1007/978-981-287-874-8_20

Thus, any vector can be written as

$$\begin{bmatrix} x \\ y \end{bmatrix} = x\,\boldsymbol{e_1} + y\,\boldsymbol{e_1}$$
$$= x\,\boldsymbol{i} + y\,\boldsymbol{j}$$

Now let's start looking at functions that map each ordered pair (x, y) into a number. Let's begin with an example. Consider the function $f(x, y) = x^2 + y^2$ defined for all x and y. Hence, for each x and y we pick, we calculate a number we can denote by z whose value is $f(x, y) = x^2 + y^2$. Using the same ideas we just used for the x–y plane, we see the set of all such triples $(x, y, z) = (x, y, x^2 + y^2)$ defines a **surface** in $\Re^3$ which is the collection of all ordered triples (x, y, z). Each of these triples can be identified with a three dimensional vector whose tail is the origin and whose head is the triple (x, y, z). We note any three dimensional vector can be written as

$$\begin{bmatrix} x \\ y \\ z \end{bmatrix} = x\,\boldsymbol{e_1} + y\,\boldsymbol{e_2} + yz\,\boldsymbol{e_3}$$
$$= x\,\boldsymbol{i} + y\,\boldsymbol{j} + z\,\boldsymbol{k}$$

where we define the special vectors used in this representation by

$$\boldsymbol{i} = \boldsymbol{e_1} = \begin{bmatrix} 1 \\ 0 \\ 0 \end{bmatrix}, \ \boldsymbol{j} = \boldsymbol{e_2} = \begin{bmatrix} 0 \\ 1 \\ 0 \end{bmatrix} \text{ and } \boldsymbol{k} = \boldsymbol{e_3} = \begin{bmatrix} 0 \\ 0 \\ 1 \end{bmatrix}$$

We can plot this surface in MatLab with fairly simple code. Let's go through how to do these plots in a lot of detail so we can see how to apply this kind of code in other situations.

20.1.1 Drawing an Annotated Surface

To draw a portion of a surface, we pick a rectangle of x and y values. To make it simple, we will choose a point (x_0, y_0) as the center of our rectangle and then for a chosen Δx and Δy and integers n_x and n_y, we set up the rectangle

$$[x_0 - n_x\Delta x, \ldots, x_0, \ldots, x_0 + n_x\Delta x] \times [y_0 - n_y\Delta y, \ldots, y_0, \ldots, y_0 + n_y\Delta y]$$

The constant x and y lines determined by this grid result in a matrix of intersections with entries (x_i, y_j) for appropriate indices i and j. We will approximate the surface by plotting the triples $(x_i, y_j, z_{ij} = f(x_i, y_j))$ and then drawing a top for each rectangle. Right now though, let's just draw this base grid. In MatLab, first setup the function we want to look at. We will choose a very simple one

Listing 20.1: Setting Up The Surface Function

```
f = @(x,y) x.^2 + y.^2;
```

Now, we draw the grid by using the function **`DrawGrid(f,delx,nx,dely,ny,x0,y0)`**. This function has several arguments as you can see and we explain them in the listing below. So we are drawing a grid centered around (0.5, 0.5) using a uniform 0.5 step in both directions. The grid is drawn at $z = 0$.

Listing 20.2: Drawing The Base Line Grid

```
% Taking the arguments in order
% f is the surface function
% delx = 0.5 is the width of the delta x
% nx = 2 is the number of steps we take right and left from
%      the base point x0
%
% dely = 0.5 is the width of the delta y
% ny = 2 is the number of steps we take right and left from
%      the base point y0
%
% x0 = 0.5
% y0 = 0.5
DrawGrid(f,0.5,2,0.5,2,0.5,0.5);
```

The resulting grid is shown in Fig. 20.1.

Make sure you play with the plot a bit. You can grab it and rotate it as you see fit to make sure you see all the detail. Right now, there is not much to see in the grid, but later when we plot the surface, the grid and other things, the ability to rotate in

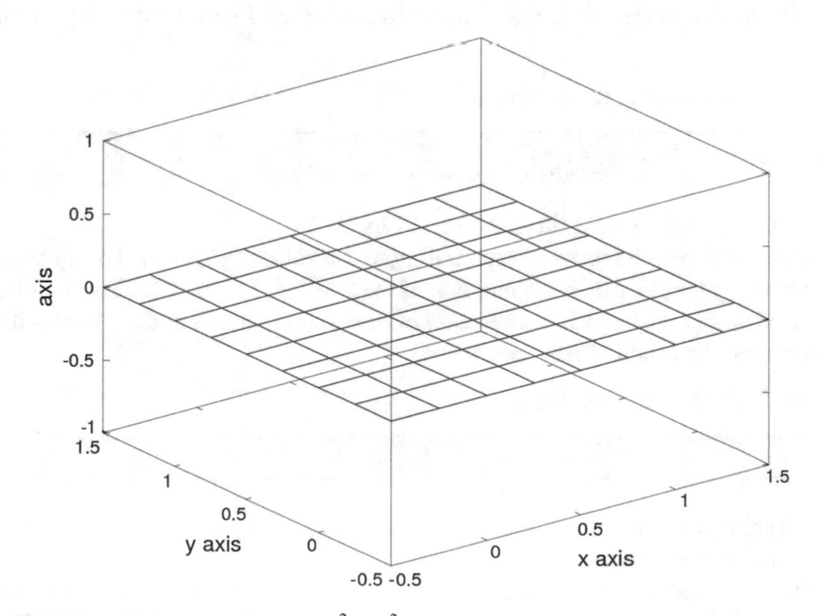

Fig. 20.1 The grid for the surface $z = x^2 + y^2$

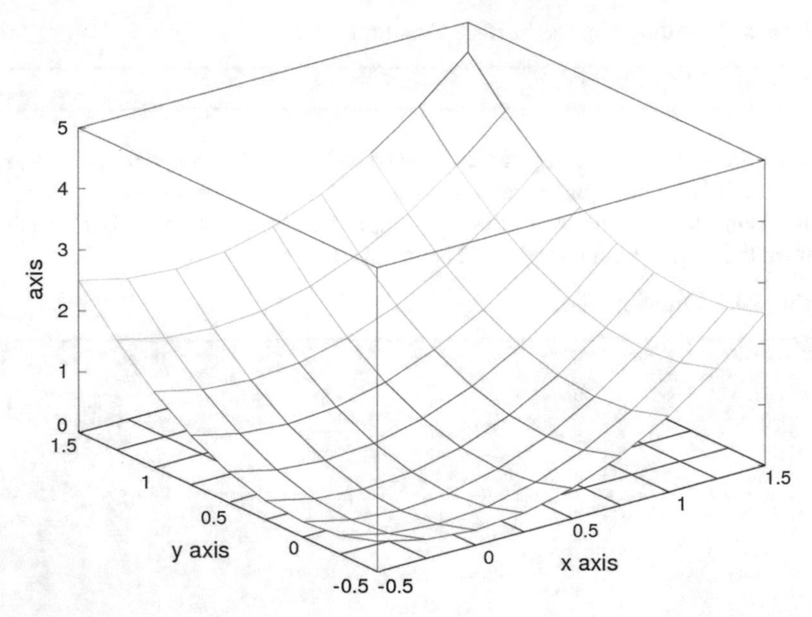

Fig. 20.2 The surface and grid for $z = x^2 + y^2$

3D is important to our understanding. So make sure you take the time to see how to do this!

To draw the surface, we find the pairs (x_i, y_j) and the associated $f(x_i, y_j)$ values and then call the **DrawMesh(f,delx,nx,dely,ny,x0,y0)** command. The meaning in the arguments is the same as in **DrawGrid** so we won't repeat them here.

Listing 20.3: Drawing the Surface as a Mesh

```
DrawMesh(f,0.5,2,0.5,2,0.5,0.5);
```

The resulting surface and grid is shown in Fig. 20.2.

Next, we draw the **traces** corresponding to the values x_0 and y_0. The x_0 trace is the function $f(x_0, y)$ which is a *function* of the two variables y and z. The y_0 trace is the function $f(x, y_0)$ which is a *function* of the two variables x and z. We plot these curves using the function **DrawTraces**.

Listing 20.4: Drawing the Traces

```
DrawTraces(f,0.5,2,0.5,2,0.5,0.5);
```

The resulting surface with grid and traces is shown in Fig. 20.3. The traces are the thick parabolas on the surface.

Next, let's add the column with the rectangular base having coordinates Lower Left (x_0, y_0), Lower Right $(x_0 + \Delta x, y_0)$, Upper Left $(x_0, y_0 + \Delta y)$ and Upper Right

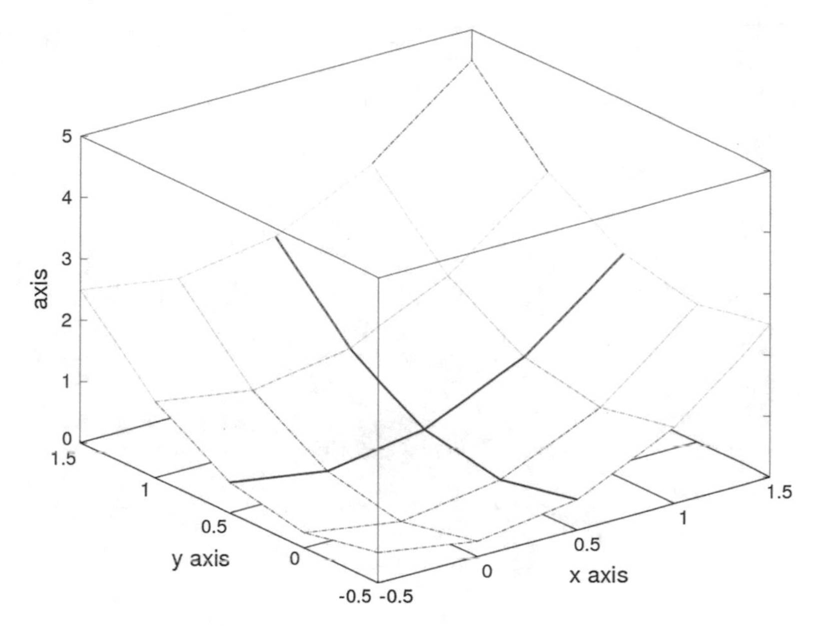

Fig. 20.3 The surface with grid and traces for $z = x^2 + y^2$

$(x_0 + \Delta x, y_0 + \Delta y)$. We draw and fill this base with the function **DrawBase**. We draw the vertical lines going from each of the four corners of the base to the surface with the code **DrawColumn** and we draw and fill the patch of surface this column creates in the full surface with the function **DrawPatch**.

Listing 20.5: Drawing The Base

```
DrawBase(f,0.5,2,0.5,2,0.5,0.5);
```

We show the added base below the surface in Fig. 20.4.

In **DrawColumn**, we draw four vertical lines from the base up to the surface. You'll note the figure is getting more crowded looking though. Make sure you grab the picture and rotate it around so you can see everything from different perspectives.

Listing 20.6: Drawing the Vertical Lines for the Base

```
DrawColumn(f,0.5,2,0.5,2,0.5,0.5);
```

We then draw the patch just like we drew the base.

Listing 20.7: Drawing The Surface Patch

```
DrawPatch(f,0.5,2,0.5,2,0.5,0.5);
```

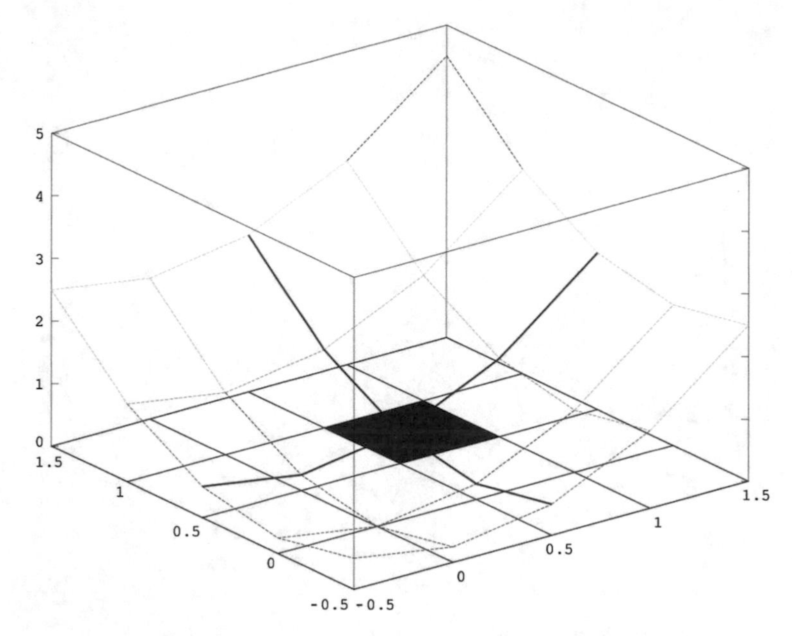

Fig. 20.4 Adding the base to the surface with grid and traces for $z = x^2 + y^2$

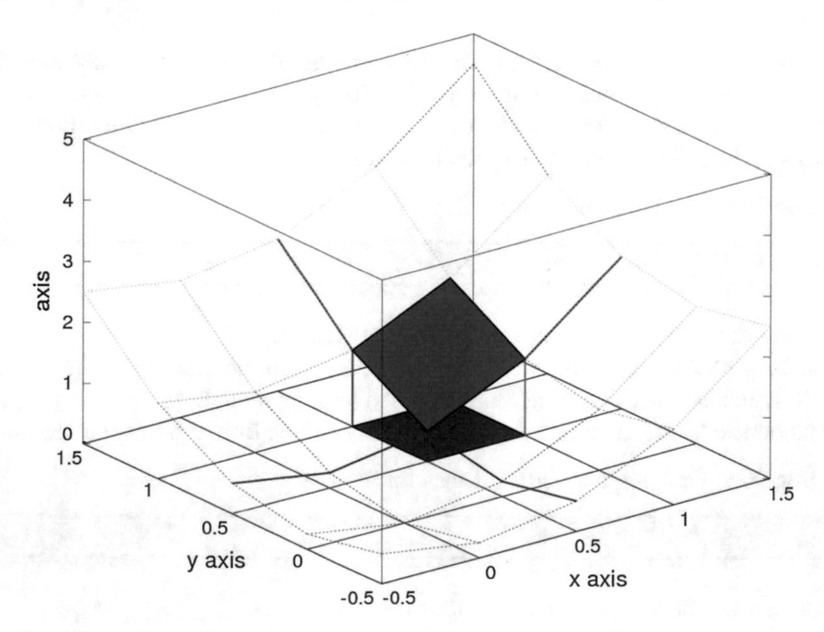

Fig. 20.5 The surface with grid and traces and column for $z = x^2 + y^2$

The resulting surface with grid and traces and with the column and patch is shown in Fig. 20.5.

We have combined all of these functions into a utility function **DrawSimpleSurface** which manages these different graphing choices using boolean variables like **DoGrid** to turn a graph on or off. If the boolean variable **DoGrid** is set to one, the grid is drawn. The code is self-explanatory so we just lay it out here. We haven't shown all the code for the individual drawing functions, but we think you'll find it interesting to see how we manage the pieces in this one piece of code. So check this out.

Listing 20.8: DrawSimpleSurface

```
function DrawSimpleSurface(f,delx,nx,dely,ny,x0,y0,domesh,dotraces,dogrid,dopatch,
    docolumn,dobase)
% f is the function defining the surface
% delx is the size of the x step
% nx is the number of steps left and right from x0
% dely is the size of the y step
% ny is the number of steps left and right from y0
% (x0,y0) is the location of the column rectangle base
% domesh = 1 if we want to do the mesh
% dogrid = 1 if we want to do the grid
% dopatch = 1 if we want the patch above the column
% dobase = 1 if we want the base of the column
% docolumn = 1 if we want the column
% dotraces = 1 if we want the traces
%
hold on
if dotraces==1
  % set up x trace for x0, y trace for y0
  DrawTraces(f,delx,nx,dely,ny,x0,y0);
end
if domesh==1 % plot the surface
  DrawMesh(f,delx,nx,dely,ny,x0,y0);
end
if dogrid==1 %plot x, y grid
  DrawGrid(f,delx,nx,dely,ny,x0,y0);
end
if dopatch==1
  % draw patch for top of column
  DrawPatch(f,delx,nx,dely,ny,x0,y0);
end
if dobase==1
  % draw patch for top of column
  DrawBase(f,delx,nx,dely,ny,x0,y0);
end
if docolumn==1
  %draw column
  DrawColumn(f,delx,nx,dely,ny,x0,y0);
end
hold off
end
```

Hence, to draw everything for this surface, we would use the session:

Listing 20.9: Drawing a simple surface

```
f = @(x,y) x.^2+y.^2;
DrawSimpleSurface(f,0.5,2,0.5,2,0.5,0.5,1,1,1,1,1,1);
```

This surface has circular cross sections for different positive values of z and it is called a *circular paraboloid*. If you used $f(x, y) = 4x^2 + 3y^2$, the cross sections for positive z would be ellipses and we would call the surface an *elliptical paraboloid*. Now this code is not perfect. However, as an exploratory tool it is not bad! Now it is time for you to play with it a bit in the exercises below.

20.1.2 Homework

Exercise 20.1.1 *Explore the surface graph of the* circular paraboloid $f(x, y) = x^2 + y^2$ *for different values of* (x_0, y_0) *and* Δx *and* Δy. *Experiment with the 3D rotated view to make sure you see everything of interest.*

Exercise 20.1.2 *Explore the surface graph of the* elliptical paraboloid $f(x, y) = 2x^2 + y^2$ *for different values of* (x_0, y_0) *and* Δx *and* Δy. *Experiment with the 3D rotated view to make sure you see everything of interest.*

Exercise 20.1.3 *Explore the surface graph of the* elliptical paraboloid $f(x, y) = 2x^2 + 3y^2$ *for different values of* (x_0, y_0) *and* Δx *and* Δy. *Experiment with the 3D rotated view to make sure you see everything of interest.*

20.2 Continuity

Let's recall the ideas of continuity for a function of one variable. Consider these three versions of a function f defined on $[0, 2]$.

$$f(x) = \begin{cases} x^2, & \text{if } 0 \leq x < 1 \\ 10, & \text{if } x = 1 \\ 1 + (x-1)^2 & \text{if } 1 < x \leq 2. \end{cases}$$

The first version is not continuous at $x = 1$ because although the $\lim_{x \to 1} f(x)$ exists and equals 1 ($\lim_{x \to 1^-} f(x) = 1$ and $\lim_{x \to 1^+} f(x) = 1$), the value of $f(1)$ is 10 which does not match the limit. Hence, we know f here has a removeable discontinuity at $x = 1$. Note continuity failed because the limit existed but the value of the function did not match it. The second version of f is given below.

$$f(x) = \begin{cases} x^2, & \text{if } 0 \leq x \leq 1 \\ (x-1)^2 & \text{if } 1 < x \leq 2. \end{cases}$$

In this case, the $\lim_{x \to 1^-} = 1$ and $f(1) = 1$, so f is continuous from the left. However, $\lim_{x \to 1^+} = 0$ which does not match $f(1)$ and so f is not continuous from the right. Also, since the right and left hand limits do not match at $x = 1$, we know

$\lim_{x \to 1}$ does not exist. Here, the function fails to be continuous because the limit does not exist. The final example is below:

$$f(x) = \begin{cases} x^2, & \text{if } 0 \leq x < 1 \\ x + (x-1)^2 & \text{if } 1 < x \leq 2. \end{cases}$$

Here, the limit and the function value at 1 both match and so f is continuous at $x = 1$. To extend these ideas to two dimensions, the first thing we need to do is to look at the meaning of the limiting process. What does $\lim_{(x,y) \to (x_0, y_0)}$ mean? Clearly, in one dimension we can approach a point x_0 from x in two ways: from the left or from the right or jump around between left and right. Now, it is apparent that we can approach a given point (x_0, y_0) in an infinite number of ways. Draw a point on a piece of paper and convince yourself that there are many ways you can draw a curve from another point (x, y) so that the curve ends up at (x_0, y_0)! We still want to define continuity in the same way; i.e. f is continuous at the point (x_0, y_0) if $\lim_{(x,y) \to (x_0, y_0)} f(x, y) = f(x_0, y_0)$. If you look at the graphs of the surface $z = x^2 + y^2$ we have done previously, we clearly see that we have this kind of behavior. There are no jumps, tears or gaps in the surface we have drawn. Let's make this formal.

Definition 20.2.1 (*Continuity*)
Let $z = f(x, y)$ be a function of the two independent variables x and y defined on some domain. At each pair (x, y) where f is defined in a circle of some finite radius r,

$$B_r(x_0, y_0) = \{(x, y) \mid \sqrt{(x - x_0)^2 + (y - y_0)^2} < r\},$$

if $\lim_{(x,y) \to (x_0, y_0)} f(x, y)$ exists and matches $f(x_0, y_0)$, we say f is continuous at (x_0, y_0).

Here is an example of a function which is not continuous at the point $(0, 0)$. Let

$$f(x, y) = \begin{cases} \frac{2x}{\sqrt{x^2+y^2}}, & \text{if } (x, y) \neq (0, 0) \\ 0, & \text{if } (x, y) = (0, 0). \end{cases}$$

If we show the limit as we approach $(0, 0)$ does not exist, then we will know f is not continuous at $(0, 0)$. If this limit exists, we should get the same value for the limit no matter what path we take to reach $(0, 0)$. Let the first path be given by $x(t) = t$ and $y(t) = 2t$. Then, as $t \to 0$, $(x(t), y(t)) \to (0, 0)$ as desired. Plugging in to f, we find for $t \neq 0$, $f(t, 2t) = 2t/\sqrt{t^2 + 4t^2} = 2/\sqrt{5}$ and hence the limit along this path is this constant value $2/\sqrt{5}$. On the other hand, along the path $x(t) = t$ and $y(t) = -3t$, for $t \neq 0$, we have $f(t, -3t) = 2/3$ which is not the same. Since the limiting value differs on two paths, the limit can't exist. Hence, f is not continuous at $(0, 0)$.

20.3 Partial Derivatives

Let's go back to our simple surface example and look at the traces again. In Fig. 20.6, we show the traces for the base point $x_0 = 0.5$ and $y_0 = 0.5$. We have also drawn vertical lines down from the traces to the x–y plane to further emphasize the placement of the traces on the surface. The surface itself is not shown as it is somewhat distracting and makes the illustration too busy.

You can generate this type of graph yourself with the function **DrawFullTraces** as follows:

Listing 20.10: Drawing a full trace

```
DrawFullTraces(f,0.5,2,0.5,2,0.5,0.5);
```

Note, that each trace has a well-defined tangent line and derivative at the points x_0 and y_0. We have

$$\begin{aligned}\frac{d}{dx} f(x, y_0) &= \frac{d}{dx}(x^2 + y_0^2) \\ &= 2x\end{aligned}$$

as the value y_0 in this expression is a constant and hence its derivative with respect to x is zero. We denote this new derivative as $\frac{\partial f}{\partial x}$ which we read as *the partial derivative*

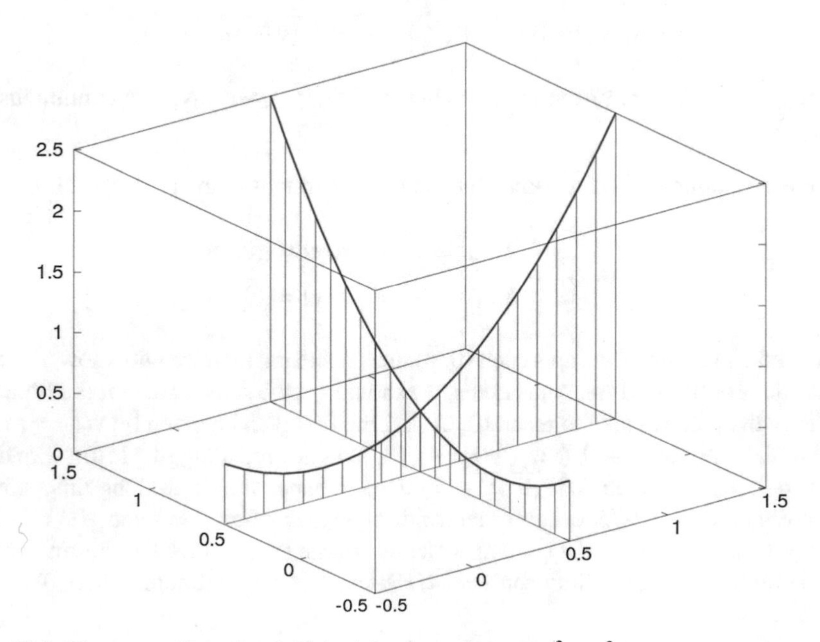

Fig. 20.6 The traces $f(x_0, y)$ and $f(x, y_0)$ for the surface $z = x^2 + y^2$ for $x_0 = 0.5$ and $y_0 = 0.5$

of f *with respect to* x. It's value as the point (x_0, y_0) is $2x_0$ here. For any value of (x, y), we would have $\frac{\partial f}{\partial x} = 2x$. We also have

$$\begin{aligned}\frac{d}{dy} f(x_0, y) &= \frac{d}{dy}(x_0^2 + y^2)\\ &= 2y\end{aligned}$$

We then denote this new derivative as $\frac{\partial f}{\partial y}$ which we read as *the partial derivative of* f *with respect to* y. It's value as the point (x_0, y_0) is then $2y_0$ here. For any value of (x, y), we would have $\frac{\partial f}{\partial y} = 2y$.

The tangent lines for these two traces are then

$$\begin{aligned}T(x, y_0) &= f(x_0, y_0) + \frac{d}{dx} f(x, y_0)\Big|_{x_0} (x - x_0)\\ &= (x_0^2 + y_0^2) + 2x_0(x - x_0)\\ T(x_0, y) &= f(x_0, y_0) + \frac{d}{dy} f(x_0, y)\Big|_{y_0} (y - y_0)\\ &= (x_0^2 + y_0^2) + 2y_0(y - y_0).\end{aligned}$$

We can also write these tangent line equations like this using our new notation for partial derivatives.

$$\begin{aligned}T(x, y_0) &= f(x_0, y_0) + \frac{\partial f}{\partial x}(x_0, y_0)\,(x - x_0)\\ &= (x_0^2 + y_0^2) + 2x_0(x - x_0)\\ T(x_0, y) &= f(x_0, y_0) + \frac{\partial f}{\partial y}(x_0, y_0)\,(y - y_0)\\ &= (x_0^2 + y_0^2) + 2y_0(y - y_0).\end{aligned}$$

We can draw these tangent lines in 3D. To draw $T(x, y_0)$, we fix the y value to be y_0 and then we draw the usual tangent line in the x–z plane. This is a copy of the x–z plane translated over to the value y_0; i.e. it is parallel to the x–z plane we see at the value $y = 0$. We can do the same thing for the tangent line $T(x, y_0)$; we fix the x value to be x_0 and then draw the tangent line in the copy of the y–z plane translated to the value x_0. We show this in Fig. 20.8. Note the $T(x, y_0)$ and the $T(x_0, y)$ lines are determined by vectors as shown below.

$$\boldsymbol{A} = \begin{bmatrix} 1 \\ 0 \\ \frac{d}{dx} f(x, y_0)\Big|_{x_0} \end{bmatrix} = \begin{bmatrix} 1 \\ 0 \\ 2x_0 \end{bmatrix} \text{ and } \boldsymbol{B} = \begin{bmatrix} 0 \\ 1 \\ \frac{d}{dy} f(x_0, y)\Big|_{y_0} \end{bmatrix} = \begin{bmatrix} 0 \\ 1 \\ 2y_0 \end{bmatrix}$$

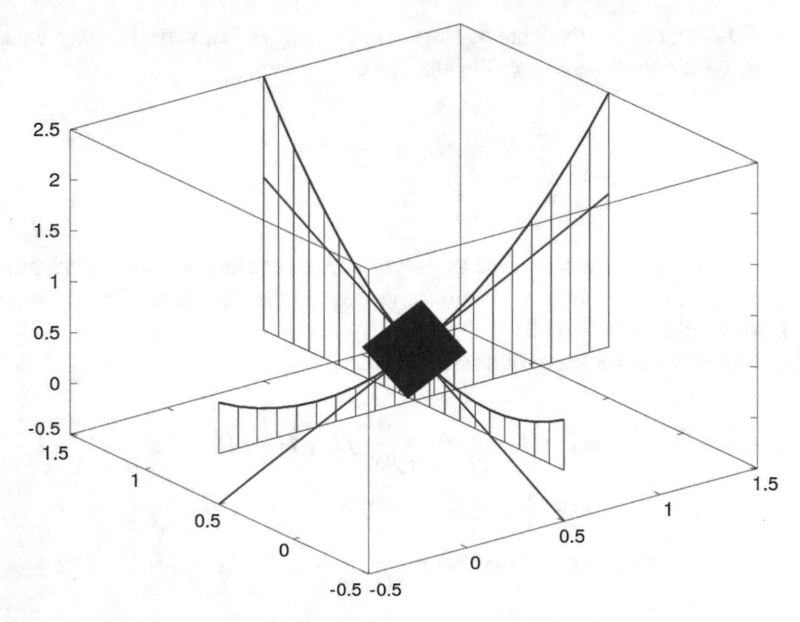

Fig. 20.7 The traces $f(x_0, y)$ and $f(x, y_0)$ for the surface $z = x^2 + y^2$ for $x_0 = 0.5$ and $y_0 = 0.5$ with added tangent lines. We have added the tangent plane determined by the tangent lines

Note that if we connect the lines determined by the vectors $\boldsymbol{A}$ and $\boldsymbol{B}$, we determine a *flat* sheet which you can interpret as a piece of paper laid on top of these two lines. Of course, we can only envision a small finite subset of this sheet of paper as you can see in Fig. 20.7. Imagine that the sheet extends infinitely in all directions! The sheet of paper we are plotting is called the **tangent plane** to our surface at the point (x_0, y_0). We will talk about this more formally later.

We use the function **DrawTangentLines** to draw this picture with tangent lines, traces and the tangent plane, This function has arguments **(f,fx,fy,delx, nx,dely,ny,r,x0,y0)**; note there are three new arguments here: **fx** which is $\partial f/\partial x$, **fy** which is $\partial f/\partial y$ and **r** which is the size of the tangent plane that is plotted. For the picture shown in Fig. 20.8, we've removed the tangent plane because the plot was getting pretty busy. We did this by commenting out the line that plots the tangent plane. It is easy for you to go into the code and add it back in if you want to play around. The MatLab command line is

Listing 20.11: Drawing Tangent Lines

```
fx = @(x,y) 2*x;
fy = @(x,y) 2*y;
%
DrawTangentLines(f,fx,fy,0.5,2,0.5,2,.3,0.5,0.5);
```

If you want to see the tangent plane as well as the tangent lines, all you have to do is look at the following lines in **DrawTangentLines.m**.

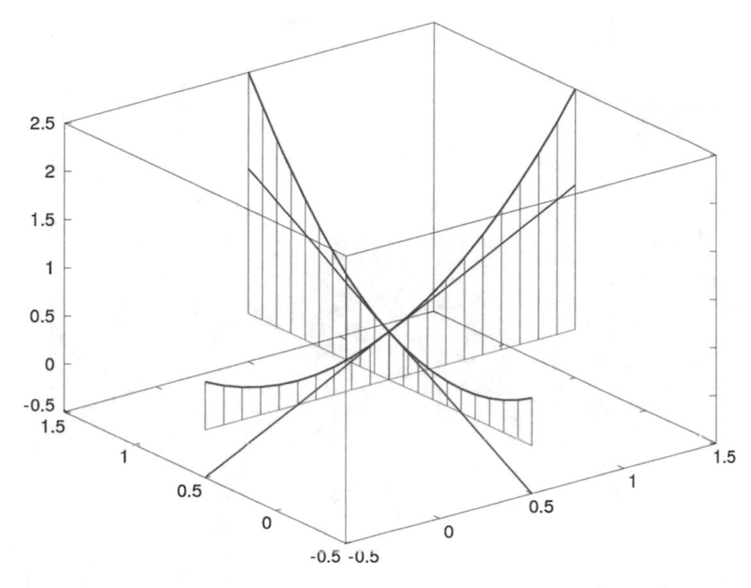

Fig. 20.8 The traces $f(x_0, y)$ and $f(x, y_0)$ for the surface $z = x^2 + y^2$ for $x_0 = 0.5$ and $y_0 = 0.5$ with added tangent lines

Listing 20.12: Drawing Tangent Lines

```
% set up a new local mesh grid near (x0,y0)
[U,V] = meshgrid(u,v)
% set up the tangent plane at (x0,y0)
W = f(x0,y0) + fx(x0,y0)*(U-x0) + fy(x0,y0)*(V-y0)
% plot the tangent plane
surf(U,V,W,'EdgeColor','blue');
```

These lines setup the tangent plane and the tangent plane is turned off if there is a % in front of **surf(U,V,W,'EdgeColor','blue');**. We edited the file to take the % out so we can see the tangent plane. We then see the plane in Fig. 20.9 as we saw before.

The ideas we have been discussing can be made more general. When we take the derivative with respect to one variable while holding the other variable constant (as we do when we find the normal derivative along a trace), we say we are taking a **partial derivative of f**. Here there are two flavors: the partial derivative with respect to x and the partial derivative with respect to y. We can now state some formal definitions and introduce the notations and symbols we use for these things. We define the process of partial differentiation carefully below.

Definition 20.3.1 (*Partial Derivatives*)
Let $z = f(x, y)$ be a function of the two independent variables x and y defined on some domain. At each pair (x, y) where f is defined in a circle of some finite radius r, $B_r(x_0, y_0) = \{(x, y) \mid \sqrt{(x - x_0)^2 + (y - y_0)^2} < r\}$, it makes sense to try to find the limits

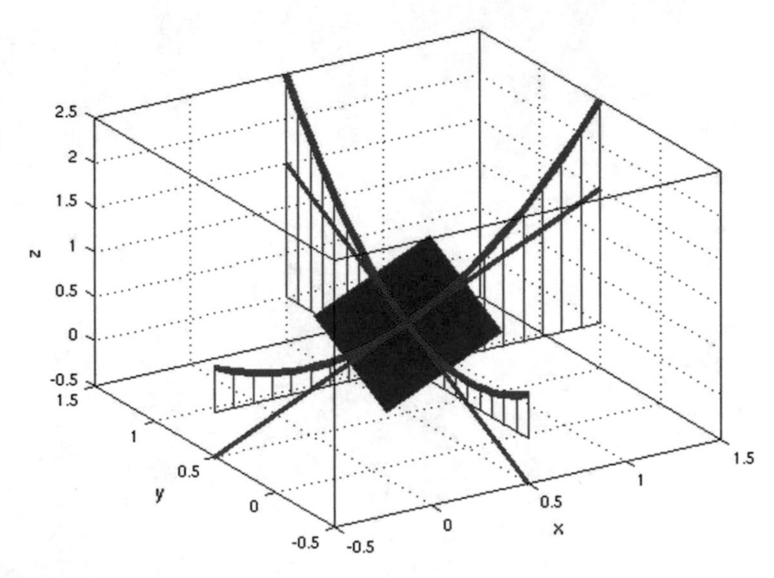

Fig. 20.9 The traces $f(x_0, y)$ and $f(x, y_0)$ for the surface $z = x^2 + y^2$ for $x_0 = 0.5$ and $y_0 = 0.5$ with added tangent lines

$$\lim_{x \rightarrow x_0, y = y_0} \frac{f(x, y) - f(x, y_0)}{x - x_0}$$
$$\lim_{x = x_0, y \rightarrow y_0} \frac{f(x, y) - f(x_0, y)}{y - y_0}$$

If these limits exists, they are called the partial derivatives of f with respect to x and y at (x_0, y_0), respectively.

Comment 20.3.1 *For these partial derivatives, we use the symbols*

$$f_x(x_0, y_0), \ \frac{\partial f}{\partial x}(x_0, y_0), \ z_x(x_0, y_0), \ \frac{\partial z}{\partial x}(x_0, y_0)$$

and

$$f_y(x_0, y_0), \ \frac{\partial f}{\partial y}(x_0, y_0), \ z_y(x_0, y_0), \ \frac{\partial z}{\partial y}(x_0, y_0)$$

Comment 20.3.2 *We often use another notation for partial derivatives. The function f of two variables x and y can be thought of as having two arguments or slots into which we place values. So another useful notation is to let the symbol $D_1 f$ be f_x and $D_2 f$ be f_y. We will be using this notation later when we talk about the* **chain rule**.

Comment 20.3.3 *It is easy to take partial derivatives. Just imagine the one variable held constant and take the derivative of the resulting function just like you did in your earlier calculus courses.*

Example 20.3.1 Let $z = f(x, y) = x^2 + 4y^2$ be a function of two variables. Find $\frac{\partial z}{\partial x}$ and $\frac{\partial z}{\partial y}$.

Solution *Thinking of* y *as a constant, we take the derivative in the usual way with respect to* x. *This gives*

$$\frac{\partial z}{\partial x} = 2x$$

as the derivative of $4y^2$ *with respect to* x *is* 0. *So, we know* $f_x = 2x$.

In a similar way, we find $\frac{\partial z}{\partial y}$. *We see*

$$\frac{\partial z}{\partial y} = 8y$$

as the derivative of x^2 *with respect to* y *is* 0. *So* $f_y = 8y$.

Example 20.3.2 Let $z = f(x, y) = 4x^2y^3$. Find $\frac{\partial z}{\partial x}$ and $\frac{\partial z}{\partial y}$.

Solution *Thinking of* y *as a constant, take the derivative in the usual way with respect to* x: *This gives*

$$\frac{\partial z}{\partial x} = 8xy^3$$

as the term $4y^3$ *is considered a "constant" here. So* $f_x = 8xy^3$.

Similarly,

$$\frac{\partial z}{\partial y} = 12x^2y^2$$

as the term $4x^2$ *is considered a "constant" here. So* $f_y = 12x^2y^2$.

Now let's do some without spelling out each step. Make sure you see what we are doing!

Example 20.3.3 $f(x, y) = \frac{x^4+1}{y^3+2}$.

Solution

$$\frac{\partial f}{\partial x} = \frac{(4x^3)}{y^3 + 2}$$

$$\frac{\partial f}{\partial y} = -\frac{(x^4 + 1)(3y^2)}{(y^3 + 2)^2}$$

Example 20.3.4 $f(x, y) = \frac{x^4y^2+2}{y^3x^5+20}$.

Solution

$$\frac{\partial f}{\partial x} = \frac{(4x^3y^2)(y^3x^5+20) - (x^4y^2+2)(5y^3x^4)}{(y^3x^5+20)^2}$$
$$\frac{\partial f}{\partial y} = \frac{(2x^4y)(y^3x^5+20) - (x^4y^2+2)(3y^2x^5)}{(y^3x^5+20)^2}$$

Example 20.3.5 $f(x, y) = \sin(x^3y+2)$.

Solution

$$\frac{\partial f}{\partial x} = \cos(x^3y+2)\,(3x^2y)$$
$$\frac{\partial f}{\partial y} = \cos(x^3y+2)\,(x^3)$$

Example 20.3.6 $f(x, y) = e^{-(x^2+y^4)}$

Solution

$$\frac{\partial f}{\partial x} = e^{-(x^2+y^4)}\,(-2x)$$
$$\frac{\partial f}{\partial y} = e^{-(x^2+y^4)}\,(-4y^3)$$

Example 20.3.7 $f(x, y) = \ln(\sqrt{x^2+2y^2})$.

Solution

$$\frac{\partial f}{\partial x} = \frac{1}{2}\,\frac{2x}{x^2+2y^2}$$
$$\frac{\partial f}{\partial y} = \frac{1}{2}\,\frac{4y}{x^2+2y^2}$$

20.3.1 Homework

These are for you: for each of these functions, find f_x and f_y.
First, functions with no cross terms.

Exercise 20.3.1 $f(x, y) = x^2 + 3y^2$.

Exercise 20.3.2 $f(x, y) = 4x^2 + 5y^4$.

Exercise 20.3.3 $f(x, y) = -3x + 2y^8$.

Next, functions with cross terms.

Exercise 20.3.4 $f(x, y) = x^2\, y^2$.

Exercise 20.3.5 $f(x, y) = 2x^3\, y^2 + 5x$.

Exercise 20.3.6 $f(x, y) = 3x\, y^2$.

Exercise 20.3.7 $f(x, y) = x^2\, y^5$.

Next, functions with fractions.

Exercise 20.3.8 $f(x, y) = \frac{x^2}{y^5}$.

Exercise 20.3.9 $f(x, y) = \frac{x^2+2y}{5x+y^3}$.

Exercise 20.3.10 $f(x, y) = \frac{x+2y}{5x+y}$.

Exercise 20.3.11 $f(x, y) = \frac{x^2+2}{5+y}$.

Now, sin and cos things.

Exercise 20.3.12 $f(x, y) = \sin(xy)$.

Exercise 20.3.13 $f(x, y) = \sin(x^2\, y)$.

Exercise 20.3.14 $f(x, y) = \cos(x + 3y)$.

Exercise 20.3.15 $f(x, y) = \sin(\sqrt{x + y})$.

Exercise 20.3.16 $f(x, y) = \cos^2(x + 4y)$.

Exercise 20.3.17 $f(x, y) = \sqrt{\sin(xy)}$.

Now, let's add ln and exp.

Exercise 20.3.18 $f(x, y) = e^{xy}$.

Exercise 20.3.19 $f(x, y) = e^{x+4y}$.

Exercise 20.3.20 $f(x, y) = e^{-3xy}$.

Exercise 20.3.21 $f(x, y) = \ln(x^2 + 4y^2)$.

Exercise 20.3.22 $f(x, y) = \ln(\sqrt{1 + xy})$.

Exercise 20.3.23 $f(x, y) = e^{\sin(3x+5y)}$.

Exercise 20.3.24 $f(x, y) = e^{\cos(3x^2+5y)}$.

Exercise 20.3.25 $f(x, y) = \ln(1 + 3x + 8y)$.

20.4 Tangent Planes

Before we discuss tangent planes to a function f again, let's digress to the ideas of planes in general in 3D. We define a plane as follows.

Definition 20.4.1 (*Planes*)
A plane in 3D through the point (x_0, y_0, z_0) is defined as the set of all points (x, y, z) so that the angle between the vectors $\boldsymbol{D}$ and $\boldsymbol{N}$ is zero where $\boldsymbol{D}$ is the vector we get by connecting the point (x_0, y_0, z_0) to the point (x, y, z). Hence, for

$$\boldsymbol{D} = \begin{bmatrix} x - x_0 \\ y - y_0 \\ z - z_0 \end{bmatrix} \text{ and } \boldsymbol{N} = \begin{bmatrix} N_1 \\ N_2 \\ N_3 \end{bmatrix}$$

the plane is the set of points (x, y, z) so that $< \boldsymbol{D}, \boldsymbol{N} >= 0$. The vector $\boldsymbol{N}$ is called the **normal vector** to the plane.

Example 20.4.1 The equation $2x + 3y - 5z = 0$ defines the plane whose normal vector is $\boldsymbol{N} = [2, 3, -5]^T$ which passes through the origin $(0, 0, 0)$.

Example 20.4.2 The equation $2(x - 2) + 3(y - 1) - 5(z + 3) = 0$ defines the plane whose normal vector is $\boldsymbol{N} = [2, 3, -5]^T$ which passes through the point $(2, 1, -3)$. Note this can be rewritten as $2x + 3y - 5z = 4 + 3 + 15 = 22$ after a simple manipulation.

Example 20.4.3 The equation $2x + 3y - 5z = 11$ corresponds to a plane with normal vector $\boldsymbol{N} = [2, 3, -5]^T$ which passes through some point (x_0, y_0, z_0). There are an infinite number of choices for the this base point: any triple which solves $2x_0 + 3y_0 - 5z_0 = 11$ will do the job. An easy way to pick one is to pick two and solve for the third. So for example, if $z_0 = 0$ and $y_0 = 4$, we find $2x_0 + 12 = 11$ which gives $x_0 = -1/2$. Thus, this plane could be rewritten as $2(x + 1/2) + 3(y - 4) - 5z = 0$.

20.4.1 The Tangent Plane to a Surface

Recall the tangent plane to a surface $z = f(x, y)$ at the point (x_0, y_0) was the plane determined by the tangent lines $T(x, y_0)$ and $T(x_0, y)$. The $T(x, y_0)$ line was determined by the vector

$$\boldsymbol{A} = \begin{bmatrix} 1 \\ 0 \\ \frac{d}{dx} f(x, y_0)\Big|_{x_0} \end{bmatrix} = \begin{bmatrix} 1 \\ 0 \\ 2x_0 \end{bmatrix}$$

and the $T(x_0, y)$ line was determined by the vector

$$\boldsymbol{B} = \begin{bmatrix} 0 \\ 1 \\ \frac{d}{dy} f(x_0, y)\Big|_{y_0} \end{bmatrix} = \begin{bmatrix} 0 \\ 1 \\ 2y_0 \end{bmatrix}$$

We know now that we can write these vectors more generally as

$$\boldsymbol{A} = \begin{bmatrix} 1 \\ 0 \\ \frac{\partial f}{\partial x}(x_0, y_0) \end{bmatrix}$$

$$\boldsymbol{B} = \begin{bmatrix} 0 \\ 1 \\ \frac{\partial f}{\partial y}(x_0, y_0) \end{bmatrix}$$

We need to find a vector perpendicular to both $\boldsymbol{A}$ and $\boldsymbol{B}$. Let's try this one: $\boldsymbol{N} = \left[-f_x(x_0, y_0),\ -f_y(x_0, y_0),\ 1\right]^T$. The dot product of $\boldsymbol{A}$ with $\boldsymbol{N}$ is

$$< \boldsymbol{A},\ \boldsymbol{N} > = 1\,(-f_x(x_0, y_0)) + 0\,(-f_y(x_0, y_0)) + (f_x(x_0, y_0))\,1 = 0.$$

and the dot product of $\boldsymbol{B}$ with $\boldsymbol{N}$ is

$$< \boldsymbol{B},\ \boldsymbol{N} > = 0\,(-f_x(x_0, y_0)) + 1\,(-f_y(x_0, y_0)) + (f_y(x_0, y_0))\,1 = 0.$$

So our $\boldsymbol{N}$ is perpendicular to both of these vectors and so we know the tangent plane to the surface $z = f(x, y)$ at the point (x_0, y_0) is then given by

$$-f_x(x_0, y_0)(x - x_0) - f_y(x_0, y_0)(y - y_0) + (z - f(x_0, y_0)) = 0.$$

This then gives the traditional equation of the tangent plane:

$$z = f(x_0, y_0) + f_x(x_0, y_0)(x - x_0) + f_y(x_0, y_0)(y - y_0). \tag{20.1}$$

We can use another compact definition at this point. We can define the **gradient** of the function f to be the vector $\boldsymbol{\nabla} f$. The gradient is defined as follows.

Definition 20.4.2 (*The Gradient*)
The gradient of the scalar function $z = f(x, y)$ is defined to be the vector $\boldsymbol{\nabla} f$ where

$$\boldsymbol{\nabla} f(x_0, y_0) = \begin{bmatrix} f_x(x_0, y_0) \\ f_y(x_0, y_0) \end{bmatrix}.$$

Note the gradient takes a scalar function argument and returns a vector answer. The word *scalar* just means the function returns a number and not a vector.

Using the gradient, Eq. 20.1 can be rewritten as

$$\begin{aligned} f(x, y) &= f(x_0, y_0) + < \nabla f, \boldsymbol{X} - \boldsymbol{X_0} > \\ &= f(x_0, y_0) + \nabla f^T (\boldsymbol{X} - \boldsymbol{X_0}) \end{aligned}$$

where $\boldsymbol{X} - \boldsymbol{X_0} = \begin{bmatrix} x - x_0 & y - y_0 \end{bmatrix}^T$. The obvious question to ask now is how much of a discrepancy is there between the value $f(x, y)$ and the value of the tangent plane?

20.4.2 Examples

Example 20.4.4 Find the gradient of $f(x, y) = x^2 + 4xy + 9y^2$ and the equation of the tangent plane to this surface at the point $(1, 2)$.

Solution

$$\nabla f(x, y) = \begin{bmatrix} 2x + 4y \\ 4x + 18y \end{bmatrix}.$$

The equation of the tangent plane at $(1, 2)$ *is then*

$$\begin{aligned} z &= f(1, 2) + f_x(1, 2)(x - 1) + f_y(1, 2)(y - 2) \\ &= 45 + 10(x - 1) + 40(y - 2) \\ &= -45 + 10x + 40y. \end{aligned}$$

Note this can also be written as $10x + 40y + z = 45$ *which is also a standard form. However, in this form, the attachment point* $(1, 2, 45)$ *is hidden from view.*

Example 20.4.5 Find the gradient of $f(x, y) = 3x^2 + 2y^2$ and the equation of the tangent plane to this surface at the point $(2, 3)$.

Solution

$$\nabla f(x, y) = \begin{bmatrix} 6x \\ 4y \end{bmatrix}.$$

The equation of the tangent plane at $(2, 3)$ *is then*

$$\begin{aligned} z &= f(2,3) + f_x(2,3)(x-2) + f_y(2,3)(y-2) \\ &= 30 + 12(x-2) + 12(y-3) \\ &= -30 + 12x + 12y. \end{aligned}$$

Note this can also be written as $12x + 12y + z = 35$ *which is also a standard form. However, in this form, the attachment point* $(2, 3, 30)$ *is hidden from view.*

20.4.3 Homework

Exercise 20.4.1 *Find the gradient of* $f(x, y) = 4x^2 + 6y^2$ *and the equation of the tangent plane to this surface at the point* $(1, 1)$.

Exercise 20.4.2 *Find the gradient of* $f(x, y) = 2x^2 + 10y^3$ *and the equation of the tangent plane to this surface at the point* $(1, 2)$.

Exercise 20.4.3 *Find the gradient of* $f(x, y) = x^2 - xy^2 + y^2$ *and the equation of the tangent plane to this surface at the point* $(2, 1)$.

Exercise 20.4.4 *Find the gradient of* $f(x, y) = -x^2 + y^2 + 4y$ *and the equation of the tangent plane to this surface at the point* $(-1, 1)$.

Exercise 20.4.5 *Find the gradient of* $f(x, y) = \sin(xy)$ *and the equation of the tangent plane to this surface at the point* $(\pi/4, -\pi/4)$.

20.4.4 Computational Results

We can use MatLab/Octave to draw tangent planes and tangent lines to a surface. Consider the function **`DrawTangentPlanePackage`**. The source code is similar to what we have done in previous functions. This time, we send in the function **`f`** and the two partial derivatives **`fx`** and **`fy`**. First, we plot the traces and draw vertical lines from the traces to the x–y plane. Note this code will not do very well on surfaces where the z values become negative! But then, this code is just for exploration and it is easy enough to alter it for other jobs. And it is a good exercise! After the traces and their *shadow lines* are drawn, we draw the tangent lines. Finally, we draw the tangent plane. The tangent plane calculation uses the partial derivatives we sent into this function as arguments.

Listing 20.13: DrawTangentPlanePackage

```
function DrawTangentPlanePackage(f,fx,fy,delx,nx,dely,ny,r,x0,y0)
% f is the function defining the surface
% delx is the size of the x step
% nx is the number of steps left and right from x0
% dely is the size of the y step
% ny is the number of steps left and right from y0
% r is the size of the drawn tangent plane
% (x0,y0) is the location of the column rectangle base
%
% set up x and y stuff
x = x0-nx*delx:delx/5:x0+nx*delx;
y = y0-ny*dely:dely/5:y0+ny*dely;
[rows,sx] = size(x);
[rows,sy] = size(y);
hold on
  % set up x trace for x = x0
  % set up y trace for y = y0
  xtrace = f(x0,y);
  ytrace = f(x,y0);
  fixedx = x0*ones(1,sx);
  fixedy = y0*ones(1,sy);
  plot3(fixedx,y,xtrace,'LineWidth',4,'Color','red');
  plot3(x,fixedy,ytrace,'LineWidth',4,'Color','red');
  % now draw x0, y0 line in xy plane
  U = [x0;x0];
  V = [y0-ny*dely;y0+ny*delx];
  W = [0;0];
  plot3(U,V,W);
  U = [y0-ny*dely;y0+ny*delx];
  V = [y0;y0];
  W = [0;0];
  plot3(U,V,W);
  % now fill in planes formed by x0, y0 lines
  for i=1:sy
    U = [x0;x0];
    V = [y(i);y(i)];
    W = [0;f(x0,y(i))];
    plot3(U,V,W,'LineWidth',1,'Color','red');
  end
  for i=1:sx
    U = [x(i);x(i)];
    V = [y0;y0];
    W = [0;f(x(i),y0)];
    plot3(U,V,W,'LineWidth',1,'Color','red');
  end
  % now draw tangent lines
  % set up new local variables centered at (x0,y0)
  TX = @(x) (f(x0,y0) + fx(x0,y0)*(x-x0));
  TY = @(y) (f(x0,y0) + fy(x0,y0)*(y-y0));
  U = [x0-nx*delx;x0+nx*delx];
  V = [y0;y0];
  W = [TX(x0-nx*delx);TX(x0+nx*delx)];
  plot3(U,V,W,'LineWidth',3,'Color','blue');
  U = [x0;x0];
  V = [y0-ny*dely;y0+ny*dely];
  W = [TY(y0-ny*dely);TY(y0+ny*dely)];
  plot3(U,V,W,'LineWidth',3,'Color','blue');
  % plot tangent plane
  u = [x0-r*delx;x0+r*delx];
  v = [y0-r*dely;y0+r*dely];
  % set up a new local mesh grid near (x0,y0)
  [U,V] = meshgrid(u,v);
  % set up the tangent plane at (x0,y0)
  w = @(u,v) f(x0,y0) + fx(x0,y0)*(u-x0) + fy(x0,y0)*(v-y0);
  W = w(U,V);
  % plot the tangent plane
  surf(U,V,W,'EdgeColor','blue');
hold off
end
```

The illustrations this code produces have already been used in Fig. 20.7. Practice with this code and draw other pictures! A typical session to generate this figure would look like

Listing 20.14: Drawing Tangent Planes

```
1 f = @(x,y) x.^2+y.^2;
  fx = @(x,y) 2*x;
  fy = @(x,y) 2*y;
  DrawTangentPlanePackage(f,fx,fy,0.5,2,0.5,2,.3,0.5,0.5);
```

20.4.5 Homework

Exercise 20.4.6 *Draw tangent lines and planes for the surface* $f(x, y) = x^2 + 3y^2$ *for various points* (x_0, y_0).

Exercise 20.4.7 *Draw tangent lines and planes for the surface* $f(x, y) = 2x^2 + 5y^2$ *for various points* (x_0, y_0). *You will need to modify the code to make this work!*

Exercise 20.4.8 *Draw tangent lines and planes for the surface* $f(x, y) = 3x^2 + 2y^2$ *for various points* (x_0, y_0). *You will need to modify the code to make this work! Make sure you try the point* $(0, 0)$.

Exercise 20.4.9 *Draw tangent lines and planes for the surface* $f(x, y) = 8x^4 + y^3$ *for various points* (x_0, y_0). *You will need to modify the code to make this work! Make sure you try the point* $(0, 0)$.

Exercise 20.4.10 *Draw tangent lines and planes for the surface* $f(x, y) = x^2 + 4y^2$ *for various points* (x_0, y_0). *You will need to modify the code to make this work! Make sure you try the point* $(0, 0)$.

20.5 Derivatives in Two Dimensions!

Let's go back to one dimensional calculus. If the function f is defined locally near x_0 that means that f is defined in a circle $B_r(x_0) = \{x : x_0 - r < x < x_0 + r\}$ for some positive value of r. In this case, we can attempt to find the usual limit as x approaches x_0 that defines the derivative of f and x_0: if this limit exists, it is called $f'(x_0)$ and

$$f'(x_0) = \lim_{x \to x_0} \frac{f(x) - f(x_0)}{x - x_0}.$$

This can be expressed in a different form. Recall that we can also use the $\epsilon - \delta$ notation to define a limit (yeah, ugly but necessary sometimes!). In this case, it means that if we choose a positive ϵ, then there is a positive δ so that

$$|x - x_0| < \delta \Longrightarrow \left| \frac{f(x) - f(x_0)}{x - x_0} - f'(x_0) \right| < \epsilon.$$

Now define the error between the function value $f(x)$ and the tangent line value $f(x_0) + f'(x_0)(x - x_0)$ to be $E(x, x_0)$. The above statement can be rewritten as

$$|x - x_0| < \delta \Longrightarrow \left| \frac{f(x) - f(x_0) - f'(x_0)(x - x_0)}{x - x_0} \right| < \epsilon.$$

Then using the definition of error, $E(x, x_0)$, we see

$$|x - x_0| < \delta \Longrightarrow \left| \frac{E(x, x_0)}{x - x_0} \right| < \epsilon.$$

This is the same as saying

$$\lim_{x \to x_0} \frac{E(x, x_0)}{x - x_0} = 0.$$

Now rewrite the inequality again to have

$$|x - x_0| < \delta \Longrightarrow \left| E(x, x_0) \right| < \epsilon |x - x_0|.$$

Since we can do this for any positive ϵ, it works for the choice $\sqrt{\epsilon}$. Hence, there is a positive δ_1 so that

$$|x - x_0| < \delta_1 \Longrightarrow \left| E(x, x_0) \right| < \sqrt{\epsilon} |x - x_0| < \sqrt{\epsilon}\, \delta_1.$$

But this work as long as $|x - x_0| < \delta_1$. So it also works if $|x - x_0| < \delta_2 = \min(\delta_1, \sqrt{\epsilon}) \leq \delta_1$! So

$$\begin{aligned} |x - x_0| < \delta_2 \Longrightarrow \left| E(x, x_0) \right| &< \sqrt{\epsilon} |x - x_0| \\ &< \sqrt{\epsilon}\, \delta_2 < \sqrt{\epsilon}\, \sqrt{\epsilon} < \epsilon. \end{aligned}$$

So we can say $\lim_{x \to x_0} E(x, x_0) = 0$ as well. This leads to the following theorem which we have already seen in the one variable part of these notes. We are stating it a bit differently here.

Theorem 20.5.1 (Error Form of Differentiability For One Variable)
If f is defined locally at x_0, then f is differentiable at x_0 if the error function $E(x, x_0) = f(x) - f(x_0) - f'(x_0)(x - x_0)$ satisfies $\lim_{x \to x_0} E(x, x_0) = 0$ and $\lim_{x \to x_0} E(x, x_0)/(x - x_0) = 0$. Conversely, if there is a number L so that the error

function $E(x, x_0) = f(x) - f(x_0) - L(x - x_0)$ satisfies the same behavior, then f is differentiable at x_0 with value $f'(x_0) = L$.

Proof If f is differentiable at x_0, we have already outlined the argument. The converse argument is quite similar. Since we know $\lim_{x \to x_0} E(x, x_0)/(x - x_0) = 0$, this tells us

$$\lim_{x \to x_0} \frac{f(x) - f(x_0) - L(x - x_0)}{x - x_0} = 0$$

or

$$\lim_{x \to x_0} \frac{\frac{f(x) - f(x_0)}{x - x_0} - L}{x - x_0} = 0.$$

But this states that f is differentiable at x_0 with value L. With this argument done, we have shown both sides of the statement are true. ■

Note if f is differentiable at x_0, f must be continuous at x_0. This follows because

$$f(x) = f(x_0) + f'(x_0)(x - x_0) + E(x, x_0)$$

and as $x \to x_0$, we have $f(x) \to f(x_0)$ which is the definition of f being continuous at x_0. Hence, we can say

Theorem 20.5.2 (Differentiable Implies Continuous: One Variable)
If f is differentiable at x_0 then f is continuous at x_0.

Proof We have sketched the argument already. ■

We can apply this idea to the partial derivatives of $f(x, y)$. As long as $f(x, y)$ is defined locally at (x_0, y_0), we can say $f_x(x_0, y_0)$ and $f_y(x_0, y_0)$ exist if and only if there are error functions $E_1(x, y, x_0, y_0)$ and $E_2(x, y, x_0, y_0)$ so that

$$f(x, y_0) = f(x_0, y_0) + f_x(x_0, y_0)(x - x_0) + E_1(x, x_0, y_0)$$
$$f(x_0, y) = f(x_0, y_0) + f_y(x_0, y_0)(y - y_0) + E_2(y, x_0, y_0)$$

with $E_1 \to 0$ and $E_1/(x - x_0) \to 0$ as $x \to x_0$ and $E_2 \to 0$ and $E_2/(y - x_0) \to 0$ as $y \to y_0$. Using the ideas we have presented here, we can come up with a way to define the differentiability of a function of two variables.

Definition 20.5.1 (*Error Form of Differentiability For Two Variables*)
If $f(x, y)$ is defined locally at (x_0, y_0), then f is differentiable at (x_0, y_0) if there are two numbers L_1 and L_2 so that the error function $E(x, y, x_0, y_0) = f(x, y) - f(x_0, y_0) - L_1(x - x_0) - L_2(y - y_0)$ satisfies $\lim_{(x,y) \to (x_0, y_0)} E(x, y, x_0, y_0) = 0$ and $\lim_{(x,y) \to (x_0, y_0)} E(x, y, x_0, y_0)/||(x - x_0, y - y_0)|| = 0$. Here, the term $||(x - x_0, y - y_0)|| = \sqrt{(x - x_0)^2 + (y - y_0)^2}$.

Note if f is differentiable at (x_0, y_0), f must be continuous at (x_0, y_0). The argument is simple:

$$f(x, y) = f(x_0, y_0) + L_1\,(x_0, y_0)(x - x_0) + L_2\,(y - y_0) + E(x, y, x_0, y_0)$$

and as $(x, y) \to (x_0, y_0)$, we have $f(x, y) \to f(x_0, y_0)$ which is the definition of f being continuous at (x_0, y_0). Hence, we can say

Theorem 20.5.3 (Differentiable Implies Continuous: Two Variables)
If f is differentiable at (x_0, y_0) then f is continuous at (x_0, y_0).

Proof We have sketched the argument already. ■

From Definition 20.5.1, we can show if f is differentiable at the point (x_0, y_0), then $L_1 = f_x(x_0, y_0)$ and $L_2 = f_y(x_0, y_0)$. The argument goes like this: since f is differentiable at (x_0, y_0), we can say

$$\lim_{(x,y)\to(x_0,y_0)} \frac{f(x, y) - f(x_0, y_0) - L_1(x - x_0) - L_2(y - y_0)}{\sqrt{(x - x_0)^2 + (y - y_0)^2}} = 0.$$

We can rewrite this using $\Delta x = x - x_0$ and $\Delta y = y - y_0$ as

$$\lim_{(\Delta x,\Delta y)\to(0,0)} \frac{f(x_0 + \Delta x, y_0 + \Delta y) - f(x_0, y_0) - L_1\Delta x - L_2\Delta y}{\sqrt{(\Delta x)^2 + (\Delta y)^2}} = 0.$$

In particular, for $\Delta y = 0$, we find

$$\lim_{(\Delta x)\to 0} \frac{f(x_0 + \Delta x, y_0) - f(x_0, y_0) - L_1\Delta x}{\sqrt{(\Delta x)^2}} = 0.$$

For $\Delta x > 0$, we find $\sqrt{(\Delta x)^2} = \Delta x$ and so

$$\lim_{\Delta x\to 0^+} \frac{f(x_0 + \Delta x, y_0) - f(x_0, y_0)}{\Delta x} = L_1.$$

Thus, the right hand partial derivative $f_x(x_0, y_0)^+$ exists and equals L_1. On the other hand, if $\Delta x < 0$, then $\sqrt{(\Delta x)^2} = -\Delta x$ and we find, with a little manipulation, that we still have

$$\lim_{(\Delta x)\to 0^-} \frac{f(x_0 + \Delta x, y_0) - f(x_0, y_0)}{\Delta x} = L_1.$$

So the left hand partial derivative $f_x(x_0, y_0)^-$ exists and equals L_1 also. Combining, we see $f_x(x_0, y_0) = L_1$. A similar argument shows that $f_y(x_0, y_0) = L_2$. Hence, we can say if f is differentiable at (x_0, y_0) then f_x and f_y exist at this point and we have

$$f(x, y) = f(x_0, y_0) + f_x(x_0, y_0)(x - x_0) + f_y(x_0, y_0)(y - y_0) + E_f(x, y, x_0, y_0)$$

where $E_f(x, y, x_0, y_0) \to 0$ and $E_f(x, y, x_0, y_0)/||(x - x_0, y - y_0)|| \to 0$ as $(x, y) \to (x_0, y_0)$. Note this argument is a pointwise argument. It only tells us that differentiability at a point implies the existence of the partial derivatives at that point.

20.6 The Chain Rule

Now that we know a bit about two dimensional derivatives, let's go for gold and figure out the new version of the chain rule. The argument we make here is very similar in spirit to the one dimensional one. You should go back and check it out! We will do this argument carefully but without tedious rigor. At least that is our hope. You'll have to let us know how we did!

We assume there are two functions $u(x, y)$ and $v(x, y)$ defined locally about (x_0, y_0) and that there is a third function $f(u, v)$ which is defined locally around $(u_0 = u(x_0, y_0), v_0 = v(x_0, y_0))$. Now assume $f(u, v)$ is differentiable at (u_0, v_0) and $u(x, y)$ and $v(x, y)$ are differentiable at (x_0, y_0). Then we can say

$$\begin{aligned}
u(x, y) &= u(x_0, y_0) + u_x(x_0, y_0)(x - x_0) + u_y(x_0, y_0)(y - y_0) + E_u(x, y, x_0, y_0) \\
v(x, y) &= v(x_0, y_0) + v_x(x_0, y_0)(x - x_0) + v_y(x_0, y_0)(y - y_0) + E_v(x, y, x_0, y_0) \\
f(u, v) &= f(u_0, v_0) + f_u(u_0, v_0)(u - u_0) + f_v(u_0, v_0)(v - v_0) + E_f(u, v, u_0, v_0)
\end{aligned}$$

where all the error terms behave as usual as $(x, y) \to (x_0, y_0)$ and $(u, v) \to (u_0, v_0)$. Note that as $(x, y) \to (x_0, y_0)$, $u(x, y) \to u_0 = u(x_0, y_0)$ and $v(x, y) \to v_0 = v(x_0, y_0)$ as u and v are continuous at the (u_0, v_0) since they are differentiable there. Let's consider the partial of f with respect to x. Let $\Delta u = u(x_0 + \Delta x, y_0) - u(x_0, y_0)$ and $\Delta v = v(x_0 + \Delta x, y_0) - v(x_0, y_0)$. Thus, $u_0 + \Delta u = u(x_0 + \Delta x, y_0)$ and $v_0 + \Delta v = v(x_0 + \Delta x, y_0)$. Hence

$$\begin{aligned}
\frac{f(u_0 + \Delta u, v_0 + \Delta v) - f(u_0, v_0)}{\Delta x} &= \frac{f_u(u_0, v_0)(u - u_0) + f_v(u_0, v_0)(v - v_0) + E_f(u, v, u_0, v_0)}{\Delta x} \\
&= f_u(u_0, v_0)\,\frac{u - u_0}{\Delta x} + f_v(u_0, v_0)\,\frac{v - v_0}{\Delta x} + \frac{E_f(u, v, u_0, v_0)}{\Delta x} \\
&= f_u(u_0, v_0)\,\frac{u_x(x_0, y_0)(x - x_0) + E_u(x, x_0, y_0)}{\Delta x} \\
&\quad + f_v(u_0, v_0)\,\frac{v_x(x_0, y_0)(x - x_0) + E_v(x, x_0, y_0)}{\Delta x} \\
&\quad + \frac{E_f(u, v, u_0, v_0)}{\Delta x} \\
&= f_u(u_0, v_0)\, u_x(x_0, y_0) + f_v(u_0, v_0)\, v_x(x_0, y_0) \\
&\quad + \frac{E_u(x, x_0, y_0)}{\Delta x} + \frac{E_v(x, x_0, y_0)}{\Delta x} + \frac{E_f(u, v, u_0, v_0)}{\Delta x}.
\end{aligned}$$

As $(x, y) \to (x_0, y_0)$, $(u, v) \to (u_0, v_0)$ and so $E_f(u, v, u_0, v_0)/\Delta x \to 0$. The other two error terms go to zero also as $(x, y) \to (x_0, y_0)$. Hence, we conclude

$$\frac{\partial f}{\partial x} = \frac{\partial f}{\partial u}\frac{\partial u}{\partial x} + \frac{\partial f}{\partial v}\frac{\partial v}{\partial x}.$$

A similar argument shows

$$\frac{\partial f}{\partial y} = \frac{\partial f}{\partial u}\frac{\partial u}{\partial y} + \frac{\partial f}{\partial v}\frac{\partial v}{\partial y}.$$

This result is known as the **Chain Rule**.

Theorem 20.6.1 (The Chain Rule)
Assume there are two functions $u(x, y)$ and $v(x, y)$ defined locally about (x_0, y_0) and that there is a third function $f(u, v)$ which is defined locally around $(u_0 = u(x_0, y_0), v_0 = v(x_0, y_0))$. Further assume $f(u, v)$ is differentiable at (u_0, v_0) and $u(x, y)$ and $v(x, y)$ are differentiable at (x_0, y_0). Then f_x and f_y exist at (x_0, y_0) and are given by

$$\begin{aligned}\frac{\partial f}{\partial x} &= \frac{\partial f}{\partial u}\frac{\partial u}{\partial x} + \frac{\partial f}{\partial v}\frac{\partial v}{\partial x}\\ \frac{\partial f}{\partial y} &= \frac{\partial f}{\partial u}\frac{\partial u}{\partial y} + \frac{\partial f}{\partial v}\frac{\partial v}{\partial y}.\end{aligned}$$

Proof We have sketched the argument already. ■

20.6.1 Examples

Example 20.6.1 Let $f(x, y) = x^2 + 2x + 5y^4$. Then if $x = r\cos(\theta)$ and $y = r\sin(\theta)$, using the chain rule, we find

$$\begin{aligned}\frac{\partial f}{\partial r} &= \frac{\partial f}{\partial x}\frac{\partial x}{\partial r} + \frac{\partial f}{\partial y}\frac{\partial y}{\partial r}\\ \frac{\partial f}{\partial \theta} &= \frac{\partial f}{\partial x}\frac{\partial x}{\partial \theta} + \frac{\partial f}{\partial y}\frac{\partial y}{\partial \theta}\end{aligned}$$

This becomes

$$\begin{aligned}\frac{\partial f}{\partial r} &= \left(2x + 2\right)\cos(\theta) + \left(20y^3\right)\sin(\theta)\\ \frac{\partial f}{\partial \theta} &= \left(2x + 2\right)\left(-r\sin(\theta)\right) + \left(20y^3\right)\left(r\cos(\theta)\right)\end{aligned}$$

You can then substitute in for x and y to get the final answer in terms of r and θ (kind of ugly though!)

Example 20.6.2 Let $f(x, y) = 10x^2y^4$. Then if $u = x^2 + 2y^2$ and $v = 4x^2 - 5y^2$, using the chain rule, we find $f(u, v) = 10u^2v^4$ and so

$$\frac{\partial f}{\partial x} = \frac{\partial f}{\partial u}\frac{\partial u}{\partial x} + \frac{\partial f}{\partial v}\frac{\partial v}{\partial x}$$
$$\frac{\partial f}{\partial y} = \frac{\partial f}{\partial u}\frac{\partial u}{\partial y} + \frac{\partial f}{\partial v}\frac{\partial v}{\partial y}$$

This becomes

$$\frac{\partial f}{\partial x} = \left(20uv^4\right)2x + \left(40u^2v^3\right)8x$$
$$\frac{\partial f}{\partial \theta} = \left(20uv^4\right)4y + \left(40u^2v^3\right)(-10y)$$

You can then substitute in for u and v to get the final answer in terms of x and y (even more ugly though!)

Example 20.6.3 In Chap. 22, we discuss a fitness function w for a model of altruism which depends on P which is the probability of giving aid and Q which is the probability of receiving aid. The model is

$$w = w_0 + b\,Q - c\,P$$

where w_0 is a baseline fitness amount. Note, the chain rule gives us

$$\frac{\partial w}{\partial P} = \frac{\partial w}{\partial P}\frac{\partial P}{\partial P} + \frac{\partial w}{\partial Q}\frac{\partial Q}{\partial P}$$
$$= -c + b\,\frac{\partial Q}{\partial P}$$

Let $\partial_P\, Q$ be denoted by r, the **coefficient of relatedness**. The parameter r is very hard to understand even though it was introduced in 1964 to study altruism. Altruism occurs if fitness increases or $\frac{\partial w}{\partial P} > 0$. So altruism occurs if $-c + br > 0$ or $rb > c$. This inequality is **Hamilton's Rule**, but what counts is we understand what these terms mean biologically.

20.6.2 Homework

Exercise 20.6.1 *Let $f(u, v) = u^2 + 5u^2v^3$ and let $u(x, y) = 2xy$ and $v(x, y) = x^2 - y^2$. Find $\partial_x f(u, v)$ and $\partial_y f(u, v)$.*

Exercise 20.6.2 *Let* $f(u, v) = 2u^2 - 5u^2v^5$ *and let* $u(s, t) = st^2$ *and* $v(s, t) = s^2 + t^4$. *Find* f_s *and* f_t.

Exercise 20.6.3 *Let* $f(u, v) = 5u^2v^3 + 10$ *and let* $u(s, t) = sin(st)$ *and* $v(s, t) = cos(st)$. *Find* f_s *and* f_t.

Exercise 20.6.4 *Let* $f(u, v) = u^2 + v^2$ *and let* $x = r\ \cos(\theta)$ *and* $y = r\ \sin(\theta)$. *Find* $\partial_r f$ *and* $\partial_\theta f$.

Chapter 21
Second Order Multivariable Calculus

We are now ready to give you a whirlwind tour of what you can call second order ideas in calculus for two variables. Or as some would say, let's drink from the fountain of knowledge with a fire hose! Well, maybe not that intense...

We will use these ideas for some practical things. We will end this chapter with a discussion of how to find minimum and maximum for our functions of two variables and then apply those ideas to the problem of finding the best straight line that fits a collection of data points. This **regression line** will be of great importance to you in your career as biologists! We will also take the time to phrase the discussion about the regression line using ideas from statistics. You definitely need to take more courses in probability and statistics so this will give you a taste of what you can do with those ideas. We will introduce the ideas of **average or mean**, **covariance** and **variance** when we work out how to find the regression line. The slope of the regression line has many important applications. We will show you one in Chap. 22 on Hamilton's Rule in evolutionary biology which roughly speaking says that altruistic traits spread throughout the population when $br > c$ where b is a term called *common good*, c is the cost of the altruistic action and r is a very difficult to understand and define term called the **coefficient of relatedness**. We will find out r is related to the slope of a particular regression line. But that is a story that is coming! Right now, we need to learn more about the underlying mathematics.

21.1 Tangent Plane Approximation Error

Now that we have the chain rule, we can quickly develop other results such as how much error we make when we approximate our surface $f(x, y)$ using a tangent plane at a point (x_0, y_0). The first thing we need is to know when a function of two variables is differentiable. Just because it's partials exist at a point is not enough to guarantee that! But we can prove that if the partials are continuous around that point, then the derivative does exist. And that means we can write the function in terms of its tangent plane plus an error. The arguments to do this are not terribly hard, but we will delay

J.K. Peterson, *Calculus for Cognitive Scientists*, Cognitive Science and Technology, DOI 10.1007/978-981-287-874-8_21

them to another class. Hint, hint—there is always something more to learn and figure out! Here is the result.

Theorem 21.1.1 (Continuous Partials Imply Differentiability)
Assume the partials of $f(x, y)$ *exist at* (x_0, y_0) *and that* f *is defined locally around* (x_0, y_0). *Further, assume the partials are continuous locally at* (x_0, y_0). *Then* f *is differentiable at* (x_0, y_0).

Proof This argument is held off to another set of notes! But if you want to see it, just ask. ∎

Now let's go back to the old idea of a tangent plane to a surface. For the surface $z = f(x, y)$ if its partials are continuous functions (they usually are for our work!) then f is differentiable and hence we know that

$$f(x, y) = f(x_0, y_0) + f_x(x_0, y_0)(x - x_0) + f_y(x_0, y_0)(y - y_0) + E(x, y, x_0, y_0)$$

and $E(x, y, x_0, y_0) \rightarrow 0$ and $E(x, y, x_0, y_0)/\sqrt{(x - x_0)^2 + (y - y_0)^2} \rightarrow 0$ as $(x, y) \rightarrow (x_0, y_0)$.

21.2 Second Order Error Estimates

We can characterize the error made when the function is replaced by its tangent plane at a point much better if we have access to what are called the second order partial derivatives of f. Roughly speaking, we take the partials of f_x and f_y to obtain the second order terms. We can make this discussion brief. Assuming f is defined locally as usual near (x_0, y_0), we can ask about the partial derivatives of the functions f_x and f_y with respect to x and y also. We define the second order partials of f as follows.

Definition 21.2.1 (*Second Order Partials*)
If $f(x, y)$, f_x and f_y are defined locally at (x_0, y_0), we can attempt to find following limits:

$$\lim_{x \rightarrow x_0, y = y_0} \frac{f_x(x, y) - f_x(x, y_0)}{x - x_0} = \partial_x(f_x)$$

$$\lim_{x = x_0, y \rightarrow y_0} \frac{f_x(x, y) - f_x(x_0, y)}{y - y_0} = \partial_y(f_x)$$

$$\lim_{x \rightarrow x_0, y = y_0} \frac{f_y(x, y) - f_y(x, y_0)}{x - x_0} = \partial_x(f_y)$$

$$\lim_{x = x_0, y \rightarrow y_0} \frac{f_y(x, y) - f_y(x_0, y)}{y - y_0} = \partial_y(f_y)$$

Comment 21.2.1 *When these second order partials exist at* (x_0, y_0), *we use the following notations interchangeably:* $f_{xx} = \partial_x(f_x)$, $f_{xy} = \partial_y(f_x)$, $f_{yx} = \partial_y(f_x)$ *and* $f_{yy} = \partial_y(f_y)$.

The second order partials are often organized into a matrix called the **Hessian**.

Definition 21.2.2 (*The Hessian*)
If $f(x, y)$, f_x and f_y are defined locally at (x_0, y_0) and if the second order partials exist at (x_0, y_0), we define the Hessian, $\boldsymbol{H}(x_0, y_0)$ at (x_0, y_0) to be the matrix

$$\boldsymbol{H}(x_0, y_0) = \begin{bmatrix} f_{xx}(x_0, y_0) & f_{xy}(x_0, y_0) \\ f_{yx}(x_0, y_0) & f_{yy}(x_0, y_0) \end{bmatrix}$$

Comment 21.2.2 *It is also possible to prove that if the first order partials are continuous locally near (x_0, y_0) then the mixed order partials f_{xy} and f_{yx} must match at the point (x_0, y_0). Most of our surfaces have this property. Hence, for these* **smooth** *surfaces, the Hessian is a symmetric matrix!*

21.2.1 Examples

Example 21.2.1 Let $f(x, y) = 2x - 8xy$. Find the first and second order partials of f and its Hessian.

Solution *The partials are*

$$\begin{aligned} f_x(x, y) &= 2 - 8y \\ f_y(x, y) &= -8x \\ f_{xx}(x, y) &= 0 \\ f_{xy}(x, y) &= -8 \\ f_{yx}(x, y) &= -8 \\ f_{yy}(x, y) &= 0. \end{aligned}$$

and so the Hessian is

$$\boldsymbol{H}(x, y) = \begin{bmatrix} f_{xx}(x, y) & f_{xy}(x, y) \\ f_{yx}(x, y) & f_{yy}(x, y) \end{bmatrix} = \begin{bmatrix} 0 & -8 \\ -8 & 0 \end{bmatrix}$$

Example 21.2.2 Let $f(x, y) = 2x^2 - 8y^3$. Find the first and second order partials of f and its Hessian.

Solution *The partials are*

$$\begin{aligned} f_x(x, y) &= 4x \\ f_y(x, y) &= -24y^2 \\ f_{xx}(x, y) &= 4 \end{aligned}$$

$$f_{xy}(x, y) = 0$$
$$f_{yx}(x, y) = 0$$
$$f_{yy}(x, y) = -48y.$$

and so the Hessian is

$$\boldsymbol{H}(x, y) = \begin{bmatrix} f_{xx}(x, y) & f_{xy}(x, y) \\ f_{yx}(x, y) & f_{yy}(x, y) \end{bmatrix} = \begin{bmatrix} 4 & 0 \\ 0 & -48y \end{bmatrix}$$

21.2.2 Homework

Exercise 21.2.1 *Let $f(x, y) = 5x - 2xy$. Find the first and second order partials of f and its Hessian.*

Exercise 21.2.2 *Let $f(x, y) = -8y + 9xy - 2y^2$. Find the first and second order partials of f and its Hessian.*

Exercise 21.2.3 *Let $f(x, y) = 4x - 6xy - x^2$. Find the first and second order partials of f and its Hessian.*

Exercise 21.2.4 *Let $f(x, y) = 4x^2 - 6xy - x^2$. Find the first and second order partials of f and its Hessian.*

21.3 Hessian Approximations

We can now explain the most common approximation result for tangent plane error. Let

$$h(t) = f(x_0 + t\Delta x, y_0 + t\Delta y)$$

as usual. Then we know we can write

$$h(t) = h(0) + h'(0)t + h''(c)\frac{t^2}{2}.$$

Using the chain rule, we find

$$h'(t) = f_x(x_0 + t\Delta x, y_0 + t\Delta y)\Delta x + f_y(x_0 + t\Delta x, y_0 + t\Delta y)\Delta y$$

and

$$
\begin{aligned}
h''(t) &= \partial_x \Big(f_x(x_0 + t\Delta x, y_0 + t\Delta y)\Delta x + f_y(x_0 + t\Delta x, y_0 + t\Delta y)\Delta y \Big) \Delta x \\
&\quad + \partial_y \Big(f_x(x_0 + t\Delta x, y_0 + t\Delta y)\Delta x + f_y(x_0 + t\Delta x, y_0 + t\Delta y)\Delta y \Big) \Delta y \\
&= f_{xx}(x_0 + t\Delta x, y_0 + t\Delta y)(\Delta x)^2 + f_{yx}(x_0 + t\Delta x, y_0 + t\Delta y)(\Delta y)(\Delta x) \\
&\quad + f_{xy}(x_0 + t\Delta x, y_0 + t\Delta y)(\Delta x)(\Delta y) + f_{yy}(x_0 + t\Delta x, y_0 + t\Delta y)(\Delta y)^2
\end{aligned}
$$

We can rewrite this in matrix–vector form as

$$
h''(t) = \begin{bmatrix} \Delta x & \Delta y \end{bmatrix} \begin{bmatrix} f_{xx}(x_0 + t\Delta x, y_0 + t\Delta y) & f_{yx}(x_0 + t\Delta x, y_0 + t\Delta y) \\ f_{xy}(x_0 + t\Delta x, y_0 + t\Delta y) & f_{yy}(x_0 + t\Delta x, y_0 + t\Delta y) \end{bmatrix} \begin{bmatrix} \Delta x \\ \Delta y \end{bmatrix}
$$

Of course, using the definition of $\boldsymbol{H}$, this can be rewritten as

$$
h''(t) = \begin{bmatrix} \Delta x \\ \Delta y \end{bmatrix}^T \boldsymbol{H}(x_0 + t\Delta x, y_0 + t\Delta y) \begin{bmatrix} \Delta x \\ \Delta y \end{bmatrix}
$$

Thus, our tangent plane approximation can be written as

$$
h(1) = h(0) + h'(0)(1 - 0) + h''(c)\frac{1}{2}
$$

for some c between 0 and 1. Substituting for the h terms, we find

$$
\begin{aligned}
f(x_0 + \Delta x, y_0 + \Delta y) &= f(x_0, y_0) + f_x(x_0, y_0)\Delta x + f_y(x_0, y_0)\Delta y \\
&\quad + \frac{1}{2} \begin{bmatrix} \Delta x \\ \Delta y \end{bmatrix}^T \boldsymbol{H}(x_0 + c\Delta x, y_0 + c\Delta y) \begin{bmatrix} \Delta x \\ \Delta y \end{bmatrix}
\end{aligned}
$$

Clearly, we have shown how to express the error in terms of second order partials. There is a point c between 0 and 1 so that

$$
E(x_0, y_0, \Delta x, \Delta y) = \frac{1}{2} \begin{bmatrix} \Delta x \\ \Delta y \end{bmatrix}^T \boldsymbol{H}(x_0 + c\Delta x, y_0 + c\Delta y) \begin{bmatrix} \Delta x \\ \Delta y \end{bmatrix}
$$

Note the error is a quadratic expression in terms of the Δx and Δy.

21.3.1 Ugly Error Estimates!

We now know for a function of two variables, $f(x, y)$, we can estimate the error made in approximating using the gradient at the given point (x_0, y_0) as follows: We know

$$f(x, y) = f(x_0, y_0) + < \nabla(f)(x_0, y_0), [x - x_0, y - y_0]^T > \\ + (1/2)[x - x_0, y - y_0]H(x_0 + c(x - x_0), y_0 + c(y - y_0))[x - x_0, y - y_0]^T$$

where $((x_0 + c(x - x_0), y_0 + c(y - y_0))$ is between (x_0, y_0) and (x, y). For convenience, we usually call this intermediate pair (x^*, y^*). Further, if we know f and its partials are bounded locally, then on the rectangle

$$R_r = [x_0 - r, x_0 + r] \times [y_0 - r, y_0 + r]$$

we can find an error estimate like this. We know

$$|f(x, y) - \nabla(f)^T(x_0, y_0)[x - x_0, y - y_0]| \leq \frac{1}{2}\Big(|f_{xx}(x^*, y^*)||\Delta x|^2 + |f_{xy}(x^*, y^*)||\Delta x||\Delta y| \\ +|f_{xy}(x^*, y^*)||\Delta x||\Delta y| + |f_{yy}(x^*, y^*)||\Delta y|^2\Big)$$

Since the biggest $|x - x_0|$ and $|y - y_0|$ can be is r, we therefore see how to find this estimate. It is easier to see this with some examples.

21.3.1.1 Examples

Here is an example with a specific function of these ideas.

Example 21.3.1 Let $f(x, y) = x^2y^4 + 2x + 3y + 10$. Estimate the tangent plane error for various local circles about $(0, 0)$.

Solution *We have*

$$f_x = 2xy^4 + 2$$
$$f_y = x^2 4y^3 + 3$$
$$f_{xx} = 2y^4$$
$$f_{xy} = f_{yx} = 8xy^3$$
$$f_{yy} = 12x^2y^2$$

So at $(x_0, y_0) = (0, 0)$, *letting E denote the error, we have*

$$x^2y^4 = 10 + < [2, 3]^T, [x - 0, y - 0]^T > + E$$

Now,

$$f_{xx} = 2y^4$$
$$f_{xy} = f_{yx} = 8xy^3$$
$$f_{yy} = 12x^2y^2$$

and in R_r, we find these estimates:

$$\max_{(x,y)\in R_r} |f_{xx}| = 2r^4$$
$$\max_{(x,y)\in R_r} |f_{xy}| = 8r^4$$
$$\max_{(x,y)\in R_r} |f_{yy}| = 12r^4.$$

Now the error is

$$\begin{bmatrix} \Delta x \\ \Delta y \end{bmatrix}^T \boldsymbol{H}(x_0 + c\Delta x, y_0 + c\Delta y) \frac{1}{2} \begin{bmatrix} \Delta x \\ \Delta y \end{bmatrix}$$
$$= f_{xx}(x^*, y^*)(\Delta x)^2 + 2f_{xy}(x^*, y^*)(\Delta x)(\Delta y) + f_{yy}(x^*, y^*)(\Delta y)^2.$$

In the box, the biggest this can be in absolute value is $\frac{1}{2}(2r^4 \times r^2 + 2 \times 8r^4 \times r^2 + 12r^4 \times r^2) = 15r^6$ *and get the largest error for the approximation is*

$$|x^2y^4 - 10 - < [2, 3]^T, [x, y]^T > | \leq 15r^6.$$

So for $r = 0.8$, *the maximum error is overestimated by* $15(0.8)^6 = 13.12$*—probably bad! For* $r = 0.4$, *the maximum error is* $15(0.4)^6 = 0.06$*—better! To make the largest error* $<10^{-4}$, *solve* $15r^6 < 10^{-4}$. *This gives* $r^6 < 6.67 \times 10^{-6}$. *Thus,* $r < 0.137$ *will do the job.*

Example 21.3.2 We can also do this sort of error estimation at another point, say $(x^*, y^*) = (1, 2)$, which, of course, is much yuckier to do!

Solution *To get these estimates, let*

$$R_r = [1 - r, 1 + r] \times [2 - r, 2 + r]$$

Again, we have

$$f_x = 2xy^4 + 2$$
$$f_y = x^2 4y^3 + 3$$
$$f_{xx} = 2y^4$$
$$f_{xy} = f_{yx} = 8xy^3$$
$$f_{yy} = 12x^2y^2$$

Then for our second order partials, in R_r,

$$\max_{(x,y)\in R_r} |f_{xx}| = 2(2 + r)^4$$
$$\max_{(x,y)\in R_r} |f_{xy}| = 8(1 + r)(2 + r)^3$$

$$\max_{(x,y)\in R_r} |f_{yy}| = 12(1+r)^2(2+r)^2.$$

Now the error is

$$\begin{bmatrix}\Delta x\\ \Delta y\end{bmatrix}^T \boldsymbol{H}(x_0 + c\Delta x,\, y_0 + c\Delta y) \begin{bmatrix}\Delta x\\ \Delta y\end{bmatrix}$$
$$= f_{xx}(x^*, y^*)(\Delta x)^2 + 2f_{xy}(x^*, y^*)(\Delta x)(\Delta y) + f_{yy}(x^*, y^*)(\Delta y)^2.$$

In the box, the biggest this can be in absolute value is

$$\begin{aligned} E &\le \frac{1}{2}(2(2+r)^4 \times r^2 + 2 \times 8(1+r)(2+r)^3 \times r^2 + 12(1+r)^2(2+r)^2 \times r^2) \\ &\le \frac{1}{2}(2(2+r)^4 r^2 + 16(2+r)^4 r^2 + 12(2+r)^4 r^2). \end{aligned}$$

as $1+r < 2+r$. *We get the largest error for the approximation is*

$$|x^2y^4 - 10 - < [2,3]^T[x,y]^T > | \le 15(2+r)^4 r^2$$

and to make the largest error in the approximation $<10^{-4}$, *we can try various* r *values to see the errors we get. An* r *of about* 0.0005 *is fine while an* r *of* 0.001 *gives too much error. The point is that to keep the tangent plane error down, we really have to stay local to the point* (1.2).

21.3.2 Homework

Exercise 21.3.1 *For the following function find its gradient and hessian and its tangent plane and error at the point* (1, 1).

$$f(x,y) = x^2 + x^4y^3$$

Exercise 21.3.2 *For the following function find its gradient and hessian and its tangent plane and error at the point* (1, 1).

$$f(x,y) = x^2 - 20y^4x^5$$

Exercise 21.3.3 *Approximate* $f(x,y) = x^2 + y^4x^5 + 3x + 4y + 25$ *near* (0, 0). *Find the* r *where the error is less than* 10^{-3}.

Exercise 21.3.4 *Approximate* $f(x,y) = 4x^4y^4x^5 + 3x + 40y + 5$ *near* (0, 0). *Find the* r *where the error is less than* 10^{-6}.

21.4 Extrema Ideas

To understand how to think about finding places where the minimum and maximum of a function to two variables might occur, all you have to do is realize it is a common sense thing. We already know that the tangent plane attached to the surface which represents our function of two variables is a way to approximate the function near the point of attachment. We have seen in our pictures what happens when the tangent plane is **flat**. This flatness occurs at the minimum and maximum of the function. It also occurs in other situations, but we will leave that more complicated event for other courses. The functions we want to deal with are quite nice and have great minima and maxima. However, we do want you to know there are more things in the world and we will touch on them only briefly.

To see what to do, just recall the equation of the tangent plane error to our function of two variables $f(x, y)$.

$$\begin{aligned} f(x, y) = f(x_0, y_0) &+ \nabla(f)(x_0, y_0)[x - x_0, y - y_0]^T \\ &+ (1/2)[x - x_0, y - y_0]H(x_0 + c(x - x_0), y_0 + c(y - y_0))[x - x_0, y - y_0]^T \end{aligned}$$

where c is some number between 0 and 1 that is different for each x. We also know that the equation of the tangent plane to $f(x, y)$ at the point (x_0, y_0) is

$$f(x, y) = f(x_0, y_0) + < \nabla f, \boldsymbol{X} - \boldsymbol{X_0} > .$$

Now let's assume the tangent plane is flat at (x_0, y_0). Then the gradient ∇f is the zero vector and we have $\frac{\partial f}{\partial x}(x_0, y_0) = 0$ and $\frac{\partial f}{\partial y}(x_0, y_0) = 0$. So the tangent plane error equation simplifies to

$$\begin{aligned} f(x, y) = f(x_0, y_0) &+ (1/2)[x - x_0, y - y_0]H(x_0 + c(x - x_0), y_0 + c(y - y_0)) \\ &\times [x - x_0, y - y_0]^T \end{aligned}$$

Now let's simplify this. The Hessian is just a 2×2 matrix whose components are the second order partials of f. Let

$$\begin{aligned} A(c) &= \frac{\partial^2 f}{\partial x^2}(x_0 + c(x - x_0), y_0 + c(y - y_0)) \\ B(c) &= \frac{\partial^2 f}{\partial x\, \partial y}(x_0 + c(x - x_0), y_0 + c(y - y_0)) = \frac{\partial^2 f}{\partial y\, \partial x}(x_0 + c(x - x_0), y_0 + c(y - y_0)) \\ D(c) &= \frac{\partial^2 f}{\partial y^2}(x_0 + c(x - x_0), y_0 + c(y - y_0)) \end{aligned}$$

Then, we have

$$f(x, y) = f(x_0, y_0) + (1/2) \begin{bmatrix} x - x_0 & y - y_0 \end{bmatrix} \begin{bmatrix} A(c) & B(c) \\ B(c) & D(c) \end{bmatrix} \begin{bmatrix} x - x_0 \\ y - y_0 \end{bmatrix}$$

We can multiply this out (a nice simple pencil and paper exercise!) to find

$$f(x, y) = f(x_0, y_0) + 1/2 \left(A(c)(x - x_0)^2 + 2B(c)(x - x_0)(y - y_0) + D(c)(y - y_0)^2 \right)$$

Now it is time to remember an old technique from high school—completing the square. Remember if we had a quadratic like $u^2 + 3uv + 6v^2$, to complete the square we take half of the number in front of the mixed term uv and square it and add and subtract it times v^2 as follows.

$$u^2 + 3uv + 6v^2 = u^2 + 3uv + (3/2)^2 v^2 - (3/2)^2 v^2 + 6v^2.$$

Now group the first three terms together and combine the last two terms into one term.

$$u^2 + 3uv + 6v^2 = \left(u^2 + 3uv + (3/2)^2 v^2 \right) + \left(6 - (3/2)^2 \right) v^2.$$

The first three terms are a *perfect square*, $(u + (3/2)v)^2$. Simplifying, we find

$$u^2 + 3uv + 6v^2 = \left(u + (3/2)v \right)^2 + (135/4)\, v^2.$$

This is called *completely the square*! Now let's do this with the Hessian quadratic we have. First, factor our the $A(c)$. We will assume it is not zero so the divisions are fine to do. Also, for convenience, we will replace $x - x_0$ by Δx and $y - y_0$ by Δy. This gives

$$f(x, y) = f(x_0, y_0) + \frac{A(c)}{2} \left((\Delta x)^2 + 2\frac{B(c)}{A(c)} \Delta x\, \Delta y + \frac{D(c)}{A(c)} (\Delta y)^2 \right).$$

One half of the $\Delta x \Delta y$ coefficient is $\frac{B(c)}{A(c)}$ so add and subtract $(B(c)/A(c))^2 (\Delta y)^2$. We find

$$\begin{aligned} f(x, y) = {} & f(x_0, y_0) \\ & + \frac{A(c)}{2} \left((\Delta x)^2 + 2\frac{B(c)}{A(c)} \Delta x\, \Delta y + \left(\frac{B(c)}{A(c)}\right)^2 (\Delta y)^2 - \left(\frac{B(c)}{A(c)}\right)^2 (\Delta y)^2 + \frac{D(c)}{A(c)} (\Delta y)^2 \right). \end{aligned}$$

Now group the first three terms together—the perfect square and combine the last two terms into one. We have

$$\begin{aligned} f(x, y) = {} & f(x_0, y_0) \\ & + \frac{A(c)}{2} \left(\left(\Delta x + \frac{B(c)}{A(c)} \Delta y \right)^2 + \left(\frac{A(c)\, D(c) - (B(c))^2}{(A(c))^2} \right) (\Delta y)^2 \right). \end{aligned}$$

Now we need this common sense result which says that if a function g is continuous at a point (x_0, y_0) and positive or negative, then it is positive or negative in a circle of radius r centered at (x_0, y_0). Here is the formal statement.

Theorem 21.4.1 (Nonzero Values and Continuity)
If $f(x_0, y_0)$ is a place where the function is positive or negative in value, then there is a radius r so that $f(x, y)$ is positive or negative in a circle of radius r around the center (x_0, y_0).

Proof This can be argued carefully using limits.

- If $f(x_0, y_0) > 0$ and f is continuous at (x_0, y_0), then if no matter now close we were to (x_0, y_0), we could find a point (x_r, y_r) where $f(x_r, y_r) = 0$, then that set of points would define a path to (x_0, y_0) and the limiting value of f on that path would be 0.
- But we know the value at (x_0, y_0) is positive and we know f is continuous there. Hence, the limiting values for all paths should match.
- So we can't find such points for all values of r. We see there will be a first r where we can't do this and so inside the circle determined by that r, f will be nonzero.
- You might think we haven't ruled out the possibility that f could be negative at some points. But the only way a continuous f could switch between positive and negative is to pass through zero. And we have already ruled that out. So f is positive inside this circle of radius r. ■

Now getting back to our problem. We have at this point where the partials are zero, the following expansion

$$\begin{aligned} f(x, y) &= f(x_0, y_0) \\ &\quad + \frac{A(c)}{2}\left(\left((\Delta x)^2 + 2\frac{B(c)}{A(c)}\Delta x\, \Delta y + \left(\frac{B(c)}{A(c)}\right)^2 (\Delta y)^2\right)\right. \\ &\qquad \left. + \left(\frac{A(c)\, D(c) - (B(c))^2}{(A(c))^2}\right)(\Delta y)^2\right). \end{aligned}$$

The algebraic sign of the terms after the function value $f(x_0, y_0)$ are completely determined by the terms which are not squared. We have two simple cases:

- $A(c) > 0$ and $A(c)\, D(c) - (B(c))^2 > 0$ which implies the term after $f(x_0, y_0)$ is positive.
- $A(c) < 0$ and $A(c)\, D(c) - (B(c))^2 > 0$ which implies the term after $f(x_0, y_0)$ is negative.

Now let's assume all the second order partials are continuous at (x_0, y_0). We know $A(c) = \frac{\partial^2 f}{\partial x^2}(x_0 + c(x - x_0), y_0 + c(y - y_0))$ and from Theorem 21.4.1, if $\frac{\partial^2 f}{\partial x^2}(x_0, y_0) > 0$, then so is $A(c)$ in a circle around (x_0, y_0). The other term $A(c)$ $D(c) - (B(c))^2 > 0$ will also be positive is a circle around (x_0, y_0) as long as

$\frac{\partial^2 f}{\partial x^2}(x_0, y_0)\, \frac{\partial^2 f}{\partial y^2}(x_0, y_0) - \frac{\partial^2 f}{\partial x \partial y}(x_0, y_0) > 0$. We can say similar things about the negative case. Now to save typing let $\frac{\partial^2 f}{\partial x^2}(x_0, y_0) = f^0_{xx}$, $\frac{\partial^2 f}{\partial y^2}(x_0, y_0) = f^0_{yy}$ and $\frac{\partial^2 f}{\partial x \partial y}(x_0, y_0) = f^0_{xy}$. So we can restate our two cases as

- $f^0_{xx} > 0$ and $f^0_{xx}\, f^0_{yy} - (f^0_{xy})^2 > 0$ which implies the term after $f(x_0, y_0)$ is positive. This implies that $f(x, y) > f(x_0, y_0)$ in a circle of some radius r which says $f(x_0, y_0)$ is a minimum value of the function locally at that point.
- $f^0_{xx} < 0$ and $f^0_{xx}\, f^0_{yy} - (f^0_{xy})^2 > 0$ which implies the term after $f(x_0, y_0)$ is negative. This implies that $f(x, y) < f(x_0, y_0)$ in a circle of some radius r which says $f(x_0, y_0)$ is a maximum value of the function locally at that point.

where, for convenience, we use a superscript 0 to denote we are evaluating the partials at (x_0, y_0). So we have come up with a great condition to verify if a place where the partials are zero is minimum or a maximum. If you think about it a bit, you'll notice we left out the case where $f^0_{xx}\, f^0_{yy} - (f^0_{xy})^2 < 0$ which is important but we will not do that in this class. That is for later courses to pick up, however it is the test for the analog of the behavior we see in the cubic $y = x^3$. The derivative is 0 but there is neither a minimum or maximum at $x = 0$. In two dimensions, the situation is more interesting of course. This kind of behavior is called a **saddle**. We have another Theorem!

Theorem 21.4.2 (Extrema Test)
If the partials of f are zero at the point (x_0, y_0), we can determine if that point is a local minimum or local maximum of f using a second order test. We must assume the second order partials are continuous at the point (x_0, y_0).

- *If $f^0_{xx} > 0$ and $f^0_{xx}\, f^0_{yy} - (f^0_{xy})^2 > 0$ then $f(x_0, y_0)$ is a local minimum.*
- *$f^0_{xx} < 0$ and $f^0_{xx}\, f^0_{yy} - (f^0_{xy})^2 > 0$ then $f(x_0, y_0)$ is a local maximum.*

The test f^0_{xx} can be replaced by the test f^0_{yy} if needed. We just don't know anything if the test $f^0_{xx}\, f^0_{yy} - (f^0_{xy})^2 = 0$. If the test gives $f^0_{xx}\, f^0_{yy} - (f^0_{xy})^2 < 0$, we have a saddle.

Proof We have sketched out the reasons for this above. ■

21.4.1 Examples

Example 21.4.1 Use our tests to show $f(x, y) = x^2 + 3y^2$ has a minimum at $(0, 0)$.

Solution *The partials here are $f_x = 2x$ and $f_y = 6y$. These are zero at $x = 0$ and $y = 0$. The Hessian at this critical point is*

$$H(x, y) = \begin{bmatrix} 2 & 0 \\ 0 & 6 \end{bmatrix} = H(0, 0).$$

as H is constant here. Our second order test says the point $(0, 0)$ *corresponds to a minimum because* $f_{xx}(0, 0) = 2 > 0$ *and* $f_{xx}(0, 0)\, f_{yy}(0, 0) - (f_{xy}(0, 0))^2 = 12 > 0$.

Example 21.4.2 Use our tests to show $f(x, y) = x^2 + 6xy + 3y^2$ has a saddle at $(0, 0)$.

Solution *The partials here are* $f_x = 2x + 6y$ *and* $f_y = 6x + 6y$. *These are zero at when*

$$\begin{aligned} 2x + 6y &= 0 \\ 6x + 6y &= 0 \end{aligned}$$

which has solution $x = 0$ *and* $y = 0$. *The Hessian at this critical point is*

$$H(x, y) = \begin{bmatrix} 2 & 6 \\ 6 & 6 \end{bmatrix} = H(0, 0).$$

as H is again constant here. Our second order test says the point $(0, 0)$ *corresponds to a saddle because* $f_{xx}(0, 0) = 2 > 0$ *and* $f_{xx}(0, 0)\, f_{yy}(0, 0) - (f_{xy}(0, 0))^2 = 12 - 36 < 0$.

Example 21.4.3 Show our tests fail on $f(x, y) = 2x^4 + 4y^6$ even though we know there is a minimum value at $(0, 0)$.

Solution *For* $f(x, y) = 2x^4 + 4y^6$, *you find that the critical point is* $(0, 0)$ *and all the second order partials are* 0 *there. So all the tests fail. Of course, a little common sense tells you* $(0, 0)$ *is indeed the place where this function has a minimum value. Just think about how it's surface looks. But the tests just fail. This is much like the curve* $f(x) = x^4$ *which has a minimum at* $x = 0$ *but all the tests fail on it also.*

Example 21.4.4 Show our tests fail on $f(x, y) = 2x^2 + 4y^3$ and the surface does not have a minimum or maximum at the critical point $(0, 0)$.

Solution *For* $f(x, y) = 2x^2 + 4y^3$, *the critical point is again* $(0, 0)$ *and* $f_{xx}(0, 0) = 4$, $f_{yy}(0, 0) = 0$ *and* $f_{xy}(0, 0) = f_{yx}(0, 0) = 0$. *So* $f_{xx}(0, 0)\, f_{yy}(0, 0) - (f_{xy}(0, 0))^2 = 0$ *so the test fails. Note the* $x = 0$ *trace is* $4y^3$ *which is a cubic and so is negative below* $y = 0$ *and positive above* $y = 0$. *Not much like a minimum or maximum behavior on this trace! But the trace for* $y = 0$ *is* $2x^2$ *which is a nice parabola which does reach its minimum at* $x = 0$. *So the behavior of the surface around* $(0, 0)$ *is not a maximum or a minimum. The surface acts a lot like a cubic. Do this in MatLab.*

Listing 21.1: The surface $f(x, y) = 2x^2 + 4y^3$

```
1 [X,Y] = meshgrid(-1:.2:1);
  Z = 2*X.^2 + 4*Y.^3;
  surf(Z);
```

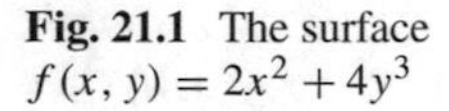

Fig. 21.1 The surface $f(x, y) = 2x^2 + 4y^3$

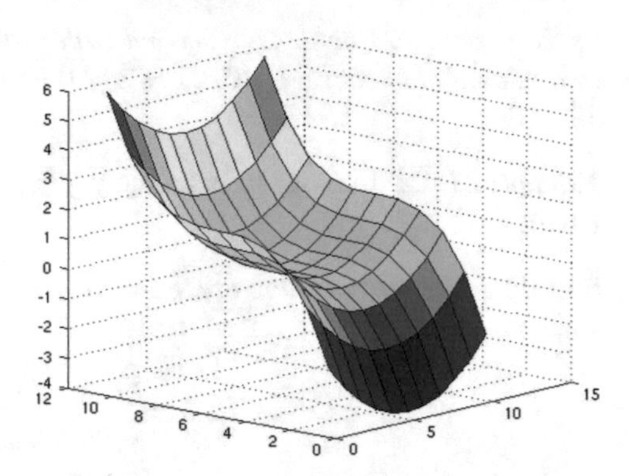

This will give you a surface. In the plot that is shown go to the tool menu and click of the rotate 3D option and you can spin it around. Clearly like a cubic! You can see the plot in Fig. 21.1.

21.4.2 Homework

Exercise 21.4.1 *Use our tests to show* $f(x, y) = 4x^2 + 2y^2$ *has a minimum at* $(0, 0)$. *Feel free to draw a surface plot to help you see what is going on.*

Exercise 21.4.2 *Use our tests to find where* $f(x, y) = 2x^2 + 3x + 3y^2 + 8y$ *has a minimum. Feel free to draw a surface plot to help you see what is going on.*

Exercise 21.4.3 *Use our tests to find where* $f(x, y) = 100 - 2x^2 + 3x - 3y^2 + 8y$ *has a maximum. Feel free to draw a surface plot to help you see what is going on.*

Exercise 21.4.4 *Use our tests to find where* $f(x, y) = 2x^2 + x + 8 + 4y^2 + 8y + 20$ *has a minimum. Feel free to draw a surface plot to help you see what is going on.*

Exercise 21.4.5 *Show our tests fail on* $f(x, y) = 6x^4 + 8y^8$ *even though we know there is a minimum value at* $(0, 0)$. *Feel free to draw a surface plot to help you see what is going on.*

Exercise 21.4.6 *Show our tests fail on* $f(x, y) = 10x^2 + 5y^5$ *and the surface does not have a minimum or maximum at the critical point* $(0, 0)$. *Feel free to draw a surface plot to help you see what is going on.*

21.5 A Regression to Regression

Now let's stop here and think about another problem. Many times we want to find the best straight line through a collection of data points. We did exactly that in our Cooling Project in Sect. 15.2. Now that we know about partial derivatives, we can look at the more general problem where the slope and the y intercept of this line need to be chosen. In the Cooling Project, we had a simpler task as we knew the y intercept had to be from our initial temperature data. So let's look at the more general problem now. It will turn out to help us understand what is happening in our altruism models.

We begin with a collection of S data pairs $\{(x_i, y_i) : 1 \le i \le S\}$. The line we want to pick has the form $y = mx + b$ for some choice of slope m and intercept b. The distance from a given data point (x_i, y_i) and our line is the usual Euclidean distance d_i given by

$$d_i = \sqrt{(mx_i + b - y_i)^2}.$$

If we want to minimize the sum of all these individual errors, we get the same result by minimizing the sum of all the errors squared. Define an error function E by

$$E = \sum_{i=1}^{S} d_i^2 = \sum_{i=1}^{S} (mx_i + b - y_i)^2.$$

We see the error function E is really a function of the two independent variables m and b. From our work with partial derivatives, we know that the minimum of a function might occur at the place where the partials $\frac{\partial E}{\partial m} = 0$ and $\frac{\partial E}{\partial b} = 0$. Since E is a sum of squares, the smallest value it could possibly have is $E = 0$ which would occur when the line passes through all the data points. However, this is not what usually happens as the data rarely aligns in that simple way. Hence, no matter how we choose the line, we typically have some data points off the line; hence, there is some error as some d_{ij} are not zero. A little common sense reasoning tells us then that the place where the tangent plane to this function is flat is indeed where the minimum value will occur. Let's get about finding where this minimum occurs. First, we need the partial derivatives.

$$\begin{aligned}
\frac{\partial E}{\partial m} &= \frac{\partial}{\partial m}\left(\sum_{i=1}^{S}(mx_i + b - y_i)^2\right) \\
&= \sum_{i=1}^{S} \frac{\partial}{\partial m}\left((mx_i + b - y_i)^2\right) \\
&= \sum_{i=1}^{S} 2\,(mx_i + b - y_i)\,x_i \\
&= 2\left(\left(m\sum_{i=1}^{S} x_i^2\right) + b\sum_{i=1}^{S} x_i - \sum_{i=1}^{S} x_i y_i\right) \qquad (21.1)
\end{aligned}$$

and

$$\begin{aligned}
\frac{\partial E}{\partial b} &= \frac{\partial}{\partial b}\left(\sum_{i=1}^{S}(mx_i + b - y_i)^2\right) \\
&= \sum_{i=1}^{S} \frac{\partial}{\partial b}\left((mx_i + b - y_i)^2\right) \\
&= \sum_{i=1}^{S} 2\,(mx_i + b - y_i)\,1 \\
&= 2\left(\left(m \sum_{i=1}^{S} x_i\right) + \sum_{i=1}^{S} b - \sum_{i=1}^{S} y_i\right) \qquad (21.2)
\end{aligned}$$

To find the minimum of our error function, we set these partials to zero and solve for m and b. This is just a bit of algebra.

$$\begin{aligned}
\frac{\partial E}{\partial m} &= 2\left(\left(m \sum_{i=1}^{S} x_i^2\right) + b \sum_{i=1}^{S} x_i - \sum_{i=1}^{S} x_i y_i\right) = 0 \\
&\Longrightarrow \\
& m \sum_{i=1}^{S} x_i^2 + b \sum_{i=1}^{S} x_i = \sum_{i=1}^{S} x_i y_i
\end{aligned}$$

The next partial condition gives

$$\begin{aligned}
\frac{\partial E}{\partial b} &= 2\left(\left(m \sum_{i=1}^{S} x_i\right) + \sum_{i=1}^{S} b - \sum_{i=1}^{S} y_i\right) = 0 \\
&\Longrightarrow m \sum_{i=1}^{S} x_i + bS = \sum_{i=1}^{S} y_i.
\end{aligned}$$

where we have used the fact that adding b up S times gives bS. Now it is time to introduce some standard notation. We encourage you all to take some real courses in probability and statistics as what we will do next is just a bare introduction! Everyone knows from high school how to average a list of test scores and so on. You just add up all the numbers and divide by the number of scores. We will start using this sort of idea now. Instead of calling this calculation an **average**, we will use a different term, **expectation**. In the back of our minds, we are thinking that the test scores are not really random (another loaded word that needs a lot more careful thought than we are putting into it here!) and indeed, follow some sort of *law*. We usually say something like the *probability* that someone gets an 80 on the exam is roughly 20 % because we have hope our professors write exams that follow a bell shaped curve with most of the student scores concentrated at the middle and the higher and lower scores falling away from that. It turns out that our usual average is a reasonable estimate of where the test scores are concentrated and so it is what we *expect*. So that is why we define the **expectation** of our collection of data points to be $\boldsymbol{E}(X)$ where $\boldsymbol{E}$ is our symbol for **expectation** and (X) is our way of saying we are finding the expected *concentration* for the data points $(X) = \{x_1, \ldots, x_S\}$. More formally, we can do this for any collection of data. So we have for the different collections of data $\boldsymbol{X}$ and $\boldsymbol{Y}$,

$$\boldsymbol{E}(\boldsymbol{X}) = \frac{1}{S} \sum_{i=1}^{S} x_i \tag{21.3}$$

$$\boldsymbol{E}(\boldsymbol{Y}) = \frac{1}{S} \sum_{i=1}^{S} y_i \tag{21.4}$$

$$\boldsymbol{E}(\boldsymbol{XY}) = \frac{1}{S} \sum_{i=1}^{S} x_i y_i \tag{21.5}$$

$$\boldsymbol{E}(\boldsymbol{X}^2) = \frac{1}{S} \sum_{i=1}^{S} x_i^2 \tag{21.6}$$

Using these ideas, we can rewrite our equations for the choice of m and b that gives us minimal error. We start with

$$m \sum_{i=1}^{S} x_i^2 + b \sum_{i=1}^{S} x_i = \sum_{i=1}^{S} x_i y_i$$
$$m \sum_{i=1}^{S} x_i + bS = \sum_{i=1}^{S} y_i.$$

and substitute in the idea of expectations to get

$$m\, S\, \boldsymbol{E}(\boldsymbol{X}^2) + b\, S\, \boldsymbol{E}(\boldsymbol{X}) = S\, \boldsymbol{E}(\boldsymbol{XY})$$
$$m\, S\, \boldsymbol{E}(\boldsymbol{X}) + bS = S\, \boldsymbol{E}(\boldsymbol{Y})$$

This is a simple system of two equations in the two unknowns b and m. Divide out the common S to find

$$m\, \boldsymbol{E}(\boldsymbol{X}^2) + b\, \boldsymbol{E}(\boldsymbol{X}) = \boldsymbol{E}(\boldsymbol{XY})$$
$$m\, \boldsymbol{E}(\boldsymbol{X}) + b = \boldsymbol{E}(\boldsymbol{Y})$$

This system is easy to solve. Multiply the bottom equation by $\boldsymbol{E}(\boldsymbol{X})$ to give

$$m\, \boldsymbol{E}(\boldsymbol{X}^2) + b\, \boldsymbol{E}(\boldsymbol{X}) = \boldsymbol{E}(\boldsymbol{XY})$$
$$m\, (\boldsymbol{E}(\boldsymbol{X}))^2 + b\, \boldsymbol{E}(\boldsymbol{X}) = \boldsymbol{E}(\boldsymbol{X})\, \boldsymbol{E}(\boldsymbol{Y})$$

and then subtract the bottom equation from the top to find m (the b terms cancel out!):

$$(E(X^2) - E(X)\,E(X))\,m = (E(XY) - E(X)\,E(Y))$$

Thus, the optimal slope is

$$m = \frac{E(XY) - E(X)\,E(Y)}{E(X^2) - E(X)\,E(X)}. \tag{21.7}$$

With m found, we can use the original top or bottom equation to solve for b. We're going to use the top one just to make you all do more algebra!

$$m\,E(X^2) + b\,E(X) = E(XY) \implies b = \frac{E(XY) - mE(X^2)}{E(X)}$$

Now plug in the optimal m to get (this is really messy algebra so we suggest you follow along with pencil and paper!)

$$\begin{aligned} b &= \frac{E(XY) - \frac{E(XY)-E(X)\,E(Y)}{E(X^2)-E(X)\,E(X)}E(X^2)}{E(X)} \\ &= \frac{\Big(E(X^2) - E(X)\,E(X)\Big)\,E(XY) - \Big(E(XY) - E(X)\,E(Y)\Big)\,E(X^2)}{E(X)\Big(E(X^2) - E(X)\,E(X)\Big)} \end{aligned}$$

Ugh!! Now looking carefully at the numerator, you will see that we have two terms that cancel each other: $E(X^2)\,E(XY) - E(X^2)\,E(XY)$. This gives

$$b = \frac{E(X)\,E(Y)\,E(X^2) - E(X)\,E(X)\,E(XY)}{E(X)\Big(E(X^2) - E(X)\,E(X)\Big)}$$

Now cancel the common $E(X)$ on the top and bottom to get

$$b = \frac{E(Y)\,E(X^2) - E(X)\,E(XY)}{E(X^2) - E(X)\,E(X)} \tag{21.8}$$

An equivalent solution (yes, it is equivalent but to see that you have to do the steps you saw above essentially in reverse—good home exercise!) that is easier to find uses the bottom equation. We find

$$b = \frac{E(X)E(Y) - m\,(E(X))^2}{E(X)} = E(Y) - m\,E(X).$$

21.5.1 Example

Example 21.5.1 Let's find the regression line for the data

$$D = \{(1.2, 2.3), (2.4, 1.9), (3.0, 4.5), (3.7, 5.2), (4.1, 3.2), (5.0, 7.2)\}.$$

Solution *We will do this in MatLab. Look at the MatLab session below which has been fully commented.*

Listing 21.2: Finding the regression line

```
% setup the data as X and Y vectors
X = [1.2;2.4;3.0;3.7;4.1;5.0];
Y = [2.3;1.9;4.5;5.2;3.2;7.2];
% get length of data
N = length(X);
% Find E(X) called EX here
EX = sum(X)/N;
% Find E(Y) called EY here
EY = sum(Y)/N
% find E(XY) called EXY here
EXY = sum(X.*Y)/N;
% find E(X^2) called EXX here
EXX = sum(X.*X)/N;
% find slope of the regression line m
m = (EXY-EX*EY)/(EXX-EX*EX);
m = 1.1824
% find the intercept b
b = EY-m*EX
b = 0.2269
% setup the regression line
regression = @(x) m*x+b;
% setup a linspace for plotting
u = linspace(1,5,21);
plot(X,Y,'o',u,regression(u));
xlabel('X');
ylabel('Y');
title('Regression line for Y vs X');
legend('Data','Regression','location','southeast');
```

In Fig. 21.2, we can see the regression line and the data plotted together. Now, for the homework problems, you will need to change the variables names and the labellings for the plots, but this is essentially all you have to do!

21.5.2 Homework

These exercises were modified from ones you can find in Sokal and Rohlf (2009) in the chapter on regression by creating completely fictional creatures and data scenarios. But we tried to make them sound very realistic! But you can look at any issue of the **Journal of Experimental Biology** for example to find lots of places where regression is used in the real world. The discussion in Sokal and Rohlf (2009) is really old as it is a Dover reprint, but it is very inexpensive and has a lot of great ideas in it. So feel free to pick it up or one more modern. The bottom line is there is a lot more we can do with these ideas and we encourage you to take additional courses in this area so you can learn more! You need to build the regression lines following the

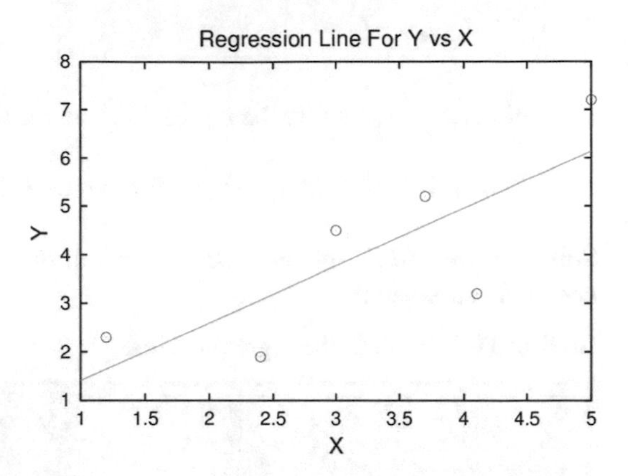

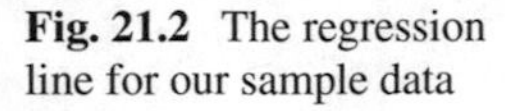
Fig. 21.2 The regression line for our sample data

MatLab template given in the example above. Make sure you label the plots nicely. Your results need to be placed in a Word doc in the usual way, nicely commented with embedded plots.

Exercise 21.5.1 *The data here has the form (Time, Temperature) where the time is the amount of time in hours that has elapsed since a mouse was inoculated with some virus and the temperature is the rabbit's temperature in degrees Fahrenheit at that time. Find the regression line for this data*

$$D = \{(24, 101.8), (32, 103.5), (48, 106.5), (56.0, 107.3), (64.0, 104.9), (80.0, 102.2)\}.$$

Exercise 21.5.2 *The data here has the form (Chromatophore density, age) where the chromatophore density is the number of chromatophore per unit area and the age is how old the squid is in months. Find the regression line for this data*

$$D = \{(3000, 3), (3650, 6), (5000, 12), (6200, 14), (6700, 15), (7300, 18)\}.$$

Exercise 21.5.3 *The data here has the form (Temperature, Calorie Expenditure) where temperature is the environmental temperature experienced by a Jewel Studded Beetle that lives only in old computer equipment in the downstairs laboratories of New Cricket University. This, of course, is the university where Jiminy Cricket's son Edgar works as a physicist. The second variable is the amount of energy the beetle uses in calories/day at that temperature which is measured in degrees Celsius. Find the regression line for this data*

$$D = \{(12.7, 67.1), (15.4, 87.3), (18.1, 200), (20.9, 190), (23.7, 200), (26.4, 210)\}.$$

Exercise 21.5.4 *The data here has the form (Spines, Neural Excitation) where spines is the number of spines on a dendrite per nm (which is known to be dependent on how active the neuron is) and neural excitation is in spikes per second. Find the regression line for this data*

$$D = \{(3000, 50), (3500, 65), (4500, 80), (6000, 100), (7000, 130), (10000, 200)\}.$$

Exercise 21.5.5 *The data here has the form (Fraction of Inhibitory Neurons, Brain Size) where the fraction of inhibitory neurons is calculated from our incomplete(!) knowledge of genomic data for a variety of species and brain size is measure in grams. Find the regression line for this data*

$$D = \{(0.1, 0.5), (0.3, 2.5), (8, 50.0), (14, 200), (20, 600), (22, 1800)\}.$$

21.6 Regression and Covariances

Now the term $\boldsymbol{E}(\boldsymbol{X}^2) - \boldsymbol{E}(\boldsymbol{X})\,\boldsymbol{E}(\boldsymbol{X})$ and $\boldsymbol{E}(\boldsymbol{XY}) - \boldsymbol{E}(\boldsymbol{X})\,\boldsymbol{E}(\boldsymbol{Y})$ occurs a lot in this kind of work. We call this kind of calculation a **covariance** and use the symbol $\boldsymbol{Cov}$ for them. The formal definitions are

$$\boldsymbol{Cov}(\boldsymbol{X}) = \boldsymbol{E}(\boldsymbol{X}^2) - \boldsymbol{E}(\boldsymbol{X})\,\boldsymbol{E}(\boldsymbol{X}) \tag{21.9}$$

$$\boldsymbol{Cov}(\boldsymbol{XY}) = \boldsymbol{E}(\boldsymbol{XY}) - \boldsymbol{E}(\boldsymbol{X})\,\boldsymbol{E}(\boldsymbol{Y}) \tag{21.10}$$

Using the idea of covariances, we see the optimal slope and intercept are given by

$$\begin{aligned} b &= \frac{\boldsymbol{E}(\boldsymbol{Y})\,\boldsymbol{E}(\boldsymbol{X}^2) - \boldsymbol{E}(\boldsymbol{X})\,\boldsymbol{E}(\boldsymbol{XY})}{\boldsymbol{Cov}(\boldsymbol{X})} \\ m &= \frac{\boldsymbol{Cov}(\boldsymbol{XY})}{\boldsymbol{Cov}(\boldsymbol{X})}. \end{aligned}$$

Finally, there is one other idea that is useful here: the idea of how our data varies from the expected value $\boldsymbol{E}(\boldsymbol{X})$. We can calculate the *expected* squared total difference as follows

$$\begin{aligned} \boldsymbol{E}((\boldsymbol{X} - \boldsymbol{E}(\boldsymbol{X}))^2) &= \frac{1}{S}\sum_{i=1}^{S}(x_i - \boldsymbol{E}(\boldsymbol{X}))^2 \\ &= \frac{1}{S}\sum_{i=1}^{S}\left(x_i^2 + 2x_i\,\boldsymbol{E}(\boldsymbol{X}) + (\boldsymbol{E}(\boldsymbol{X}))^2\right) \\ &= \frac{1}{S}\sum_{i=1}^{S}x_i^2 - 2\,\frac{\boldsymbol{E}(\boldsymbol{X})}{S}\sum_{i=1}^{S}x_i + \frac{1}{S}\sum_{i=1}^{S}(\boldsymbol{E}(\boldsymbol{X}))^2 \\ &= \boldsymbol{E}(\boldsymbol{X}^2) - 2\,(\boldsymbol{E}(\boldsymbol{X}))^2 + (\boldsymbol{E}(\boldsymbol{X}))^2 \\ &= \boldsymbol{E}(\boldsymbol{X}^2) - (\boldsymbol{E}(\boldsymbol{X}))^2 = \boldsymbol{Cov}(\boldsymbol{X}). \end{aligned}$$

This calculation gives us what is called the **variance** of our data and you should learn more about this tool in other courses as it is extremely useful. Alas, our needs are quite limited in this course, so we just need to mention it. So we have another definition. The **variance** is denoted by the symbol $\boldsymbol{Var}$ and defined by

$$\boldsymbol{Var(X)} = \boldsymbol{E((X - E(X))^2)} = \boldsymbol{E(X^2)} - \boldsymbol{(E(X))^2} = \boldsymbol{Cov(X)}. \quad (21.11)$$

Note that the variance $\boldsymbol{Var(X)}$ is exactly the same as the covariance $\boldsymbol{Cov(X)}$! We can now see that there is an interesting connection between the covariance of X and Y in terms of the covariance and variance of X. Let's denote the slope m obtaining by our optimization strategy by $\boldsymbol{m(X, Y)}$ so we always remember is a function of our data pairs. A summary of our work is in order. We have found the optimal slope and y intercept for our data is given by m and b, respectively, where

$$\begin{aligned} \boldsymbol{m(X, Y)} &= \frac{\boldsymbol{Cov(XY)}}{\boldsymbol{Cov(X)}} \\ \boldsymbol{b(X, Y)} &= \frac{\boldsymbol{E(Y)\,E(X^2) - E(X)\,E(XY)}}{\boldsymbol{Cov(X)}} \\ &= \boldsymbol{E(Y) - m(X, Y)\,E(X)} \end{aligned}$$

We call the optimal slope $\boldsymbol{m(X, Y)}$ the slope of the **regression** of Y on X. This is something we can measure so it is an estimate of how the variable y changes with respect to x. We now know a fair bit of calculus, so we can think of $\boldsymbol{m(X, Y)}$ as an estimate of either $\frac{dy}{dx}$ or $\frac{\partial y}{\partial x}$ which is a really useful idea. Then, we notice that

$$\begin{aligned} \boldsymbol{Cov(X, Y)} &= \boldsymbol{Var(X)}\,\frac{\boldsymbol{Cov(X, Y)}}{\boldsymbol{Var(X)}} \\ &= \boldsymbol{Var(X)\,m(X, Y)}. \end{aligned}$$

Thus, the covariance of x and y is proportional to the slope of the regression line of Y on X with proportionality constant given by the variance $\boldsymbol{Var(X)}$ which, of course, is the same as the covariance of X with itself, $\boldsymbol{Cov(X, X)}$.

21.6.1 Example

Example 21.6.1 Let's find the Cov(X) = Var(X) and Cov(XY) for the data

$$D = \{(1.2, 2.3), (2.4, 1.9), (3.0, 4.5), (3.7, 5.2), (4.1, 3.2), (5.0, 7.2)\}.$$

Solution *We will do this in MatLab. Look at the MatLab session below which has been fully commented.*

Listing 21.3: Finding the regression line

```
% setup the data as X and Y vectors
X = [1.2;2.4;3.0;3.7;4.1;5.0];
Y = [2.3;1.9;4.5;5.2;3.2;7.2];
% get length of data
N = length(X);
% Find E(X) called EX here
EX = sum(X)/N;
% Find E(Y) called EY here
EY = sum(Y)/N
% find E(XY) called EXY here
EXY = sum(X.*Y)/N;
% find E(X^2) called EXX here
EXX = sum(X.*X)/N;
% find Cov(X), here COVX
COVX = EXX-EX*EX; % here COVX = 1.4956
% Find COV(X,Y), here COVXY
COVXY =  EXY - EX*EY  % here COVXY = 1.7683
```

21.6.2 Homework

These exercises here are the same ones you did in Sect. 21.5 but this time you need to find the covariances like we did in the example. Your results need to be placed in a Word doc in the usual way.

Exercise 21.6.1 *The data here has the form (Time, Temperature) where the time is the amount of time in hours that has elapsed since a mouse was inoculated with some virus and the temperature is the rabbit's temperature in degrees Fahrenheit at that time. Find the regression line for this data*

$$D = \{(24, 101.8), (32, 103.5), (48, 106.5), (56.0, 107.3), (64.0, 104.9), (80.0, 102.2)\}.$$

Exercise 21.6.2 *The data here has the form (Chromatophore density, age) where the chromatophore density is the number of chromatophore per unit area and the age is how old the squid is in months. Find the regression line for this data*

$$D = \{(3000, 3), (3650, 6), (5000, 12), (6200, 14), (6700, 15), (7300, 18)\}.$$

Exercise 21.6.3 *The data here has the form (Temperature, Calorie Expenditure) where temperature is the environmental temperature experienced by a Jewel Studded Beetle that lives only in old computer equipment in the downstairs laboratories of New Cricket University. This, of course, is the university where Jiminy Cricket's son Edgar works as a physicist. The second variable is the amount of energy the beetle uses in calories/day at that temperature which is measured in degrees Celsius. Find the regression line for this data*

$$D = \{(12.7, 67.1), (15.4, 87.3), (18.1, 200), (20.9, 190), (23.7, 200), (26.4, 210)\}.$$

Exercise 21.6.4 *The data here has the form (Spines, Neural Excitation) where spines is the number of spines on a dendrite per nm (which is known to be dependent on how active the neuron is) and neural excitation is in spikes per second. Find the regression line for this data*

$$D = \{(3000, 50), (3500, 65), (4500, 80), (6000, 100), (7000, 130), (10000, 200)\}.$$

Exercise 21.6.5 *The data here has the form (Fraction of Inhibitory Neurons, Brain Size) where the fraction of inhibitory neurons is calculated from our incomplete(!) knowledge of genomic data for a variety of species and brain size is measure in grams. Find the regression line for this data*

$$D = \{(0.1, 0.5), (0.3, 2.5), (8, 50.0), (14, 200), (20, 600), (22, 1800)\}.$$

Reference

R. Sokal, F.J. Rohlf, *Introduction to Biostatistics* (Dover Publications Inc, New York, 2009)

Chapter 22
Hamilton's Rule in Evolutionary Biology

We are now ready to look at another interesting model from evolutionary biology. Recall in Chap. 2, we discusses how a phenotype of type **A** might spread through a population. We didn't say much about what the phenotype **A** might have been, but what we were really thinking about was an allele that codes for the high level trait we might call **altruistic behavior**. So type **A** is the allele that codes for **altruism** and type **B** the allele that represents the lack of **altruistic** behavior. Note our discussion is about characteristics of human populations that are very high level—not at all like our previous efforts on low level things like protein transcription. Our discussions here are based on some of the material in Chap. 3 of McElreath and Boyd (2007) and use different notation and emphasis than what is in there. However, that book has a lot more detail and covers many more issues; we are just going to focus on a small part of the study of altruistic behavior as it is an excellent vehicle into how to study a topic that is very difficult to form a quantitative model for. Again, think of the major issues we face in cognitive modeling: what are emotions, what is depression and other cognitive disorders and what is insight? These are important things to think about and extremely hard to model. But we can get started into learning how to think about such tough questions by tacking another one like viability which is the study of altruistic behavior. When you read about major problems in biology, one that pops up is how altruistic behavior could evolve in a population. Key work was done by W. Hamilton (1963, 1964). You should know that Hamilton had a very hard time publishing those papers as the blend of mathematics and biology he used was very strange to his fellow biologists! But those papers are now widely recognized as being important. Hence, in this work, you can't be dismayed if your peer group doesn't *get* what you are trying to do. Just keep plugging away is our motto! Anyway, you should look at those papers as you have now learned enough mathematics to follow quite a bit of them. Our discussion in this chapter just touches on all these interesting things.

Before we go on, we want to take a minute and discuss the philosophy of this course again. We have been working hard to show you that to develop **insight** into complicated problems requires you to think hard and deep about the biology and behavior you see in your data and so forth and then use your training in mathematics

J.K. Peterson, *Calculus for Cognitive Scientists*, Cognitive Science and Technology, DOI 10.1007/978-981-287-874-8_22

and computational techniques to help you build simple models that confirm or deny your suspicions. This is a very iterative process and we go back and forth between our different versions as we hone in on a better understanding. But the key really is to learn how to **think** for yourself.

Now in the sections that follow these remarks, we are going to explore the idea of altruistic behavior and how it might be a winning thing to pass on the future generations in a variety of ways using a number of different modeling paradigms. It is a great example to show you how those of us who model keep trying to find better and better ways to **understand**.

22.1 How Do We Define Altruism

We could define **altruism** as a behavior which reduces the **fitness** of the person who has this behavior but increases the **fitness** of the person who benefits from this altruistic behavior. However, another notion of **altruism** is that it is a behavior which reduces the fitness of the person who engages in it *relative* to the people who benefit from it. So there are two types of altruism: absolute and relative. Let's design a simple model to help us think about this more clearly. Let's also call the person or individual in the population that uses some behavior an **actor**.

22.1.1 A Shared Common Good Model

We have some animals that live in a pool: following McElreath, these are salamanders. Each salamander can choose to *poop* in the pool or walk outside of the pool to do it. McElreath's use of *poop* here was inspired; the abstraction of this is that there is a closed environment where a population of creatures has a behavior that hurts the environment they share and so hurts what you might call some measure of their *fitness* in some way. Any one of these creatures can help all of them by moving out of their closed environment to do this action and thereby save all of them from the negative consequences. But you have to admit, all that abstraction makes so much more sense when it is attached to salamanders pooping in their pool! For them the act of walking outside the pool is a **cost** to the salamander that does it. But every creature in the pond shares the common benefit of having cleaner water. Consider the following matrix for interaction of two salamanders, Salamander 1 and Salamander 2. This table records the fitness consequences of each of four possible interactions. In the interaction of two salamanders, each has a 50 % chance of either cooperating by going out of the pool to poop or staying in the pool to poop. There are four types of interactions.

- they both go outside the pool to poop is called V(walk out, walk out).
- the salamander 1 is nice and walks outside the pool to poop but salamander 2 poops in the pool. Call this V(walk out, stay in).
- salamander one poops in the pool and salamander two walks out of the pool to poop. Call this V(stay in, walk out).
- salamander one and salamander two both poop in the pool. Call this V(stay in, stay in).

Since going outside of the pool to poop is an action that helps the group, we can call this **cooperation**. If a salamander stays in the pool to poop, the salamander goes against the group benefit; we say the salamander **defects**. The general form of our matrix is

	Salamander Two	
Salamander One	**Go outside of the pool to poop**	**Stay in the pool to poop**
Go outside of the pool to poop	V(walk out, walk out)	V(walk out, stay in)
Stay in the pool to poop	V(stay in, walk out)	V(stay in, stay in)

Let B be the total benefit for the population; we call this the **group benefit**. We let c be the private cost to a salamander to leaving the pool to poop. Next, we have to decide what these four fitness consequences should be.

- V(walk out, walk out): since this option is chosen 50 % of the time the average group benefit seen is $B/2$. However, Salamander one uses this option at the same time salamander two uses it. So there is an additional $B/2$ that comes from that choice. The cost of going outside to poop is always c so the total fitness for player one is upgraded by $B - c$.
- V(stay in, walk out): since Salamander one stays in the pool there is no private cost incurred. Thus, the upgrade in fitness for Salamander one is $B/2$.
- V(walk out, stay in): since Salamander one goes out of the pool to poop, the fitness increase Salamander one sees is $B/2 - c$. There is no effect from Salamander two's choice on Salamander One's fitness.
- V(stay in, stay in): If both salamanders poop in the pool, this does not change the group benefit, so the fitness increase of salamander one should be 0.

We can summarize what we have said into a table as shown below. The table shows the fitness consequences to Salamander one. This is called the **Payoff Matrix for Salamander One**.

	Salamander Two	
Salamander One	**Go outside of the pool to poop**	**Stay in the pool to poop**
Go outside of the pool to poop	$B - c$	$B/2 - c$
Stay in the pool to poop	$B/2$	0

When does the action to be altruistic and help the group spread throughout the population? This happens when the best choice for each salamander to make is the one that increases the overall good. In our example above, if it was true that $B/2 - c > 0$, then we have the other terms are all larger than $B/2 - c$ as

$$B - c = B/2 + B/2 - c > B/2 - c$$
$$B/2 > B/2 - c$$

In this case, the *best* choice to make is always that both salamander's cooperate for the group benefit and go outside the poop to poop. So this action spreads throughout the population and becomes dominant.

22.1.1.1 Example

Example 22.1.1 Let $B = 3$ and $c = 1$ in our salamander model. Write down the payoff matrix and determine if the action to go outside the pool, i.e. be altruistic, spreads throughout the population.

Solution *For these values, the payoff matrix is*

	Salamander Two	
Salamander One	*Go outside of the pool to poop*	*Stay in the pool to poop*
Go outside of the pool to poop	2	1/2
Stay in the pool to poop	3/2	0

and since $B/2 - c = 1/2 > 0$, this action will spread throughout the population.

Example 22.1.2 Let $B = 2$ and $c = 0.8$ in our salamander model. Write down the payoff matrix and determine if the action to go outside the pool, i.e. be altruistic, spreads throughout the population.

Solution *For these values, the payoff matrix is*

	Salamander Two	
Salamander One	*Go outside to poop*	*Stay in to poop*
Go outside to poop	1.2	0.2
Stay in to poop	1.0	0

and since $B/2 - c = 0.2 > 0$, this action will spread throughout the population.

Example 22.1.3 Let $B = 2$ and $c = 1.3$ in our salamander model. Write down the payoff matrix and determine if the action to go outside the pool, i.e. be altruistic, spreads throughout the population.

Solution *For these values, the payoff matrix is*

	Salamander Two	
Salamander One	***Go outside to poop***	***Stay in to poop***
Go outside to poop	0.7	−0.3
Stay in to poop	1.0	0

and since $B/2 - c = -0.3 < 0$, *this action will not spread throughout the population.*

22.1.1.2 Homework

Exercise 22.1.1 *Let* $B = 4$ *and* $c = 3$ *in our salamander model. Write down the payoff matrix and determine if the action to go outside the pool, i.e. be altruistic, spreads throughout the population.*

Exercise 22.1.2 *Let* $B = 10$ *and* $c = 4$ *in our salamander model. Write down the payoff matrix and determine if the action to go outside the pool, i.e. be altruistic, spreads throughout the population.*

Exercise 22.1.3 *Let* $B = 2$ *and* $c = 1$ *in our salamander model. Write down the payoff matrix and determine if the action to go outside the pool, i.e. be altruistic, spreads throughout the population.*

Exercise 22.1.4 *Let* $B = 5$ *and* $c = 2$ *in our salamander model. Write down the payoff matrix and determine if the action to go outside the pool, i.e. be altruistic, spreads throughout the population.*

22.1.2 The Abstract Version

Now let's make this a bit more abstract. Label the cooperative action where the salamander goes out of the pool to poop as **C** and the action where the salamander goes against the group benefit and poops in the pool by **D**. Then our matrix can be written more succinctly as

	Salamander Two	
Salamander One	**C**	**D**
C	$B - c$	$B/2 - c$
D	$B/2$	0

Note that our simple (and silly) example could have been about something else. It works as long as there is a shared common good and a private cost for helping the group. For example, it is well known that if a squirrel is on the ground and sees a predator, often the squirrel will stand on its feet and utter loud piercing cries to alerts its mates of the danger. Of course, this has a great personal cost: the squirrel that calls out the alarm increases it own risk greatly! So the squirrel example fits into our common good model nicely too. So you don't have to think about *poop* but we figured we could grab your attention by starting with it! In general, instead of salamanders in our example, we would call them **agents** and we would rewrite our table as

	Agent Two	
Agent One	**C**	**D**
C	$B - c$	$B/2 - c$
D	$B/2$	0

Finally, we define the following four things:

- $\boldsymbol{V(C, C)}$: payoff or fitness consequence to agent one's choice of **C** given that agent two chooses **C**.
- $\boldsymbol{V(C, D)}$: payoff or fitness consequence to agent one's choice of **C** given that agent two chooses **D**.
- $\boldsymbol{V(D, C)}$: payoff or fitness consequence to agent one's choice of **D** given that agent two chooses **C**.
- $\boldsymbol{V(D, D)}$: payoff or fitness consequence to agent one's choice of **D** given that agent two chooses **D**.

which are just like the $\boldsymbol{V}$'s we defined earlier. Then the general **fitness** matrix or **payoff** matrix is

	Agent Two	
Agent One	**C**	**D**
C	$\boldsymbol{V(C, C)}$	$\boldsymbol{V(C, D)}$
D	$\boldsymbol{V(D, C)}$	$\boldsymbol{V(D, D)}$

The total fitness consequence to a choice of $\boldsymbol{D}$ is then called $\boldsymbol{W(D)}$ and is given by

$$\boldsymbol{W(D)} = \boldsymbol{W_0} + \boldsymbol{p}\, \boldsymbol{V(D, C)} + (1 - \boldsymbol{p})\, \boldsymbol{V(D, D)}.$$

where in general $\boldsymbol{p}$ is the probability the choice of $\boldsymbol{C}$ is made and $1 - \boldsymbol{p}$, the probability the other choice of $\boldsymbol{D}$ is made. Also the term $\boldsymbol{W_0}$ is a *baseline* fitness level the population has before any agent choices have been made. Similarly, the fitness consequence to a choice of $\boldsymbol{C}$ is then called $\boldsymbol{W(C)}$ and is given by

$$\boldsymbol{W(C)} = \boldsymbol{W_0} + \boldsymbol{p}\, \boldsymbol{V(C, C)} + (1 - \boldsymbol{p})\, \boldsymbol{V(C, D)}.$$

We can simplify this a bit more (well, we think it simplifies but then we like abstraction!) using a little matrix and vector multiplication. The definitions above can be written more succinctly as

$$\begin{bmatrix} \boldsymbol{W(C)} \\ \boldsymbol{W(D)} \end{bmatrix} = \begin{bmatrix} \boldsymbol{V(C,C)} & \boldsymbol{V(C,D)} \\ \boldsymbol{V(D,C)} & \boldsymbol{V(D,D)} \end{bmatrix} \begin{bmatrix} \boldsymbol{p} \\ 1-\boldsymbol{p} \end{bmatrix} + \begin{bmatrix} \boldsymbol{W_0} \\ \boldsymbol{W_0} \end{bmatrix}$$

So for our example, we have

$$\begin{aligned} \begin{bmatrix} \boldsymbol{W(C)} \\ \boldsymbol{W(D)} \end{bmatrix} &= \begin{bmatrix} B-c & B/2-c \\ B/2 & 0 \end{bmatrix} \begin{bmatrix} 0.5 \\ 0.5 \end{bmatrix} + \begin{bmatrix} \boldsymbol{W_0} \\ \boldsymbol{W_0} \end{bmatrix} \\ &= \begin{bmatrix} \boldsymbol{W_0} + (0.5)(B-c) + (0.5)(B/2-c) \\ \boldsymbol{W_0} + (0.5)(B/2) + (0.5)(0) \end{bmatrix} \end{aligned}$$

We conclude

$$\begin{aligned} \boldsymbol{W(C)} &= \boldsymbol{W_0} + (0.5)(B-c) + (0.5)(B/2-c) \\ \boldsymbol{W(D)} &= \boldsymbol{W_0} + (0.5)(B/2). \end{aligned}$$

Now for cooperation to spread through the population, the change in difference $\boldsymbol{W(C)} - \boldsymbol{W(D)}$ should be positive. Note this change is just the amount above the baseline $\boldsymbol{W_0}$ and so

$$\begin{aligned} \boldsymbol{W(C)} - \boldsymbol{W(D)} &= (0.5)(B-c) + (0.5)(B/2-c) - (0.5)(B/2) \\ &= B/4 + 2\,(B/2-c) - B/4 = 2\,(B/2-c). \end{aligned}$$

This difference is positive if we assume $B/2 - c > 0$. Hence, we have our first inkling that this relationship, $B/2 > c$ is an important one that is linked to the spread of altruistic behavior throughout the population. Indeed, it implies that the benefit to cooperation is always increasing!

22.1.2.1 Examples

Example 22.1.4 Determine if action $\boldsymbol{C}$ will spread throughout the population for $\boldsymbol{V(C,C)} = 3$, $\boldsymbol{V(C,D)} = 2$, $\boldsymbol{V(D,C)} = 1$ and $\boldsymbol{V(D,D)} = 0.5$ when the probability of action $\boldsymbol{C}$ is $p = 0.5$.

Solution *The payoff matrix is*

	Agent Two	
Agent One	**C**	**D**
C	3	2
D	1	0.5

The payoff for action $\boldsymbol{C}$ *is then*

$$\begin{aligned} \boldsymbol{W(C)} &= \boldsymbol{W_0} + \boldsymbol{p}\, \boldsymbol{V(C, C)} + (1 - \boldsymbol{p})\, \boldsymbol{V(C, D)} \\ &= \boldsymbol{W_0} + (0.5)\,(3) + (0.5)\,(2.0) = \boldsymbol{W_0} + 2.50. \end{aligned}$$

and the payoff for action $\boldsymbol{D}$ *is*

$$\begin{aligned} \boldsymbol{W(D)} &= \boldsymbol{W_0} + \boldsymbol{p}\, \boldsymbol{V(D, C)} + (1 - \boldsymbol{p})\, \boldsymbol{V(D, D)} \\ &= \boldsymbol{W_0} + (0.5)\,(1) + (0.5)(0) = \boldsymbol{W_0} + 0.5. \end{aligned}$$

The difference $\boldsymbol{W(C)} - \boldsymbol{W(D)}$ *is* 2.0 *and so action* $\boldsymbol{C}$ *will spread.*

Example 22.1.5 Determine if action $\boldsymbol{C}$ will spread throughout the population for $\boldsymbol{V(C, C)} = 2$, $\boldsymbol{V(C, D)} = -2$, $\boldsymbol{V(D, C)} = 1$ and $\boldsymbol{V(D, D)} = 0$ when the probability of action $\boldsymbol{C}$ is $p = 0.5$.

Solution *The payoff matrix is*

	Agent Two	
Agent One	C	D
C	2	−2
D	1	0

The payoff for action $\boldsymbol{C}$ *is then*

$$\begin{aligned} \boldsymbol{W(C)} &= \boldsymbol{W_0} + \boldsymbol{p}\, \boldsymbol{V(C, C)} + (1 - \boldsymbol{p})\, \boldsymbol{V(C, D)} \\ &= \boldsymbol{W_0} + (0.5)\,(2) + (0.5)\,(-2) = \boldsymbol{W_0} + 0.0. \end{aligned}$$

and the payoff for action $\boldsymbol{D}$ *is*

$$\begin{aligned} \boldsymbol{W(D)} &= \boldsymbol{W_0} + \boldsymbol{p}\, \boldsymbol{V(D, C)} + (1 - \boldsymbol{p})\, \boldsymbol{V(D, D)} \\ &= \boldsymbol{W_0} + (0.5)\,(1) + (0.5)(0) = \boldsymbol{W_0} + 0.5. \end{aligned}$$

The difference $\boldsymbol{W(C)} - \boldsymbol{W(D)}$ *is* −0.5 *and so action* $\boldsymbol{C}$ *will not spread.*

22.1.2.2 Homework

Exercise 22.1.5 *Determine if action* $\boldsymbol{C}$ *will spread throughout the population for* $\boldsymbol{V(C, C)} = 1$, $\boldsymbol{V(C, D)} = 2$, $\boldsymbol{V(D, C)} = 1$ *and* $\boldsymbol{V(D, D)} = 0$ *when the probability of action* $\boldsymbol{C}$ *is* $p = 0.5$.

Exercise 22.1.6 *Determine if action* $\boldsymbol{C}$ *will spread throughout the population for* $\boldsymbol{V(C, C)} = 4$, $\boldsymbol{V(C, D)} = -3$, $\boldsymbol{V(D, C)} = 2$ *and* $\boldsymbol{V(D, D)} = 0$ *when the probability of action* $\boldsymbol{C}$ *is* $p = 0.5$.

Exercise 22.1.7 *Determine if action $\boldsymbol{C}$ will spread throughout the population for $\boldsymbol{V}(\boldsymbol{C}, \boldsymbol{C}) = 3$, $\boldsymbol{V}(\boldsymbol{C}, \boldsymbol{D}) = 2$, $\boldsymbol{V}(\boldsymbol{D}, \boldsymbol{C}) = 1$ and $\boldsymbol{V}(\boldsymbol{D}, \boldsymbol{D}) = 0$ when the probability of action $\boldsymbol{C}$ is $p = 0.4$.*

Exercise 22.1.8 *Determine if action $\boldsymbol{C}$ will spread throughout the population for $\boldsymbol{V}(\boldsymbol{C}, \boldsymbol{C}) = 2$, $\boldsymbol{V}(\boldsymbol{C}, \boldsymbol{D}) = -1$, $\boldsymbol{V}(\boldsymbol{D}, \boldsymbol{C}) = 1$ and $\boldsymbol{V}(\boldsymbol{D}, \boldsymbol{D}) = 0$ when the probability of action $\boldsymbol{C}$ is $p = 0.6$.*

22.2 Hamilton's Rule

Now let's talk about what is called Hamilton's Rule (Hamilton 1964). Hamilton had the insight that **kinship** could lead to altruism evolving. As McElreath says (McElreath and Boyd 2007) on p. 78,

> He reasoned that an allele that codes for altruism could selectively help other copies of itself if altruistic behavior was preferentially directed toward kin.

In our cooperate/ defection model interaction in the last section, we assumed random sampling of cooperative/defecting individuals from the population. So the chance of an agent being cooperative or defecting was the same. This is what determined our fitness calculations. It turns out **non random** interactions are the key to the formation of altruism. Let

- $\boldsymbol{P}(\boldsymbol{A}|\boldsymbol{A})$ be the probability an altruist **A** is paired with another altruist **A**.
- $\boldsymbol{P}(\boldsymbol{A}|\boldsymbol{N})$ be the probability an altruist **A** is paired with a non altruist **N**.
- $\boldsymbol{P}(\boldsymbol{N}|\boldsymbol{A})$ be the probability a non altruist **N** is paired with an altruist **A**.
- $\boldsymbol{P}(\boldsymbol{N}|\boldsymbol{N})$ be the probability a non altruist **N** is paired with another non altruist **N**.

We also need the fitness values for these interactions.

- $\boldsymbol{V}(\boldsymbol{A}|\boldsymbol{A})$ is the fitness change for the **A** and **A** pairing.
- $\boldsymbol{V}(\boldsymbol{A}|\boldsymbol{N})$ is the fitness change for the **A** and **N** pairing.
- $\boldsymbol{V}(\boldsymbol{N}|\boldsymbol{A})$ is the fitness change for the **N** and **A** pairing.
- $\boldsymbol{V}(\boldsymbol{N}|\boldsymbol{N})$ is the fitness change for the **N** and **N** pairing.

We can then calculate fitness like usual with these values. The fitness $\boldsymbol{W}(\boldsymbol{A})$ and $\boldsymbol{W}(\boldsymbol{N})$ are given by

$$\boldsymbol{W}(\boldsymbol{A}) = w_0 + \boldsymbol{P}(\boldsymbol{A}|\boldsymbol{A})\ \boldsymbol{V}(\boldsymbol{A}|\boldsymbol{A}) + \boldsymbol{P}(\boldsymbol{A}|\boldsymbol{N})\ \boldsymbol{V}(\boldsymbol{A}|\boldsymbol{N})$$
$$\boldsymbol{W}(\boldsymbol{N}) = w_0 + \boldsymbol{P}(\boldsymbol{N}|\boldsymbol{A})\ \boldsymbol{V}(\boldsymbol{N}|\boldsymbol{A}) + \boldsymbol{P}(\boldsymbol{N}|\boldsymbol{N})\ \boldsymbol{V}(\boldsymbol{N}|\boldsymbol{N})$$

Note, before we had two probabilities p and $1 - p$ which applied to both the cooperative and defective choices. Now we allow each possible interaction to be governed by its own probability. So we have added further abstraction and more possibilities.

This allows us to model more things! Now let's go back and think about our previous cooperate/defecting strategy analysis. If the agents are not being chosen randomly with equal probability, we have to rethink how we calculated our payoff matrix. Let b be the benefit gain to the population for an altruistic act; we call this the **group benefit**. We let c be the private cost to a altruist. We have to decide what these four fitness consequences should be.

- $\boldsymbol{V(A|A)}$: this pairing is not a random choice anymore, so the change in fitness to agent one is just the group benefit b minus its cost c: hence, this value is $b-c$ now.
- $\boldsymbol{V(A|N)}$: since this pairing is not a random choice anymore, so the change in fitness to agent one is just its cost c: hence, this value is $-c$ now.
- $\boldsymbol{V(N|A)}$: since agent two's action adds group benefit, the payoff to agent one is b.
- $\boldsymbol{V(N|N)}$: there is no change in fitness for now. So the value of the payoff to agent one is 0.

The payoff matrix for the altruistic agent is then

	Agent Two	
Agent One	**A**	**N**
A	$\boldsymbol{V(A\|A)}$	$\boldsymbol{V(A\|N)}$
N	$\boldsymbol{V(N\|A)}$	$\boldsymbol{V(N\|N)}$

=

	Agent Two	
Agent One	**A**	**N**
A	$b-c$	$-c$
N	b	0

The fitness calculations are then

$$\begin{aligned}\boldsymbol{W(A)} &= w_0 + \boldsymbol{P(A|A)}\,\boldsymbol{V(A|A)} + \boldsymbol{P(A|N)}\,\boldsymbol{V(A|N)}\\ &= w_0 + (b-c)\,\boldsymbol{P(A|A)} - c\,\boldsymbol{P(A|N)}\\ \boldsymbol{W(N)} &= w_0 + \boldsymbol{P(N|A)}\,\boldsymbol{V(N|A)} + \boldsymbol{P(N|N)}\,\boldsymbol{V(N|N)}\\ &= w_0 + b\,\boldsymbol{P(N|A)}.\end{aligned}$$

But we also know that $\boldsymbol{P(A|A)} + \boldsymbol{P(A|N)} = 1$, so we can simplify a bit to

$$\begin{aligned}\boldsymbol{W(A)} &= w_0 + (b-c)\,\boldsymbol{P(A|A)} - c\,(1 - \boldsymbol{P(A|A)})\\ &= w_0 + b\,\boldsymbol{P(A|A)} - c.\end{aligned}$$

Altruistic behavior will increase in frequency when $\boldsymbol{W(A)} > \boldsymbol{W(N)}$. Note

$$\boldsymbol{W(A)} > \boldsymbol{W(N)} \Longrightarrow w_0 + b\,\boldsymbol{P(A|A)} - c > w_0 + b\,\boldsymbol{P(N|A)}.$$

With a bit of algebra, we find altruism increases when

$$\boldsymbol{W(A)} > \boldsymbol{W(N)} \Longrightarrow b\,(\boldsymbol{P(A|A)} - \boldsymbol{P(N|A)}) > c.$$

We are almost done. Now let's get to Hamilton's insight. Hamilton introduced what he called the **coefficient of relatedness**, **r**, to the mix. There is a chance **r** that individuals possess the same allele due to common descent. He then modified the probabilities like this: he let

- $\boldsymbol{P(A|A)}$ be the probability an altruist **A** is paired with another altruist **A**. This is now

$$\boldsymbol{P(A|A)} = r \times 1 + (1 - r) \times p$$

  which says altruistics interact with their kin with probability 1 and with everyone else at probability $1 - r$. Here p is the frequency of the altruistic allele in the population.
- $\boldsymbol{P(A|N)}$ be the probability an altruist **A** is paired with a non altruist **N** which is

$$\boldsymbol{P(A|N)} = r \times 0 + (1 - r) \times (1 - p)$$

  as kin don't interact with non kin and $1 - p$ is the frequency of non altruists in the population.
- $\boldsymbol{P(N|A)}$ be the probability a non altruist **N** is paired with an altruist **A**. We have

$$\boldsymbol{P(N|A)} = r \times 0 + (1 - r) \times (1 - p)$$

  as kin don't interact with non kin and again, $1 - p$ is the frequency of non altruists in the population.
- $\boldsymbol{P(N|N)}$ be the probability a non altruist **N** is paired with another non altruist **N**. This is

$$\boldsymbol{P(N|N)} = r \times 1 + (1 - r) \times (1 - p)$$

  where non altruist kin interact with their own kin with probability 1 and with the rest of the population with probability $1 - p$.

We already know altruism spreads or increases if

$$\boldsymbol{W(A)} > \boldsymbol{W(N)} \Longrightarrow b\,(\boldsymbol{P(A|A)} - \boldsymbol{P(N|A)}) > c.$$

Plugging in our new probabilities, we have altruism increases if

$$\boldsymbol{W(A)} > \boldsymbol{W(N)} \Longrightarrow b\,(r + (1 - r)p - (1 - r)p) > c \Longrightarrow r\,b > c.$$

This relationship is Hamilton's insight. Of course, it is not clear at all what **r** should mean! The rest of this chapter gives various answers to how **r** might be interpreted.

22.2.1 Example

Example 22.2.1 Determine if altruism increases when $\boldsymbol{P(A|A)} = 0.6$, $\boldsymbol{P(N|A)} = 0.4$, $b = 2$ and $c = 1$.

Solution *We know altruism increases when*

$$\boldsymbol{W(A)} > \boldsymbol{W(N)} \Longrightarrow b\,(\boldsymbol{P(A|A)} - \boldsymbol{P(N|A)}) > c.$$

Here, we have $(2)\ ((0.6) - (0.4)) = 0.4$ *which is less than* $c = 1$. *So altruism doesn't increase.*

Example 22.2.2 Determine if altruism increases when the coefficient of relatedness $r = 0.3$, the benefit to the group $b = 3$ and the cost of altruism $c = 1$.

Solution *We know altruism increases when* $r\ b > c$ *so this is a simple calculation. We have* $r\ b = 0.9$ *which is less than* $c = 1$. *So altruism does not increase.*

22.2.2 Homework

Exercise 22.2.1 *Determine if altruism increases when* $\boldsymbol{P(A|A)} = 0.5$, $\boldsymbol{P(N|A)} = 0.5$, $b = 3$ *and* $c = 2$.

Exercise 22.2.2 *Determine if altruism increases when* $\boldsymbol{P(A|A)} = 0.4$, $\boldsymbol{P(N|A)} = 0.6$, $b = 1$ *and* $c = 1$.

Exercise 22.2.3 *Determine if altruism increases when the coefficient of relatedness* $r = 0.4$, *the benefit to the group* $b = 2$ *and the cost of altruism* $c = 0.6$.

Exercise 22.2.4 *Determine if altruism increases when the coefficient of relatedness* $r = 0.7$, *the benefit to the group* $b = 4$ *and the cost of altruism* $c = 1.2$.

22.3 Gene Survival

Now let's extend the work we did on viability selection to a multiple gene situation. In a population, each individual i has a fitness w_i and in that individual the frequency of allele **A** is p_i. As an example, we could have an organism in which the allele **A** either occurs or it does not. So if individual i has $p_i = 1$, this means that individual has allele **A**. If another individual j does not have allele **A**, it would have $p_j = 0$. The situation could be more complicated; the organism could have the allele **A** occur with frequency 0, $1/2$ or 1 instead of the simpler model which had only two states, 0 and 1. Hence, in a population of n individuals, we should look at the average frequency $\bar{p}$ and average fitness $\bar{w}$ of allele **A** in the population. The fitness of allele **A** is individual i is just $w_i\, p_i$ and so we can calculate

$$\bar{p} = (1/n) \sum_{i=1}^{n} p_i$$

$$\bar{w} = (1/n) \sum_{i=1}^{n} w_i\, p_i$$

Recall the value of **A** in our earlier viability selection model was the probability the **A** phenotype survives to adulthood. In this new case, it will be interpreted as the average fitness of the allele **A**. The average fitness of allele **A** in the population is thus

$$\boldsymbol{V_A} = \frac{\bar{w}}{\bar{p}} = \frac{(1/n) \sum_{i=1}^{n} w_i\, p_i}{(1/n) \sum_{i=1}^{n} p_i} = \frac{\sum_{i=1}^{n} w_i\, p_i}{\sum_{i=1}^{n} p_i}$$

Let's go back and look at the viability selection model we started this book with. Recall, we assume there are only two phenotypes, type **A** and type **B**. Now we focus on the phenotype that arises from an individual i having a frequency p_i of the allele **A**. We lump all other phenotypes into the type **B**. We define

- $\boldsymbol{N_A}(t)$ is the number of individuals of type **A** and $\boldsymbol{N_B}(t)$ is the number of individuals of **B** in a given generation.
- $\boldsymbol{N}(t)$ is the number of individuals in the population and this number changes each generation as

$$\boldsymbol{N}(t) = \boldsymbol{N_A}(t) + \boldsymbol{N_B}(t).$$

We also keep track of the fraction of individuals in the population that are type **A** or **B**. This fraction is also called the **frequency** of type **A** and **B** respectively. As usual, we define

$$\boldsymbol{P_A(t)} = \frac{\boldsymbol{N_A(t)}}{\boldsymbol{N_A(t) + N_B(t)}}$$

$$\boldsymbol{P_B(t)} = \frac{\boldsymbol{N_B(t)}}{\boldsymbol{N_A(t) + N_B(t)}}$$

We let the average fitness of type **A** be $\boldsymbol{V_A}$ and the average fitness of the type **B** be $\boldsymbol{V_B}$, Then the fitness of type **A** is $\boldsymbol{V_A}\, \boldsymbol{N_A}(t)$ and the fitness of type **B** is $\boldsymbol{V_B}\, \boldsymbol{N_B}(t)$. We see the frequency of type **A** must be

$$\boldsymbol{f_A(t)} = \frac{\boldsymbol{V_A}\, \boldsymbol{N_A}(t)}{\boldsymbol{V_A}\, \boldsymbol{N_A}(t) + \boldsymbol{V_B}\, \boldsymbol{N_B}(t)}.$$

Now divide top and bottom of this fraction by $\boldsymbol{N_A(t)} + \boldsymbol{N_B(t)}$ (this is messy but at this point, we think you can do it yourself on scratch paper!) to find

$$\boldsymbol{f_A(t)} = \frac{\boldsymbol{V_A}\ \boldsymbol{P_A}(t)}{\boldsymbol{V_A}\ \boldsymbol{P_A}(t) + \boldsymbol{V_B}\ \boldsymbol{P_B}(t)}.$$

Note the original $\boldsymbol{P_A}$ and $\boldsymbol{P_B}$ could be written using $\boldsymbol{V(A)} = 0.5$ and $\boldsymbol{V(B)} = 0.5$; i.e. both type **A** and **B** are equally likely. Thus we could have written

$$\boldsymbol{P_A(t)} = \frac{0.5\ \boldsymbol{N_A(t)}}{0.5\ \boldsymbol{N_A(t)} + 0.5\ \boldsymbol{N_B(t)}}$$
$$\boldsymbol{P_B(t)} = \frac{0.5\ \boldsymbol{N_B(t)}}{0.5\ \boldsymbol{N_A(t)} + 0.5\ \boldsymbol{N_B(t)}}$$

Hence, the frequency $\boldsymbol{f_A(t)}$ is really the change in fitness for type **A** in the population due to the mixture of the allele for **A** throughout the population due to the frequencies p_i which are no longer necessarily all 0.5. So we will call $\boldsymbol{f_A(t)}$, the new fitness at generation $t + 1$ and write

$$\boldsymbol{P_A(t+1)} = \frac{\boldsymbol{V_A}\ \boldsymbol{P_A}(t)}{\boldsymbol{V_A}\ \boldsymbol{P_A}(t) + \boldsymbol{V_B}\ \boldsymbol{P_B}(t)}.$$

Now let $\boldsymbol{W_{av}}(t)$ be the average fitness in the population $\boldsymbol{V_A}\ \boldsymbol{P_A}(t) + \boldsymbol{V_B}\ \boldsymbol{P_B}(t)$. Then rewriting again, we have

$$\boldsymbol{P_A(t+1)} = \frac{\boldsymbol{V_A}\ \boldsymbol{P_A}(t)}{\boldsymbol{W_{av}}(t)}.$$

Now subtract $\boldsymbol{P_A}(t)$ to get

$$\boldsymbol{\Delta P}(t) = \boldsymbol{P_A(t+1)} - \boldsymbol{P_A(t)} = \frac{\boldsymbol{V_A}\ \boldsymbol{P_A}(t)}{\boldsymbol{W_{av}}(t)} - \boldsymbol{P_A(t)}.$$

Multiplying through by $\boldsymbol{W_{av}}(t)$, we get to our final form.

$$\boldsymbol{\Delta P}(t)\ \boldsymbol{W_{av}}(t) = \boldsymbol{V_A}\ \boldsymbol{P_A}(t) - \boldsymbol{P_A(t)}\ \boldsymbol{W_{av}}(t).$$

22.3.1 Back to Covariance!

Now we will use what we know about covariance to help us understand what is happening in this multiple gene setting. Let's also drop the (t)'s for convenience. We let $n = \boldsymbol{N_A} + \boldsymbol{N_B}$ too. We know here the average frequency and fitness of **A** is then

$$\boldsymbol{P_A} = \frac{1}{n} \sum_{i=1}^{n} p_i$$

$$\boldsymbol{W_{av}} = \frac{1}{n} \sum_{i=1}^{n} w_i \; p_i$$

and the average fitness is

$$\boldsymbol{V_A} = \frac{\sum_{i=1}^{n} w_i \; p_i}{\sum_{i=1}^{n} p_i}$$

and so

$$\begin{aligned} \boldsymbol{\Delta P \; W_{av}} &= \left(\frac{\sum_{i=1}^{n} w_i \; p_i}{\sum_{i=1}^{n} p_i} \right) \frac{1}{n} \sum_{i=1}^{n} p_i - \frac{1}{n} \sum_{i=1}^{n} p_i \; \frac{1}{n} \sum_{i=1}^{n} w_i \\ &= \frac{1}{n} \sum_{i=1}^{n} w_i \; p_i - \frac{1}{n} \sum_{i=1}^{n} p_i \; \frac{1}{n} \sum_{i=1}^{n} w_i \\ &= \boldsymbol{E(WP) - E(W)\; E(P) = Cov(WP)}. \end{aligned}$$

We conclude

$$\boldsymbol{\Delta P \; W_{av} = Cov(WP) = Var(P)\, m(P, W)}, \tag{22.1}$$

where $\boldsymbol{m(P, W)}$ is the slope of the regression line $w = mp + b$ that fits the w versus p data. Recall it gives us an estimate of how the average fitness w depends on the frequency p of the type **A** allele and hence is an estimate of $\frac{dw}{dp}$. We will use this equation to figure out more about when altruism spreads throughout the population.

22.4 Altruism Spread Under Additive Fitness

Now let's use Eq. 22.1 to see how altruism might spread under the assumption that we have additive fitness. This means that if you are helped twice, your fitness doubles. Most biologists and ecologists think this is **not** true as helping, like most actions, suffers from diminishing returns. But the algebra is much simpler in this case so let's give it a go.

We assume that the individual fitness w_i satisfies the relationship

$$w_i = w_0 + b \; y_i - c \; h_i \tag{22.2}$$

where b is the value of increase in fitness due to a cooperative action and c is the cost of a cooperative action. The numbers y_i and h_i are interpreted as the *probability an individual i benefits from another individual's cooperative act* and *the probability of individual i is cooperative and incurs a cost*, respectively. If we interpret these probabilities in terms of our allele model, we would say y_i is the probability individual i gains a benefit from having p_i alleles of type **A** and h_i is the probability individual i has a loss or cost associated with having the allele frequency. From Eq. 22.1, we know the fitness of the **A** phenotype will increase if $\boldsymbol{Cov(WP)} > 0$. So we need to calculate $\boldsymbol{Cov(WP)}$.

$$\boldsymbol{Cov(WP)} = \boldsymbol{E(WP)} - \boldsymbol{E(W)}\,\boldsymbol{E(P)}.$$

We see

$$\begin{aligned}
\boldsymbol{E(WP)} &= \frac{1}{n}\sum_{i=1}^{n}(w_0 + by_i - ch_i)p_i \\
&= w_0\frac{1}{n}\sum_{i=1}^{n} p_i + b\,\frac{1}{n}\sum_{i=1}^{n} y_i\,p_i - c\frac{1}{n}\sum_{i=1}^{n} h_i\,p_i \\
&= w_0 + b\,\boldsymbol{E(YP)} - c\,\boldsymbol{E(HP)}.
\end{aligned}$$

and

$$\begin{aligned}
\boldsymbol{E(W)} &= \frac{1}{n}\sum_{i=1}^{n}(w_0 + by_i - ch_i) \\
&= w_0\frac{1}{n}\sum_{i=1}^{n} + b\,\frac{1}{n}\sum_{i=1}^{n} y_i - c\frac{1}{n}\sum_{i=1}^{n} h_i \\
&= w_0 + b\,\boldsymbol{E(Y)} - c\,\boldsymbol{E(H)}.
\end{aligned}$$

Combining, we have

$$\begin{aligned}
\boldsymbol{Cov(WP)} &= w_0 + b\,\boldsymbol{E(YP)} - c\,\boldsymbol{E(HP)} - (w_0 + b\,\boldsymbol{E(Y)} - c\,\boldsymbol{E(H)})\,\boldsymbol{E(P)} \\
&= b\,(\boldsymbol{E(YP)} - \boldsymbol{E(Y)}\,\boldsymbol{E(P)}) - c\,(\boldsymbol{E(HP)} - \boldsymbol{E(H)}\,\boldsymbol{E(P)}) \\
&= b\,\boldsymbol{Cov(YP)} - c\,\boldsymbol{Cov(HP)}.
\end{aligned}$$

For this allele to spread throughout the population, we want $\boldsymbol{Cov(WP)} > 0$ and so this happens when

$$b\,\boldsymbol{Cov(YP)} - c\,\boldsymbol{Cov(HP)} > 0$$

or

$$b\left(\frac{\boldsymbol{Cov(YP)}}{\boldsymbol{Cov(HP)}}\right) > c$$

In our first model of a cooperative action, we found fitness increased if $B\ (1/2) > c$. Hence, we have a bit more general result now. Our 1/2 came from an argument based on each agent having an equal opportunity of being cooperative or not—hence, the 1/2 indicating 50 % either way. Now we are more general. We see the term $\boldsymbol{Cov(YP)/Cov(HP)}$ looks like Hamilton's **r**! But we still don't know much about how to find it.

22.4.1 Example

Example 22.4.1 If $\boldsymbol{Cov(YP)/Cov(HP)} = 0.7$, the benefit $b = 2.4$ and the cost of altruism $c = 0.8$, does altruism spread?

Solution *We calculate* $b\left(\frac{Cov(YP)}{Cov(HP)}\right) - c = (2.4)\ (0.7) - 0.8 = 1.68 - 0.8 > 0$. *So altruism does spread.*

Example 22.4.2 If $\boldsymbol{Cov(YP)/Cov(HP)} = 0.3$, the benefit $b = 1.4$ and the cost of altruism $c = 1.2$, does altruism spread?

Solution *We calculate* $b\left(\frac{Cov(YP)}{Cov(HP)}\right) - c = (1.4)\ (0.3) - 1.2 = -0.78 < 0$. *So altruism does not spread.*

22.4.2 Homework

Exercise 22.4.1 *If* $\boldsymbol{Cov(YP)/Cov(HP)} = 0.95$, *the benefit* $b = 1.1$ *and the cost of altruism* $c = 0.9$, *does altruism spread?*

Exercise 22.4.2 *If* $\boldsymbol{Cov(YP)/Cov(HP)} = 1.7$, *the benefit* $b = 2.9$ *and the cost of altruism* $c = 4.0$, *does altruism spread?*

Exercise 22.4.3 *If* $\boldsymbol{Cov(YP)/Cov(HP)} = 0.25$, *the benefit* $b = 5.0$ *and the cost of altruism* $c = 1.5$, *does altruism spread?*

Exercise 22.4.4 *If* $\boldsymbol{Cov(YP)/Cov(HP)} = 0.1$, *the benefit* $b = 1.8$ *and the cost of altruism* $c = 1.5$, *does altruism spread?*

22.5 Altruism Spread Under Additive Genetics

Now let's use Eq. 22.1 to see how altruism might spread under the assumption that we have additive genetics. We are going to reinterpret our model setup a bit, so although it is similar to what we have already done, note some of the underlying structure is changed. We will be assuming that we have a genotype which corresponds to a phenotype that is altruistic. In our earlier notation, we identified our genotypes with a fraction p_i of the alleles of a gene in the individual i in the population. So the genotypes that code for the altruism phenotype corresponds to some fractions also.

Indeed, there might be more than one frequency that gives rise to altruistic behavior phenotype. Let's call this phenotype for altruism P_A where the A stands for *altruism*. Whether or not this phenotype will spread through the population will depend on whether or not there is some sort of correlation between P_A and the other phenotypes which we will call P_B. Let's assume the fractions p_i which correspond to the possible frequencies of the alleles can take on a finite number of values: $\{f_1, \ldots, f_M\}$. For example, in a haploid model, the choices are $p = 1$ or $p = 0$ as there are only two possibilities. In a diploid organism, p could be 0, 1/2 or 1 and other organisms can have a richer set of possible frequencies. Let's assume a large population so we can always find N samples of altruistic phenotypes and N samples of the others. Let's assume in a sample, the altruistic phenotype and other phenotypes take on the frequencies

$$
\begin{aligned}
P_A &= \{p_1^A, \ldots, p_N^A\} \\
P_B &= \{p_1^B, \ldots, p_N^B\}
\end{aligned}
$$

Next, we assume that the individual probability of helping the common good due to the a phenotype which depends linearly on the genotype. Hence, we let helping by denoted by h^A and define

$$h_i^A = a + k\, p_i^A \tag{22.3}$$

where a is the amount of helping that occurs even when an individual has a frequency of 0. The slope term k indicates the impact that gene i has on helping. If the frequency is low, then the amount of helping is low also. We also assume the probability of individual i benefits from an altruistic individual's cooperative act can also be modeling linearly. So we assume

$$y_i^A = a + k\, p_i^B \tag{22.4}$$

The baseline amount of benefit to individual i is still a so that is why we set up the linear model using a as the intercept. We use the same slope, k, because since the slope term k indicates the impact that gene i has on helping, this is the same impact that it should have on the benefit. Now these assumptions are also ruled out by most biological situations. We will leave it to you to think more about this as you take more biology courses and learn more about genetics.

We have the fitness equation

$$
\begin{aligned}
w_i^A &= w_0 + b\, y_i^A - c\, h_i^A \\
&= w_0 + b\,(a + k\, p_i^A) - c\,(a + k\, p_i^B).
\end{aligned}
$$

We already know from our discussion using the additive fitness model that altruism spreads when

$$b\left(\frac{\boldsymbol{Cov(Y^A\ P^A)}}{\boldsymbol{Cov(H^A\ P^A)}}\right) > c$$

So we need to calculate $\boldsymbol{Cov(H^A\ P^A)}$. We know

$$\boldsymbol{Cov(H^A\ P^A)} = \boldsymbol{E(H^A\ P^A)} - \boldsymbol{E(H^A)}\ \boldsymbol{E(P^A)}.$$

We calculate

$$\begin{aligned}
\boldsymbol{E(H^A\ P^A)} &= \frac{1}{N}\sum_{i=1}^{N}(a + kp_i^A)p_i^A \\
&= a\frac{1}{N}\sum_{i=1}^{N} p_i^A + k\,\frac{1}{N}\sum_{i=1}^{N}\ p_i^A\, p_i^A \\
&= a\boldsymbol{E(P^A)} - k\boldsymbol{E((P^A)^2)}.
\end{aligned}$$

Further

$$\begin{aligned}
\boldsymbol{E(H^A)} &= \frac{1}{N}\sum_{i=1}^{N}(a + kp_i^A) \\
&= a + k\,\frac{1}{N}\sum_{i=1}^{N}\ p_i^A \\
&= a + k\boldsymbol{E(P^A)}.
\end{aligned}$$

Combining, we have

$$\begin{aligned}
\boldsymbol{Cov(H^A\ P^A)} &= a\boldsymbol{E(P^A)} + k\boldsymbol{E((P^A)^2)} - (a + k\boldsymbol{E(P^A)})\ \boldsymbol{E(P^A)} \\
&= k\ \left(\boldsymbol{E((P^A)^2)} + \boldsymbol{E(P^A)}\ \boldsymbol{E(P^A)}\right) \\
&= k\ \boldsymbol{Cov(P^A)} = k\ \boldsymbol{Var(P^A)}.
\end{aligned}$$

The calculations for $\boldsymbol{E(Y^A\ P^A)}$ are similar. We find, since each y_i^A depends on p_i^B, that

$$\begin{aligned}
\boldsymbol{E(Y^A\ P^A)} &= \frac{1}{N}\sum_{i=1}^{N}(a + kp_i^B)p_i^A \\
&= a\frac{1}{N}\sum_{i=1}^{N} p_i^A + k\,\frac{1}{N}\sum_{i=1}^{N}\ p_i^A\, p_i^B \\
&= a\boldsymbol{E(P^A)} + k\ \boldsymbol{E(P^A\ P^B)}.
\end{aligned}$$

Also,

$$\begin{aligned} \boldsymbol{E}(\boldsymbol{Y}^A) &= \frac{1}{N}\sum_{i=1}^{N}(a + kp_i^B) \\ &= a + k\,\frac{1}{N}\sum_{i=1}^{N} p_i^B \\ &= a + k\boldsymbol{E}(\boldsymbol{P}^B). \end{aligned}$$

Combining

$$\begin{aligned} \boldsymbol{Cov}(\boldsymbol{Y}^A\ \boldsymbol{P}^A) &= a\boldsymbol{E}(\boldsymbol{P}^A) + k\,\boldsymbol{E}((\boldsymbol{P}^A\ \boldsymbol{P}^B) - (a + k\boldsymbol{E}(\boldsymbol{P}^B))\,\boldsymbol{E}(\boldsymbol{P}^A) \\ &= k\,\left(\boldsymbol{E}((\boldsymbol{P}^A\ \boldsymbol{P}^B) - \boldsymbol{E}(\boldsymbol{P}^B)\,\boldsymbol{E}(\boldsymbol{P}^A))\right) \\ &= k\,\boldsymbol{Cov}(\boldsymbol{P}^A\ \boldsymbol{P}^B). \end{aligned}$$

For this allele to spread throughout the population, we want $b\left(\frac{\boldsymbol{Cov}(\boldsymbol{Y}^A\ \boldsymbol{P}^A)}{\boldsymbol{Cov}(\boldsymbol{H}^A\ \boldsymbol{P}^A)}\right) > c$ and so this happens when

$$b\ k\,\boldsymbol{Cov}(\boldsymbol{P}^A\ \boldsymbol{P}^B) - c\ k\,\boldsymbol{Var}(\boldsymbol{P}^A) > 0$$

or dividing out the common k,

$$b\left(\frac{\boldsymbol{Cov}(\boldsymbol{P}^A\ \boldsymbol{P}^B)}{\boldsymbol{Var}(\boldsymbol{P}^A)}\right) > c$$

But we know that we can rewrite this using the regression slope of P^A on P^B as

$$\boldsymbol{m}(\boldsymbol{P}^A, \boldsymbol{P}^B) = \frac{\boldsymbol{Cov}(\boldsymbol{P}^A\ \boldsymbol{P}^B)}{\boldsymbol{Var}(\boldsymbol{P}^A)}.$$

Hence, our condition for the spread of altruism is

$$b\,\boldsymbol{m}(\boldsymbol{P}^A, \boldsymbol{P}^B) > c$$

We finally have a way to compute Hamilton's **r**. It is found as the slope of the regression of $\boldsymbol{P}^B$ on $\boldsymbol{P}^A$! We might even have a way to get that from data!!

22.5.1 Examples

Example 22.5.1 If the slope of the regression line of $\boldsymbol{P}^B$ on $\boldsymbol{P}^A$ is 0.2, the benefit $b = 1.0$ and the altruism cost is $c = 3$, does altruism spread?

Solution *We calculate* $b\,\boldsymbol{m}(\boldsymbol{P}^A, \boldsymbol{P}^B) - c = (1.0)(0.2) - 3 < 0$ *and so altruism does not spread.*

Example 22.5.2 If the slope of the regression line of $\boldsymbol{P}^B$ on $\boldsymbol{P}^A$ is 1.2, the benefit $b = 2.0$ and the altruism cost is $c = 1.8$, does altruism spread?

Solution *We calculate* $b\,\boldsymbol{m}(\boldsymbol{P}^A, \boldsymbol{P}^B) - c = (2.0)(1.2) - 1.8 > 0$ *and so altruism does spread.*

22.5.2 Homework

Exercise 22.5.1 *If the slope of the regression line of* $\boldsymbol{P}^B$ *on* $\boldsymbol{P}^A$ *is* 1.1, *the benefit* $b = 1.4$ *and the altruism cost is* $c = 0.8$, *does altruism spread?*

Exercise 22.5.2 *If the slope of the regression line of* $\boldsymbol{P}^B$ *on* $\boldsymbol{P}^A$ *is* 0.45, *the benefit* $b = 2.5$ *and the altruism cost is* $c = 1.1$, *does altruism spread?*

Exercise 22.5.3 *If the slope of the regression line of* $\boldsymbol{P}^B$ *on* $\boldsymbol{P}^A$ *is* 0.6, *the benefit* $b = 4.0$ *and the altruism cost is* $c = 1.9$, *does altruism spread?*

Exercise 22.5.4 *If the slope of the regression line of* $\boldsymbol{P}^B$ *on* $\boldsymbol{P}^A$ *is* 0.5, *the benefit* $b = 3.0$ *and the altruism cost is* $c = 1.8$, *does altruism spread?*

22.6 The Optimization Approach

The final approach is just another way to get insight. This time we will yet again modify how we set up the model and use calculus to derive Hamilton's rule. However, note the mathematics does not tell you how to find **r** itself. Our previous discussions help us with that—and they did not use calculus at all. Just a lot of thinking...We will assume that the fitness function is given by

$$w = w_0 + b\, P_{receive} - c\, P_{aid}$$

where P_{aid} is the probability of giving aid and $P_{receive}$ is the probability of receiving aid. The fitness function is always grows if its derivative is positive. Calculating, we have

$$\begin{aligned} \frac{dw}{dP_{aid}} &= \frac{\partial w}{\partial P_{aid}} + \frac{\partial w}{\partial P_{receive}} \frac{dP_{receive}}{dP_{aid}} \\ &= -c + b\, \frac{dP_{receive}}{dP_{aid}}. \end{aligned}$$

But we know we can estimate $\frac{dP_{receive}}{dP_{aid}}$ by $\boldsymbol{m(P_{aid}, P_{receive})}$, the slope of the regression of $P_{receive}$ on P_{aid}. The derivative is positive when

$$b\, \boldsymbol{m(P_{aid}, P_{receive})} > c.$$

Since, we know $P_{receive} \approx \boldsymbol{\alpha} + \boldsymbol{m(P_{aid}, P_{receive})}\, P_{aid}$, where $\boldsymbol{\alpha}$ is the regression intercept, we see

$$\begin{aligned} w &\approx w_0 + b\,(\boldsymbol{\alpha} + \boldsymbol{m(P_{aid}, P_{receive})}\, P_{aid}) - c\, P_{aid} \\ &\approx w_0 + b\,\boldsymbol{\alpha} + P_{aid}\,(b\,\boldsymbol{m(P_{aid}, P_{receive})} - c). \end{aligned}$$

We see we maximize fitness by letting $P_{aid} \rightarrow 1$ as long as $b\,\boldsymbol{m(P_{aid}, P_{receive})} - c > 0$. So we are led to our usual condition that $b\,\boldsymbol{m(P_{aid}, P_{receive})} > c$! We also see that it is only our previous complicated discussions that connect the $\frac{dP_{receive}}{dP_{aid}}$ to the slope of the regression line. So the mathematics comes to the same result but needs to be tempered by a lot of biology!

Now that you have studied the basics of the theory behind altruism using a variety of approaches, you might look at a recent paper using these ideas (Smith et al. 2010). The paper focuses on the problem Hamilton's theory has shen selection is strong and fitness effects are not additive. In the discussion, there is a fair bit of mathematical terminology thrown around. You should be able to follow it now with the training you have received so far!

22.6.1 Example

Example 22.6.1 If $P_{receive} \approx 0.2 + 0.5\, P_{aid}$, the benefit $b = 1.4$ and the cost of altruism is $c = 1.3$, does altruism spread?

Solution *We check* $b\,\boldsymbol{m(P_{aid}, P_{receive})} - c = (1.4)(0.5) - 1.3 < 0$ *and so altruism does not spread.*

Example 22.6.2 If $P_{receive} \approx 0.45 + 1.3\, P_{aid}$, the benefit $b = 2.4$ and the cost of altruism is $c = 3.0$, does altruism spread?

Solution *We check* $b\,\boldsymbol{m(P_{aid}, P_{receive})} - c = (2.4)(1.3) - 3.0 < 3.12 - 3.0 > 0$ *and so altruism does spread.*

22.6.2 Homework

Exercise 22.6.1 *If* $P_{receive} \approx 0.32 + 0.9\, P_{aid}$, *the benefit* $b = 1.9$ *and the cost of altruism is* $c = 2.1$, *does altruism spread?*

Exercise 22.6.2 *If* $P_{receive} \approx 0.12 + 1.5\ P_{aid}$, *the benefit* $b = 1.2$ *and the cost of altruism is* $c = 1.0$, *does altruism spread?*

Exercise 22.6.3 *If* $P_{receive} \approx 0.44 + 0.75\ P_{aid}$, *the benefit* $b = 1.8$ *and the cost of altruism is* $c = 0.4$, *does altruism spread?*

Exercise 22.6.4 *If* $P_{receive} \approx 0.03 + 0.0.8\ P_{aid}$, *the benefit* $b = 1.4$ *and the cost of altruism is* $c = 0.7$, *does altruism spread?*

References

W.D. Hamilton, The evolution of altruistic behaviour. Am. Natural. **97**, 354–356 (1963)

W.D. Hamilton, The genetical evolution of social behavior. J. Theoret. Biol. **7**, 1–52 (1964)

R. McElreath, R. Boyd, *Mathematical Models of Social Evolution: A Guide for the Perplexed* (University of Chicago Press, Chicago, 2007)

J. Smith, D. Van Dyken, P. Zee, A generalization of Hamilton's rule for the evolution of microbial cooperation. Science **25**, 1700–1703 (2010)

Part IV
Summing It All Up

Chapter 23
Final Thoughts

So how do we become a good user of the three fold way of mathematics, science and computer tools? We believe it is primarily a question of deep respect for the balance between these disciplines. The basic idea is that once we abstract from biology or some other science how certain quantities interact, we begin to phrase these interactions in terms of mathematics. It is very important to never forget that once the mathematical choices have been made, the analysis of the mathematics alone will lead you to conclusions which may or may not be biologically relevant. You must always be willing to give up a mathematical model or a computer science model if it does not lead to useful insights into the original science.

We can quote from Lisa Randall (2005, pp. 70–71). She works in Particle Physics which is the experimental arm that gives us data to see if string theory or loop gravity is indeed a useful model of physical reality. It is not necessary to know physics here to get the point. Notice what she says:

> The term "model" might evoke a small scale battleship or castle you built in your childhood. Or you might think of simulations on a computer that are meant to reproduce known dynamics – how a population grows, for example, or how water moves in the ocean. Modeling in particle physics is not the same as either of these definitions. Particle physics models are guesses at alternate physical theories that might underlie the standard model...Different assumptions and physical concepts distinguish theories, as do the distance or energy scales at which a theory's principles apply. Models are a way at getting at the heart of such distinguishing features. They let you explore a theory's potential implications. If you think of a theory as general instructions for making a cake, a model would be a precise recipe. The theory would say to add sugar, a model would specify whether to add half a cup or two cups.

Now substitute *Biological* for *Particle Physics* and so forth and you can get a feel for what a model is trying to do. Of course, biological models are much more complicated than physics ones!

The primary message of this course is thus to teach you to think deeply and carefully. The willingness to attack hard problems with multiple tools is what we need in young scientists. We hope this course teaches you a bit more about that.

We believe that we learn all the myriad things we need to build reasonable models over a lifetime of effort. Each model we design which pulls in material from disparate

J.K. Peterson, *Calculus for Cognitive Scientists*, Cognitive Science and Technology, DOI 10.1007/978-981-287-874-8_23

areas of learning enhances our ability to develop the kinds of models that give insight. As Raymond Pierrehumbert (2010, p. xi) says about climate modeling

> When it comes to understanding the whys and wherefores of climate, there is an infinite amount one needs to know, but life affords only a finite time in which to learn it... It is a lifelong process. [We] attempt to provide the student with a sturdy scaffolding upon which a deeper understanding may be built later.
>
> The climate system [and other biological systems we may study] is made up of building blocks which in themselves are based on elementary ... principles, but which have surprising and profound collective behavior when allowed to interact on the [large] scale. In this sense, the "climate game" [the biological modeling game] is rather like the game of Go, where interesting structure emerges from the interaction of simple rules on a big playing field, rather than complexity in the rules themselves.

So welcome to the journey!

References

R. Pierrehumbert, *Principles of Planetary Climate* (Cambridge University Press, Cambridge, 2010)

L. Randall, *Warped Passages: Unraveling the Mysteries of the Universe's Hidden Dimensions* (Harper Collins, New York, 2005)

Part V
Advise to the Beginner

Chapter 24
Background Reading

To learn more about how the combination of mathematics, computational tools and science have been used with profit, we have some favorite books that have tried to do this. Reading these books have helped us design biologically inspired algorithms and models and has inspired us in many ways

- Evolving Brains (Allman 1999): you should read about how complicated nervous systems might have evolved. Understanding cognitive dysfunctions like depression depend on knowing this kind of stuff.
- An Introduction to Systems Biology: Design Principles of Biological Circuits (Alon 2006): this is a great book on how mathematics, computer science and science can be used together to figure out how things tick. Now that you have had this course, you can profitably read most of this book on your own. You can amaze your friends!
- Information in the Brain: A Molecular Perspective (Black 1991): this is a fantastic book, now out of print but you can find it used, which helps you understand how the way animals process information means that neural signals alter both hardware and software constantly. You can all read this book right now too!
- Creativity and Unpredictability (Boden 1995): did you ever think about how to understand creativity? This book will help you think about it in new ways and it is highly recommended.
- A dynamical systems model for Parkinson's disease (Connolly et al. 2000): you have learned a fair bit about modeling using differential equations. Try looking at this paper. You'll be able to understand quite a bit of it!
- Animal Physiology: Mechanisms and Adaptations (Eckert et al. 1998): a great first read on physiology that is not human centric. So it makes you generalize which is good!
- Cells, Embryos and Evolution: Towards a Cellular and Developmental Understanding of Phenotypic Variation and Evolutionary Adaptability (Gerhart and Kirschner 1997): this is a wonderful book that looks at low level and high level biological processing and tries to find patterns and abstractions that can be used to give

J.K. Peterson, *Calculus for Cognitive Scientists*, Cognitive Science and Technology, DOI 10.1007/978-981-287-874-8_24

insight. You can read a lot of this now—not all—but it shows you what to shoot for!

- A simulation model of long-term climate, livestock and vegetation interactions on communal range lands in the semi-arid Succulent Karoo, Namaqualand, South Africa (Hahn et al. 2005): who says we have to limit our modeling to biology? Here is a great article on modeling in a different arena. But many of the tools we have learned work just fine. Pay attention to that. It is why learning the mathematics is a good career helper. It gives you a tool you can use all over the place!
- Complex Worlds from Simpler Nervous Systems (Harland and Jackson 2004): This book models how a lot of small animals explore the world with their sensory systems. You would be surprised how much we can figure out by studying these things with our tools!
- The Evolution of Communication (Hauser 1996): this is book all about how animals have evolved to process sensory information. You can read most of it right now.
- Molecules and Cognition: The Latterday Lessons of Levels, Language and lac: Evolutionary Overview of Brain Structure and Function in Some Vertebrates and Invertebrates (Miklos 1993) this is an old article but very helpful in seeing the big picture!
- Principles of Planetary Climate (Pierrehumbert 2010): here is where you take the stuff you know from this class and start putting it to work modeling planetary climates. Really interesting and the book is not tied just to the earth!

All have helped us to see the *big picture* and they will help you too. And there are more!!

References

J. Allman, *Evolving Brains* (Scientific American Library, New York, 1999)

U. Alon, *An Introduction to Systems Biology: Design Principles of Biological Circuits*, CRC Mathematical and Computational Biology (Chapman & Hill, Boca Raton, 2006)

I. Black, *Information in the Brain: A Molecular Perspective*, A Bradford Book (MIT Press, Cambridge, 1991)

M. Boden, Creativity and unpredictability. Stanf. Humanity Rev. **4**(2), 123–140 (1995)

C. Connolly, J. Burns, M. Jog, A dynamical systems model for Parkinson's disease. Biol. Cybern. **83**, 47–59 (2000)

R. Eckert, D. Randall, G. Augustine, *Animal Physiology: Mechanisms and Adaptations*, 3rd edn. (W. H. freeman and Company, New York, 1998)

J. Gerhart, M. Kirschner, *Cells, Embryos and Evolution: Towards a Cellular and Developmental Understanding of Phenotypic Variation and Evolutionary Adaptability* (Blackwell Science, Oxford, 1997)

B. Hahn, F. Richardson, M. Hoffman, R. Roberts, S. Todd, P. Carrick, A simulation model of long-term climate, livestock and vegetation interactions on communal rangelands in the semi-arid Succulent Karoo, Namaqualand, South Africa. Ecol. Model. **183**, 211–230 (2005)

D. Harland, R. Jackson, Portia perceptions: the umwelt of an araneophagic jumping spider, in *Complex Worlds from Simpler NervousSystems*, A Bradford Book, ed. by F. Prete (MIT Press, Cambridge, 2004), pp. 5–40

M. Hauser, *The Evolution of Communication*, A Bradford Book (MIT Press, Cambridge, 1996)

G. Miklos, Molecules and Cognition: The Latterday Lessons of Levels, Language and lac: Evolutionary Overview of Brain Structure and Function in Some Vertebrates and Invertebrates. J. Neurobiol. **24**(6), 842–890 (1993)

R. Pierrehumbert, *Principles of Planetary Climate* (Cambridge University Press, Cambridge, 2010)

Glossary

A

Antiderivative The antiderivative of a function f is any function F which is differentiable and satisfies $F'(t) = f(t)$ at all points in the domain of f, p. 115.

B

Biological Modeling This is the study of biological systems using a combination of mathematical, scientific and computational approaches, p. 3.

C

Cancer Model A general model of cancer based on TSG inactivation is as follows. The tumor starts with the inactivation of a **TSG** called **A**, in a small compartment of cells. A good example is the inactivation of the **APC** gene in a colonic crypt, but it could be another gene. Initially, all cells have two active alleles of the **TSG**. We will denote this by $A^{+/+}$ where the superscript $+/+$ indicates both alleles are active. One of the alleles becomes inactivated at mutation rate u_1 to generate a cell type denoted by $A^{+/-}$. The superscript $+/-$ tells us one allele is inactivated. The second allele becomes inactivated at rate $\hat{u}_2$ to become the cell type $A^{-/-}$. In addition, $A^{+/+}$ cells can also receive mutations that trigger **CIN**. This happens at the rate u_c resulting in the cell type $A^{+/+CIN}$. This kind of a cell can inactivate the first allele of the **TSG** with normal mutation rate u_1 to produce a cell with one inactivated allele (i.e. a $+/-$) which started from a CIN state. We denote these cells as $A^{+/-CIN}$. We can also get a cell of type $A^{+/+CIN}$ when a cell of type $A^{+/-}$ receives a mutation which triggers **CIN**. We will assume this happens at the

J.K. Peterson, *Calculus for Cognitive Scientists*, Cognitive Science and Technology, DOI 10.1007/978-981-287-874-8

same rate u_c as before. The $\boldsymbol{A}^{+/-CIN}$ cell then rapidly undergoes **LOH** at rate $\hat{u}_3$ to produce cells of type $\boldsymbol{A}^{-/-CIN}$. Finally, $\boldsymbol{A}^{-/-}$ cells can experience **CIN** at rate u_c to generate $\boldsymbol{A}^{-/-CIN}$ cells. The first allele is inactivated by a point mutation. The rate at which this occurs is modeled by the rate u_1. We assume the mutations governed by the rates u_1 and u_c are **neutral**. This means that these rates do not depend on the size of the population N. The events governed by $\hat{u}_2$ and $\hat{u}_3$ give what is called **selective advantage**. This means that the size of the population size does matter. Using these assumptions, we therefore model $\hat{u}_2$ and $\hat{u}_3$ as

$$\hat{u}_2 = N\, u_2$$

and

$$\hat{u}_3 = N\, u_3.$$

where u_2 and u_3 are neutral rates. The mathematical model is then setup as follows. Let

$X_0(t)$ is the probability a cell in cell type $A^{+/+}$ at time t.
$X_1(t)$ is the probability a cell in cell type $A^{+/-}$ at time t.
$X_2(t)$ is the probability a cell in cell type $A^{-/-}$ at time t.
$Y_0(t)$ is the probability a cell in cell type $A^{+/+CIN}$ at time t.
$Y_1(t)$ is the probability a cell in cell type $A^{+/-CIN}$ at time t.
$Y_2(t)$ is the probability a cell in cell type $A^{-/-CIN}$ at time t.

We can then derive rate equations to be

$$\begin{aligned} X_0' &= -(u_1 + u_c)\, X_0 \\ X_1' &= u_1\, X_0 - (u_c + N\, u_2)\, X_1 \\ X_2' &= N\, u_2\, X_1 - u_c\, X_2 \\ Y_0' &= u_c\, X_0 - u_1\, Y_0 \\ Y_1' &= u_c\, X_1 + u_1\, Y_0 - N\, u_3\, Y_1 \\ Y_2' &= N\, u_3\, Y_1 + u_c\, X_2 \end{aligned}$$

We are interested in analyzing this model over a typical human life span of 100 years, p. 390.

Cauchy Fundamental Theorem of Calculus Let G be any antiderivative of the Riemann integrable function f on the interval $[a, b]$. Then $G(b) - G(a) = \int_a^b f(t)\, dt$, p. 159.

Continuity A function f is continuous at a point p if for all positive tolerances ϵ, there is a positive δ so that $\mid f(t) - f(p) \mid < \epsilon$ if t is in the domain of f and $\mid t - p \mid < \delta$. You should note continuity is something that is only defined at a point and so functions in general can have very few points of continuity. Another way of defining the continuity of f at the point p is to say the $\lim_{t \to p} f(t)$ exists and equals $f(p)$, p. 64.

D

Differentiability A function f is differentiable at a point p if there is a number L so that for all positive tolerances ϵ, there is a positive δ so that

$$\mid \frac{f(t) - f(p)}{t - p} - L \mid < \epsilon \quad \text{if } t \text{ is in the domain of } f \text{ and } \mid t - p \mid < \delta$$

You should note differentiability is something that is only defined at a point and so functions in general can have very few points of differentiability. Another way of defining the differentiability of f at the point p is to say the $\lim_{t \to p} \frac{f(t)-f(p)}{t-p}$ exists. At each point p where this limit exists, we can define a new function called the derivative of f at p. This is usually denoted by $f'(p)$ or $\frac{df}{dt}(p)$, p. 87.

E

Exponential growth Some biological systems can be modeled using the idea of exponential growth. This means the variable of interest, x, has growth proportional to its rate of change. Mathematically, this means $x' \propto rx$ for some proportionality constant r, p. 207.

F

Function approximation Let $f : [a, b] \rightarrow \Re$ be continuous on $[a, b]$ and be at least twice differentiable on (a, b). For a given p in $[a, b]$, for each x, there is at least one point c, between p and x, so that $f(x) = f(p) + f'(p)(x - p) + (1/2)f''(c)(x - p)^2$. The $f(p) + f'(p)(x - p)$ is called the **first order Taylor Polynomial** for f at p and we denote it by $P_2(x; p)$. The point p is again called the **base point**. Note we are approximating $f(x)$ by the linear function $f(p) + f'(p)(x - p)$ and the error we make is $E_1(x, p) = (1/2)f''(c)(x - p)$. We can extend this result to quadratic terms and obtain a similar result: given p in $[a, b]$, for each x, there is at least one point c, between p and x, so that $f(x) = f(p) + f'(p)(x-) + (1/2)f''(p)(x - p)^2 + (1/6)f'''(c)(x - p)^3$. The quadratic $f(p) + f'(p)(x-) + (1/2)f''(p)(x - p)^2$ is called the **second order Taylor Polynomial** for f at p and we denote it by $P_2(x; p)$. The point p is again called the **base point**. Note we are approximating $f(x)$ by the quadratic $f(p) + f'(p)(x-) + (1/2)f''(p)(x - p)^2$ and the error we make is $E_2(x, p) = (1/6)f'''(c)(x - p)$.
Hence, for a function with sufficient differentiability, we can approximate its behavior using a straight line called the tangent line which has local error h^2 where h measures the distance between the x arguments or we can use a quadratic approximation. The quadratic approximation has a local error h^3, p. 286.

Fundamental Theorem of Calculus Let f be Riemann Integrable on $[a, b]$. Then the function F defined on $[a, b]$ by $F(x) = \int_a^x f(t)\, dt$ satisfies

1. F is continuous on all of $[a, b]$.
2. F is differentiable at each point x in $[a, b]$ where f is continuous and $F'(x) = f(x)$, p. 156.

H

Half life The amount of time it takes a substance x to lose half its original value under exponential decay. It is denoted by $t_{1/2}$ and can also be expressed as $t_{1/2} = \ln(2)/r$ where r is the decay rate in the differential equation $x'(t) = -r\, x(t)$, p. 220.

Hamilton's Rule Hamilton was a theoretical biologist who had the insight that **kinship** could lead to altruism evolving. In a standard cooperate/ defection model interaction, random sampling is assumed for cooperative/ defecting individuals from the population. So the chance of an agent being cooperative or defecting was the same. This is what determines the fitness calculations. It turns out non random interactions are the key to the formation of altruism. Hamilton introduced what he called the **coefficient of relatedness**, **r**, to these models. There is a chance **r** that individuals who possess the same allele due to common descent. He then modified the probabilities in the cooperative/ defection model to include kinship probabilities. Specifically, he let

- $\boldsymbol{P(A|A)}$ be the probability an altruist **A** is paired with another altruist **A**. This is now

$$\boldsymbol{P(A|A)} = r \times 1 + (1 - r) \times p$$

which says altruistics interact with their kin with probability 1 and with everyone else at probability $1 - r$. Here p is the frequency of the altruistic allele in the population.

- $\boldsymbol{P(A|N)}$ be the probability an altruist **A** is paired with a non altruist **N** which is

$$\boldsymbol{P(A|N)} = r \times 0 + (1 - r) \times (1 - p)$$

as kin don't interact with non kin and $1 - p$ is the frequency of non altruists in the population.

- $\boldsymbol{P(N|A)}$ be the probability a non altruist **N** is paired with an altruist **A**. We have

$$\boldsymbol{P(N|A)} = r \times 0 + (1 - r) \times (1 - p)$$

as kin don't interact with non kin and again, $1 - p$ is the frequency of non altruists in the population.

- $\boldsymbol{P(N|N)}$ be the probability a non altruist **N** is paired with another non altruist **N**. This is

$$\boldsymbol{P(N|N)} = r \times 1 + (1-r) \times (1-p)$$

where non altruist kin interact with their own kin with probability 1 and with the rest of the population with probability $1 - p$.

Plugging in these new probabilities, we find altruism increases if

$$\boldsymbol{W(A)} > \boldsymbol{W(N)} \Longrightarrow b\,(r + (1-r)p - (1-r)p) > c \Longrightarrow r\,b > c.$$

where $\boldsymbol{W(A)}$ and $\boldsymbol{W(N)}$ are the payoffs for the two choices A and N. You should look at the text for more details. Then, the big question is what does r mean? p. 461.

L

Linear approximations We know we can approximate a function g at a point x_0 using a tangent line. The tangent line to the differentiable function g at the point x_0 is given by $T(x, x_0)$ where

$$T(x, x_0) = g(x_0) + g'(x_0)\,(x - x_0)$$

and the error at $x_0 + h$ would be

$$\begin{aligned} E(h) &= g(x_0 + h) - T(x_0 + h, x_0) \\ &= g(x_0 + h) - g(x_0) - g'(x_0)\,h. \end{aligned}$$

We also know from the Taylor Theorem with remainder type results the absolute error is always given by p. 315.

$$\begin{aligned} E(x - x_0) &= g(x) - T(x, x_0) \\ &= g(x) - g(x_0) - g'(x_0)\,(x - x_0) \\ &= g''(c)/2\,h^2, \end{aligned}$$

Logistic Model Another first order differential equation that is of interest is the one known as the Logistics Differential Equation. This has the form

$$\begin{aligned} u'(t) &= \alpha\,u(t)\,(L - u(t)), \\ u(0) &= u_0 \end{aligned}$$

The positive parameter L is called the carrying capacity and the positive parameter α controls how quickly the solution approaches the carrying capacity L. We will see the solutions to this problem either start out above L and decay asymptotically to L from above or they begin below L and grow up to L from below. We assume we have $0 < u_0 < L$ or $u_0 > L$. We can think of the rate of change of u as determined by the balance between a growth term and a decay or loss term.

- Let $u'_{growth} = \alpha\, L\, u$ and $u'_{decay} = -\alpha\, u^2$. Then

$$\begin{aligned} u' &= u'_{growth} + u'_{decay} \\ &= \alpha\, L\, u - \alpha\, u^2 \\ &= \alpha\, u\, (L - u). \end{aligned}$$

- The growth term could be a birth rate model where new members of the population grow following an exponential growth model.
- The decay term is an example of how to model interaction between populations. Given two populations $x(t)$ and $y(t)$, we can model their interaction with the product of the population sizes. Often the decay rate of change of a populations x and y would then have interaction components

$$\begin{aligned} x'_{I,\, decay} &\propto x(t)\; y(t) \\ y'_{I,\, decay} &\propto x(t)\; y(t) \end{aligned}$$

- The decay term is an example of how to model interaction between populations. Given two populations $x(t)$ and $y(t)$, we can model their interaction with the product of the population sizes. Often the decay rate of change of a populations x and y would then have interaction components

$$\begin{aligned} x'_{I,\, decay} &\propto x(t)\; y(t) \\ y'_{I,\, decay} &\propto x(t)\; y(t) \end{aligned}$$

  You see this kind of population interaction in the Predator–Prey model and the Simple Disease model also. Here, since there is only one variable and so this is called **self interaction**. In general, a self interaction decay term would be modeled by

$$x'_{SI,\, decay} = -a\, x(t)\, x(t) = -ax^2(t)$$

  So for logistic models, the decay component is a **self-interaction** term, p. 259.

P

Partial derivatives Let $z = f(x, y)$ be a function of the two independent variables x and y defined on some domain. At each pair (x, y) where f is defined in a circle of some finite radius r, $B_r(x_0, y_0) = \{(x, y) \mid \sqrt{(x - x_0)^2 + (y - y_0)^2} < r\}$, it makes sense to try to find the limits

$$\lim_{x \to x_0, y = y_0} \frac{f(x, y) - f(x, y_0)}{x - x_0}$$
$$\lim_{x = x_0, y \to y_0} \frac{f(x, y) - f(x_0, y)}{y - y_0}$$

If these limits exists, they are called the partial derivatives of f with respect to x and y at (x_0, y_0), respectively. For these partial derivatives, we use the symbols

$$f_x(x_0, y_0), \ \frac{\partial f}{\partial x}(x_0, y_0), \ z_x(x_0, y_0), \ \frac{\partial z}{\partial x}(x_0, y_0)$$

and

$$f_y(x_0, y_0), \ \frac{\partial f}{\partial y}(x_0, y_0), \ z_y(x_0, y_0), \ \frac{\partial z}{\partial y}(x_0, y_0)$$

We often use another notation for partial derivatives. The function f of two variables x and y can be thought of as having two arguments or slots into which we place values. So another useful notation is to let the symbol $D_1 f$ be f_x and $D_2 f$ be f_y, p. 411.

Primitive The primitive of a function f is any function F which is differentiable and satisfies $F'(t) = f(t)$ at all points in the domain of f, p. 115.

Protein Modeling The gene $\boldsymbol{Y}$ is a string of nucleotides (**A**, **C**, **T** and **G**) with a special starting string in front of it called the **promoter**. The nucleotides in the gene $\boldsymbol{Y}$ are *read* three at a time to create the amino acids which form the protein $\boldsymbol{Y^*}$ corresponding to the gene. The process is this: a special **RNA** polymerase, **RNAp**, which is a complex of several proteins, binds to the promoter region. Once **RNAp** binds to the promoter, messenger **RNA**, **mRNA**, is synthesized that corresponds to the specific nucleotide triplets in the gene $\boldsymbol{Y}$. The process of forming this **mRNA** is called **transcription**. Once the **mRNA** is formed, the protein $\boldsymbol{Y^*}$ is then made. The protein creation process is typically regulated. A single **regulator** works like this. An activator called $\boldsymbol{X}$ is a protein which increases the rate of **mRNA** creation when ti binds to the promoter. The activator $\boldsymbol{X}$ switches between and active and inactive version due to a signal $\boldsymbol{S_X}$. We let the active form be denoted by $\boldsymbol{X^*}$. If $\boldsymbol{X^*}$ binds in front of the promoter, **mRNA** creation increases implying an increase in the creation of the protein $\boldsymbol{Y^*}$ also. Once the signal $\boldsymbol{S_X}$ appears, $\boldsymbol{X}$ rapidly transitions to its state $\boldsymbol{X^*}$, binds with the front of the promoter and protein $\boldsymbol{Y^*}$ begins to accumulate. We let β denote the rate of protein accumulation which is

constant once the signal S_X begins. However, proteins also degrade due to two processes:

- proteins are destroyed by other proteins in the cell. Call this rate of destruction α_{des}.
- the concentration of protein in the cell goes down because the cell grows and therefore its volume increases. Protein is usually measured as a concentration and the concentration goes down as the volume goes up. Call this rate α_{dil}—the *dil* is for *dilation*.

The net or total *loss of protein* is called α and hence

$$\alpha = \alpha_{des} + \alpha_{dil}$$

The net rate of change of the protein concentration is then our familiar model

$$\frac{dY^*}{dt} = \underbrace{\beta}_{\text{constant growth}} - \underbrace{\alpha\, Y^*}_{\text{loss term}}$$

We usually do not make a distinction between the gene Y and its transcribed protein Y^*. We usually treat the letters Y and Y^* as the same even though it is not completely correct. Hence, we just write as our model

$$Y' = \beta - \alpha\, Y$$
$$Y(0) = Y_0$$

and then solve it using the integrating factor method even though, strictly speaking, Y is the gene! p. 239.

R

Response time The amount of time it takes a protein x to rise from its current concentration half way to its steady state value β/α. It is denoted by t_R and can also be expressed as $t_R = \ln(2)/\alpha$ where α is the decay rate in the differential equation $x'(t) = -\alpha\, x(t) + \beta$, p. 242.
In a protein synthesis model

$$y'(t) = -\alpha\, y(t) + \beta$$
$$y(0) = y_0$$

the time it takes the solution to go from its initial concentration y_0 to a value halfway between the initial amount and the steady state value is called the

response time. It is denoted by t_r and $t_r = \ln(2)/\alpha$ so it is functionally the same as the **half life** in an exponential decay model, p. 246.

Riemann Integral If a function on the finite interval $[a, b]$ is bounded, we can define a special limit which, if it exists, is called the Riemann Integral of the function on the interval $[a, b]$. Select a finite number of points from the interval $[a, b]$, $\{t_0, t_1, \ldots, t_{n-1}, t_n\}$. We don't know how many points there are, so a different selection from the interval would possibly gives us more or less points. But for convenience, we will just call the last point t_n and the first point t_0. These points are not arbitrary—t_0 is always a, t_n is always b and they are ordered like this:

$$t_0 = a < t_1 < t_2 < \cdots < t_{n-1} < t_n = b$$

The collection of points from the interval $[a, b]$ is called a Partition of $[a, b]$ and is denoted by some letter—here we will use the letter **P**. So if we say P is a partition of $[a, b]$, we know it will have $n + 1$ points in it, they will be labeled from t_0 to t_n and they will be ordered left to right with strict inequalities. But, we will not know what value the positive integer n actually is. The simplest Partition P is the two point partition $\{a, b\}$. Note these things also:

1. Each partition of $n + 1$ points determines n subintervals of $[a, b]$
2. The lengths of these subintervals always adds up to the length of $[a, b]$ itself, $b - a$.
3. These subintervals can be represented as

$$\{[t_0, t_1], [t_1, t_2], \ldots, [t_{n-1}, t_n]\}$$

   or more abstractly as $[t_i, t_{i+1}]$ where the index i ranges from 0 to $n - 1$.
4. The length of each subinterval is $t_{i+1} - t_i$ for the indices i in the range 0 to $n - 1$.

Now from each subinterval $[t_i, t_{i+1}]$ determined by the Partition P, select any point you want and call it s_i. This will give us the points s_0 from $[t_0, t_1]$, s_1 from $[t_1, t_2]$ and so on up to the last point, s_{n-1} from $[t_{n-1}, t_n]$. At each of these points, we can evaluate the function f to get the value $f(s_j)$. Call these points an **Evaluation Set** for the partition P. Let's denote such an evaluation set by the letter E. If the function f was nice enough to be positive always and continuous, then the product $f(s_i) \times (t_{i+1} - t_i)$ can be interpreted as the area of a rectangle; in general, though, these products are not areas. Then, if we add up all these products, we get a sum which is useful enough to be given a special name: the Riemann sum for the function f associated with the Partition P and our choice of evaluation set $E = \{s_0, \ldots, s_{n-1}\}$. This sum is represented by the symbol $S(f, P, E)$ where the things inside the parenthesis are there to remind us that this sum depends on our choice of the function f, the partition P and the evaluation set E. The Riemann sum is normally written as

$$S(f, P, E) = \sum_{i \in P} f(s_i)\,(t_{i+1} - t_i)$$

and we just remember that the choice of P will determine the size of n. Each partition P has a maximum subinterval length—let's use the symbol $\| P \|$ to denote this length. We read the symbol $\| P \|$ as the **norm** of P. Each partition P and evaluation set E determines the number $S(f, P, E)$ by a simple calculation. So if we took a collection of partitions P_1, P_2 and so on with associated evaluation sets E_1, E_2 etc., we would construct a sequence of real numbers $\{S(f, P_1, E_1),\ S(f, P_2, E_2), \ldots, S(f, P_n, E_n), \ldots\}$. Let's assume the norm of the partition P_n gets smaller all the time; i.e. $\lim_{n \to \infty} \| P_n \| = 0$. We could then ask if this sequence of numbers converges to something. What if the sequence of Riemann sums we construct above converged to the same number I no matter what sequence of partitions whose norm goes to zero and associated evaluation sets we chose? Then, we would have that the value of this limit is *independent* of the choices above. This is what we mean by the **Riemann Integral** of f on the interval $[a, b]$. If there is a number I so that

$$\lim_{n \to \infty} S(f, P_n, E_n) = I$$

no matter what sequence of partitions $\{P_n\}$ with associated sequence of evaluation sets $\{E_n\}$ we choose as long as $\lim_{n \to \infty} \| P_n \| = 0$, we say that the Riemann Integral of f on $[a, b]$ exists and equals the value I. The value I is dependent on the choice of f and interval $[a, b]$. So we often denote this value by $I(f, [a, b])$ or more simply as, $I(f, a, b)$. Historically, the idea of the Riemann integral was developed using area approximation as an application, so the summing nature of the Riemann Sum was denoted by the 16th century *letter S* which resembled an elongated or stretched letter *S* which looked like what we call the integral sign $\int$. Hence, the common notation for the Riemann Integral of f on $[a, b]$, when this value exists, is $\int_a^b f$. We usually want to remember what the independent variable of f is also and we want to remind ourselves that this value is obtained as we let the norm of the partitions go to zero. The symbol dt for the independent variable t is used as a reminder that $t_{i+1} - t_i$ is going to zero as the norm of the partitions goes to zero. So it has been very convenient to add to the symbol $\int_a^b f$ this information and use the augmented symbol $\int_a^b f(t)\,dt$ instead. Hence, if the independent variable was x instead of t, we would use $\int_a^b f(x)\,dx$. Since for a function f, the name we give to the independent variable is a matter of personal choice, we see that the choice of variable name we use in the symbol $\int_a^b f(t)\,dt$ is very arbitrary. Hence, it is common to refer to the independent variable we use in the symbol $\int_a^b f(t)\,dt$ as the dummy variable of integration, p. 151.

S

Second order partials We can characterize the error made when a function of two variables is replaced by its tangent plane at a point better if we have access to the second order partial derivatives of f. To find these, we take the partials of f_x and f_y to obtain the second order terms. Assuming f is defined locally as usual near (x_0, y_0), we can ask about the partial derivatives of the functions f_x and f_y with respect to x and y also. We define the second order partials of f as follows: if $f(x, y)$, f_x and f_y are defined locally at (x_0, y_0), we can attempt to find following limits:

$$\lim_{x \to x_0, y = y_0} \frac{f_x(x, y) - f_x(x, y_0)}{x - x_0} = \partial_x(f_x)$$
$$\lim_{x = x_0, y \to y_0} \frac{f_x(x, y) - f_x(x_0, y)}{y - y_0} = \partial_y(f_x)$$
$$\lim_{x \to x_0, y = y_0} \frac{f_y(x, y) - f_y(x, y_0)}{x - x_0} = \partial_x(f_y)$$
$$\lim_{x = x_0, y \to y_0} \frac{f_y(x, y) - f_y(x_0, y)}{y - y_0} = \partial_y(f_y)$$

When these second order partials exist at (x_0, y_0), we use the following notations interchangeably: $f_{xx} = \partial_x(f_x)$, $f_{xy} = \partial_y(f_x)$, $f_{yx} = \partial_y(f_x)$ and $f_{yy} = \partial_y(f_y)$. The second order partials are often organized into a matrix called the **Hessian**, $\boldsymbol{H}$, as follows p. 430:

$$\boldsymbol{H}(x_0, y_0) = \begin{bmatrix} f_{xx}(x_0, y_0) & f_{xy}(x_0, y_0) \\ f_{yx}(x_0, y_0) & f_{yy}(x_0, y_0), \end{bmatrix}$$

Smoothness Smoothness is a term we use to describe the behavior of a function. The lowest level of smoothness is that the function is continuous on a domain of interest and higher levels of smoothness add the first, second and so on orders of derivatives to the function, p. 65.

T

Tangent plane The tangent plane to the surface $z = f(x, y)$ at the point (x_0, y_0) is given by

$$-f_x(x_0, y_0)(x - x_0) - f_y(x_0, y_0)(y - y_0) + (z - f(x_0, y_0)) = 0.$$

This then gives the traditional equation of the tangent plane p. 417:

$$z = f(x_0, y_0) + f_x(x_0, y_0)(x - x_0) + f_y(x_0, y_0)(y - y_0),$$

Tangent plane error We can characterize the error made when a function of two variables is replaced by its tangent plane at a point better if we have access to the second order partial derivatives of f. The value of f at the point $(x_0+\Delta x, y_0+\Delta y)$ can be expresses as follows:

$$f(x_0 + \Delta x, y_0 + \Delta y) = f(x_0, y_0) + f_x(x_0, y_0)\Delta x + f_y(x_0, y_0)\Delta y + \frac{1}{2}\begin{bmatrix}\Delta x\\ \Delta y\end{bmatrix}^T \boldsymbol{H}(x_0 + c\Delta x, y_0 + c\Delta y)\begin{bmatrix}\Delta x\\ \Delta y\end{bmatrix}$$

where c between 0 and 1 so that the tangent plane error is given by p. 432

$$E(x_0, y_0, \Delta x, \Delta y) = \frac{1}{2}\begin{bmatrix}\Delta x\\ \Delta y\end{bmatrix}^T \boldsymbol{H}(x_0 + c\Delta x, y_0 + c\Delta y)\begin{bmatrix}\Delta x\\ \Delta y,\end{bmatrix}$$

Index

J.K. Peterson, *Calculus for Cognitive Scientists*, Cognitive Science and Technology, DOI 10.1007/978-981-287-874-8

D

Printed in the United States
By Bookmasters